Birkhäuser

Current topics on deformation monitoring and modelling, geodynamics and natural hazards

Edited by
Pablo J. González
Gerardo Herrera
Francisco Luzón
Pietro Tizzani

Previously published in *Pure and Applied Geophysics*
(PAGEOPH), Volume 172, No. 11, 2015

 Birkhäuser

Editors

Pablo J. González
COMET. School of Earth
and Environment
University of Leeds
LS2 9JT, Leeds
United Kingdom

Gerardo Herrera
Geohazards InSAR laboratory
and modeling group
Área de Riesgos Geológicos
Departamento de Investigación
Geocientífica, Instituto Geológico y
Minero de España, Calle Alenza 1
28050 Madrid, Spain

Francisco Luzón
Department of Applied Physics
University of Almería
04120 Almería
Spain

Pietro Tizzani
Istituto per il Rilevamento
Elettromagnetico dell'Ambiente
Consiglio Nazionale delle Recerca
IREA - CNR
Via Diocleziano 328
80124 Napoli, Italy

ISBN 978-3-0348-0956-6   ISBN 978-3-0348-0957-3 (eBook)
DOI 10.1007/978-3-0348-0957-3

Library of Congress Control Number: 2015955822

Springer Basel Heidelberg New York Dordrecht London
© Springer Basel 2016

Cover illustration: The cover picture depicts the Pico do Fogo volcano and was taken by Pablo J. González in June 2015.

Cover design: deblik, Berlin.

Printed on acid-free paper

Springer Basel AG is part of Springer Science+Business Media

www.springer.com

Contents

Pure Appl. Geophys. 172 (2015), 2961–2964
© 2015 Springer Basel
DOI 10.1007/s00024-015-1176-9

▌Pure and Applied Geophysics

Current Topics on Deformation Monitoring and Modelling, Geodynamics and Natural Hazards: Introduction

Pablo J. González,[1] Gerardo Herrera,[2] Francisco Luzón,[3] and Pietro Tizzani[4]

The term natural hazard includes a wide range of diverse physical processes. A non-exhaustive list should include geophysical (e.g., earthquakes, tsunamis, landslides, volcanic eruptions); hydrological (e.g., floods, droughts); shallow processes (e.g., ground subsidence and collapse); atmospheric (e.g., cyclones, tornadoes, hail, lighting); and biophysical (e.g., wildfires) (Gill and Malamud 2014). In this book (should be topical issue), we emphasize the geophysical and shallow processes natural hazards, although the concept is larger by definition. In particular over the last decades, the investigation on natural hazards has greatly advanced by the widespread use of technology and influx of ideas from multiple earth sciences and related disciplines. In this regard, a current revision of advances in the field of monitoring and modelling of surface deformation processes and, in general geodynamics, with emphasis on the natural hazards was a clear necessity.

In 2013 at the 13th congress of the International Association of Mathematical Geology, we envisioned a special session dedicated to the most recent advances in the study of the Natural Hazards with emphasis on the use of modern geodetic techniques. Based on the positive experience, we proposed to Springer-Birkhäuser, the edition of a topical issue. The current book (should be issue) represents selected papers presented at the session "Deformation modelling, Geodynamics and Natural Hazards", chaired by José Fernández and Pablo J. González. We, the editors, are greatly indebted to Dr. José Fernández for the help during the organization of the session and the early stages of completion of the current issue. Our main duty was to help the discussion and foster the dissemination of the latest trends in leading problems on geodynamics and natural hazards.

Reducing the risk associated with the natural hazards related to earthquakes, volcanic eruptions, land subsidence or landslides is one of the most challenging problems confronting the Earth Science community. For example, earthquakes release accumulated elastic strain energy, radiating energy from an expanding rupture surface area. The growing process is poorly known and the accumulation of energy is slow enough to be difficult to capture. Ground deformation is a powerful tool to investigate such complex and challenging processes, with the aim of not only qualitative assessment, but model it for a deeper understanding, leading eventually to future forecast of hazardous activities. This topical issue represents an attempt to compile a broad view on current topics on the use of geodetic data, enhanced with other techniques to capture the inner working of Geodynamics and Natural Hazard processes. The book consists of 14 chapters grouped in 3 main topics: near-surface; structures and seismotectonics; and volcanic processes.

In many instances, natural and anthropogenic geohazards overlap, indicating different processes; of particular interest due to its prevalence is the occurrence of landslide and land subsidence phenomena. In

---

[1] COMET, Institute of Geophysics and Tectonics, School of Earth and Environment, University of Leeds, Leeds LS2 9JT, UK. E-mail: pabloj.glez@gmail.com

[2] Geohazards InSAR Laboratory and Modeling Group, Área de Riesgos Geológicos, Departamento de Investigación Geocientífica, Instituto Geológico Y Minero de España, Calle Alenza 1, 28050 Madrid, Spain.

[3] Department of Applied Physics, University of Almería, 04120 Almería, Spain.

[4] Istituto Per Il Rilevamento Elettromagnetico Dell'Ambiente, Consiglio Nazionale Delle Recerca (IREA—CNR), Via Diocleziano, 328 Naples, Italy.

this topical issue, we start by grouping a large set of studies illustrating examples ranging from observational to numerical modelling approaches of near-surface processes. Near-surface hazards are receiving large attention due to its impact on safety of human settlements and infrastructures. In Cigna et al. the Great London (UK) area has been imaged and analysed using radar interferometry within the PanGeo project. Ground motion estimates and geological data were combined to generate a catalogue of geohazards ranging from compaction of river thames fluvial deposits, slope instabilities, land subsidence due to groundwater management and aquifer changes, and geotechnical works such as subway and electricity tunnelling. In the next paper, displacement estimates derived from the application of Permanent Scatterers (PS). Interferometry to ERS (1992–2000) and ENVISAT (2002–2005) images have been integrated with in situ geological information in the City of Roma to produce a Ground Stability Layer (GSL). The GSL identifies in Rome numerous polygons enclosing areas where geohazards have been pointed out by PS data and/or in situ surveys, including: landslides, collapsible grounds, compressible grounds, groundwater abstraction, mining, made (What is made Ground?) ground, tectonic movements, volcanic inflation/deflation. The following paper (perhaps give the name of the first author?) is dedicated to an application of conventional and advanced DInSAR to detect mining subsidence in the Upper Silesian Coal Basin (Southern Poland) exploiting multi-sensor SAR imaginary. Conventional DInSAR from ALOS-PALSAR and TerraSAR-X images was used to detect the fastest displacements and to monitor the underground mining front in a period of several months. Advanced DInSAR analysis (PSInSAR and SqueeSAR) from ERS, ENVISAT and TerraSAR-X imaginary was useful to detect residual displacement around the subsidence bowl even after the closure of the mines.

Bianchini et al. use C- and X-band radar Persistent Scatterer Interferometry to evaluate the spatial and temporal movements on a known landslide. This paper contains a wealth of in situ field measurements that validate and support the spaceborne motion data. The paper addresses the complex issue of delineating and identifying the most active unstable areas to help

risk-mitigation measures. The San Fratello landslide test case provides an ideal place to test the technology and assess the potential for future sliding events, similar to the February 2010. With the initiation of such large datasets, the logical next step is to improve existing modelling approaches. In Castaldo et al. an inverse numerical modelling of slow landslides is performed from advanced DInSAR time series derived from 20 years of ERS-1/2 and ENVISAT satellite acquisitions. The proposed methodology allows for automatically searching the physical parameters that characterize the landslide behaviour. This approach is validated on the slow Ivancich landslide (Assisi, central Italy), which behaviour is simulated through a two-dimensional time-dependent finite element model. Comparison between the model results and DInSAR measurements and in situ data reveals that the creep model is suitable to describe the kinematic (not kinematical) evolution of the landslide. To conclude with near-surface processes section (land subsidence and landslides), Notti et al. proposed a set of low-level user-oriented indexes to interpret multi-temporal Differential Synthetic Aperture Radar (SAR) Interferometry (DInSAR and PSI).

In the present volume, contributions broadly relevant to the different geodesy branches as gravimetric and ground deformation measurements in A variety of geodynamic contexts were enclosed. In the case of large regional low-rate subsidence phenomenon, Camacho et al. present a new methodological work to adapt a published inversion approach (growth method) to the case of an alluvial valley (sedimentary stratification, with density increase downward). More specifically, the paper presents a gravimetric study of the low Andarax valley that is an alluvial basin, close to its river mouth, is located in the extreme south of the province of Almería. In an additional study, Conejo-Martín et al. proposed a novel methodology to determine the electromagnetic velocity of the ground due to a combination of LIDAR Digital Terrain Model with Ground Penetrating Radar profiles. This technique is suitable for use in shallow underground spaces with access from surface, as other natural cavities, archaeological cavities, sewerage systems, drainpipes, etc. It has been applied over some cavities used for underground wine cellars in Atauta (Soria, Spain), and allowed the correct detection of the inner structures of cavities in high resolution and with great accuracy.

At higher sampling rates, the ground deformation is better characterized by seismological methods. In this topical issue, three papers are dedicated to understand Natural Hazards with emphasis in seismotectonic aspects. Induced and natural seismicity is a research subject with profound societal implications. There is a clear gap about different studied regions with much of the current research focused on the US and European cases, but Sarychikhina et al. present a study on ground deformation and seismicity with the aim to separate the origin of the observed seismicity in Mexicali Valley, Mexico. Levelling surveys, radar interferometry and seismic catalogues illustrate a complex setting where active tectonics and human activity interact. Untangling the contributions from the fluid extraction at the Cerro Prieto geothermal field and activity along tectonic faults ensures future research in this area, exploiting spaceborne and ground-based instruments (tiltmeters and creepmeters). Jiménez et al. faced the challenge to determine useful earthquake sources and attenuation parameters around the Itoiz dam, the hydrological infrastructure in Northern Spain, using low magnitude events. Novel inverse methods allow the authors to obtain estimates of moments, source radii and stress drops. Empirical and theoretical relationships show a good agreement between moment and magnitude. Evaluation of seismic activity around Itoiz is a relevant case due to the observed correlation with the water-level variations. Finally, Carmona et al. embrace the identification of internal waves on smaller ocean basins. The authors take advantage of the special setting of the Northern Moroccan seismicity and a dense seismological network in Southern Spain (the Red Sísmica de Andalucía) to record late seismic arrivals. Those late arrivals decayed in amplitude with distance from the coast and have relatively constant delays of 85 s with respect to the P-wave. Those evidences suggested that T-waves were responsible for the late arrivals, usually observed in large oceanic basins. This study shows the first evidence of such waves in a small closed sea, such as the Alboran Sea.

In volcanic context, three contributions were enclosed in the volume. Fernández et al. present an overview of geodetic volcano research in the Canary Islands (Spain). (No: is proposed). The authors describe the research in volcano geodetic monitoring carried out in the Canary Islands and the results obtained and consider for each epoch the two main constraints existing: the level of volcanic activity in the archipelago and the limitations of the techniques available at the time. The second enclosed contribution is relevant to the ground deformation prior to the 2010 eruption of Mt. Sinabung. In particular, González et al. used differential interferometric synthetic aperture radar (DInSAR) obtained from Japanese ALOS-PALSAR radar imagery, between 05 January 2007 and 31 August 2010. InSAR time series processing results detected significant ground deformation (subsidence) at several locations in the Karo plateau, and uplift at the summit area of Mt. Sinabung of hydrothermal origin preceding a phreatic eruption. D'Auria et al. proposed a novel approach to retrieve the full stress field based on ground deformation and seismological estimates. The inversion of the surface ground deformation independently, is consistent with a planar crack source, located at a depth of about 2.56 (perhaps just 2.5, this could not be given with such detail?) km underneath the centre of the caldera. The addition of regional background stress field helped to delineate the full stress field into a variable component (of volcanic origin) and a weak NNE-SSW extension due to constant tectonic forces. This methodology highlights the resolving power of the joint use of seismological and geodetic data for volcano monitoring.

This monographic topical issue does not thematically exhaust all directions that have been taken by the earth science community to understand the Natural Hazards and Geodynamics. However, it does provide an essential view of significant issues in the observation and modelling of geophysical, geodetic and shallow hazardous processes, which should prompt further research and interpretations from the current trend of multidisciplinary approaches.

This topical issue has been made possible due to a dedicated group of people. The editors of the book offer to all of them their sincere gratitude. In particular, we would like to acknowledge Dr. Renata Dmowska, who persevered in the effort to complete the work. Instrumental was Ms. Priyanka Ganesh who helped immensely with the technical editorial work, making our work easier and smoother. We thank all the crew at the editorial offices of Springer-Birkhäuser, who greatly helped at different stages (Katherina Steinmetz, Clemens Heine, Barbara Hellriegel and Thomas Tschech), our apologies to anyone missing in this brief list. We deeply thank all authors for their contributions and, in particular those leading manuscripts that unfortunately could not be accepted. Finally, we deeply acknowledge the assistance of the following reviewers [listed in alphabetical order]: J.M. Azañón, B. Benjumea, A.G. Camacho, A. Cannata, F. Cigna, J. Díaz, Y. Fialko,

P.J. González, F. Guglielmino, J. Fernández, J.A. Fernández-Merodo, Z. Li, A. Manconi, M. Motagh, U. Niethammer, M. Palano, S. Pepe, E. Poyiadji, F. Raspini, M. Sanabria, F. Sánchez-Sesma, M.A. Santoyo, S. Samsonov, D.A. Schmidt, V. Singhroy, R. Stephen, P. Teatini, K.F. Tiampo and R. Tomás.

REFERENCE

GILL, J. C., and MALAMUD, B. D. (2014), Reviewing and visualizing the interactions of natural hazards, Rev. Geophys., 52, 680–722, doi: 10.1002/2013RG000445.

(Published online October 8, 2015)

Pure Appl. Geophys. 172 (2015), 2965–2995
© 2014 The Author(s)
This article is published with open access at Springerlink.com
DOI 10.1007/s00024-014-0927-3

❙ **Pure and Applied Geophysics**

Natural and Anthropogenic Geohazards in Greater London Observed from Geological and ERS-1/2 and ENVISAT Persistent Scatterers Ground Motion Data: Results from the EC FP7-SPACE PanGeo Project

Francesca Cigna,[1] Hannah Jordan,[1] Luke Bateson,[1] Harry McCormack,[2] and Claire Roberts[2]

Abstract—We combine geological data and ground motion estimates from satellite ERS-1/2 and ENVISAT persistent scatterer interferometry (PSI) to delineate areas of observed natural and anthropogenic geohazards in the administrative area of Greater London (United Kingdom). This analysis was performed within the framework of the EC FP7-SPACE PanGeo project, and by conforming to the interpretation and geohazard mapping methodology extensively described in the Production Manual (cf. http://www.pangeoproject.eu). We discuss the results of the generation of the PanGeo digital geohazard mapping product for Greater London, and analyse the potential of PSI, geological data and the PanGeo methodology to identify areas of observed geohazards. Based on the analysis of PSI ground motion data sets for the years 1992–2000 and 2002–2010 and geology field campaigns, we identify 25 geohazard polygons, covering a total of ~650 km². These include not only natural processes such as compaction of deposits on the River Thames flood plain and slope instability, but also anthropogenic instability due to groundwater management and changes in the Chalk aquifer, recent engineering works such as those for the Jubilee Line Extension project and electricity tunnelling in proximity to the River Thames, and the presence of made ground. In many instances, natural and anthropogenic observed geohazards overlap, therefore indicating interaction of different processes over the same areas. In terms of ground area covered, the dominant geohazard is anthropogenic land subsidence caused by groundwater abstraction for a total of ~300 km², followed by natural compression of River Thames sediments over ~105 km². Observed ground motions along the satellite line-of-sight are as high as +29.5 and −25.3 mm/year, and indicate a combination of land surface processes comprising ground subsidence and uplift, as well as downslope movements. Across the areas of observed geohazards, urban land cover types from the Copernicus (formerly GMES) EEA European Urban Atlas, e.g., continuous and discontinuous urban fabric and industrial units, show the highest average velocities away from the satellite sensor, and the smallest standard deviations (~0.7–1.0 mm/year). More rural land cover types such as agricultural, semi-natural and green areas reveal the highest

spatial variability (up to ~4.4 mm/year), thus suggesting greater heterogeneity of observed motion rates within these land cover types. Areas of observed motion in the PSI data for which a geological interpretation cannot be found with sufficient degree of certainty are also identified, and their possible causes discussed. Although present in Greater London, some geohazard types such as shrink–swell clays and ground dissolution are not highlighted by the interpretation of PSI annual motion rates. Reasons for absence of evidence of the latter in the PSI data are discussed, together with difficulties related to the identification of good radar scatterers in landsliding areas.

Key words: Geohazard, persistent scatterer interferometry, ground motion, InSAR, monitoring, land surface processes.

1. Introduction

Geohazards and their impacts in the United Kingdom (UK) have long been discussed in the literature. Gibson *et al.* (2013) analyse aspects related to management of landslide hazards in an environment considered as low-risk, but where the financial loss from such a hazard is likely to be in excess of £10 million per year. Farrant and Cooper (2008) investigate geological properties of soluble rocks, and report on karstic features observed in Carboniferous limestone, chalk and Permo-Triassic gypsum and halite. Flooding and storms occurring in vulnerable floodplains and coastal areas have large economic impacts, and single hydro-meteorological events have caused damage of over £3 billion (Pitt 2008). Predominantly affecting the southeast of the country, volume changes of clay soils and mudrocks in response to variations in moisture content are considered the cause of the largest financial impact in the UK, with costs up to £500 million in a single year

[1] British Geological Survey, Natural Environment Research Council, Nicker Hill, Keyworth, Nottinghamshire NG12 5GG, UK. E-mail: fcigna@bgs.ac.uk; hann@bgs.ac.uk; lbateson@bgs.ac.uk
[2] CGG, NPA Satellite Mapping Ltd., Crockham Park, Edenbridge, Kent TN8 6SR, UK. E-mail: harry.mccormack@cgg.com; claire.roberts@cgg.com

(JONES and TERRINGTON 2011). The correlation between geotechnical and mineralogical factors and the shrink–swell susceptibility of the UK has been analysed and discussed widely (e.g., JONES and JEFFERSON 2012), and major relationships between the number of subsidence claims due to shrinkage and historical records of both precipitation and average temperatures have been found (HARRISON et al. 2012).

The British Geological Survey (BGS) has undertaken natural geohazard susceptibility mapping for Great Britain, and produced the GeoSure data set (BOOTH et al. 2010; WALSBY 2008). Developed at a scale of 1:50,000, this data set provides information about potential natural ground movement resulting from collapsible deposits, compressible ground, landslides (GIBSON et al. 2013), running sand, shrink–swell (HARRISON et al. 2012) and soluble rocks (COOPER et al. 2011). Susceptibility is classified using an A (lowest) to E (highest) rating for each of these six geohazard types (BGS 2014).

Depending on their causes, geohazards can be observed within areas of high natural geohazard susceptibility or even areas where susceptibility is low. For instance, land subsidence induced by anthropogenic activities such as mining or tunnelling can occur in areas with low susceptibility to ground compaction, where motions at the surface can occur in response to artificial changes in the in situ stress induced by the excavation and removal of subsurface material. Similarly, changes in the pore water pressure due to groundwater abstraction for domestic use, or water levels control during engineering and mining works, can alter local conditions in the effective stress and result in ground surface motions. Surface evidence of these processes in the UK was recently investigated by processing satellite synthetic aperture radar (SAR) imagery and interpreting their derived ground motion data together with a range of geological layers and information (e.g., BANTON et al. 2013; BATESON et al. 2014; CULSHAW et al. 2006; LEIGHTON et al. 2013; SOWTER et al. 2013). A feasibility study with nationwide coverage has also been carried out by CIGNA et al. (2014) to assess the potential of ERS-1/2 and ENVISAT C-band SAR-based imaging of the landmass of Great Britain, by analysing archive data availability, simulating geometric distortions and modelling land cover control on the success of SAR,

interferometric SAR (InSAR) and persistent scatterer interferometry (PSI) applications.

For the area of London and Thames estuary, the DEFRA/EA R&D Land Levels project analysed land level changes using ERS-1/2 and ENVISAT PSI ground motion data from March 1997 to December 2005. A variety of regional to local-scale controlling factors on the rates of these changes were found, ranging from near-surface to deep-seated mechanisms, and with a variety of temporal scales, from <10 to over 100,000 years duration (ALDISS et al. 2014; BINGLEY et al. 2007). Geological interpretation of spatial correlations between the larger variations in satellite data and various geological data sets allowed the identification of 'domains' of approximately uniform vertical ground velocity and the identification of major lineaments within the data distribution. This project examined the data at up to 1:20,000 scale, whereas higher spatial scales and smaller variations in the PSI motion data were not interpreted (ALDISS et al. 2014).

Areas of observed and potential ground instability at the reference scale of 1:10,000 have been depicted for 52 of the largest towns across 27 countries within the European Union by the validated geohazard layers generated by the Geological Surveys of Europe in the framework of the PanGeo project (http://www.pangeoproject.eu). PanGeo was funded in 2011–2014 by the European Commission under the Space Theme, Seventh Framework Programme (FP7-SPACE), and led by Fugro—Nigel Press Associates (NPA), now CGG—NPA Satellite Mapping Ltd. Among these 52 towns, London (CIGNA et al. 2013c) and Stoke-On-Trent (JORDAN et al. 2013) were selected as targets for the UK, and the BGS was responsible for the generation of the respective geohazard layers. The objective of PanGeo was to enable free and open access to geohazard information in Europe in support of Copernicus, the European Earth Observation Programme. For each PanGeo town, the Geological Survey of the respective country has generated: (1) a polygon-wise ground stability layer (GSL) showing location, extent and typology of the observed and potential geohazards, and (2) the geohazard description (GHD) document, a supporting report that describes in detail the geological setting and places of interest affected by each geohazard, the

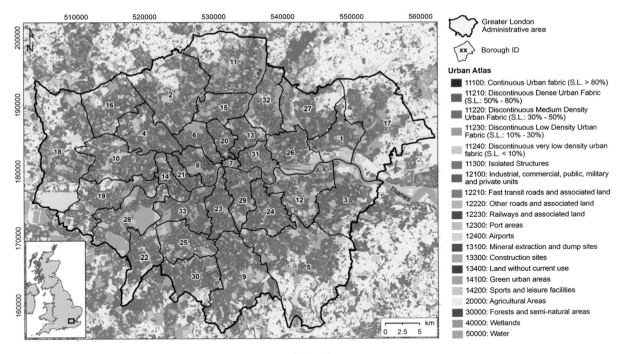

Figure 1

Land cover from the GMES EEA European Urban Atlas for Greater London administrative area and Boroughs; *S.L.* sealing layer. Borough IDs: *1* Barking and Dagenham, *2* Barnet, *3* Bexley, *4* Brent, *5* Bromley, *6* Camden, *7* City and County of the City of London, *8* City of Westminster, *9* Croydon, *10* Ealing *11* Enfield, *12* Greenwich, *13* Hackney, *14* Hammersmith and Fulham, *15* Haringey, *16* Harrow, *17* Havering, *18* Hillingdon, *19* Hounslow, *20* Islington, *21* Kensington and Chelsea, *22* Kingston upon Thames, *23* Lambeth, *24* Lewisham, *25* Merton, *26* Newham, *27* Redbridge, *28* Richmond upon Thames, *29* Southwark, *30* Sutton, *31* Tower Hamlets, *32* Waltham Forest, *33* Wandsworth. British National Grid; Projection: Transverse Mercator; Datum: OSGB 1936. GMES EEA Urban Atlas © EEA 2013, Directorate-General Enterprise and Industry (DG-ENTR), Directorate-General for Regional Policy

confidence and any additional evidence associated with the interpretation. All PanGeo GSL and GHD products have been made freely accessible and usable via a portal based on OneGeology Europe infrastructure that can be accessed at: http://pangeo.brgm-rec.fr/pangeoportal, or visualised in Google Earth via the PanGeo coverage map on: http://www. pangeoproject.eu/eng/coverage_map. Integration of the GSL with the Copernicus Land Theme's Urban Atlas (EC 2011) shows the land cover and use classes influenced by such hazards, and supports the end-users in the management of hazards and induced risks within the concerned areas.

Geological and other geospatial layers and information for the towns analysed within PanGeo were integrated with ground motion data generated from the processing of long temporal stacks of satellite SAR imagery, using PSI approaches (e.g., Crosetto *et al.* 2010). When compared against other monitoring data, the results of PSI techniques can achieve accuracies up to the level of a few mm/year, depending on surface characteristics of the processed area, quantity and quality of the input SAR imagery, and quality of the PSI processing. Findings of the European Space Agency (ESA) Terrafirma Validation study for the Alkmaar and Amsterdam sites in The Netherlands have shown that the observed overall accuracy of PSI average annual velocity was 1.0–1.8 mm/year (RMSE) for ERS-1/2 and ENVISAT data when compared against levelling (Crosetto *et al.* 2008; Hanssen *et al.* 2008).

In this paper, we show the results of the generation of the PanGeo digital geohazard mapping product for Greater London. The total investigated site corresponds to the administrative area and extends $\sim$1,580 km^2 (Fig. 1). This area includes the City of London and 32 other surrounding boroughs, within which a total population of more than 8

7

million inhabitants was censused in 2011, corresponding to an average of $\sim 5{,}200$ inhabitants/km^2.

We briefly describe the PanGeo methodology based on the Production Manual (BATESON et al. 2012) and illustrate the available input data for Greater London in Sect. 2. These include newly processed ERS-1/2 and ENVISAT PSI data sets showing the ground motion history of London over the last two decades. In Sect. 3, building upon the results of the PanGeo products generation, we analyse the potential of PSI, geological data and the PanGeo methodology to delineate areas of observed geohazards. Some types of geohazards that were not highlighted by the interpretation of annual motion rates from the PSI data, such as shrink–swell clays and dissolution, are considered in Sect. 3.4, where the reasons for absence of evidence of the latter in the PSI average motion velocities are also discussed, together with difficulties related to the identification of good radar scatterers in landsliding areas, where land cover exerts significant control on the success of conventional PSI techniques. Areas of potential instability due to natural geohazards have not been mapped for the UK PanGeo towns, as these are already delineated for all Great Britain at the 1:50,000 scale through the GeoSure data sets, for which further information is provided in Sect. 3.4.

2. Interpretation Methodology and Input Data

The interpretation and geohazard mapping approach employed for the delineation of observed geohazards in Greater London, conforms to the step-by-step methodology that is extensively described by BATESON et al. (2012) in the PanGeo Production Manual. The latter is a freely downloadable document that was distributed to the Geological Surveys to support the generation of their GSL by following instructions and procedures in accordance with the PanGeo Product Specification (BATESON 2013), in addition to specific training workshops and related material, which can be accessed at: http://www.pangeoproject.eu/eng/educational.

All the GSL polygons are attributed with classifications and hazard categories compliant with the Natural Risk Zones data specification of INSPIRE

(EC-JRC, 2013), which are also used in the project portal to provide a summary of the geohazard identified within the area. Geohazard categories considered by PanGeo are 'Deep Ground Motions', 'Natural Ground Instability', 'Natural Ground Movement', 'Anthropogenic Ground Instability', 'Other' (i.e., their geological explanation does not fit into any of the previous categories) and 'Unknown' (i.e., of which a geological interpretation cannot be found with sufficient degree of certainty), and each includes a number of hazard types. These conform to the Glossary of terms for PanGeo, available within the Product and Service Specification report of the project (BATESON 2013). A measure of the confidence in the interpretation is attributed to each geohazard, by using a scale of 'low', 'medium' and 'high' depending on the number of data sets used in the interpretation, or 'external' for those geohazards imported from an external source (e.g., landslide inventory). The determination method refers to the main information source that has been used to identify the geohazard, and is classified in PanGeo as: 'Observed in PSI', 'Observed in Other Deformation Measurement' (e.g., levelling, GPS), 'Observed in Geology Field Campaigns' or 'Potential'. Full details and a step-by-step methodological approach are described in the PanGeo Production Manual by BATESON et al. (2012).

The identification of geohazards in Greater London was performed through combined interpretation of geological, geomorphologic, land use and other geospatial layers available at the BGS (cf. CIGNA et al. 2013c), together with satellite PSI ground motion data for the 18-year long period between 1992 and 2010 (see Sect. 2.1). Background input data used to map geohazards include Ordnance Survey (OS) topographic maps at 1:10,000–1:50,000 scales, 0.25-m resolution aerial photographs, 5-m resolution NEXTMap$^\circledR$ DEM, the Digital Geological Map of Great Britain (DiGMapGB) at 1:625,000 to 1:10,000 scales (SMITH, 2013), the Superficial Deposits Thickness Model (SDTM) at 1:50,000 scale (LAWLEY and GARCIA-BAJO 2009), the National Landslide Database (NLD; FOSTER et al. 2012) and Karst Database (FARRANT and COOPER 2008), and groundwater pumping records from recent surveys carried out by the Environment Agency (EA 2007, 2010).

Table 1

Main characteristics of the PSI data sets employed for the generation of the GSL of Greater London

Data stack	No. scenes	Period (day/mo/yr)	Master scene	Georeference accuracy (m)	Reference point (lat, long)	PS coherence threshold	PSI processing area			GSL area	
							Area (km^2)	No. PS	PS density (PS/km^2)	No. PS	PS density (PS/km^2)
ERS-1/2 SAR ascending	27	19/06/ 1992–31/ 07/2000	13/01/ 1997	10	51.552°N, – 0.113°E	0.53	2,500	730,254	292	615,950	386
ENVISAT ASAR descending	45	13/12/ 2002–17/ 09/2010	11/05/ 2007	10	51.554°N, – 0.101°E	0.49	2,350	838,939	336	712,236	446

The results of the projects ESA GSE Terrafirma for the London H3 Modelled Product (BATESON *et al.* 2009; ESA 2009), and DEFRA/EA R&D Land Levels project results (ALDISS *et al.* 2014; BINGLEY *et al.* 2007) were also incorporated into the interpreted geohazard polygons of the GSL. The latter include average vertical ground motion information for 'domains' of uniform land level change that were identified based on an absolute gravimetry (AG) and GPS-aligned, ERS-ENVISAT-combined PSI product covering the period between 1997 and 2005, and processed by CGG—NPA Satellite Mapping, Ltd.

2.1. *ERS-1/2 and ENVISAT Interferometric Point Target Analysis*

To generate two new data sets of PSI ground motion data for Greater London, we employed the GAMMA SAR and Interferometry software, and in particular, the interferometric point target analysis (IPTA) algorithm, developed at GAMMA Remote Sensing and Consulting AG in Switzerland (WERNER *et al.* 2003). IPTA exploits the spatial and temporal characteristics of interferometric phase signatures of ground targets that exhibit point-like scattering behaviour and remain coherent over the monitored period, to estimate their ground motion velocities, time histories, terrain heights, and relative atmospheric path delays. This technique was recently used to monitor a variety of geological processes and man-made geohazards, including landsliding, ground subsidence, deep-level mining and structural instability (e.g., GIGLI *et al.* 2012; STROZZI and AMBROSI 2007; TEATINI *et al.* 2007; WEGMULLER *et al.* 2010). As input

for the IPTA processing, we used the following data stacks of C-band, VV polarized SAR imagery acquired from sun-synchronous near-polar orbits and with 35 days nominal repeat cycle (Table 1):

1. 27 ERS-1 and ERS-2 SAR scenes acquired between 19/06/1992 and 31/07/2000 in ascending mode, along the satellite track 201; and
2. 45 ENVISAT advanced SAR (ASAR) Image Mode IS2 scenes acquired between 13/12/2002 and 17/09/2010 in descending mode, along the satellite track 51.

The inclination of both ERS-1/2 and ENVISAT satellite ground tracks at the SAR scene centre was ~14° with respect to the S–N axis at the latitude of Greater London, and the incidence angle of the employed sensor modes was ~23° measured from the vertical direction. This means that the employed LOS were able to estimate purely vertical motions as ~92 % of their actual amount, E–W motions as ~38 %, and N–S as only ~9 %.

The processing followed the iterative methodology described by WERNER *et al.* (2003), and was carried out based on a selected number of candidate points in the radar imagery that were persistent over the observation time period and dominated the backscattering within the resolution pixels. The 90 m resolution Shuttle Radar Topography Mission (SRTM) Digital Surface Model (DSM) by NASA-JPL was used to simulate the initial topographic phase components, and a simple linear model of phase variation through time was chosen to extract phase signals relating to ground displacements.

9

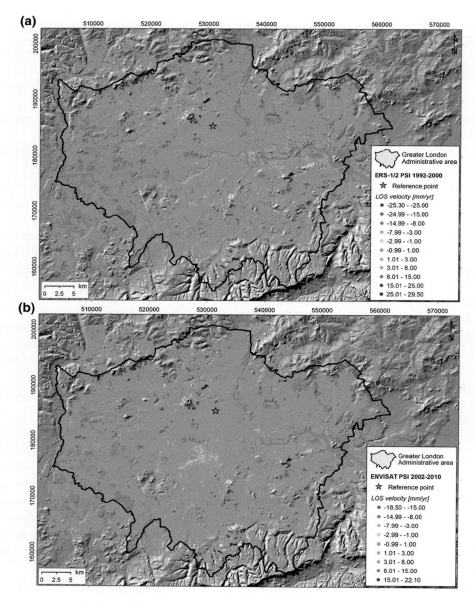

Figure 2
Average motion velocities in **a** 1992–2000 and **b** 2002–2010 for Greater London, estimated along the line-of-sight of, respectively, ERS-1/2 satellites in ascending mode and ENVISAT in descending mode. Refer to Table 1 for detailed processing statistics. *Green PS* are considered stable with respect to the reference point location, whereas *yellow to dark red PS* indicate motions recorded away from the sensor, and *light to dark blue* indicate motions towards the sensor. PSI data are overlapped onto shaded relief of NEXTMap® DTM at 50 m resolution. British National Grid; Projection: Transverse Mercator; Datum: OSGB 1936. NEXTMap® Britain © 2003, Intermap Technologies Inc., All rights reserved

Step-wise iterative processing using height corrections, linear motion rates, standard deviations and residual phases, allowed us to progressively improve the different phase components, and to extract a total of 730,254 ERS-1/2 persistent scatterers (PS), corresponding to an average density of 292 PS/km^2 across the ∼2,500 km^2 processing area, and 386 PS/km^2 within the administrative area of Greater London. The processing of the ENVISAT stack identified 838,939 targets, hence 336 PS/km^2

over the respective processing area of $\sim 2,350$ km^2, and 446 PS/km^2 in Greater London (Fig. 2). Despite the difference in the number of scenes populating the two stacks, the observed number and density of targets over the GSL area are similar, and amount to ~ 400 PS/km^2 per data set. The somewhat greater density observed for the ENVISAT results is likely due to the larger number of scenes composing the ENVISAT stack, as opposed to the ERS stack. Indeed, higher numbers of input scenes generally result in PS data sets with both higher quality and reliability, and denser networks of good reflectors. In this particular case, the different coherence thresholds used to extract the final set of radar targets (i.e., 0.53 for ERS and 0.49 for ENVISAT) clearly had an impact on the final number of points, with the ENVISAT data set more likely to have more scatterers due to the lower threshold employed. Bearing in mind that the selection of the optimal coherence thresholds is a trade-off between the quality and the number of point targets composing the resulting data set, for the area of Greater London, a higher coherence threshold during the ERS-1/2 processing was chosen. This was done in order to minimise the inclusion of lower quality targets in the final results, accounting for the smaller number of ERS-1/2 ascending mode SAR scenes available to perform the IPTA processing with respect to the more populated ENVISAT ASAR stack.

Reference points for the PS data sets were identified at similar locations, i.e., WGS84 51.552°N, −0.113°E for ERS-1/2 and 51.554°N, −0.101°E for ENVISAT (see green stars in Fig. 2a, b). The locations were selected accounting for both the interferometric phase stability of the PS candidates and the geological setting of this sector of the area, which was considered a good site to reference all ground motion data to.

For over 95 % of the PS targets found within the GSL area, the standard deviations of the estimated annual velocity along the satellite LOS are between 0.09 and 1.09 mm/year in the ERS-1/2 data set, and between 0.17 and 1.13 mm/year in the ENVISAT data set (by assuming the data are normally distributed). Taking these values into account, we considered the PS points showing annual deformation velocities along the LOS in the range of ±1.0 mm/year as 'stable' (i.e., green PS in Fig. 2a, b).

From the observation of average annual velocities across the administrative area, it is apparent that although the two PSI data sets revealed a general stability at the regional scale over both periods of 1992–2000 and 2002–2010, some areas show significant motions away from the satellite. In most cases, these are located along the Thames valley, and in the Fulham, Battersea and Clapham areas (Fig. 2). Minimum and maximum annual velocities observed within Greater London along the satellite LOS are −25.3 and +29.5 mm/year in the ERS-1/2 data set (1992–2000), and −18.5 and +22.1 mm/year in the ENVISAT data set (2002–2010). The distribution of average velocities also confirms the absence of regional trends or wide scale shifts that could have resulted from inappropriate selection of the reference location.

It is worth noting that the accuracy of ERS-1/2 and ENVISAT data sets can be assessed by comparing the resulting PSI ground motion velocities and time series against continuous GPS stations that operated in the study region during the same time intervals. For instance, vertical motion histories from GPS stations of the NERC-funded British Isles continuous GNSS Facility (BIGF; http://www.bigf.ac.uk) could be considered. This specific analysis would allow estimation of the reference accuracy of our results in Greater London, and correction of potential shifts due to the reference point selection and tilts that were artificially removed during the processing, though it is beyond the scope of this paper to analyse this aspect further.

3. Results and Discussion

Using the methodology described in Sect. 2, we identified a total of 25 geohazard polygons over Greater London, covering a total of ~ 700 km^2, or ~ 650 km^2 if excluding overlapping geohazards (Fig. 3; Table 2). In most cases, observed geohazards are identified as a single-part polygon, whereas in the case of landslides, the areas of motion are grouped into multi-part polygons sharing the same set of standardised PanGeo attributes.

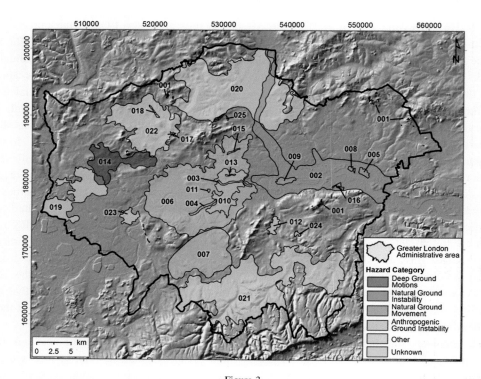

Figure 3
PanGeo Ground Stability Layer of Greater London: observed geohazards classified by Hazard Category and overlapped onto shaded relief of NEXTMap® DTM at 50 m resolution. *Labels* indicate the last three digits of the INSPIRE polygon IDs. Refer to Table 2 for detailed information and PSI ground motion statistics for each observed geohazard. British National Grid; Projection: Transverse Mercator; Datum: OSGB 1936. NEXTMap® Britain © 2003, Intermap Technologies Inc., All rights reserved

A range of geohazard types are observed, including both natural processes such as compaction of deposits on the River Thames flood plain and anthropogenic instability due to water abstraction and recent engineering works. There are nine areas of observed anthropogenic geohazards in total, including both the 'Anthropogenic Ground Instability' and the 'Other' hazard categories (see Sect. 3.2). Natural geohazards include three polygons (see Sect. 3.1) of 'Natural Ground Instability', 'Natural Ground Movement' and 'Deep Ground Motion', whereas 13 have unknown causes (see Sect. 3.3). In terms of ground area covered, the dominant geohazard is anthropogenic land subsidence caused by groundwater abstraction for a total of ~ 300 km^2 (see Sect. 3.2.2), followed by natural compression of River Thames sediments over ~ 105 km^2 (see Sect. 3.1.1). In many instances, geohazards of different categories and types overlap, thus indicating interaction of different processes (cf. Fig. 3).

As regards confidence in the interpretation, six high, six medium, 12 low and one external confidence level polygons are identified. As for the determination method, only one mapped polygon corresponds to geohazards observed by geology field campaigns (i.e., landslides from the NLD and DiGMapGB mass movement layer), whereas the remaining 24 correspond to areas observed in PSI data, 18 of which show subsidence and six of which show uplift. Ground motion statistics from ERS-1/2 and ENVISAT PSI data for all observed geohazards are summarized in Table 2. For each geohazard polygon, the maximum, minimum and average observed velocity of all PS points identified within its boundary are computed for both monitoring periods, with an understanding that different levels of homogeneity (or heterogeneity) can be observed within the various polygons; these are indicated by the observed standard deviation. The latter quantifies the dispersion of the annual velocity values for all the PS included

Table 2

Observed geohazards in Greater London: INSPIRE ID, hazard category and type, motion type, determination method, confidence level, total area, and ground motion statistics for each observed geohazard

INSPIRE ID (polygon ID)	Hazard Category	Hazard Type	Motion type	Determination method	Confidence level	Area (km²)	Observation date Start	Observation date End	ERS-1/2 data from 1992 to 2000 No. PS	LOS velocity (VEL) (mm/year) Avg	Min	Max	SD	ENVISAT data from 2002 to 2010 No. PS	LOS velocity (VEL) (mm/year) Avg	Min	Max	SD
PGGH_London_001	Natural Ground Instability	Landslide	Down slope	Observed Geology Field Campaigns	External	0.52	14/10/2010[a]		3	−3.32	−11.64	+1.05	7.20	5	−1.36	−1.87	−0.14	0.69
PGGH_London_002	Natural Ground Movement	Compressible Ground	Subsidence	Observed PSI	High	104.27	19/06/1992	17/09/2010	33,597	−0.47	−19.19	+29.51	1.78	39,972	−0.51	−12.85	+9.09	1.00
PGGH_London_003	Anthropogenic Ground Instability	Underground Construction	Subsidence	Observed PSI	High	1.32	19/06/1992	17/09/2010	883	−0.92	−15.89	+18.92	1.86	1,428	−1.16	−4.13	+2.16	0.74
PGGH_London_004	Anthropogenic Ground Instability	Underground Construction	Subsidence	Observed PSI	High	1.39	19/06/1992	17/09/2010	870	−1.93	−14.04	+20.10	1.78	1,305	−1.81	−6.04	+0.62	0.64
PGGH_London_005	Anthropogenic Ground Instability	Made Ground	Subsidence	Observed PSI	High	1.38	19/06/1992	17/09/2010	317	−3.23	−15.99	+5.57	3.45	400	−2.78	−10.64	+2.57	2.52
PGGH_London_006	Anthropogenic Ground Instability	Groundwater Abstraction	Subsidence	Observed PSI	High	146.65	13/12/2002	17/09/2010	101,138	+0.19	−19.45	+22.47	1.15	130,813	−0.81	−9.22	+5.48	0.64
PGGH_London_007	Anthropogenic Ground Instability	Groundwater Abstraction	Subsidence	Observed PSI	High	47.17	19/06/1992	17/09/2010	19,769	−0.14	−25.28	+20.11	1.16	23,477	−0.47	−8.43	+4.98	0.62
PGGH_London_008	Anthropogenic Ground Instability	Made Ground	Subsidence	Observed PSI	Medium	0.57	19/06/1992	17/09/2010	261	−1.69	−11.92	+17.98	2.38	371	−1.72	−7.27	+1.39	1.32
PGGH_London_009	Other	Other	Uplift	Observed PSI	Medium	2.19	19/06/1992	31/07/2000	377	+1.44	−2.36	+20.46	2.40	1,033	−0.96	−6.24	+1.79	1.04
PGGH_London_010	Other	Other	Uplift	Observed PSI	Medium	21.29	19/06/1992	31/07/2000	17,747	+0.45	−16.28	+20.90	1.12	23,651	−0.96	−7.73	+4.72	0.68
PGGH_London_011	Unknown	Unknown	Subsidence	Observed PSI	Medium	0.14	19/06/1992	17/09/2010	225	−1.34	−14.92	+0.37	1.15	218	−1.74	−4.83	+1.11	0.63
PGGH_London_012	Unknown	Unknown	Subsidence	Observed PSI	Medium	2.80	13/12/2002	17/09/2010	1,774	−0.34	−11.52	+19.87	1.20	2,109	−1.21	−5.56	+1.72	0.70
PGGH_London_013	Unknown	Unknown	Subsidence	Observed PSI	Medium	0.23	13/12/2002	17/09/2010	302	+0.43	−2.33	+7.09	0.63	476	−1.64	−5.38	+2.26	0.94
PGGH_London_014	Deep Ground Motions	Tectonic Movements	Uplift	Observed PSI	Low	26.63	19/06/1992	31/07/2000	10,028	+0.40	−19.48	+22.23	1.22	15,049	−0.18	−7.74	+5.81	0.69
PGGH_London_015	Anthropogenic Ground Instability	Underground Construction	Subsidence	Observed PSI	Low	0.38	13/12/2002	17/09/2010	272	+0.07	−4.55	+5.28	0.81	286	−1.09	−5.69	+4.76	0.85
PGGH_London_016	Unknown	Unknown	Subsidence	Observed PSI	Low	0.65	19/06/1992	17/09/2010	428	−1.57	−12.76	+0.54	1.36	307	−1.59	−12.51	+1.33	1.51
PGGH_London_017	Unknown	Unknown	Subsidence	Observed PSI	Low	0.19	19/06/1992	17/09/2010	164	−0.85	−9.40	+9.12	1.67	165	−1.33	−9.72	+1.36	1.17
PGGH_London_018	Unknown	Unknown	Subsidence	Observed PSI	Low	0.44	13/12/2002	17/09/2010	325	−0.40	−20.12	+7.31	1.48	377	−1.10	−3.59	+2.31	0.66
PGGH_London_019	Unknown	Unknown	Subsidence	Observed PSI	Low	32.21	13/12/2002	17/09/2010	7,045	+0.15	−17.52	+17.12	1.27	15,585	−0.74	−9.83	+4.90	0.69

Table 2 continued

INSPIRE ID (polygon ID)	Hazard		Motion type	Determination method	Confidence level	Area (km²)	Observation date		ERS-1/2 data from 1992 to 2000					ENVISAT data from 2002 to 2010				
	Category	Type					Start	End	No. PS	LOS velocity (VEL) (mm/year)				No. PS	LOS velocity (VEL) (mm/year)			
										Avg	Min	Max	SD		Avg	Min	Max	SD
PGGH_London_020	Unknown	Unknown	Uplift	Observed PSI	Low	127.16	19/06/1992	31/07/2000	59,918	+0.48	−19.47	+26.47	1.25	68,741	−0.09	−11.01	+8.43	0.62
PGGH_London_021	Unknown	Unknown	Uplift	Observed PSI	Low	126.85	19/06/1992	17/09/2010	52,695	+0.51	−22.62	+22.48	1.09	53,582	−0.10	−9.98	+5.65	0.55
PGGH_London_022	Unknown	Unknown	Subsidence	Observed PSI	Low	54.45	19/06/1992	17/09/2010	24,058	−0.28	−22.44	+18.81	1.15	28,561	−0.22	−9.72	+6.82	0.64
PGGH_London_023	Unknown	Unknown	Uplift	Observed PSI	Low	4.10	19/06/1992	31/07/2000	1,993	+0.60	−18.95	+17.63	1.11	2,375	−0.26	−5.06	+3.73	0.54
PGGH_London_024	Unknown	Unknown	Subsidence	Observed PSI	Low	0.52	13/12/2002	17/09/2010	202	+0.00	−4.01	+3.05	0.63	263	−1.15	−5.29	+1.03	0.64
PGGH_London_025	Unknown	Unknown	Subsidence	Observed PSI	Low	0.13	19/06/1992	17/09/2010	74	−0.59	−4.24	+8.63	1.60	31	−1.21	−3.97	+1.19	1.08

[a] Publication date of the DiGMapGB-50 V6.2 mass movements layer

within the boundaries of each geohazard polygon with respect to their spatial average. Low standard deviations indicate homogeneity, and high standard deviations indicate heterogeneity.

The Copernicus EEA European Urban Atlas (UA) shows that land use within the region is typified by continuous or discontinuous dense to medium density urban fabric, with sparse industrial and commercial units, and extended port areas present along the River Thames (EC 2011). Table 3 summarizes the total areas of each of the 20 UA land cover types present in Greater London, the fraction of these that are covered by PanGeo observed geohazard polygons, and respective ground motion velocity statistics during 1992–2000 and 2002–2010 based on the ERS-1/2 and ENVISAT PS data sets. Areas of observed geohazards mainly involve discontinuous urban fabric, with ∼200 km² dense (UA code 11210) and ∼94 km² medium density (UA code 11220) fabric. Industrial and commercial units are also widely affected by geohazards, with a total extent of ∼88 km² UA code 12100 land cover polygons intersected by observed geohazards. Areas of continuous and discontinuous urban fabric, industrial, port areas and roads show the highest average velocities away from the satellite sensor and the smallest standard deviations (i.e., ∼1.0 mm/year in the ERS-1/2, and ∼0.7 mm/year in the ENVISAT data) across the UA polygons covered by observed geohazards. On the other hand, more rural land cover types, such as agricultural, semi-natural and green areas, mineral extraction and dump sites and forests, generally reveal the highest standard deviations across the UA polygons (up to ∼4.4 mm/year in the ERS-1/2, and ∼1.0–1.6 mm/year in the ENVISAT data), thus suggesting greater spatial variability of observed motions within these land cover types. Both subsidence and uplift are observed for the various UA types, although these are related to different geohazard categories and types, as discussed in the following sections.

It is worth noting that the London GSL provides information on geohazards identified from PSI data and geology field campaigns. The geohazard polygons within the GSL, therefore, represent geohazards observed by interpreting these input layers, and accounting for their specific temporal reference and

Table 3

Areal and velocity statistics for UA land cover types covered by PanGeo observed motions in Greater London

UA code	UA land cover type	Area (km²)			ERS-1/2 data from 1992 to 2000					ENVISAT data from 2002 to 2010				
		Greater London (A)	Observed geohazards (B)	(B)/(A) (%)	No. PS	LOS velocity (VEL) (mm/year)				No. PS	LOS velocity (VEL) (mm/year)			
						Avg	Min	Max	SD		Avg	Min	Max	SD
11100	Continuous urban fabric (S.L. >80 %)	36.6	25.7	70	28,888	0.11	−21.89	+18.92	1.04	36,229	−0.70	−7.30	+5.80	0.67
11210	Discontinuous dense urban fabric (S.L. 50–80 %)	432.2	199.2	46	131,540	0.19	−25.28	+21.57	1.10	155,559	−0.47	−12.28	+7.09	0.67
11220	Discontinuous medium density urban fabric (S.L. 30–50 %)	207.9	93.9	45	45,539	0.44	−21.09	+26.47	1.11	50,465	−0.18	−11.01	+5.72	0.59
11230	Discontinuous low density urban fabric (S.L. 10–30 %)	28.6	10.5	37	4,643	0.45	−18.08	+20.73	1.23	5,399	−0.16	−9.42	+5.04	0.60
11240	Discontinuous very low density urban fabric (S.L. <10 %)	7.0	2.4	34	1,191	−0.05	−14.93	+21.25	1.45	1,621	−0.36	−5.32	+2.14	0.72
11300	Isolated structures	1.2	0.1	6	28	0.38	−0.45	+1.22	0.42	30	−0.11	−1.40	+1.37	0.62
12100	Industrial, commercial, public, military and private units	170.3	87.9	52	63,971	0.13	−22.44	+22.48	1.31	76,504	−0.45	−12.51	+9.09	0.84
12210	Fast transit roads and associated land	2.4	0.5	21	127	−0.18	−15.17	+10.48	2.35	169	−0.65	−7.83	+3.06	1.27
12220	Other roads and associated land	120.9	59.1	49	14,855	0.08	−19.37	+22.12	1.62	26,530	−0.51	−12.85	+6.82	0.79
12230	Railways and associated land	18.4	9.6	52	3,563	−0.05	−17.98	+20.34	1.94	5,355	−0.56	−9.98	+6.50	0.92
12300	Port areas	12.1	11.9	99	6,018	−0.70	−18.99	+20.93	1.87	7,164	−0.49	−8.20	+4.54	1.08
12400	Airports	16.3	8.2	50	1,543	0.01	−12.43	+17.12	1.41	2,555	−0.80	−9.53	+4.15	0.82
13100	Mineral extraction and dump sites	5.4	3.1	57	50	−0.07	−13.59	+18.54	3.97	74	−0.44	−7.29	+4.52	1.48
13300	Construction sites	3.1	2.3	74	1,386	0.27	−18.02	+19.43	1.24	136	−0.69	−4.19	+3.79	1.36
13400	Land without current use	2.2	1.3	59	184	−0.25	−13.75	+7.70	2.86	134	−0.93	−6.79	+0.84	1.35
14100	Green urban areas	146.0	46.3	32	2,313	−0.21	−18.72	+22.20	2.62	2,771	−0.64	−7.07	+4.90	0.95
14200	Sports and leisure facilities	116.6	39.9	34	2,665	−0.05	−19.47	+21.61	2.42	3,933	−0.47	−8.43	+4.88	0.92
20000	Agricultural areas, semi-natural areas and wetlands	189.8	25.8	14	527	−0.60	−17.97	+21.38	3.66	708	−0.77	−10.57	+4.09	1.64
30000	Forests	40.3	6.4	16	84	0.39	−17.09	+17.83	4.42	64	−0.37	−3.97	+1.10	0.85
50000	Water bodies	37.4	29.3	78	687	0.03	−19.19	+29.51	4.38	992	−0.43	−4.15	+3.37	0.72

The total area of each UA type in Greater London and the respective fraction affected by PanGeo observed geohazards is shown. Observed average, minimum, maximum and standard deviation for the ERS-1/2 and ENVISAT LOS ground motion velocities during 1992–2000 and 2002–2010 are also summarized for each UA land cover type

spatial scales and resolutions. This aspect is discussed in detail in Sect. 3.4. The PanGeo product also needs to be used in conjunction with geohazard susceptibility maps indicating areas of potential geohazards. As mentioned above, for Great Britain, these are provided by the BGS through the GeoSure data set (Booth et al. 2010; Walsby 2008) for the six natural ground movement types of collapsible deposits, compressible ground, landslides, running sand, shrink–swell and soluble rocks.

3.1. Observed Natural Geohazards

Topography, geomorphology and geology of Greater London are controlled by the presence of the drainage network formed by the River Thames and its tributaries. This network is associated with an alluvial tract that lies at about 10 mOD in the west of the area, falling towards sea level to the east of the district. Gently sloping valley sides rising to approximately 30 mOD border the riparian zones. The north-eastern sector is characterised by a dissected plateau at about 100 mOD, whilst to the south of the River Thames, the land rises gently southwards. Alluvium, till, marine, glaciofluvial and river terrace deposits of Quaternary age are mapped within the administrative area. Interfluves in the north-west of the area are formed of dissected London Clay capped, in places, by fine-grained sands of the Bagshot Formation. Sparse outliers of glaciofluvial deposits are also present. The ground rises to a dissected plateau of till at about 100 mOD in the north-east of the study area. South of the River Thames, the land rises gently across the London Clay towards the southern extremity of the district, where white to grey Chalk is present at surface. Clays and some sands and gravels of the Lambeth Group and silts and sands of the Thanet Sand Formation are present at the surface between the areas of London Clay and Chalk (Ellison et al. 2004).

The presence of alluvium in the river flood plain, extensive areas of clays at the surface, lithologies containing loosely packed sandy layers, slopes prone to landsliding, and deep-seated tectonic structures make this area particularly susceptible to natural hazards. Observed natural geohazards, based on the two PSI ground motion data sets for 1992–2000 and

2002–2010, and geology field campaigns, encompass the three main categories of 'Natural Ground Movement', 'Natural Ground Instability' and 'Deep Ground Motions' (Table 2), and in Greater London include a total of three PanGeo geohazard polygons, classified as 'Compressible Ground', 'Landslide' and 'Tectonic Movements', respectively, according to the PanGeo Geohazard Glossary (Bateson 2013).

3.1.1 Compressible Ground

Centred upon the River Thames and its tributary, the River Lea in the east of Greater London, is a 101 km² , low-lying area with typically gentle relief, identified in PanGeo as geohazard polygon 'PGGH_London_002' and indicating ground motion caused by compressible deposits (Fig. 4). The observed ground motion extends 20 km inland from its most easterly limit at the administrative boundary near Erith, and at its widest, reaches 6 km, diverging from the channel of the River Thames by a maximum of 3 km. Landmarks such as the London Docklands, Millennium dome, Olympic Park, Cutty Sark, London City Airport, Thames Barrier and the Blackwall Tunnel are present in this area, and regions of water, port areas, discontinuous dense urban fabric and discontinuous medium density urban fabric predominate, although land cover within this area is generally varied.

The area of instability coincides with extensive areas of relatively thick deposits of Holocene alluvium in the flood plain and salt marshes of the Rivers Lea and Thames. These overly deposits are of the London Clay Formation, Lambeth Group, Thanet Sand Formation and the Seaford Chalk Formation and Newhaven Chalk Formation. The Holocene deposits are susceptible to progressive subsidence from compaction, drying and resulting compression.

Analysis of average motion velocities from 1992 to 2010 for the areas of the Hornchurch, Rainham, Aveley and Wennington Marshes, where the thickness of the superficial deposits is up to ~40 m, reveals that the PSI-derived motion velocities increase up to −15 mm/year, with increasing deposit thickness and presence of made ground (Fig. 4). On the other hand, there seems to be no significant correlation between sediment thickness and the

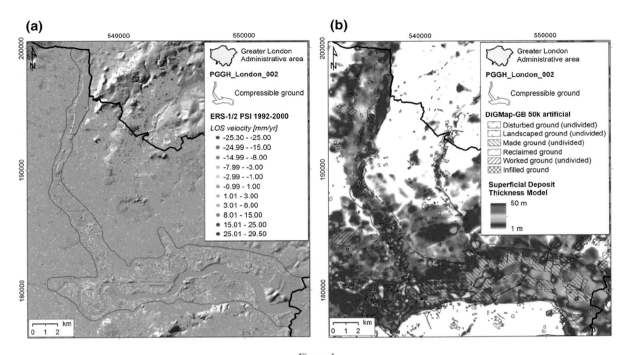

Figure 4

a Average motion velocities from 1992 to 2000 for polygon PGGH_London_002, estimated along the line-of-sight of ERS-1/2 satellites in ascending mode, overlapped onto the shaded relief of NEXTMap® DTM at 50 m resolution. **b** Artificial deposits from DiGMapGB at 1: 50,000 scale, onto Superficial Deposit Thicknesses derived from the BGS Superficial Deposit Thickness Model (SDTM). Refer to Table 2 for detailed information and PSI ground motion statistics. British National Grid; Projection: Transverse Mercator; Datum: OSGB 1936. NEXTMap® Britain © 2003, Intermap Technologies Inc., All rights reserved. Geological materials © NERC, All rights reserved

amount of subsidence based on PSI velocity values in the upper parts of river catchments including the River Thames upstream of Tower Bridge, where alluvium is <5 m thick. This confirms observations by BINGLEY *et al.* (2007).

The identified geohazard polygon corresponds well with domains '5F' and '6D' identified by ALDISS *et al.* (2014). Ground motion in the domains identified by ALDISS *et al.* (2014) reached vertical motion velocities in the order of −1.30 ± 0.95 mm/year along the River Lea and −1.99 ± 1.87 mm/year along the Thames in the period from 1997 to 2005. This means that the spatial average velocity of all PS targets within the boundaries of these domains was −1.30 (domain '5F') and −1.99 mm/year (domain '6D'), with observed deviations from these values equal to 0.95 and 1.87 mm/year, respectively. The latter indicate that for 68.2 % of the PS targets within the boundaries of these two domains, the observed annual velocities were, in turn, in the ranges of −2.25 to −0.35 mm/year, and −3.86 to −0.12 mm/year

during 1997–2005 (by assuming normal distribution of the PS velocities within the polygon boundary). The average LOS velocities observed by analysing our ERS-1/2 and ENVISAT IPTA targets within the mapped geohazard polygon boundary are −0.47 ± 1.77 mm/year during 1992–2000, and −0.51 ± 1.00 mm/year during 2002–2010 (Table 2). These values indicate that the majority of the targets reveal annual velocities in the range of −2.24 to +1.30 mm/ year from 1992 to 2000, and −1.51 to +0.49 mm/ year from 2002 to 2010. A maximum negative LOS PS velocity of −19.19 mm/year is achieved between 1992 and 2000, amounting to a maximum total displacement of 153.5 mm along the LOS over the 8-year period. By assuming a purely vertical motion direction for this area, the projection of the LOS values to the vertical direction can be performed by simply rescaling LOS observations by a factor of 1.09 (by diving the LOS values by the cosine of the 23° incidence angle). This rescaling results in observed spatial averages within the polygon boundary of

−0.51 ± 1.93 mm/year from 1992 to 2000, and −0.56 ± 1.09 mm/year from 2002 to 2010.

3.1.2 Landslides

Areas of observed landslides in Greater London cover ∼0.53 km² and are mainly located in the Havering, Barnet, Ealing, Greenwich, and Richmond upon Thames Boroughs. These have been categorized in the multi-part geohazard PanGeo polygon 'PGGH_London_001', and include 37 individual landslide deposits. The latter were mapped by BGS at 1:10,000 scale based on geology field campaigns and digital stereoscopic aerial photo interpretation, digital field data capture, terrestrial and airborne LiDAR, and differential GPS using a multi-stage methodology (EVANS *et al.* 2013), and recorded in the mass movement layer of the DiGMapGB (BECKEN and GREEN 2000; SMITH 2013), and the NLD (FOSTER *et al.* 2012). Mapped landslide deposits include both phenomena active at the time of survey, and older, inactive and relict landslides that are identified based upon the identification of certain morphological and sedimentological characteristics, and not necessarily on the observation of motion.

The majority of landslide features in Greater London occur on deposits of the London Clay Formation, often in close proximity to the boundary with overlying, more-permeable units. Landsliding mechanisms within the area vary from flows to multiple, successive rotational slides (FORSTER *et al.* 2003), and the ages of the features range between old (<1,000 years) and recent (<100 years). The Claygate Member of the London Clay is particularly prone to failure, and possesses a high plasticity and high water content on account of water-bearing sand layers and the uppermost deposits of the underlying London Clay (ELLISON *et al.* 2004; FORSTER 1997; SUMBLER 1996). In addition, where the Claygate Member is overlain by water-bearing sand in the Bagshot Formation, spring lines may develop, potentially raising pore-water pressure in material below. Many London Clay slopes steeper than 3° are covered by a veneer of head composed of redeposited London Clay including the Claygate Member, and may potentially be unstable (ELLISON *et al.* 2004).

Analysis of the ERS-1/2 and ENVISAT PSI data distribution for landslides in Greater London shows extremely low densities to absence of radar targets within the landslide deposit areas (Table 2). A total of only three ERS and five ENVISAT PS were identified for the 37 landslide deposits. To the south of the River Thames, within the Greenwich Borough, PSI data for one landslide deposit record velocities of −1.87, −1.63, −1.51 mm/year between 2002 and 2010 (Fig. 5a). Ground motion due to landsliding has been observed on the flanks of Shooters Hill, Eltham Common. The hill itself is capped by Pleistocene sand and gravel of the Stanmore Gravel Formation, Crag Group (Fig. 5b). The landslide deposits possess maximum elevations of ∼100 m a.s.l., variable aspects (15°–105°) and widths ranging between 74 and 205 m. Four occur in the London Clay Formation and one in the Claygate Member. The observed LOS velocities for the three PS targets mentioned above correspond, respectively, to 4–5 mm/year if re-projected along the steepest slope direction of the respective locations, and suggest that part of the deposit still shows signs of activity.

In the Richmond-upon-Thames Borough, seven landslide deposits situated along the western edge of Richmond Park are mapped in the DiGMapGB mass movement data set (Fig. 5d). The area is low-lying with typically gentle relief. However, slopes in the immediate vicinity of the landslides reach approximately 20 %. The largest landslide feature possesses a width of 338 m, whilst maximum elevations of each of the polygons range between 48 and 24 mOD. Landslide failures occurred within the London Clay Formation of the Thames Group. Upslope of the majority of the deposits lies the Black Park Gravel Member of the Thames Valley Formation, and the close proximity of this more permeable unit to the failed areas suggests that hydrological regime and pore water pressure may influence ground stability in the area. PSI data show ground motion velocity of −11.63 mm/year between 1992 and 2000 for one PS located within one landslide deposit (Fig. 5c). Another PS located only 11 m away from the feature records −6.17 mm/year within the same time period, and indicates presence of motion outside the deposit, thus suggesting possible enlargement of the geohazard polygon,

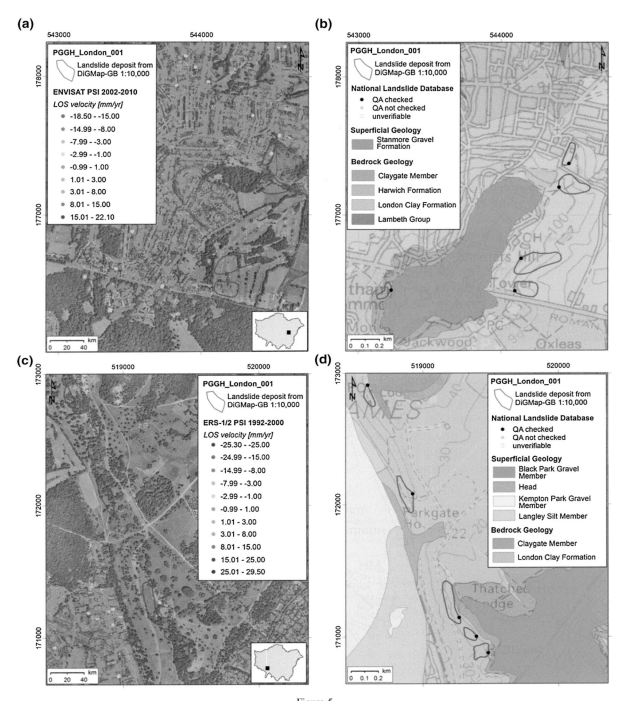

Figure 5
Examples of observed landslides in Greater London, delineated as PanGeo polygon PGGH_London_001: Average motion velocities in
a 2002–2010 for landslide deposits in the Richmond upon Thames Borough and in **c** 1992–2000 for landslide deposits in the Greenwich
Borough, estimated along the line-of-sight of ENVISAT in descending mode, and ERS-1/2 in ascending mode, respectively. Surface geology
from the DiGMapGB at 1:50,000 scale for landslides in the **b** Richmond upon Thames and the **d** Greenwich Boroughs. PSI data are
overlapped onto aerial photographs, whereas surface geology onto OS topographic map at 1:50,000 scale. *Insets* location within Greater
London. Refer to Table 2 for detailed information and PSI ground motion statistics. British National Grid; Projection: Transverse Mercator;
Datum: OSGB 1936. Geological materials © NERC, All rights reserved. OS data © Crown Copyright and database rights 2013. Aerial
photography © UKP/Getmapping Licence No. UKP2006/01

which will be verified by analysis of geomorphological data and other field evidences.

Full details about all landslides deposits and PSI observations in other boroughs can be found in the GHD report for London, i.e., CIGNA et al. (2013c).

3.1.3 Tectonic Processes

Based upon the analysis of the ERS-1/2 and ENVISAT PS data, the regional gravity field of the area of London and ground motion domains of average vertical velocity from the DEFRA/EA joint project Land Levels (BINGLEY et al. 2007), we mapped a 26.63 km^2 area of observed deep ground motions related to tectonic movements. This area was identified as PanGeo geohazard 'PGGH_London_014' (Table 2), is centred upon the Greenholt area of west London, and was attributed a low confidence due to the level of uncertainty in the delineation of its overall extension.

Within the pre-Mesozoic basement, the Midlands Microcraton underlies the geohazard polygon. This terrane is characterised by the occurrence of Proterozoic rocks at relatively shallow depths, recorded recent isostatic uplift, and was delineated in terms of the generalised domain 'G1' by ALDISS et al. (2014). Gravity data that were processed using gravity stripping to the base of the Mesozoic succession,

enhanced variations in the regional Bouguer gravity anomaly field, which largely relates to Palaeozoic or Proterozoic age geological formations beneath the Chalk Group (ALDISS et al. 2014). For domain 'G1', a gravity 'high' (i.e., where the mass of underlying rock is greater than average) suggests presence of relatively dense rocks close to the surface, and that deep-seated tectonic structures could have causative relationship with the observed ground motions (Fig. 6b).

The ERS-1/2 PS results for 1992–2000 identify ground uplift with +0.40 ± 1.22 mm/year LOS velocity for the PS targets within the polygon boundary (Fig. 6a), and an observed maximum of +22.23 mm/year, which corresponds to a total movement of 18 cm towards the satellite sensor over the monitored interval. Analysis of ground motions between 2002 and 2010 reveals a general change in the deformation trend of the geohazard polygon. Subsidence is recorded by the ENVISAT PS results, with −0.18 ± 0.69 mm/year for the PS within the polygon, and −7.74 mm/year observed peak velocities, due to either an inversion of the motion trend or, more likely, the presence of another geohazard type, overlapped onto (and thus masking) the existing uplift. Comparison with ground motion domains from ALDISS et al. (2014) suggests that the unstable area corresponds with domain '1' identified from AG-

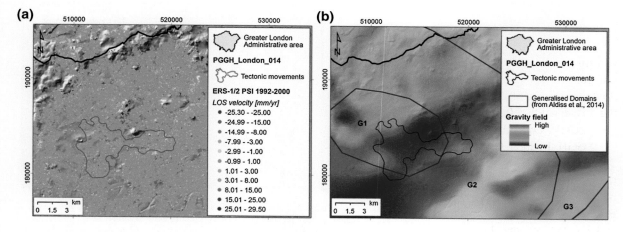

Figure 6

a Average motion velocities from 1992 to 2000 for PGGH_London_014, estimated along the line-of-sight of ERS-1/2 in ascending mode; and **b** generalised vertical ground velocity domains and gravity field stripped to base of Mesozoic succession [modified from ALDISS et al. (2014)], overlapped onto shaded relief of NEXTMap® DTM at 50 m resolution. Refer to Table 2 for detailed information and PSI ground motion statistics, and Fig. 3 for location of this polygon within Greater London. British National Grid; Projection: Transverse Mercator; Datum: OSGB 1936. Geological materials © NERC, All rights reserved. NEXTMap® Britain © 2003, Intermap Technologies Inc., All rights reserved

aligned and GPS-aligned estimates of vertical velocity. These authors show that ground motion in this area reached a maximum velocity of +8.26 mm/year from 1997 to 2005, whilst the average velocity for the domain was +0.25 mm/year.

3.2. Observed Anthropogenic Geohazards

The surface geology in Greater London has been modified throughout the centuries by several anthropogenic factors that widely influence the local and regional patterns of ground motions; for example, engineering works and groundwater management. Ground motions related to anthropogenic factors have long been studied through the analysis of PSI data, including land motions induced by engineering works (e.g., ASTRIUM-GEO 2014; BERARDINO et al. 2000; KRIVENKO et al. 2012) and groundwater management (e.g., AMELUNG et al. 1999; BELL et al. 2008; CIGNA et al. 2012a; GALLOWAY et al. 1998; HERRERA et al. 2009).

Observed anthropogenic geohazards in Greater London include a total of nine PanGeo geohazard polygons, classified as 'Underground Construction', 'Made Ground', 'Groundwater Abstraction' and 'Other' (Table 2), according to the PanGeo Geohazard Glossary (BATESON 2013).

3.2.1 Underground Construction

Areas of observed geohazard associated with anthropogenic ground instability due to underground construction in Greater London consist of three polygons in the City of Westminster, Lambeth and Southwark areas of London ('PGGH_London_003'), the Wandsworth ('PGGH_London_004') and Islington ('PGGH_London_015') Boroughs (Fig. 7).

'PGGH_London_003' covers a linear area of 1.32 km^2 (Table 2) and crosses the River Thames in the region of Westminster Bridge, where a number of landmarks are present, such as sections of the British Rail Network, Waterloo Train Station, Buckingham Palace, the London Eye, the Tower of London and the Tate Modern Art Gallery. This area is a low-lying river flood plain with elevations generally in the range of 5 to 10 mOD, and the bedrock geology is characterised by the London Clay

Formation. Superficial deposits in the area include alluvium, peat, Kempton Park Gravel Formation, Langley Silt Member and Taplow Gravel Formation.

The unstable area indicated by the PSI data from both 1992 to 2000, and to a lesser degree, 2002 to 2010, corresponds with the location of the Jubilee Line Extension, which was constructed between 1993 and 1999 (BURLAND et al. 2001; PAGE 1995). In particular, the polygon area coincides with the 6 km-long line branch running between the Green Park and Bermondsey stations, opened at the end of 1999 (Fig. 7b). It is suggested that the motion observed from the PS data is due to ground compaction following underground engineering works of the Jubilee Line Extension project, and removal of subsurface material, which altered the support for the overlying terrain. STANDING and BURLAND (2006) report on tunnelling volume losses measured during construction of the tunnels for this line, and observe that losses higher than 3 % were measured in Westminster and in St James's Park, south of the lake, while north of St James's Park, losses were generally lower than 2 % as expected.

PSI data sets show motions away from the satellite sensor, indicating that land subsidence occurred during both time intervals, with LOS velocity for the PS targets within the polygon of -0.92 ± 1.86 mm/year during 1992–2000, and -1.16 ± 0.74 mm/year in the ENVISAT PS data from 2002 to 2010 (Fig. 7a). Maximum PS velocities estimated along the satellite LOS are approximately -15.9 mm/year from 1992 to 2000, amounting to a maximum total displacement of 13 cm over the 8-year period. Although average velocities decrease to < -4.1 mm/year during 2002–2010, motions due to the underground works are still identifiable from the ENVISAT monitoring data, and are discernible from the compaction of the alluvium affecting a wider sector of the city and due to groundwater abstraction (see Sect. 3.2.2). This geohazard polygon also corresponds well with domain '6C' identified by ALDISS et al. (2014).

A similar pattern in the PS ground motion data was observed for a 1.34 km^2, south-west trending, 4.5 km long area to the south of the River Thames in Wandsworth Borough ('PGGH_London_004'; Fig. 7c, d), in close proximity to the Battersea Park,

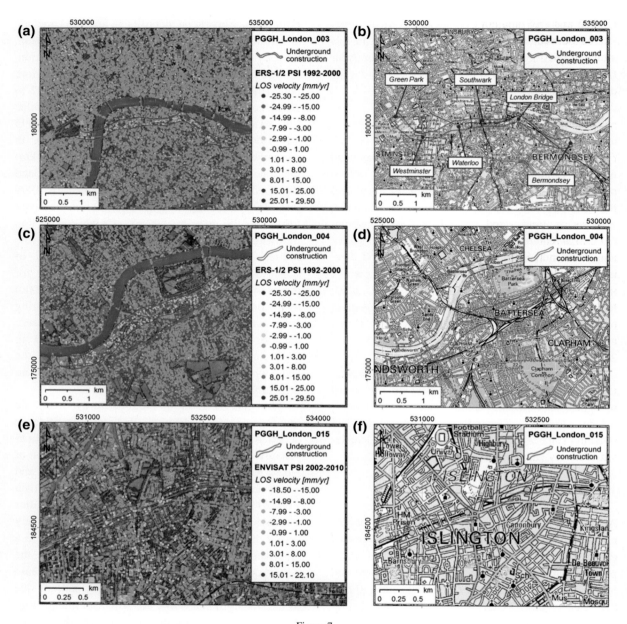

Figure 7

a, c, e Average LOS motion velocities from the ERS-1/2 and ENVISAT PSI results and **b, d, f** OS topographic map at 1:50,000 scale for PanGeo polygons PGGH_London_003 (**a, b**), PGGH_London_004 (**c, d**) and PGGH_London_015 (**e, f**), depicting areas of observed geohazards due to underground constructions in Greater London. PSI data are overlapped onto aerial photographs. Stations of the Jubilee Line Extension track between the stations Green Park and Bermondsey are indicated in **b**. Refer to Table 2 for detailed information and PSI ground motion statistics, and Fig. 3 for location of these polygons within Greater London. British National Grid; Projection: Transverse Mercator; Datum: OSGB 1936. Aerial photography © UKP/Getmapping Licence No. UKP2006/01. OS data © Crown Copyright and database rights 2013

Power, and Queenstown Road Stations, New Covent Garden Market and Wandsworth Bridge. This low-lying area of instability is characterised by deposits of the Langley Silt Member and Kempton Park Gravel Formation that, towards the centre and south of the polygon, abut Holocene alluvium.

The unstable area coincides with the route of the A3205 between Nine Elms and Wandsworth, along which tunnelling works for electricity cables were carried out between 1997 and 2005 and are the likely cause of the observed ground motion (BINGLEY et al. 2007). Observed PS LOS velocity within the geohazard polygon is -1.93 ± 1.78 mm/year from 1992 to 2000, with a peak of -14.0 mm/year achieved by a few isolated PS, corresponding to a movement of -110 mm over the monitored interval (Fig. 7c). Up to -8.0 mm/year are observed in the inner sector of the polygon lying along the axis of the A3205, whereas motion velocity decreases to -1.0 to -3.0 mm/year towards the boundaries of the polygon, with increasing distance from the track of the underground excavation. Our results confirm observations by ALDISS et al. (2014), and depict domain '6A' identified by these authors for 1997–2005, as moving at -2.1 ± 1.3 mm/year on average within the domain boundary. Motion velocities estimated by the ENVISAT PS data decrease to -1.81 ± 0.64 mm/year during 2002–2010, and peaks of no more than -6.04 mm/year are observed. During this time period, velocities within the area of PGGH_London_004 are not distinguishable from those affecting the larger surrounding area, and are better attributed to alternative sources of motion (see Sect. 3.2.2).

The third area of underground construction identified by the PSI data extends 0.38 km^2 within the Islington Borough, and follows the track of the Channel Tunnel Rail Link (CTRL) or High Speed 1 (HS1), the UK high speed rail link between the Channel Tunnel and London St Pancras International, opened in full in November 2007 (cf. 'PGGH_London_015'; Fig. 7e, f; Table 2). The geohazard polygon includes a 2-km long portion of the CTRL tunnel under the built-up areas of London between Caledonian Rd and Barnsbury to the West (before the line emerges on the surface and arrives at St Pancras) and Canonbury to the East, before Stratford station.

Variable motion rates are observed within the polygon area, with most targets moving away from the satellite sensor, and only a few moving towards the sensor. The ERS-1/2 PS data within the geohazard polygon boundary show LOS velocity of -0.07 ± 0.81 mm/year during 1992–2000, whereas

significant acceleration is observed in the ENVISAT data from 2002 to 2010, when values increase to -1.09 ± 0.85 mm/year, with peaks of -5.69 mm/year along the CTRL track (Fig. 7f). We relate the observed motions with ground subsidence resulting from the construction of the track of the CTRL between Caledonian Rd and Barnsbury and Canonbury. Indeed, the above track segment was built during the same time frame of the motions estimated over the ENVISAT data set, and possibly exerted local control on ground stability by inducing compaction of the sediments above the tunnel along its track.

Ground subsidence is also observed over a wider area around this polygon, as revealed by the presence of several PS showing up to -3 mm/year average motion velocities outside the geohazard polygon boundaries (Fig. 7e). This motion is identified by PanGeo polygon 'PGGH_London_006', and is due to groundwater abstraction (see Sect. 3.2.2).

3.2.2 Groundwater Abstraction and Rise

Vast areas of ground motions due to groundwater management are revealed in Greater London by the PSI data and analysis of water level records from the Environment Agency (EA). These concern both areas undergoing land subsidence induced by water pumping and decreased ground water levels in the aquifers, and water rise due to recovery of the historical levels. Ground levels tend to change in response to water levels; for instance, by subsiding when water levels fall, and uplifting when it recovers, as a direct effect of changes in the pore water pressure, and consequently, the effective stress acting on the terrain.

A total of four geohazard polygons belonging to groundwater management are delineated in London, and these are classified as 'Groundwater Abstraction' when showing subsidence, and 'Other' when showing ground uplift.

The wider geohazard polygon relating to this category refers to a large area of 146.65 km^2, which encompasses portions of 13 London boroughs including those of Islington, City of Westminster, Wandsworth, Lambeth and Southwark, and numerous landmarks such as Hyde Park, St James Palace, the City of Westminster, Wimbledon Common and

London Bridge (cf. 'PGGH_London_006'; Table 2). Over 80 % of this area is underlain by the London Clay Formation, and the deposits of the Lambeth Group and Thanet Sand Formation are present only in the south-eastern sector. LOS velocity from the PS data within the geohazard polygon boundary reveal -0.81 ± 0.64 mm/year during 2002–2010 and $+0.19 \pm 1.15$ mm/year from 1992 to 2000 for this area, with peak negative (-19.45 mm/year) and positive LOS velocities ($+22.47$ mm/year) achieved in the ERS-1/2 ascending data set. The ENVISAT PSI results display a more consistent trend in ground motion away from the satellite during 2002–2010 (Fig. 8a), and the wide area and range of lithologies over which this negative motion operates suggests a relatively deep-seated cause for the motion.

Monitoring of groundwater levels by the EA (2010) reveals a period of groundwater abstraction within the study area between 2000 and 2010, and a fall in groundwater levels by as much as 22 m has been recorded in the centre of the polygon area (Fig. 8b).

Velocity trends for the PS data from 1992 to 2000 are more variable. A relatively distinct area of positive velocities can be seen centred around Lambeth, and can be attributed to groundwater recharge. This has been delineated as polygon 'PGGH_London_010' in PanGeo, and covers a total of 21.29 km^2 and portions of nine London boroughs, including those of the City of Westminster, Lambeth, Southwark and the City of London. LOS velocity of all the ERS-1/2 PS targets within the polygon is $+0.45 \pm 1.12$ mm/year during 1992–2000 (Fig. 9a),

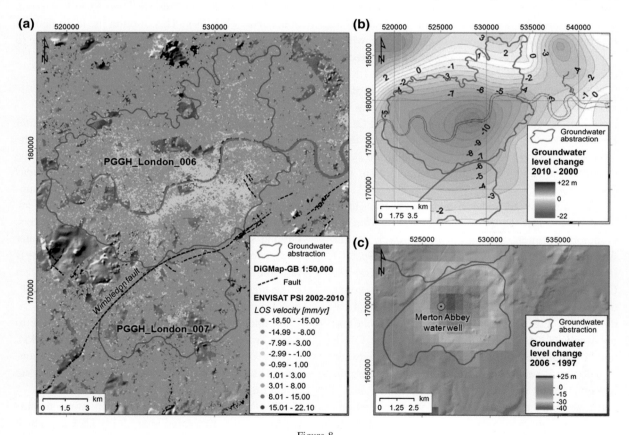

Figure 8
a Faults from the DiGMapGB at 1:50,000 scale and average motion velocities from 2002 to 2010, estimated along the line-of-sight of ENVISAT in descending mode for areas of observed groundwater abstraction; and **b, c** groundwater level changes between 2000 and 2010 [modified from EA (2010)] and between 1997 and 2006 [modified from EA (2007)], overlapped onto shaded relief of NEXTMap® DTM at 50 m resolution. Refer to Table 2 for detailed information and PSI ground motion statistics, and Fig. 3 for location of these polygons within Greater London. British National Grid; Projection: Transverse Mercator; Datum: OSGB 1936. Geological materials © NERC, All rights reserved. NEXTMap® Britain © 2003, Intermap Technologies Inc., All rights reserved

with maximum observed velocities of +20.90 mm/year. Monitoring of groundwater levels by the EA (2007) reveals a period of groundwater recharge within the study area between 1996 and 2001 (Fig. 9b). This is particularly the case for the northern section of the polygon area, where values approaching +3.39 m for the period 1997 to 2006 can be seen. Groundwater recharge in these areas facilitates uplift by increasing pore water pressures.

Another area of observed land subsidence due to groundwater abstraction has been delineated and classified as PanGeo polygon 'PGGH_London_007' (Table 2; Fig. 8a, c). This area covers a total of 47.17 km^2 and includes portions of six Boroughs of Greater London, namely Merton, Sutton, Croydon, Lambeth, Wandsworth and Kingston upon Thames. The bedrock geology of the area is dominated by the London Clay Formation, with small sectors where clays, silt and sand of the Lambeth Group crop out.

Several groundwater wells are located across this geohazard polygon and the observed subsidence is thought to be related to increased ground water abstraction at these locations. Groundwater levels were lowered by 39 m between January 1997 and January 2006 (EA 2007), due to abstraction at the Merton Abbey public water supply well, which is one of a number of sites in this part of the London area where water is taken from the Chalk (i.e., the principal aquifer of the London Basin) at depths in excess of 70 m. Groundwater level maps over 1997–2006 also record rates of groundwater level changes of the order of −2 to −5 m/year from 1996 to 2002.

ERS-1/2 and ENVISAT PS mainly show very low motion rates (in the range ± 1 mm/year) and several zones moving away from the satellite sensor at rates of a few mm/year. LOS velocity of the ERS-1/2 PS within the polygon is −0.14 ± 1.16 mm/year during

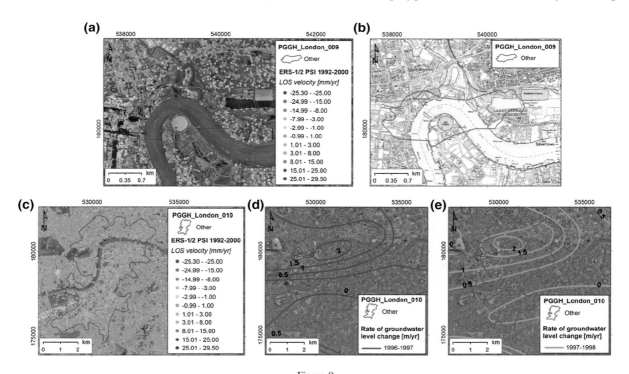

Figure 9

a Average motion velocities from 1992 to 2000, estimated along the line-of-sight of ERS-1/2 in ascending mode, and **b** OS topographic map at 1:25,000 scale for PanGeo polygon PGGH_London_009. **c** Average motion velocities from 1992 to 2000 estimated along the line-of-sight of ERS-1/2 in ascending mode, and **d**, **e** rates of groundwater level change recorded in 1996–1997 and 1997–1998 [modified from EA (2007)]. Refer to Table 2 for detailed information and PSI ground motion statistics, and Fig. 3 for location of these polygons within Greater London. **a**, **d**, and **e** are overlapped onto aerial photographs, whereas **c** onto shaded relief of NEXTMap® DTM at 50 m resolution. British National Grid; Projection: Transverse Mercator; Datum: OSGB 1936. Aerial photography © UKP/Getmapping Licence No. UKP2006/01. OS data © Crown Copyright and database rights 2013. NEXTMap® Britain © 2003, Intermap Technologies Inc., All rights reserved

1992–2000, with peaks of −25.28 mm/year in the northern sector of the polygon, around Tooting (Fig. 8a). In this area, the fastest water table decrease (−10 m/year) was observed in 2001–2002 (EA 2007). LOS velocity decreases to −0.47 ± 0.62 mm/year in the ENVISAT data from 2002 to 2010 and no more than −8.43 mm/year are observed; however, the areas revealing subsidence in this period appear wider than from 1992 to 2000, and zones moving at higher and consistent rates during 2002–2010 are concentrated in the central sector around Mitcham.

The north-west edge of this geohazard polygon is bounded by the Wimbledon Fault. In this area, it appears that faults parallel to the Wimbledon Fault are exerting control on local subsidence patterns, as major lineaments in average velocity distribution are aligned to these faults (BATESON et al. 2009). It is also noteworthy that the width of the Thames floodplain increases markedly downstream of the Wimbledon Fault, as shown by the outcrop of the Holocene deposits. Our PSI data confirm the observations for domain '5A' identified by ALDISS et al. (2014), who found velocities of −1.55 ± 0.83 mm/year for this area during 1997–2005, with the largest subsidence rates centred close to the Merton Abbey public water supply well. Ground motions in this area were attributed to groundwater abstraction from the above water well, and were investigated further in the framework of the ESA Terrafirma project, via the production of the Terrafirma—London H3 Modelled Product (BATESON et al. 2009). Quantitative analysis and modelling of the relationship between ground motion rates and groundwater pumping from the Merton Abbey water well showed agreement between groundwater modelling results and observed ground motions over the Merton Abbey area.

Lying within the large area of compaction of the Holocene alluvium (see Sect. 3.1.1), land uplift due to groundwater aquifer recharge is also observed for a 2.19 km^2 area crossing the Tower Hamlets, Greenwich and Newham Boroughs of Greater London, and including the far end of the Greenwich Peninsula in South East London (cf. 'PGGH_London_009'; Table 2). This area is low-lying river flood plain with elevations in the range of 2 to 14 mOD, and maximum values reached in the north-western sector, around Blackwall. The bedrock geology is dominated by the London Clay Formation and clays, silt and sand of the Lambeth Group, whereas superficial deposits in the area consist mostly of alluvium of the River Thames. Made, infilled and landscaped ground (undivided) is found in this area, and the thickness of the superficial deposits is typically 10–15 m, with maximum values of 30 m in the area of the Blackwall Stairs.

PSI data indicate uplift of the polygon area from 1992 to 2000 with LOS motion rates of +1.44 and ± 2.40 mm/year within the geohazard polygon boundary, whereas during the more recent data set of ground motion data from 2002 to 2010, the uplift cannot be distinguished and subsidence is observed with −0.96 ± 1.04 mm/year (Fig. 9c). During 1992–2000 although the PS data show that most of the Canary Wharf area underwent subsidence, the eastern part of that area, around West India Dock and Blackwall Basin underwent uplift and a quite sharp demarcation between uplifting and subsiding ground across the middle of the Canary Wharf area can be observed. Areas of uplift are also seen around the north end of the Blackwall tunnel and extending over to the end of Royal Victoria Dock. The observed ground uplift is thought to be due to groundwater changes during construction of some blocks on Canary Wharf during the late 1980s (e.g., One Canada Square) and associated local disturbance in local water levels. Indeed, to allow the construction works to be performed, the block basements were surrounded by cofferdams, and groundwater level pumped down temporarily. After the construction, ceased groundwater pumping likely resulted in local aquifer recharge and consequent ground uplift.

It is worth noting that despite the urban fabric and presence of radar reflective structures, no PSI points are found over the area of the Millennium Dome in the ERS data set (1992–2000), likely due to significant land cover changes and construction works performed during the 1990s, before the opening of the Dome to the public in 2000 for the Millennium Experience.

3.2.3 Made Ground

Areas undergoing subsidence due to consolidation of artificial ground and compaction of underlying deposits are identified in Greater London and delineated in the geohazard polygons 'PGGH_London_005' and

'PGGH_London_008' (Table 2; Fig. 10). These are both located in the southern sector of the Havering Borough adjacent to the River Thames, and lie within identified geohazard polygon of compressible ground along the flood plain of the River Lea and Thames (see Sect. 3.1.1), in a sector of the Thames flood plain where areas of made ground are largely present.

One of these mapped geohazards is located within the Hornchurch Marshes, and includes the Fairview Industrial Park and car compounds and some business centres along the Marsh, Barlow and Creek Ways. This represents observed ground motion related to subsidence of recent (∼1940s) made ground, and covers 1.38 km^2 low-lying river flood plain with gentle relief and elevations in the range of 2 to 10 mOD. This area is characterised by the presence of the London Clay Formation and the Lambeth Group, which are overlain by alluvium and tidal river or creek deposits. These are generally susceptible to progressive subsidence from compaction, drying and resulting compression.

The geohazard mapped via PanGeo coincides with an area of made ground (undivided), where the thickness of the superficial deposits is between 10 and 25 m (Fig. 10c). For this area, most ERS-1/2 and ENVISAT PS targets show motion away from the satellite sensor, and indicate subsidence in both time intervals 1992–2000 and 2002–2010, with consistent contrast in average velocities between this and adjacent areas. LOS velocity within the polygon is −3.23 ± 3.45 mm/year during 1992–2000, with maximum observed velocities of −15.99 mm/year in the western sector of the polygon area. Velocity decreases to −2.78 ± 2.52 mm/year in the ENVISAT descending data set from 2002 to 2010, and no more than −10.64 mm/year are observed (Fig. 10a, b).

Another similar area of compacting made ground due to the presence of artificial ground (undifferentiated) overlying the Holocene alluvium has been delineated nearby. This covers 0.57 km^2 and includes sections of the Dagenham motor works and abuts the Dagenham Power Station and Horse Shoe Corner. In this area, the maximum total superficial deposit thickness (including superficial geology and artificial ground) of 21.64 m is reached in the centre of the geohazard polygon (Fig. 10c).

A distinct concentration of PS is visible here when compared to the neighbouring areas, and a significant contribution from the made ground is considered likely with respect to the compaction of the River Thames alluvium, which affects the area at a larger scale. ERS-1/2 and ENVISAT data highlight LOS velocity within the polygon of −1.69 ± 2.38 mm/year during 1992–2000, with maximum observed negative velocities of −11.92 mm/year, and −1.72 ± 1.32 mm/year from 2002 to 2010, with peak velocities of −7.27 mm/year (Table 2; Fig. 10a, b). For this area, the difference in ground motion between other areas of artificial ground within the vicinity is likely due to differing dates of development. For example, the Dagenham Motor Works to the east of the polygon was developed in 1935. Any ground motion related to this older development is, therefore, likely to have slowed with time as the artificial deposits settle.

3.3. Unknown Geohazards

A number of areas showing ground motion but with a level of uncertainty in their related causes was observed (cf. 'Unknown' hazard types in Table 2). For these geohazard polygons, although clear spatial evidence of the presence of land motion was revealed by the PSI data, it was difficult to attribute with confidence a type or category due to the absence of validation with external data or information.

Geohazards of unknown origin and cause are, for instance, found in north-west London in the Harrow, Barnet and Brent Boroughs. At the intersection of the latter, 11 km north of the River Thames, a narrow 0.44 km^2 polygon following the line of the Edgware Road (A5) from junction with B461 to Annesley Avenue for a total of 2.39 km is observed (cf. 'PGGH_London_018'; Fig. 11a). Subsidence is clearly visible between 2002 and 2010, and LOS velocity is −1.10 ± 0.66 mm/year in the ENVISAT data set, with peak of −3.59 mm/year. Given the close association with the local transport network and concentrated temporal nature of the motion, the cause is likely to be anthropogenic activity in the area. As the area is underlain by the London Clay Formation, which is a known shrink–swell-prone lithology, altered drainage during 2002–2010 may be a possible

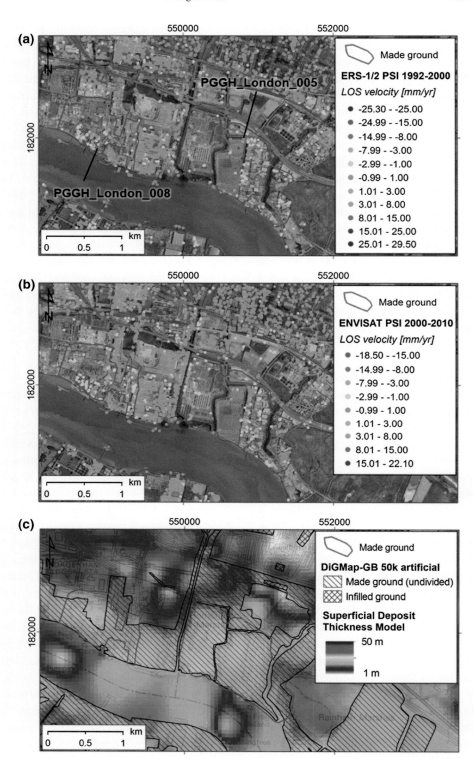

◄
Figure 10
Average motion velocities from 1992 to 2000 (**a**) and from 2002 to 2010 (**b**), estimated along the line-of-sight of ERS-1/2 in ascending mode and ENVISAT descending mode, respectively, overlapped onto aerial photographs, for areas of observed made ground in Greater London. **c** Artificial deposits from BGS DiGMapGB at 1:50,000 scale, onto Superficial Deposit Thicknesses derived from the BGS Superficial Deposit Thickness Model (SDTM) and OS topographic map at 1:50,000 scale. Refer to Table 2 for detailed information and PSI ground motion statistics, and Fig. 3 for location of these polygons within Greater London. British National Grid; Projection: Transverse Mercator; Datum: OSGB 1936. Aerial photography © UKP/Getmapping Licence No UKP2006/01. OS data © Crown Copyright and database rights 2013

cause. In this instance, the improvement of drainage from the road network or improved buried utilities will reduce water leaks and consequent swell of the underlying clay, though no direct proof of this hypothesis is available.

To the north-west of this area, another small area of observed ground motions is found (cf. 'PGGH_London_017'; Fig. 11b). This coincides with the Staples Corner, a major road junction of London built in the 1960s and consisting of two linked roundabouts and flyovers that connect the A406/North Circular Road (crossing North London, and linking W and E London) with the A5 Edgware Road, and the start of the M1 motorway. LOS velocity of the PS data within the polygon was -0.85 ± 1.67 mm/year during 1992–2000, and -1.33 ± 1.17 mm/year during 2002–2010. ERS-1/2 PS moving at -3.3 to -3.7 mm/year from 1992 to 2000 are found over the A406 flyover, and at up to -9.1 mm/year over the JVC, London Group and Aquarius Business Parks (north of the A406/North Circular Road). ENVISAT data show more consistently distributed motions all over the geohazard polygon, and up to -3.0 mm/year of along the infrastructure of the roundabout to the M1 and the business parks north of the A406/North Circular Road, and -3.4 mm/year along the A5 Edgware Road.

Serious damage to the road infrastructure and nearby buildings was caused by the explosion of a Provisional IRA van bomb underneath the A406 flyover and near the junction in the early 1990s, and the junction was temporarily closed for reconstruction and repair works. The format of the junction was modified during the reconstruction works and an additional slip road onto the M1 from the east

was added. Although no definite causative relationship of the observed ground motions was identified for this polygon, it is worth considering a possible correlation with these events. Indeed, the engineering works for the reconstruction and repairing of the junction might have been followed by structural and ground settlement that was imaged by the satellite data.

Another area of ground motion with unknown causes is centred on the Sloane Sq. London Underground station, which is served by the District and Circle Lines and is between South Kensington and Victoria (cf. 'PGGH_London_011'; Fig. 11c). This area is generally low-lying with elevations in the range of 11 to 16 mOD, and lies over alluvium and sediments of the River Westbourne that ran southwards towards the Thames through Hyde Park as the Serpentine Lake, and originally crossed by the Knight's Bridge at Knightsbridge. At Sloane Sq., the River Westbourne now flows over the Circle and District Line platforms inside a large iron conduit suspended from girders that was built in the 1850s when Belgravia, Chelsea and Paddington were developed. PS data indicating a general pattern of subsidence within this area during 1992–2010, with LOS velocity of -1.34 ± 0.37 mm/year during 1992–2000, and -1.74 ± 0.63 mm/year during 2002–2010. This area corresponds with domain '6B' identified by ALDISS et al. (2014), who estimated up to -5.1 mm/year at the centre of the unstable area. Historical records document groundwater flood incidents due to heavy rainfall and sewer surcharge occurred in 2006 and 2007 in the Sloane Sq. and the Notting Hill London Underground stations (HALCROW 2011), and preliminary Flood Risk Assessments and Flood Risk Areas for the Royal Borough of Kensington and Chelsea from EA indicate widespread vulnerability to surface water flooding across the entire Borough. Although for Sloane Sq. the extension of the critical drainage area coincides approximately with the location of the observed ground motions, no direct correlation between the latter and the above events has been identified.

Details and PSI observations for all unknown geohazards found in Greater London are discussed within the GHD report for London, i.e., CIGNA et al. (2013c).

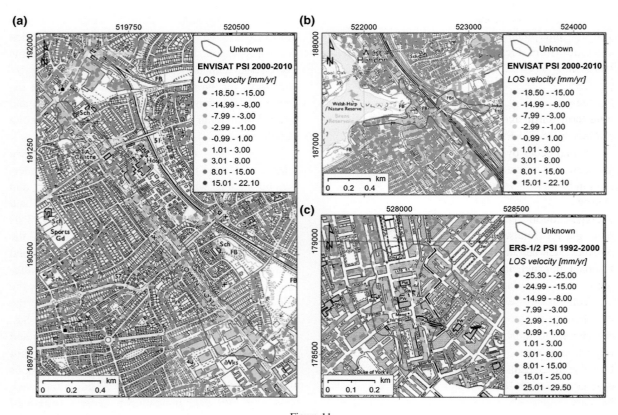

Figure 11

Examples of observed geohazards of unknown category in Greater London: PanGeo polygons **a** PGGH_London_018, **b** PGGH_London_017, and **c** PGGH_London_011. Average motion velocities in **a**, **b** 2002–2010 and **c** 1992–2000 estimated along the lines-of-sight of ENVISAT descending mode and ERS-1/2 in ascending mode, respectively. PSI data are overlapped onto OS topographic maps at **a**, **b** 1:25,000 and **c** 1:10,000 scales. Refer to Table 2 for detailed information and PSI ground motion statistics, and Fig. 3 for location of these polygons within Greater London. British National Grid; Projection: Transverse Mercator; Datum: OSGB 1936. OS data © Crown Copyright and database rights 2013

3.4. Discussion

PSI ground motion data for Greater London help to place the capabilities of these remote sensing techniques into a wider context, and analyse their potential to detect surface motions related to near-surface geological processes and the surface expression of deeper-seated motions and dynamics.

From the typologies of ground motions that we were able to detect and delineate with confidence using PSI, it becomes apparent that InSAR and PSI with such input as ERS and ENVISAT SAR data are generally sensitive to slow (up to a few tens of mm/year), relatively constant through time, ground motions, such as compressible ground and the effects of ground water abstraction and rise. For instance, we

observe in Greater London what we expected along the River Thames and other flood plains (see Sect. 3.1.1), and the consequences of groundwater level changes in the main aquifer are also evident across the investigated area (see Sect. 3.2.2). There are, however, geohazards that affect London for which the PSI data encountered difficulties in depicting.

It has been observed and largely discussed that Palaeogene clays of the London Clay and Lambeth Group and other Mesozoic and Tertiary clay soils and mudrocks are susceptible to natural shrinkage and swelling induced by variations in moisture content induced by changes in environmental conditions (e.g., HARRISON *et al.* 2012; JONES and JEFFERSON

2012). This type of geohazard has not been identified via the interpretation of average ground motion velocities from the PSI data for Greater London. We believe that this is partly due to the difficulties of the ERS-1/2 and ENVISAT PSI data in distinguishing the generally low rates of motion associated with long-term shrinkage (e.g., HOBBS *et al.* 2014; and references therein) with respect to other land processes that occur over vast areas (e.g., compaction of alluvium). Moreover, the usually seasonally variable motion history of shrinkage and swelling clays cannot clearly be highlighted by the interpretation of the sole average motion velocities. Indeed, detailed analysis of single PS time series would be required to verify whether such seasonal variations are depicted by the PSI data, but such an analysis by visual inspection for more than hundreds of thousands of targets is unfeasible. A few studies have tried to overcome this limitation and to ease the task of the radar interpreters to identify nonlinear components within large volumes of PSI data sets (BERTI *et al.* 2013; CIGNA *et al.* 2012b; TAPETE and CASAGLI 2013). These approaches will be tested and assessed with our data in Greater London to verify whether observed geohazards relating to shrink–swell clays can be incorporated into upgraded version of the PanGeo product.

Another geohazard that is apparently missed by the PSI data concerns dissolution processes. Although this geohazard has been both observed in the field and recorded in the National Karst Database (FARRANT and COOPER 2008) and in terms of susceptibility via the GeoSure data set, for instance where the white Cretaceous Chalk is present in the southern sector of Greater London, no PanGeo polygons are associated to this particular geohazard. This is due to the fact that dissolution processes and associated motions are generally faster than the maximum potential motion that PSI techniques can estimate before encountering phase unwrapping problems (i.e., 15 cm/year for ERS or ENVISAT data with 35 days repeat cycle). As of today, only in a few cases InSAR-based studies have been successful to image land motions associated with karstic features, and these mainly relate to conventional InSAR applications, and only in a few instances to multi-interferogram techniques (e.g., CASTAÑEDA *et al.* 2009; CLOSSON *et al.* 2005; FERRETTI

et al. 2004; GUTIÉRREZ *et al.* 2011), to the best of our knowledge.

With regard to landsliding, very few PS targets were found across the investigated area within the mapped landslide deposits or in close proximity to landslide deposits as mapped in the DiGMapGB. This evidence can be mostly explained by analysing the typical land covers of landslide deposits in Greater London, which mainly shows green urban areas, forests and agricultural land covers, and only in a few instances, small areas of urban fabric of generally low density [cf. also CIGNA *et al.* (2013c)]. These land cover types are generally affected by significant temporal decorrelation and strong variations of the interferometric phase, which prevent good radar reflectors to be identified within the radar imagery stacks (e.g., CIGNA *et al.* 2013a, b; COLESANTI and WASOWSKI 2006). Rapid ground motions are a further possible explanation for low PS density; however, we believe that this factor does not play a role for the landslide processes mapped within Greater London, due to their age and recent state of inactivity as mapped in the database and geological maps (see Sect. 3.1.2). The state of inactivity of these landslides also results in the absence of apparent evidence of surface motion in the PSI data for the majority of the mapped landslide deposits in Greater London. These often refer to inactive (e.g., stabilized or relict) phenomena that are depicted by the PSI average velocities as stable or undergoing almost null motions [cf. also CIGNA *et al.* (2013c)].

The identification of geohazards for the generation of the PanGeo products was, for Greater London, largely focussed on EO data and the analysis of PSI ground motion data that covers the period 1992–2010. Therefore, the products and polygons do not claim to be an exhaustive representation of all geohazards affecting the administrative area. This is mainly due to the temporal coverage of the PSI data sets that can only look as far back as the SAR archive data allows (to the beginning of the 1990s).

Further need to integrate the PanGeo geohazards products is found in the representation of areas susceptible to the various geohazards, where ground motions are potential, but have not necessarily occurred. As mentioned above, geohazard susceptibility mapping has been undertaken by the BGS for the entire landmass of Great Britain, and is available through the BGS's commercial GeoSure data set (BOOTH et al. 2010; WALSBY 2008). Thus, the use of the PanGeo GSL for Greater London in conjunction with areas of potential geohazards identified through GeoSure is highly recommended.

4. Conclusions

We have mapped a range of interacting natural and manmade geohazards within the administrative area of Greater London, by combining ground motion data derived from the IPTA processing of ERS-1/2 and ENVISAT SAR images acquired between 1992 and 2010, with a variety of geological and other geospatial layers.

Areas of observed geohazards that were identified via the PanGeo standardised methodology cover a total of over 40 % of the investigated area (i.e., ~ 650 km^2 out of a total of $\sim 1,580$ km^2), and range from natural compaction of the Holocene deposits of the River Thames flood plain, to land subsidence and heave resulting from groundwater management for engineering works or domestic use, and changes in the groundwater levels in the main aquifer of the London Basin. Observed ground motions indicate a combination of land surface processes comprising ground subsidence and uplift, as well as down slope movements, and minimum and maximum observed LOS velocities are -25.3 and $+29.5$ mm/year during 1992–2010.

Integration of the geohazard mapping with the Copernicus EEA European Urban Atlas has revealed greater spatial variability of observed motion velocities within non-urban land cover types such as agricultural, semi-natural and green areas, when compared to observations for continuous and discontinuous urban fabric and industrial units that seem to show the smallest standard deviations in their annual velocity statistics.

We have also analysed difficulties in the identification of land processes relating to the shrink–swell of clayey deposits that are based mainly upon interpreting just the velocity and not the time series spatio-temporally, and to ground dissolution and collapse, the associated difficulties of which mainly relate to technological constraints due to temporal decorrelation. Challenges relating to the detection of good radar targets for PSI analysis in landsliding affected rural areas are also discussed. Future research will focus on analysing further PSI data for this area with specific regard to these geohazards; for instance, to ascertain whether seasonal variations of ground levels due to shrinking–swelling clays have been depicted by the motion time series of the radar targets identified across Greater London, or to verify whether precursors of ground collapses have been recorded by the PSI data in areas subject to dissolution processes. Further research is already being carried out at BGS to test and assess new processing techniques to improve the radar target coverage and density in non-urban areas. The latter has been recognized as a priority for areas like the UK, where land cover exerts significant control on the success of interferometric studies using C-band SAR imagery (BATESON et al. 2014; CIGNA et al. 2013a, 2014).

Acknowledgments

The research leading to these results has received funding from the European Union's Seventh Framework Programme (FP7/2007–2013) under grant agreement no. 262371. The authors would like to thank Ren Capes, formerly PanGeo Project Coordinator at CGG—NPA Satellite Mapping, for launching, leading and promoting the project with dedication from 2011 to 2013, and Prof. Stuart Marsh, formerly at BGS, for chairing the Service Design and Validation Panel in PanGeo and his enthusiastic contribution to the project. Lee Jones at BGS is acknowledged for his support with the discussion on expansive soils, and Don Aldiss, formerly at BGS, for aspects concerning local geology of the London area. F. Cigna, H. Jordan and L. Bateson publish with the permission of the Executive Director of the British Geological Survey (BGS), Natural Environment Research Council (NERC). PanGeo Data: Copyright © 2012. Used and/or reproduced with the permission of the rights holders who participated in the EC FP7 PanGeo Project (262371) and European Environment Agency. Details of the rights holders and the terms of the licence to use PanGeo project data can be found at http://www.pangeoproject.eu. Ordnance Survey data © Crown Copyright and database rights 2013. Geological materials © NERC, All rights reserved. GMES EEA Urban Atlas © EEA 2013, Directorate-General Enterprise and Industry (DG-ENTR), Directorate-General for Regional Policy. NEXT-Map® Britain © 2003, Intermap Technologies Inc., All rights reserved. Aerial photography © UKP/Getmapping Licence No. UKP2006/01.

REFERENCES

ALDISS, D., BURKE, H., CHACKSFIELD, B., BINGLEY, R., TEFERLE, N., WILLIAMS, S., BLACKMAN, D., BURREN, R. and PRESS, N., 2014. *Geological interpretation of current subsidence and uplift in the London area, UK, as shown by high precision satellite-based surveying.* Proceedings of the Geologists' Association, *125*(1): 1–13.

AMELUNG, F., GALLOWAY, D.L., BELL, J.W., ZEBKER, H.A. and LACZNIAK, R.J., 1999. *Sensing the ups and downs of Las Vegas: InSAR reveals structural control of land subsidence and aquifer-system deformation.* Geology, *27*(6): 483–486.

ASTRIUM-GEO, 2014. Underground Construction in Budapest: TerraSAR-X-based analysis of surface movement phenomena related to underground construction in Budapest Available at: http://www.astrium-geo.com/en/140-underground-construction-in-budapest, Accessed on: 07/03/2014.

BANTON, C., BATESON, L., McCORMACK, H., HOLLEY, R., WATSON, I., BURREN, R., LAWRENCE, D. and CIGNA, F., 2013. Monitoring post-closure large scale surface deformation in mining areas, Mine Closure 2013, Eighth International Conference on Mine Closure. Australian Centre for Geomechanics, Perth, Eden Project, Cornwall, UK.

BATESON, L., 2013. PanGeo—D3.3 Product and Service Specification. Version 2.3, 51 pp. Available at: http://www.pangeoproject.eu/sites/default/files/pangeo_other/D3.3-Product-Specification-v1-of-4th-Aug-2011[1].pdf. Accessed on: 30/06/2014.

BATESON, L., CIGNA, F., BOON, D. and SOWTER, A., 2014. The application of the Intermittent SBAS (ISBAS) InSAR method to the South Wales Coalfield, UK. International Journal of Applied Earth Observation and Geoinformation 34:249–257.

BATESON, L., CUEVAS, M., CROSETTO, M., CIGNA, F., SCHIJF, M. and EVANS, H., 2012. PANGEO: enabling access to geological information in support of GMES: deliverable 3.5 production manual. Version 1. Available at: http://www.pangeoproject.eu/sites/default/files/pangeo_other/D3.5-PanGeo-Production-Manual-v1.3.pdf, Accessed on: 07/03/2014.

BATESON, L.B., BARKWITH, A.K.A.P., HUGHES, A.G. and ALDISS, D., 2009. Terrafirma: London H-3 Modelled Product. Comparison of PS data with the results of a groundwater abstraction related

subsidence Model. British Geological Survey Commissioned Report, OR/09/032, 47 pp.

BECKEN, K.H. and GREEN, C.A., 2000. *DiGMap: a digital geological map of the United Kingdom.* Earthwise, *16*:8–9.

BELL, J.W., AMELUNG, F., FERRETTI, A., BIANCHI, M. and NOVALI, F., 2008. *Permanent scatterer InSAR reveals seasonal and long-term aquifer-system response to groundwater pumping and artificial recharge.* Water Resources Research, *44*(2):W02407.

BERARDINO, P., FORNARO, G., FRANCESCHETTI, G., LANARI, R., SANSOSTI, E. and TESAURO, M., 2000. Subsidence effects inside the city of Napoli (Italy) revealed by differential SAR interferometry, Geoscience and Remote Sensing Symposium, 2000. Proceedings. IGARSS 2000. IEEE 2000 International, pp. 2765–2767 vol.6.

BERTI, M., CORSINI, A., FRANCESCHINI, S. and IANNACONE, J.P., 2013. *Automated classification of Persistent Scatterers Interferometry time series.* Nat. Hazards Earth Syst. Sci., *13*(8): 1945–1958.

BGS, 2014. GeoSure: National Ground Stability Data. Available at: http://www.bgs.ac.uk/products/geosure/home.html, Accessed on: 07/03/2014.

BINGLEY, R., TEFERLE, F.N., ORLIAC, E.J., DODSON, A.H., WILLIAMS, S.D.P., BLACKMAN, D.L., BAKER, T.F., RIEDMANN, M., HAYNES, M., ALDISS, D.T., BURKE, H.C., CHACKSFIELD, B.C. and TRAGHEIM, D.G., 2007. Absolute Fixing of Tide Gauge Benchmarks and Land Levels: Measuring Changes in Land and Sea Levels around the coast of Great Britain and along the Thames Estuary and River Thames using GPS, Absolute Gravimetry, Persistent Scatterer Interferometry and Tide Gauges. Joint DEFRA/EA Flood and Coastal Erosion Risk Management R&D Programme. Available at: http://nora.nerc.ac.uk/1493/1/Absolutefixing.pdf, Accessed on: 07/03/2014.

BOOTH, K.A., DIAZ DOCE, D., HARRISON, M. and WILDMAN, G., 2010. User Guide for the British Geological Survey GeoSure data set. British Geological Survey Internal Report, OR/10/066, 17 pp.

BURLAND, J.B., STANDING, J.R. and JARDINE, F.M., 2001. Building response to tunnelling: Case studies from the construction of the Jubilee Line Extension, London. Volume 1: projects and methods.

CASTAÑEDA, C., GUTIÉRREZ, F., MANUNTA, M. and GALVE, J.P., 2009. *DInSAR measurements of ground deformation by sinkholes, mining subsidence, and landslides, Ebro River, Spain.* Earth Surface Processes and Landforms, *34*(11): 1562–1574.

CIGNA, F., BATESON, L., JORDAN, C. and DASHWOOD, C., 2013a. Nationwide monitoring of geohazards in Great Britain with InSAR: Feasibility mapping based on ERS-1/2 and ENVISAT imagery. 2013 IEEE International Geoscience and Remote Sensing Symposium (IGARSS 2013), pp. 672–675.

CIGNA, F., BATESON, L., JORDAN, C. and DASHWOOD, C., 2014. *Simulating SAR geometric distortions and predicting Persistent Scatterer densities for ERS-1/2 and ENVISAT C-band SAR and InSAR applications: nationwide feasibility assessment to monitor the landmass of Great Britain with SAR imagery.* Remote Sensing of Environment, *152*, 441–466, doi:10.1016/j.rse.2014.06.025.

CIGNA, F., BIANCHINI, S. and CASAGLI, N., 2013b. *How to assess landslide activity and intensity with Persistent Scatterer Interferometry (PSI): The PSI-based matrix approach.* Landslides, *10*(3): 267–283.

CIGNA, F., JORDAN, H. and BATESON, L., 2013c. GeoHazard Description for London, Version 2.0. Available at: http://www.pangeoproject.eu/pdfs/english/london/Geohazard-Description-london.pdf, 194 pp. Accessed on: 07/03/2014.

CIGNA, F., OSMANOĞLU, B., CABRAL-CANO, E., DIXON, T.H., ÁVILA-OLIVERA, J.A., GARDUÑO-MONROY, V.H., DEMETS, C. and WDO-WINSKI, S., 2012a. *Monitoring land subsidence and its induced geological hazard with Synthetic Aperture Radar Interferometry: a case study in Morelia, Mexico*. Remote Sensing of Environment, *117*: 146–161.

CIGNA, F., TAPETE, D. and CASAGLI, N., 2012b. *Semi-automated extraction of Deviation Indexes (DI) from satellite Persistent Scatterers time series: tests on sedimentary volcanism and tectonically-induced motions*. Nonlin. Processes Geophys, *19*: 643–655.

CLOSSON, D., KARAKI, N.A., KLINGER, Y. and HUSSEIN, M.J., 2005. *Subsidence and sinkhole hazard assessment in the Southern Dead Sea area, Jordan*. Pure and Applied Geophysics, *162*(2): 221–248.

COLESANTI, C. and WASOWSKI, J., 2006. *Investigating landslides with space-borne Synthetic Aperture Radar (SAR) interferometry*. Engineering Geology, *88*(3–4): 173–199.

COOPER, A.H., FARRANT, A.R. and PRICE, S.J., 2011. *The use of karst geomorphology for planning, hazard avoidance and development in Great Britain*. Geomorphology, *134*(1–2): 118–131.

CROSETTO, M., MONSERRAT, O., ADAM, N., PARIZZI, A., BREMMER, C., DORTLAND, S., HANSSEN, R.F., & VAN LEIJEN, F.J. (2008). Validation of existing processing chains in TerraFirma stage 2—Final Report. GMES TERRAFIRMA ESRIN/Contract no. 19366/05/I-E. (15 pp.).

CROSETTO, M., MONSERRAT, O., IGLESIAS, R. and CRIPPA, B., 2010. *Persistent Scatterer Interferometry: Potential, Limits and Initial C- and X-band Comparison*. Photogrammetric Engineering and Remote Sensing, *76*(9): 1061–1069.

CULSHAW, M., TRAGHEIM, D., BATESON, L. and DONNELLY, L., 2006. Measurement of ground movements in Stoke-on-Trent (UK) using radar interferometry. In: M. Culshaw, H. Reeves, I. Jefferson and T. Spink (Editors), 10th Congress of the International Association for Engineering Geology and the Environment, IAEG2006. Geological Society, London, Nottingham, UK, pp. 1–10.

EA, 2007. Groundwater Levels in the Chalk-Basal Sands Aquifer of the London Basin. Unpublished report.

EA, 2010. Management of the London Basin Chalk Aquifer. Status Report 2010. Environment Agency of England and Wales, Thames Region Report.

EC-JRC, 2013. D2.8.III.12 Data Specification on Natural Risk Zones—Technical Guidelines. Available at: http://inspire.jrc.ec.europa.eu/documents/Data_Specifications/INSPIRE_DataSpecification_NZ_v3.0.pdf, Accessed on: 07/03/2014.

EC, 2011. Mapping Guide for a European Urban Atlas. Available at: http://www.eea.europa.eu/data-and-maps/data/urban-atlas/, 31 pp. Accessed on: 07/03/2014.

ELLISON, R.A., WOODS, M.A., ALLEN, D.J., FORSTER, A., PHARAOH, T.C. and KING, C., 2004. Geology of London. Memoir of the British Geological Survey, Sheets 256 (North London), 257 (Romford), 270 (South London), 271 (Dartford) (England and Wales). 114 pp.

ESA, 2009. The Terrafirma Atlas. Terrain-motion across Europe. A compendium of results produced by the European Space Agency GMES service element project Terrafirma 2003–2009. Available at: http://esamultimedia.esa.int/multimedia/publications/TerrafirmaAtlas/pageflip.html, 94 pp. Accessed on: 07/03/2014.

EVANS, H., PENNINGTON, C., JORDAN, C. and FOSTER, C., 2013. Mapping a Nation's Landslides: A Novel Multi-Stage Methodology. In: C. Margottini, P. Canuti and K. Sassa (Editors),

Landslide Science and Practice. Springer Berlin Heidelberg, pp. 21–27.

FARRANT, A.R. and COOPER, A.H., 2008. *Karst geohazards in the UK: the use of digital data for hazard management*. Quarterly Journal of Engineering Geology and Hydrogeology, *41*(3): 339–356.

FERRETTI, A., BASILICO, M., NOVALI, F. and PRATI, C., 2004. Possibile utilizzo di dati radar satellitari per individuazione e monitoraggio di fenomini di sinkholes. In: S. Nisio, S. Panetta and L. Vita (Editors), Stato dell'arte sullo studio dei fenomeni di sinkholes e ruolo delle amministrazioni statali e locali nel governo del territorio, APAT, Roma (2004), pp. 331–340.

FORSTER, A., 1997. The Engineering Geology of the London Area 1:50,000 Geological sheets 256, 257, 270, 271. British Geological Survey Technical Report WN/97/27.

FORSTER, A., WILDMAN, G. and POULTON, C., 2003. Landslide potential modelling of North London. British Geological Survey Internal Report. IR/03/122R.

FOSTER, C., PENNINGTON, C.V.L., CULSHAW, M.G. and LAWRIE, K., 2012. *The national landslide database of Great Britain: development, evolution and applications*. Environmental Earth Sciences, *66*(3): 941–953.

GALLOWAY, D.L., HUDNUT, K.W., INGEBRITSEN, S.E., PHILLIPS, S.P., PELTZER, G., ROGEZ, F. and ROSEN, P.A., 1998. *Detection of aquifer system compaction and land subsidence using interferometric synthetic aperture radar, Antelope Valley, Mojave Desert, California*. Water Resources Research, *34*(10): 2573–2585.

GIBSON, A.D., CULSHAW, M.G., DASHWOOD, C. and PENNINGTON, C.V.L., 2013. *Landslide management in the UK—the problem of managing hazards in a 'low-risk' environment*. Landslides, *10*(5): 599–610.

GIGLI, G., FRODELLA, W., MUGNAI, F., TAPETE, D., CIGNA, F., FANTI, R., INTRIERI, E. and LOMBARDI, L., 2012. *Instability mechanisms affecting cultural heritage sites in the Maltese Archipelago*. Natural Hazards and Earth System Sciences, *12*(6): 1883–1903.

GUTIÉRREZ, F., GALVE, J.P., LUCHA, P., CASTAÑEDA, C., BONACHEA, J. and GUERRERO, J., 2011. *Integrating geomorphological mapping, trenching, InSAR and GPR for the identification and characterization of sinkholes: A review and application in the mantled evaporite karst of the Ebro Valley (NE Spain)*. Geomorphology, *134*(1–2): 144–156.

HALCROW, 2011. Surface Water Management Plan for Royal Borough of Kensington & Chelsea, Drain London. The Royal Borough of Kensington and Chelsea. SWMP Halcrow Report for RBKC, v0.10. Aug 2011.

HANSSEN, R.F., VAN LEIJEN, F.J., VAN ZWIETEN, G.J., BREMMER, C., DORTLAND, S. and KLEUSKENS, M., 2008. Validation of existing processing chains in Terrafirma stage 2. Product validation: validation in the Amsterdam and Alkmaar area. Terrafirma project report. Accessed on: 26/05/2008.

HARRISON, A.M., PLIM, J.F.M., HARRISON, M., JONES, L.D. and CULSHAW, M.G., 2012. *The relationship between shrink–swell occurrence and climate in south-east England*. Proceedings of the Geologists' Association, *123*(4): 556–575.

HERRERA, G., FERNÁNDEZ, J.A., TOMÁS, R., COOKSLEY, G. and MULAS, J., 2009. *Advanced interpretation of subsidence in Murcia (SE Spain) using A-DInSAR data—modelling and validation*. Nat. Hazards Earth Syst. Sci., *9*(3): 647–661.

HOBBS, P.R.N., JONES, L.D., KIRKHAM, M.P., ROBERTS, P., HASLAM, E.P. and GUNN, D.A., 2014. *A new apparatus for determining the shrinkage limit of clay soils*, Géotechnique, *64*(3): 195–203.

JONES, L.D. and JEFFERSON, I., 2012. Expansive soils. In: J. Burland, T. Chapman, H. Skinner and M. Brown (Editors), ICE Manual of Geotechnical Engineering. Volume 1: Geotechnical Engineering Principles, Problematic Soils and Site Investigation. ICE Publishing, London, UK, pp. 413–441.

JONES, L.D. and TERRINGTON, R., 2011. *Modelling Volume Change Potential in the London Clay*. Quarterly Journal of Engineering Geology and Hydrogeology, *44*(1): 109–122.

JORDAN, H., CIGNA, F. and BATESON, L., 2013. GeoHazard Description for Stoke-On-Trent, Version 1.0. Available at: http://www.pangeoproject.eu/pdfs/english/stoke/Geohazard-Description-stoke.pdf, 122 pp. Accessed on: 07/03/2014.

KRIVENKO, A., RIEDEL, B., NIEMEIER, W., SCHINDLER, S., HEEK, P., MARK, P. and ZIEM, E., 2012. Application of Radar Interferometry for Monitoring a Subway Construction Site. In: W.e.a. Niemeier (Editor), GeoMonitoring 2012. TU Braunschweig, pp. 67–85.

LAWLEY, R. and GARCIA-BAJO, M., 2009. The National Superficial Deposit Thickness Model (version 5). Information Products Programme. British Geological Survey Internal Report OR/09/049. Available at: http://nora.nerc.ac.uk/8279/1/SDTM_V5_UserGuide.pdf, 18 pp. Accessed on: 07/03/2014.

LEIGHTON, J.M., SOWTER, A., TRAGHEIM, D., BINGLEY, R.M. and TEFERLE, F.N., 2013. *Land motion in the urban area of Nottingham observed by ENVISAT-1*. International Journal of Remote Sensing, *34*(3): 982–1003.

PAGE, D.P., 1995. *Jubilee Line Extension*. Quarterly Journal of Engineering Geology and Hydrogeology, *28*(2): 97–104.

PITT, M., 2008. The Pitt review: learning lessons from the 2007 floods, June 2008. Cabinet Office, Met.

SMITH, A., 2013. Digital Geological Map of Great Britain, information notes, 2013, Nottingham, UK. Available at: http://nora.nerc.ac.uk/502315/, 54 pp. Accessed on: 07/03/2014.

SOWTER, A., BATESON, L., STRANGE, P., AMBROSE, K. and SYAFIUDIN, M., 2013. *DInSAR estimation of land motion using intermittent coherence with application to the South Derbyshire and Leicestershire coalfield*. Remote Sensing Letters, *4*(10): 979–987.

STANDING, J.R. and BURLAND, J.B., 2006. *Unexpected tunnelling volume losses in the Westminster area, London*, Géotechnique, pp. 11–26.

STROZZI, T. and AMBROSI, C., 2007. SAR Interferometric Point Target Analysis and Interpretation of Aerial Photographs for Landslides Investigations in Ticino, Southern Switzerland, ENVISAT Symposium 2007, Montreux, Switzerland.

SUMBLER, M.A., 1996. British Regional Geology: London and the Thames Valley (4th Edition). London, UK.

TAPETE, D. and CASAGLI, N., 2013. Testing Computational Methods to Identify Deformation Trends in RADARSAT Persistent Scatterers Time Series for Structural Assessment of Archaeological Heritage. In: B. Murgante, S. Misra, M. Carlini, C. Torre, H.-Q. Nguyen, D. Taniar, B. Apduhan and O. Gervasi (Editors), Computational Science and Its Applications—ICCSA 2013. Lecture Notes in Computer Science. Springer Berlin Heidelberg, pp. 693–707.

TEATINI, P., STROZZI, T., TOSI, L., WEGMÜLLER, U., WERNER, C. and CARBOGNIN, L., 2007. *Assessing short- and long-time displacements in the Venice coastland by synthetic aperture radar interferometric point target analysis*. Journal of Geophysical Research: Earth Surface, *112*(F1): F01012.

WALSBY, J.C., 2008. Geosure: a bridge between geology and decision making. In: D.G.E. Liverman, C.P.G. Pereira and B. Marker (Editors), Communicating Environmental Geoscience. Geological Society, London, UK, pp. 81–87.

WEGMULLER, U., WALTER, D., SPRECKELS, V. and WERNER, C., 2010. *Nonuniform Ground Motion Monitoring With TerraSAR-X Persistent Scatterer Interferometry*. Geoscience and Remote Sensing, IEEE Transactions on, *48*(2): 895–904.

WERNER, C., WEGMULLER, U., STROZZI, T. and WIESMANN, A., 2003. Interferometric point target analysis for deformation mapping. 2003 IEEE International Geoscience and Remote Sensing Symposium (IGARSS 2003), vol. 7, pp. 4362–4364.

(Received March 11, 2014, revised August 6, 2014, accepted August 20, 2014, Published online September 20, 2014)

Reprinted from the journal

Pure Appl. Geophys. 172 (2015), 2997–3028
© 2015 Springer Basel
DOI 10.1007/s00024-015-1066-1

Geohazards Monitoring in *Roma* from InSAR and In Situ Data: Outcomes of the PanGeo Project

VALERIO COMERCI,[1] EUTIZIO VITTORI,[1] CARLO CIPOLLONI,[1] PIO DI MANNA,[1] LUCA GUERRIERI,[1] STEFANIA NISIO,[1] CLAUDIO SUCCHIARELLI,[2] MARIA CIUFFREDA,[2] and ERIKA BERTOLETTI[2]

Abstract—Within the PanGeo project (financed by the European Commission under the 7th Framework Program), the Geological Survey of Italy (*ISPRA*) and the Urban Planning Department of the City of *Roma* developed a geodatabase and map of the geological hazards for the territory of *Roma*, integrating remotely sensed data (PSInSAR—Permanent Scatterer Interferometry Synthetic Aperture Radar) and in situ geological information. Numerous thematic layers, maps and inventories of hazards (e.g., landslides, sinkholes, cavities), geological and hydrogeological data added to historical and recent urbanization information were compared to the permanent scatterer (PS) data from the European Remote Sensing satellites (ERS-1/2, 1992–2000) and ENVISAT (2002–2005) descending scenes, in order to produce a ground stability layer (GSL). Based on the PS data, most of the territory appears stable (almost 70 % of PS velocities are within ±1 mm/year). About 14 % of the PSs show positive line-of-sight (LOS) velocities (measured along the LOS of the satellite) between 1 and 3 mm/year and more than 2 % exceed 3 mm/year; more than 11 % of PSs show negative LOS velocities between −1 and −3 mm/year, while about 3 % exceed −3 mm/year (with tens of the PSs showing velocities over −20 mm/year). The GSL is comprised of polygons or multi-polygons (multipart polygons grouping individual polygons under a single identifier geohazard) enclosing areas where geohazards have been pointed out by PS data and/or in situ surveys (observed instabilities), and by polygons enclosing areas potentially affected by geohazards, based on the available knowledge of the territory (potential instabilities). In *Roma*'s GSL, 18 multi-polygons (covering ca. 600 km^2) related to observed instabilities have been outlined, where ground movements could be detected through InSAR data or where landslides and sinkholes are known to have occurred. Other 13 multi-polygons (covering nearly 900 km^2) concern areas where the potential occurrence of geohazards was inferred by combining geological and/or geothematic data (potential instabilities). The geohazards mapped in *Roma* have been: landslides, collapsible grounds, compressible grounds, groundwater abstraction, mining, man-made ground, tectonic movements, and volcanic inflation/deflation. The lattermost is the likely cause of the significant uplift observed in the Alban Hills area. However, this paper focuses on two more currently impending hazards: subsidence and sinkholes. In general, sinkhole-prone areas (areas of dense underground cavities) are hard to discern from satellite data, but can be revealed by ruling out other potential causes of observed ground movement based on in situ data. Subsiding zones are effectively detected by the available PSInSAR dataset over a total extent of about 60 km^2, mostly overlapping the recent alluvial areas of the Tiber and its tributaries. PSs show a very different behaviour inside and outside the historical centre. Inside, loading by anthropogenic construction and man-made ground since ancient times has led to an almost complete consolidation of the recent river deposits, marked by modest to absent subsidence. In contrast, outside, subsidence clearly stands out, with negative LOS velocities that, although generally within several mm/year, can locally exceed −25 mm/year. PS data have provided motion information at both a regional and local scale (up to the scale of a single building). Closer to the sea, in the Tiber delta area, velocities increase, especially above recently reclaimed marsh areas, rich in peat and organic clays. Velocities can change significantly over short distances, as in the international airport area, reflecting the local stratigraphic setting. The same occurs in the two subsiding areas located within the Alban Hills volcanic complex, which is generally affected by ground uplift. As a whole, PSInSAR ground motion velocities offer a significant contribution to susceptibility and hazard recognition studies. In particular, such a method provides a fast and effective tool available to local authorities to monitor ground and building behaviour, possibly allowing for timely prevention activities, especially when coupled with appropriate in situ knowledge.

Key words: Geohazards, PSInSAR, *Roma*, PanGeo, Ground stability layer.

1. Introduction

In the last two decades, several projects have developed monitoring methods and mapping of natural hazards in Europe, most of them financed by the European Commission (EC), e.g., *MUSCL (monitoring urban subsidence, cavities and landslides by remote sensing)*, *LESSLOSS (Risk Mitigation for*

[1] ISPRA (Istituto Superiore per la Protezione e la Ricerca Ambientale), Geological Survey of Italy, via Vitaliano Brancati 48, 00144 Roma, Italy. E-mail: valerio.comerci@isprambiente.it

[2] Department of Urban Planning, Roma Capitale, via del Turismo 30, 00144 Roma, Italy.

Earthquakes and Landslides), *SafeLand (Living with landslide risk in Europe: Assessment, effects of global change, and risk management strategies)*, *DORIS (An advanced downstream service for the detection, mapping, monitoring, and forecasting of ground deformations)*, *LAMPRE (Landslide Modelling and tools for vulnerability assessment Preparedness and Recovery management)*, *SHARE (Seismic Hazard Harmonization in Europe)*, etc.

Between 2011 and 2014, the *PanGeo* project, a *GMES (Global Monitoring for Environment and Security)-Copernicus* service has been financed by the EC (Seventh Framework Program). Unlike previous projects, PanGeo was aimed at providing geohazard information at a city scale based on satellite data (Permanent Scatterer Interferometry Synthetic Aperture Radar, PSInSAR), interpreted by the national geological surveys of 27 European Union countries, by comparing them with in situ geological, geothematic and geohazard data (BATESON *et al.* 2012). Now, PanGeo provides users with open access to geohazard maps across 52 European cities and towns (representing approximately 13 % of the EU population), down to a mapping scale of 1:10,000. Therefore, PanGeo offers to local authorities a valuable information tool for land use management.

The PanGeo products, built in strict consistency with the last version of the INSPIRE Data Specification related to the Natural Risk Zone theme (INSPIRE DS-D2.8/III-12 2013), include a ground stability layer (GSL), which maps, using attributed vector polygons, all the areas of a given town that are affected by terrain motion (ground instability), and a geohazard description (GHD), a document that illustrates the geological reasons for the discovered motions. The areas of mapped ground instability are differentiated into areas of observed instability and areas of potential instability. The former includes zones where the instability has been detected by all types of direct or indirect observation, e.g., by PSI, by other types of deformation measurement devises and techniques or from geology field surveys and campaigns. The areas of potential instability include zones identified as having the potential for ground instability using the available geological, geomorphological and auxiliary data provided by the geological surveys.

Generally, geohazards are the geomorphological, geological, or environmental processes, phenomena, and conditions that are potentially dangerous or pose a level of threat to human life, health and property, or to the environment (KOMAC and ZORN 2013). Geohazards include subaerial and submarine processes, such as earthquakes, volcanic eruptions, floods, erosion, subsidence, sinkholes, landslides and tsunamis; in some cases, geohazards can also be induced by human activities. In PanGeo, the geohazards taken into consideration are natural and man-made phenomena that make the ground unstable, including: landslides, volcanic inflation or deflation, tectonics, compressible ground, collapsible deposits, man-made ground, underground constructions, mining, groundwater abstraction and recharge, shrinking and swelling clays, liquefaction, and soluble rocks (flooding is not considered).

The PSInSAR technique, patented by the Polytechnic University of Milan (POLIMI) (FERRETTI *et al.* 2000a, 2001), allows characterization of ground motion with millimetre accuracy, based on the identification of permanent scatterers (PSs), which are points on the ground that consistently backscatter stable radar signals to a satellite sensor, allowing recognition of phase differences between successive acquisitions. The radar wave-lengths of the most common sensors range from one to a few centimetres, so that the measurement precision can be submillimetric. The PSInSAR techniques provide usable information where backscatterers can be repeatedly recorded, requiring they retain a sufficient signal coherence over a sufficiently long time span. Typically, they correspond to buildings, metallic objects, exposed rocks, etc. Thus, in general, PSs are very efficient at monitoring hazards related to slowly evolving ground deformation in urban areas and areas covered by a dense network of infrastructures, which are also areas of higher risk.

This technique has already proved successful in many applications (e.g., FERRETTI *et al.* 2000a; HILLEY *et al.* 2004; HOOPER *et al.* 2004; DIXON *et al.* 2006; MEISINA *et al.* 2006, 2008; HERRERA *et al.* 2009a; CIGNA *et al.* 2012a, b; RIGHINI *et al.* 2012).

In this paper, we present some results of the *Roma* case study, carried out by the Italian Geological

Survey (ISPRA) in collaboration with the Urban Planning Department of the City of *Roma*.

We have elaborated a GSL for *Roma* (and a GHD; COMERCI *et al.* 2013) that integrates PSInSAR data (ERS-1/2 1992–2000 and ENVISAT 2002–2005 scenes) with geological and geothematic data (BERTOLETTI *et al.* 2013). We already analysed the same PSInSAR dataset in the framework of the Terrafirma project (http://esamultimedia.esa.int/multimedia/publications/TerrafirmaAtlas/pageflip.html).

This paper focuses on two geohazards: subsidence and sinkholes, showing representative case studies. The former, one of the most widespread in the GSL, was chosen because the PanGeo methodology followed proved to be highly effective. On the other hand, sinkhole hazards offer a good opportunity to exploit, at best, integration of SAR and in situ data to address its elusiveness. Although application of InSAR techniques provides valid results in the study of landslides (e.g., FARINA *et al.* 2006; CIGNA *et al.* 2013; HERRERA *et al.* 2013), our PSInSAR dataset proved to be poorly effective in detecting slope movements. Actually, the slopes known to be affected by landslides show a general lack of backscattering, mostly due to vegetation.

The large amount of available in situ data (after their homogenization) has sometimes enabled one to overcome the limits of the PSInSAR technique, i.e., lack of data on vegetated and bare areas (on this argument see CIGNA *et al.* 2011; TAPETE *et al.* 2012a).

Several case-studies of subsidence are here illustrated, regarding the areas of *Viale Giustiniano Imperatore* and *Fosso di Vallerano* (sited along two tributaries of Tiber), the Tiber delta (in particular, inside the *Leonardo da Vinci* airport compound), and two sectors, near *Rocca di Papa* and *Grottaferrata*, located on the slopes of the Alban Hills volcanic complex, generally affected by ground uplift. Finally, we describe a susceptibility study, based on the integration of PS and in situ data, regarding the probability of sinkhole formation above a warren of underground artificial cavities in the *Caffarella–Appio Latino* and *Prenestino-Labicano* boroughs.

In recent years, a number of papers showed the usefulness of radar interferometry as a tool for monitoring urban areas (e.g., HERRERA *et al.* 2009b, c, 2010, 2012; BRU et al. 2013; TOSI *et al.*, 2013;

EZQUERRO *et al.* 2014; SANABRIA *et al.* 2014). As regards the Roman area, some studies focused on the *Viale Giustiniano Imperatore* area, where subsidence causes a particularly significant hazard (MANUNTA *et al.* 2008; STRAMONDO *et al.* 2008; FORNARO *et al.* 2010, 2014; ZENI *et al.* 2011; ARANGIO *et al.* 2013), and on the city centre (MANUNTA *et al.* 2008; ZENI *et al.* 2011; TAPETE *et al.* 2012b; CIGNA *et al.* 2014), by applying different interferometric techniques, e.g., IPTA (interferometric point target analysis; WERNER *et al.* 2003; WEGMÜLLER *et al.* 2004), SqueeSAR (FERRETTI *et al.* 2011), SBAS (Small BAseline Subset)-DInSAR (differential synthetic aperture radar interferometry) (GABRIEL *et al.* 1989; ROSEN *et al.* 2000; BERARDINO *et al.* 2002; MORA *et al.* 2003; LANARI *et al.* 2004; PRATI *et al.* 2010), differential SAR tomography (REIGBER and MOREIRA 2000; FORNARO *et al.* 2012) and CAESAR (FORNARO *et al.* 2013, 2015).

2. *Geological background of Roma*

The city of *Roma*, the capital of Italy and the UNESCO World Heritage Site, is located in the central-western portion of the Italian peninsula (Latium region), where the river Aniene joins the Tiber, which is the third main river of Italy (length 405 km). With around 2.8 million residents in an area of nearly 1300 km^2, *Roma* is the most populated town in Italy. Although the city centre is about 24 kilometres inland from the Tyrrhenian Sea, the city extends southward to the very shore with its Ostia district. So, the altitude of *Roma* ranges from the sea level to its highest point at 139 m a.s.l. (Monte Mario Hill).

The area covered by the GSL of *Roma* (Fig. 1) amounts to around 1940 km^2 and corresponds to the extent of the *Roma* Municipality with its 15 administrative districts. To this area, are added the territories of some smaller municipalities surrounding *Roma* and falling inside its larger urban zone (LUZ), as defined in the Urban Atlas 2006 (EEA 2011). These are the whole territories of the municipalities of Fiumicino (where the main airport of *Roma* is located), Pomezia, Ariccia, Albano, Castel Gandolfo, Marino, Ciampino, Frascati, Grottaferrata, Monte

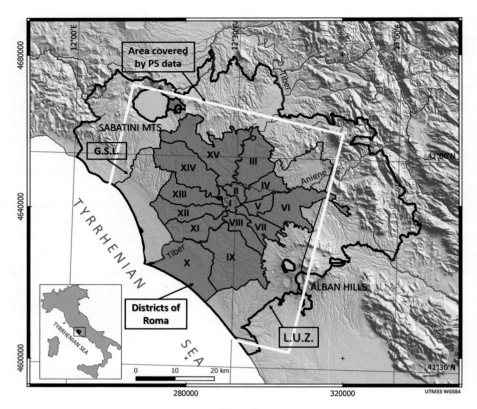

Figure 1
Location of the study area corresponding to the GSL (ground stability layer), which is inside the large urban zone (LUZ). The limits of the area covered by PS data are also indicated. *I* to *XV* are the districts of *Roma*

Porzio, Guidonia, and portions of Ardea, Genzano, Rocca di Papa, Montecompatri, Zagarolo, Gallicano and Tivoli. The total population reaches more than 3.2 million inhabitants (2011 census), corresponding to a density of ca. 1600 inhabitants/km^2, mostly concentrated in the city and the other main satellite towns: Fiumicino, Tivoli, Pomezia, and the many boroughs on the pleasant foothills of the Alban Hills volcanic area.

2.1. Stratigraphic and tectonic evolution of the study region

The geology and stratigraphy of *Roma* and surroundings is relatively well known and detailed (HEIKEN *et al.* 2007; FUNICIELLO 1995; FUNICIELLO *et al.* 2008), thanks to many boreholes drilled for public and private constructions (e.g., VENTRIGLIA 2002) and high resolution mapping. Most of *Roma*'s GSL is covered by three geological sheets

at a scale of 1:50,000 of the CARG (*CARtografia Geologica*) project (374 *Roma*, 373 *Cerveteri* and 387 *Albano*). Outside the CARG Sheets, geo-lithological maps at a scale of 1:100,000 cover the GSL (AMANTI *et al.* 2007), accessible online at http://sgi.isprambiente.it/geoportal/catalog/content/project/litologica.page.

The study area lies in a wide nearly flat landscape extending from the southwestern flank of the Central Apennine chain to the Tyrrhenian Sea coast, bounded by volcanic apparatuses and structural highs (Fig. 2). Its origin is linked to the Neogene extensional tectonic evolution of the Tyrrhenian Sea—Apennines boundary (MALINVERNO and RYAN 1986; PATACCA *et al.* 1992). A neo-autochthon marine sedimentation has filled this subsiding area since the Late Messinian. However, during the Plio—Quaternary, interplay between climatic changes, with their alternating depositional and erosive phases, and extensional tectonics and related volcanism, caused a complex

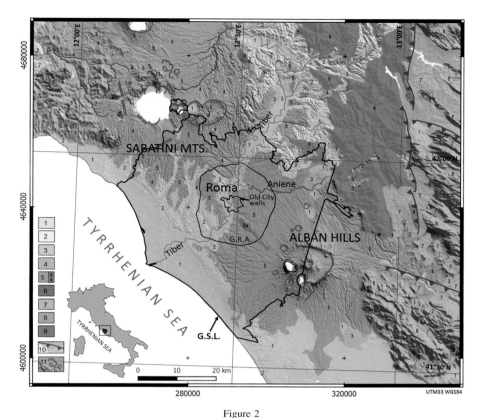

Figure 2

Simplified geological map of the surroundings of Roma (geological units from the OneGeology Project, scale 1:500,000: http://sgi. isprambiente.it/geoportal/catalog/content/project/onegeology.page). *1* Holocene—Latest Pleistocene sedimentary units; *2* Pleistocene sedimentary units; *3* Pleistocene travertines; *4* mainly tefritic pyroclastic flow and hydromagmatic units (Middle Pleistocene); *5* Mainly latitic-trachitic pyroclastic flow and hydromagmatic units (Middle-Late Pleistocene); *5a* Capo di Bove lava flow (277 kyr); *6* Acid (Riolites, riodacides) lavas and ignimbrites (Early Quaternary); *7* Miocene flysch; *8* Meso-Cenozoic carbonate platform units (Latium-Abruzzi facies); *9* Meso-Cenozoic pelagic carbonate units (Sabina facies); *10* Main thrust fronts; *11* Caldera rims. G.R.A. (*Grande Raccordo Anulare*) is the Great Ring Road

suite of geological features (FUNICIELLO and GIORDANO 2008a, b; MATTEI *et al.* 2008; PAROTTO 2008).

The deep, graben-like structure is mostly made of Meso—Cenozoic basin to shelf carbonate and flysch facies units. After the Miocene northeast-verging shortening that originated the Apennines thrust and fold belt, these units sunk during the Neogene, affected by the extensional tectonics following the opening of the Tyrrhenian Sea (BARTOLE 1984). The roof of this basement is now from a few hundred (structural highs) to more than one thousand meters deep (FUNICIELLO and PAROTTO 2001). In the wide depressions, Late Pliocene marine *grey-blue* clays are deposited (Mt. Vaticano formation); they are the oldest sedimentary outcroppings in *Roma*, representing the aquitard of all the hydrogeological units

(Fig. 3). Made of well-consolidated marly clays and silty clays, with rare sand intercalations, they crop out at various spots in town (often hidden by recent constructions), but mainly along a structural ridge parallel to the right bank of the Tiber river, at the base of the Mt. Mario, Vaticano and Gianicolo hills (FUNICIELLO and GIORDANO 2008a, b). To the east and west of this fault-bounded ridge, these clays lie at various depths, being particularly thick in tectonic lows of the bedrock, as in the Circus Maximus area (at least 870 m) (MARRA and ROSA 1995).

An erosive phase, corresponding to the first big glaciation marking the passage to the Pleistocene, interrupted this marine cycle. Two subsequent marine transgression cycles followed, dating to the Early Pleistocene. Mainly made of sandy units (Mt. Mario

41

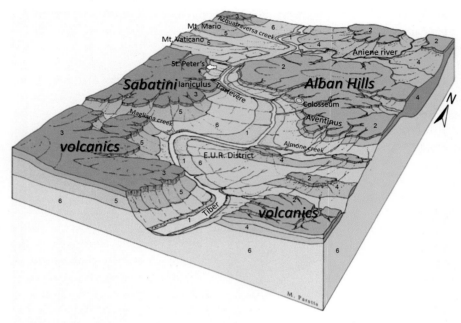

Figure 3

Highly schematic paleomorphology of the Roma area during the last sea level lowstand (20–22 kyr; from PAROTTO 2008, modified). Well evident are the incision of the pre-volcanic marine and continental basement and the caps of mostly ignimbrites from the Sabatini and Alban Hills volcanic centres, roughly separated by the Tiber valley. Nowadays, a large part of the valley floor is filled by post-glacial, mainly Holocene, fluvial and palustrine deposits and man-made ground. Also, the Tiber river bed now shows a different system of meanders. The Aniene river and the Acquatraversa, Almone and Magliana creeks are the main tributaries of the Tiber in the city area. *1* Late Pleistocene-Holocene alluvial deposits of Tiber and its tributaries; *2* Volcanic products of the Alban Hills volcanic district (Middle-Late Pleistocene); *3* Volcanic products of the Sabatini volcanic district (mainly ignimbrites; Middle Pleistocene); *4* Acquatraversa Fm. (fluvial gravel and sand of Paleo-Tiber, Early-Middle Pleistocene); *5* Mt. Mario Fm. (sandy-silty clays and grey to yellowish sands, Early Pleistocene); *6* Mt. Vaticano Fm. (overconsolidated grey-blue clays, Late Pliocene)

formation), they indicate a much shallower marine environment, partly due to uplift of the basement and partly to the main eustatic sea level fluctuations during this period (MARRA *et al.* 1995; MARRA and ROSA 1995; COSENTINO *et al.* 2009). These marine deposits crop out in *Roma* in the hills west of the Tiber.

The change from marine to coastal, then to continental sedimentation, is evident toward the end of the Early Pleistocene in the fluvial–lacustrine facies deposits carried by the Paleo-Tiber and its tributaries (BOZZANO *et al.* 2000). Such strong paleogeographic change took place due to general cooling and concurrent basement uplift. Sea regression induced erosion of the Plio-Pleistocene bedrock through downcutting of the drainage network. The mainly continental deposits related to this activity (PAROTTO 2008, and references therein) start with a basal gravel overlain by fluvial and delta facies

(Acquatraversa formation; Fig. 3), with fluvial-palustrine sandy clays and frequent peat beds (MARRA and ROSA 1995; FACCENNA *et al.* 1995; MARRA *et al.* 1995).

Since about 0.6 Ma, this evolving landscape was widely modified by the activity of two main volcanic districts near *Roma*, the Alban Hills to the southeast and the Sabatini Mounts to the northwest. Between 500 and 1000 km^3 of alkalin-potassic pyroclastic flows, surges, airfall deposits and lavas erupted from these volcanic districts from the Middle to Upper Pleistocene, during several paroxysmal events (DE RITA et al. 1994, 1995; FUNICIELLO and PAROTTO 2001; GIORDANO 2008).

With regard to the Alban District, the most ancient known deposits (561–355 kyr) are represented by the Tuscolano-Artemisio highly explosive phase (DE RITA *et al.* 1995), made prevalently of pyroclastic flows and subordinately of lavas, which can be seen in several places in the eastern side of the town, being probably one of the main causes of migration of the Paleo-Tiber to its present-day position against the western structural high comprised of Pliocene sediments (Fig. 3). The Tuscolano-Artemisio Phase closed with the collapse of the caldera. Afterwards, no more than 2 km^3 of volcanic material erupted, a very small volume compared to the 280 km^3 of the preceding phase. A strato-volcano grew inside the caldera and a remarkably long leucitic lava flow escaped from the caldera (*Lava di Capo di Bove*—277 kyr; Fig. 2) following the future trace of the ancient Appian Way down to near the Tiber, representing afterwards the main source of paving stones for *Roma* until the last century. The last Hydromagmatic Phase (up to 37 kyr) involved some eccentric craters in the southwestern sector of the volcanic edifice (MARRA and ROSA 1995; KARNER et al. 2001). According to FUNICIELLO et al. (2003), there is archaeological and stratigraphic evidence of nearly historical volcanic activity, possibly confirmed by some descriptions in the Annals of Livy. Actually, thermal anomalies, gas emissions, seismic activity and the ongoing uplift testify that this volcanic apparatus is far from extinct.

The most ancient Sabatini deposits in *Roma* (548 kyr) are pyroclastic flows, encountered in the underground of the old city. The first big explosive eruption of the Bracciano sector, followed by a collapse of its caldera (449 kyr), was characterized by red tuffs with black scoriae. Yellow tuffs (285 kyr) closed the activity of this sector with the final collapse of the caldera (MARRA et al. 1998). The most recent known activity is about 250 kyr old, with pyroclastic flows and hydromagmatic products from a secondary centre (MARRA and ROSA 1995; KARNER et al. 2001). Despite strong positive thermal anomalies, there is no further evidence of residual activity nowadays.

Continental deposits are interlayered with the volcanic products, mostly fluvial and lacustrine deposits (including diatomites and travertine) with resedimented levels of volcanics. They are temporally separated by erosive phases, corresponding to the glacial periods. However, the cap of relatively hard volcanic deposits (ignimbrites) has substantially protected from erosion the softer uplifted marine and continental sediments underneath.

As a result, the deep incision of the drainage network (Fig. 3) following the Last Glacial sea level drop (22–20 kyr ago) has shaped a system of tabular hills bounded by cliffs and steep slopes. The Pliocene bedrock was locally eroded down to about 50 m below the sea level in *Roma*. The subsequent sea rise (end of the Pleistocene-Holocene) has sensibly reduced the gradients of the drainage system, determining the meandering of the Tiber and the development of a flood plain greatly broadening seaward and of a delta with wide humid environments (marshes and swamps). The valleys carved into the volcanic deposits were filled by soft alluvial sediments, mainly sands, silts and clay often rich in organic matter, recently buried in town by artificial filling that may reach a thickness of 20 m (BELLOTTI et al. 1994, 1995; MARRA and ROSA 1995; AUTORITÀ DI BACINO DEL FIUME TEVERE 2004; CAMPOLUNGHI et al. 2008; PAROTTO 2008). A large part of these Holocene alluvial sediments are still today affected by subsidence, as described in the following text.

2.2. Overview of the main geohazards in Roma

The stratigraphic and structural backbone of the region enclosing the GSL and its most recent climatically driven geomorphological evolution, as shaped by the Tiber River and its tributaries, had a great influence on the ancient history of *Roma* and still determines the distribution and types of natural hazards.

Roma lies where the Tiber River valley narrows considerably crossing the western outer slopes of the Alban Hills, more or less where their deposits come in contact with the similar products of the Sabatini volcanic field located to the northwest.

The first Romans chose to settle on the ample, nearly flat surfaces at the hilltops (Fig. 4), away from the periodic river floods, which represented, and still do, the major source of hazard for the town. When the first urban sprawl took place (during the Roman age), the reclamation of areas on the Tiber's left bank proceeded with infill of the marsh areas. A system of culverts draining into the *Cloaca Maxima* (main sewer) was implemented to drain the valley between *Capitolium*, *Palatinus* and *Velia*: another system of culverts was implemented between *Palatinus* and

SYNOPSIS XLVI TABVLARVM

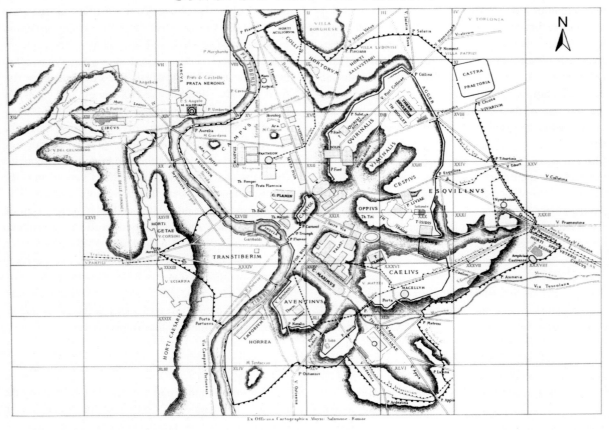

Figure 4
Simplified morphology of Roma with the Seven Hills and the old city walls (see location in Fig. 2) (modified, from LANCIANI 1901)

Aventinus in order to allow the realization of the *Circus Maximus*. Some morphological elements were even removed or completely modified, as is the case of the morphological saddle between the *Quirinalis* and *Capitolium* hills and the *Velia* hill just before the *Colosseum*, cancelled in the modern age (1930) to realize the *Fori Imperiali* street (Fig. 4).

In recent times, the flood-prone areas along the Tiber, especially seaward, have been densely occupied, also thanks to a system of protective measures and extensive reclamation works since *Roma* became the capital of the newly formed Italian state in 1870 (BERSANI and BENCIVENGA 2001). This urban sprawl above the very recent soft alluvial sediments has led to a substantial increase of risk, related, above all, to subsidence and differential compaction, enhanced by the often thick layer of man-made ground. The seismic hazard in town is related to earthquakes with both local epicentres (Alban Hills area, with generally modest magnitudes but shallow depths) and epicentres in the Apennines, where magnitudes can be close to or even exceed 7 (MOLIN *et al.* 1995; historical seismic catalogues in BOSCHI *et al.* 1997; ROVIDA *et al.* 2011). Seismic amplification is expected to be severe in the zones of recent fluvial deposition and artificial fill.

Another type of hazard is posed by the pervasive network of underground, often quite shallow, artificial cavities. Mostly, they are mines of volcanic soft, easy to quarry, construction material (*Pozzolana*), which are sometimes affected by sinkhole phenomena.

Finally, a far from trivial volcanic hazard comes from the unrest of the Alban Hills, as shown by uplift, seismicity, thermal and gas anomalies, and potentially the historical manifestations of activity (FUNICIELLO *et al.* 2003; FUNICIELLO and GIORDANO 2010).

Information concerning urban development and the local morphology before the numerous artificial changes of the last 130 years can also be retrieved from historical maps (e.g., ARCHIVIO STORICO CAPITOLINO 2002).

3. Datasets

3.1. Geographical and geological dataset

The background input data used to support the geological interpretation of each geohazard within the GSL include (Table 1): topographic maps, geological, structural and lithological maps, maps and inventories of specific hazards (landslides, sinkholes, cavities), hydrogeological data, characteristics of historical and recent urbanization and human activities (thickness of anthropic deposits, quarries and mines distribution, roads and railways network, tunnelling). The geographic reference systems of the different datasets were manifold; they were all transformed into the WGS84 reference system to enable allow layers to overlap at their best, considering the uncertainties due to their different scale and origin.

3.2. Permanent scatterers interferometry dataset

The PanGeo consortium has provided ISPRA terrain motion velocities covering a period of 13 years between 1992 and 2005, derived from ERS-1/2 descending scenes acquired in 1992–2000 and ENVISAT descending scenes acquired in 2002–2005. Multi-temporal interferometric processing of this satellite radar imagery, carried out by TRE (TeleRilevamentoEuropa) by means of the patented POLIMI PS Technique (PSInSAR^{TM}; FERRETTI *et al.* 2000a, 2001), has provided motion time-histories and mean velocities of a dense cluster of PS reflection points along the LOS of the satellites. With the nominal repeat cycle of both ERS-1/2 and ENVISAT

data being 35 days, the resulting time-series are characterized by a ca. monthly temporal sampling. Details about the number of PSs, the extension of the total processed area and the average PS density are summarized in Table 2.

A satisfactory quality characterizes the ERS-1/2 descending dataset (standard deviation (SD) of the mean PS velocities: between 0.26 and 0.89 mm/ year), likely due to the good number of scenes available and employed for the processing (65 scenes). Lower quality and reliability characterize the ENVISAT descending dataset (for a discussion on processing methodologies to improve data analysis see NOTTI *et al.* 2014), due to the low number of images with which it was generated (13 scenes) and consequent lower coherence and higher SD (between 0.8 and 6.6 mm/year).

The orbit inclination of both ERS-1/2 and ENVISAT is about 12° with respect to the south-north axis at the latitude of *Roma*, and the look angle of the employed sensor modes is about 23° from the zenith. For these reasons, the PS measurements expressed with respect to LOS are very sensitive to vertical motion, but almost blind with respect to north–south deformation and slightly influenced by east–west components. RASPINI *et al.* (2012) analysed the subsidence ground motion components in a sedimentary basin infilled by Quaternary fine-grained deposits, evidencing the predominance of the vertical direction. The relationship between the velocities along the LOS (V_{LOS}) and along the vertical direction (V_V) (assuming null the horizontal components) is represented by the following formula:

$$V_V = V_{LOS}/\cos\vartheta = V_{LOS} \times 1.086$$

where ϑ is the satellite look angle (23°). For the subsidence cases illustrated in this paper occurring in areas occupied by alluvial deposits, because of the geological setting similar to that described in RASPINI *et al.* (2012), it is possible to multiply the LOS velocities by a factor of 1.086 to obtain vertical velocities.

The distribution of identified PSs and their average motion velocities are shown in Figs. 5 and 6. Taking into account the characteristics of the two

Table 1

Spatial datasets used for the interpretation of ground motions and the definition of geohazards

Geohazard	Dataset	Description	Scale	Source	Observed/ Potential	Coverage
Landslide	IFFI—Italian Landslides Inventory	Landslide polygons	1:10,000	ISPRA	Observed	GSL
	Roma landslide DB	Landslide polygons/points	1:10,000	ISPRA	Observed	Roma Capitale Municipality
	Susceptibility landslide map	Roma landslide DB + scarp steepness	1:10,000	ISPRA	Potential	Roma Capitale Municipality
Collapsible ground	Sinkhole DB	Areas affected by sinkholes	1:10,000	ISPRA	Observed	GSL
	Cavities and collapse DB	Areas affected by cavities and collapses	1:20,000	VENTRIGLIA (2002)	Potential	Roma Capitale Municipality
	Cavities	Cavities DB	1:20,000	Sovraintendenza Beni Culturali Roma	Potential	Roma Capitale Municipality
	Cavities probability map	Susceptibility map based on documented collapses, sinkholes and ancient roman mines	1:20,000	SCIOTTI et al.(2000)	Potential	Roma Capitale Municipality
Compressible Ground	Geology 10 k Roma	Source of distribution of alluvial deposits, peat and silt deposits, sand and gravel deposits, etc.	1:10,000	ISPRA; AUTORITÀ DI BACINO DEL FIUME TEVERE (2004)	Potential	Roma Capitale Municipality
	Geology CARG 25 k	Source of distribution of alluvial deposits outside Roma municipality	1:25,000	ISPRA	Potential	GSL
	Lithology 100 k	Geo-lithological map at scale 1:100,000 derived from the geological map of Italy at the same scale	1:100,000	ISPRA	Potential	GSL
Underground Construction	Underground train network	Existing and in progress underground lines	1:10,000	Roma Capitale	Potential	Roma Capitale Municipality
	Tunnelling	Galleria Giovanni XXIII	1:10,000	Roma Capitale	Potential	Roma Capitale Municipality
Tectonic Movements	Faults	Faults taken from CARG project maps	1:25,000	ISPRA	Potential	GSL
Mining	Quarries and Mines distribution	Mining inventory	1:25,000	Roma Capitale	Potential	Roma Capitale Municipality
	Geology CARG 25 k	Quarries and mines reported in the CARG db	1:25,000	ISPRA	Potential	GSL
Groundwater Abstraction	Water wells	Inventory of water boreholes for private or public use	1:10,000	Roma Capitale	Potential	Roma Capitale Municipality
	Map of piezometric variation	Water table oscillation of aquifers in *Roma* in the last 20 years (4 classes).	1:10,000	Autorità di Bacino del Fiume Tevere	Potential	Roma Capitale Municipality
Made Ground	Man-made deposit thickness	Anthropic deposit thickness: in the GSL, only thickness >5 meters has been considered	1:25,000	CORAZZA and MARRA (1995), VENTRIGLIA (2002), FUNICIELLO et al. (2008)	Potential	Roma Capitale Municipality
	Railway network	Local and regional railway network	1:10,000	Roma Capitale	Potential	GSL
	Road network	Main roads and highways network	1:10,000	Roma Capitale	Potential	GSL
	Geology CARG 25,000	Source for alluvial deposits outside the *Roma* municipality	1:25,000	ISPRA	Potential	GSL

Table 2

Main characteristics of the PS dataset employed for generation of the Roma GSL

Data stack	Number of scenes	Period	Ground resolution (m)	PSI processing area			GSL (*Roma* area)		
				Area (km^2)	Nr. PS	PS density (PS/km^2)	Area (km^2)	Nr. PS	PS density (PS/km^2)
ERS-1/2 descending	65	21/04/1992–29/12/2000	26 range 6–30 azimuth	2600	184,531	71	1970	164,963	83.7
ENVISAT descending	13	3/01/2002–1/07/2005	30	2600	14,905	5.7	1970	13,717	7

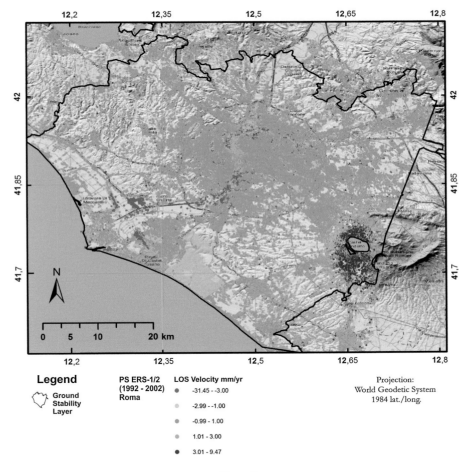

Figure 5

ERS-1/2 (1992–2000) PS distribution in the *Roma* GSL

datasets and the associated SDs, the PSs showing annual velocities in the range of ±1.0 mm/year (color-coded green) are considered stable. Yellow to red PSs indicate motions away from the sensor (subsidence), while light to dark blue indicate motions toward the sensor (uplift).

Most of the town was generally stable during the whole period of 1992–2005, but some sectors showed significant motions away from the satellite, in most cases, located along the Tiber and Aniene river valleys. Significant motions toward the satellite characterize the volcanic area of the Alban Hills.

47

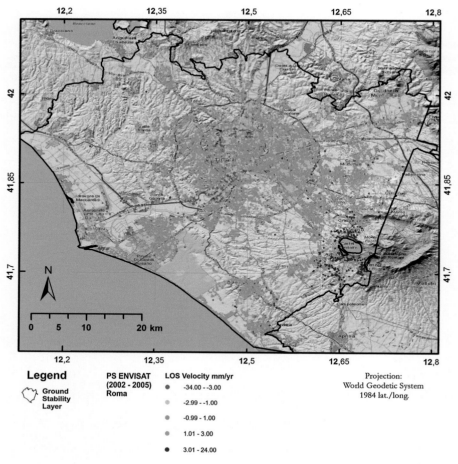

Figure 6
ENVISAT (2002–2005) PS distribution in the *Roma GSL*

Minimum and maximum annual LOS velocities estimated for the area covered by the GSL are −31.4 and 9.6 mm/year (corresponding to −34.1 and 10.4 mm/year in the vertical direction) in the ERS-1/2 descending dataset (1992–2000), and −34.0 and 24.3 mm/year (corresponding to −36.9 and 26.4 mm/year in the vertical direction) in the ENVISAT descending dataset (2003–2005). Basically, for the purposes of PanGeo, the ERS-1/2 dataset has been utilized, while the ENVISAT dataset has been mostly used for cross-checking.

For representation purposes, five ranges of mean velocities were set, as in Table 3, which also gives the PS distribution in each range.

Even if most of the territory appears stable (almost 70 % of PS velocities are within 1 mm/year), more than 25 % of the PSs show LOS velocities

Table 3

Distribution of PS annual LOS velocities (ERS-1/2)

PS annual velocity range (mm/year)	Number of PSs in velocity range	Percent (%)
−31.45 to −3.00	5500	2.96
−2.99 to −1.00	21,121	11.43
−0.99 to +1.00	127,883	69.28
+1.01 to +3.00	25,992	14.06
+3.01 to +9.47	4235	2.27

(positive or negative) between 1 and 3 mm/year, and more than 5 % exceed 3 mm/year (with tens of PSs showing velocities over −20 mm/year). According to these results, there are wide areas deserving attention, being affected by significant ground motion. They are discussed in some details in the next section. In the

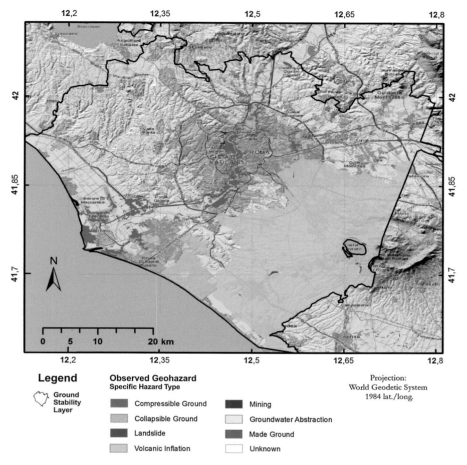

Figure 7
Ground Stability Layer of *Roma*: Observed geohazards. Polygons enclose areas where ground motions have been detected by PSInSAR data and/or where geohazards are known to occur. Zoomed images available in PanGeo portal (http://www.pangeoproject.eu)

mapping, there are also single PSs showing anomalous velocities compared to the surroundings, supposedly pertaining to noise in the dataset or to single buildings or structures. Evidently, the causes of such behaviour deserve verification, which, nevertheless, will be not be further dealt with here, being outside the scope of this paper.

4. Results of the project in Roma and case studies

In *Roma*'s GSL, 18 multi-polygons (a multipart polygon is a grouping of individual polygons under a single identifier geohazard) bound areas affected by observed geohazards (Fig. 7): they enclose areas (totalling 590 km^2) where ground motions have been detected through PSInSAR data or where landslides and sinkholes (according to datasets of Table 1) are known to have occurred. Observed geohazards include both natural processes (compaction of sediments of the Tiber and Aniene rivers and their tributaries, slope instabilities, uplift in the volcanic area of the Alban Hills) and anthropogenic instability due to water abstraction, land reclamation, extensive man-made ground to fill recent alluvial valleys, and ancient and recent engineering works (e.g., underground excavations). Table 4 provides some statistics on the identified polygons. A measure of the confidence in the interpretation of each polygon is also given, based on the number and quality of contributing datasets: in particular, *medium* is assigned based on a single dataset of sufficient reliability or on

Table 4

Statistics on area coverage and PS data for observed geohazards with respect to the GSL

Type of hazard	Landslide	Compressible ground	Collapsible ground	Groundwater abstraction	Man-made ground	Mining	Volcanic inflation	Unknown
Total Area (km^2)	1.10	56.71	5.84	5.11	1.33	0.66	517.60	2.43
Coverage % of GSL	0.06	2.93	0.30	0.26	0.07	0.03	26.71	0.12
Determination method[a]	A	B	A	B	B	B	B	B
Confidence level	High (medium in some cases)	High	Medium	High	High	High	High	High
Number of PSs	304	9257	568	639	134	66	13,161	598
Average PS velocity	0.52	−3.15	−0.61	−1.67	−2.69	−1.66	0.97	−1.4
PS velocity range	−6.4 to 5.2	−25.8 to 3.9	−14.9 to 2.7	−16.1 to 2.6	−22.5 to 0.7	−7.8 to 0.9	−14.9 to 7.5	−31.4 to 1.1

Velocities in mm/year

[a] A: Observed in geological field campaigns; B: Observed in PS data

more than one dataset of lower reliability, while *high* applies when the interpretation is supported by more than one dataset of sufficient reliability or by at least one dataset with high reliability.

Further, 13 multi-polygon areas were defined, where the *potential* occurrence of geohazards could be inferred by combining geological and/or geothematic and auxiliary data (888 km^2). A map of the polygons enclosing areas affected by potential geohazards (out of the scope of this paper) is available on the PanGeo portal (www.pangeoproject.eu). Mapped areas affected by observed or potential instability enclose more than 1475 km^2, corresponding to about 76 % of the entire GSL. The largest mapped polygon (518 km^2) corresponds to the observed (by PS data) uplift of the Alban Hills volcanic system (Fig. 7), followed by the potentially compressible ground represented by the Holocene and historical alluvial deposits of the Tiber and its tributaries (288 km^2; see PanGeo web portal), delimited based on in situ information (see Table 1). In Fig. 7, the "Compressible Ground" polygons enclose only areas (about 57 km^2; see Table 4) occupied by such unconsolidated deposits, where lowering movements have been actually observed through PSInSAR data.

While ground subsidence corresponding to unconsolidated sediments was clearly evidenced by PSInSAR data, we cannot say the same about other geohazards. The often abrupt and large surface displacement of landslides or sinkholes, sometimes also coupled with vegetation cover, frequently cause loss

of coherence in the radar signal, hindering identification of these phenomena by means of the PSInSAR technique. Moreover, radar sensitivity to displacement is also limited corresponding to steep slopes and north-facing surfaces (CASAGLI *et al.* 2010). Therefore, the information provided by PanGeo about these geohazards is derived principally from the in situ databases (see Table 1).

4.1. Compressible ground and subsidence

Almost all the PSs showing significant subsidence velocity (mean value faster than −3 mm/year along the LOS) fall in the flood plain areas occupied by Holocene and recent alluvial deposits of the Tiber, Aniene and their tributaries, mapped as "Compressible Ground" in the GSL (Figs. 5, 6 and 7; see also MANUNTA *et al.* 2008; ZENI *et al.* 2011; ARANGIO *et al.* 2013). The flood areas are generally low-lying with mixed land-use, including, according to the Urban Atlas, major categories of continuous and discontinuous urban fabric, industrial, commercial, public, military and private units, agricultural + seminatural areas + wetlands and green urban areas. Places of interest within the polygon include sections of the historical centre (Piazza Cavour, Via Frattina), the Isola Tiberina and the Trastevere and Testaccio boroughs. The subsidence is a consequence of natural consolidation of recent soft sediments, locally enhanced by anthropogenic factors (e.g., buildings and man-made ground load, water abstraction). While maximum PS LOS velocities are larger than −25 mm/year (−27 mm/year in the vertical direction), most of such velocities are within a few mm/year, in the considered time interval.

In the historical centre, it is worth noting the presence of large reclaimed areas with thick layers of man-made ground. Notably, the alluvial-palustrine areas are affected by fast subsidence both north and south of the city walls. Here, land reclamation and intensive construction started in the last decades of the XIX century (north) or only after World War II (south). Instead, the same types of sediments inside the city centre appear markedly stable (Fig. 8) or are only slowly subsiding (LOS velocities lower than −3 mm/year). Likely, they have nearly completed their compaction cycle, having already beibng

subjected to loading since Roman to baroque ages. According to their geotechnical properties (Bozzano *et al.* 2000; Campolunghi *et al.* 2007), the Holocene Tiber alluvial deposits in the city centre should have almost totally completed their consolidation, after two millennia of loading (Funiciello *et al.* 2004, 2005). Moreover, geotechnical prospections on relatively old man-made deposits gave local relative density values higher than 60–70 %, inducing consideration of their geomechanical behaviour being prevalently frictional (Corazza *et al.* 1999), with friction angles up to 40° (Campolunghi *et al.* 2008). The red dots (negative velocities higher than −3 mm/year) dispersed also inside the city walls offer likely evidence of structural/foundation deficiencies of single buildings. Previous studies have already

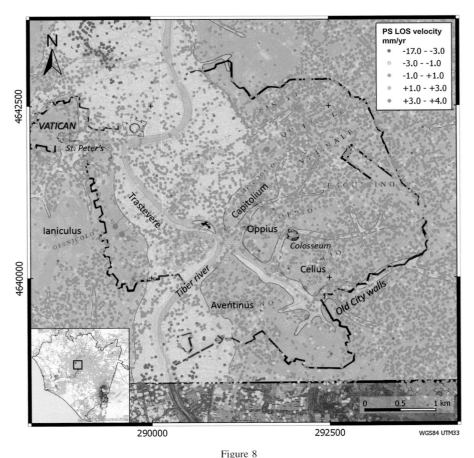

Figure 8
PSInSAR data for 1992–2000 (ERS1/2 descending) overlapped on the early historical hydrography of the city centre according to Corazza and Lombardi (1995). 2008 aerial ortophoto of *Regione Lazio* in the background. *Light blue* alluvial and palustrine areas. Although subsidence clearly marks the Tiber valley floor, velocities are very small in the most ancient occupation areas and increase moving toward and outside the old city walls. Location in *inset*

Reprinted from the journal

described local deformations occurring in a generally stable historical centre, related to single buildings or structures (MANUNTA *et al.* 2008; TAPETE *et al.* 2012b), or even portions of buildings (ZENI *et al.* 2011).

4.1.1 Viale Giustiniano Imperatore area

Worthy of mention is the subsidence affecting the *Viale Giustiniano Imperatore* area, in the Saint Paul's district, where Late Pliocene gravel and Middle-Late Pleistocene deposits, mainly composed of resedimented volcanic sediments, are overlain by Holocene alluvial deposits of the *Fosso* (creek) of *Grotta Perfetta* (Fig. 9). These soft deposits are characterized by a very low shear stress resistance

due to the abundance of organic material. The whole alluvial body is generally subjected to subsidence involving buildings and subsurface water mains and sewers. The subsidence rates, in some cases exceeding 10 mm/year, affect mostly 7 to 8-story building blocks around 50–60 years old (Fig. 10), and have been interpreted as a long-lasting primary consolidation, a consequence of the peculiar stratigraphic setting of the *Grotta Perfetta* site (STRAMONDO *et al.* 2008), coupled with poor foundation design (e.g., CAMPOLUNGHI *et al.* 2008). In fact, the analyses conducted afterwards have revealed that most of the damaged buildings have foundation piles not reaching the rigid Plio-Pleistocene layer, but partly or completely floating inside poorly consolidated alluvial deposits, with their bearing capacity relying only

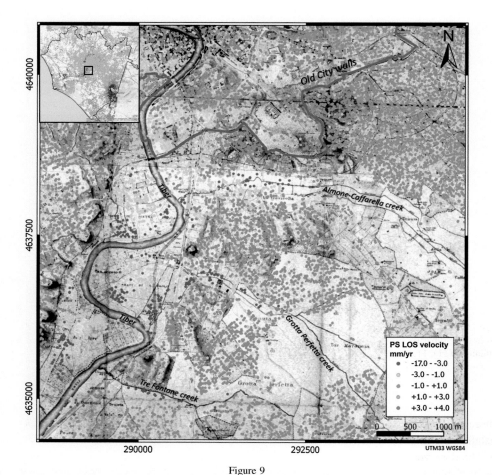

Figure 9

PSInSAR data for 1992–2000 (ERS1/2 descending) overlapped on the XIX century hydrography of the region south of the city centre according to a map realized around 1860 (ARCHIVIO STORICO CAPITOLINO 2000). Subsidence clearly marks the Tiber valley floor (cf. Fig. 8). Here, velocities correlate well with the hydrographic network now largely buried under the urban sprawl started in the 50′. Location in *inset*

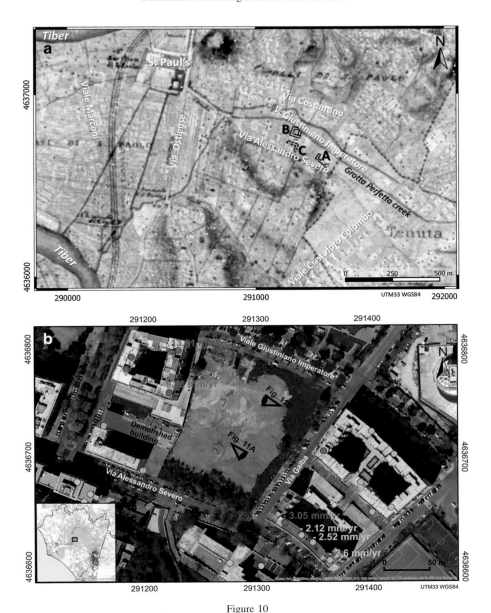

Figure 10

Viale Giustiniano Imperatore area. **a** Zooming of Fig. 9, with the XIX century trace of the Fosso (creek) of Grotta Perfetta. **a**, **b** indicate the same buildings in Figs. 10b and 11; **c** indicates the location of the building demolished in 2005. PS legend as in Fig. 9. **b** The building on the lower right (building of Fig. 11a) shows a rather homogeneous subsidence rate on its southwest façade. Instead, the building in the *upper part* of the figure (building of Fig. 11b) shows higher velocities in its northern edge, facing the center of the paleo-river valley, now corresponding to the trace of Viale Giustiniano Imperatore. The markers labelled "Fig. 11a, b" identify viewpoints of buildings in Fig. 11a, b. The building demolished in 2005 is also indicated. The *inset* shows the location with respect to the GSL

on their lateral resistance. This has led to locally severe translational and rotational movements (CEC-CHINI 2005). The observed sinking is not uniform, due also to the variable thickness of the Late Pleistocene—Holocene alluvial deposits, up to 45–50 m in the central part of the paleo-river bed and in the

presence of less compressible sandy paleo-channel beds. The complex geometry, due to the river meandering, of alluvial sand and clay beds and thick lenses of marsh peat and organic clays causes an uneven distribution of building settling, as shown by the rotations affecting the buildings of Figs. 10b and

11, and can impose anomalous stresses and strains to the reinforced concrete frames.

MANUNTA et al. (2008) have shown a deformation velocity (detected by 1995–2002 ERS-1/2 data) higher in the middle of the paleovalley and tapering toward both edges, locally corresponding to Via Alessandro Severo and Via Costantino (Fig. 10). CAMPOLUNGHI et al. (2008) described three different types of subsidence behaviour, based on the depth of the "bedrock". In the marginal zone of the paleovalley there are rotations and/or differential lowering: buildings rest partially on the concrete slab or only in part on piles entering the bedrock. In the intermediate zone the buildings are affected by the variable thickness of the highly compressible deposits, but less than in the marginal zone. In the central zone, instead, the constant bedrock depth produces a uniform building lowering. Moreover, the hydrogeological setting and the absence of a structured drainage network system induce a rise of the watertable in the man-made ground, causing local piping phenomena, which contributes to destabilizing of buildings. CINTI et al. (2008) published a detailed cross-section of the southern border of the valley, based on a number of boreholes, radiocarbon datings and geophysical sounding, that shows subhorizontal layers of Holocene clays and peat on a Middle Pleistocene to upper Pliocene substratum of gravel and compacted clay. The rapid thickening of the Holocene deposits is proposed to be related to recent offset on northwest-southeast trending faults. Also, this cross-section cannot provide an account of the rotation of the building in Figs. 10b and 11a, caused by very local stratigraphic conditions, suggesting the need for a 3D view of the sedimentation pattern.

The PS velocities in Fig. 10b prove the capability of the PSInSAR technique to detect displacements, even between different parts of a single building. Our results are in agreement with those presented by ZENI et al. (2011) and ARANGIO et al. (2013) in the same zone, based on the ERS-1/2 and ENVISAT 1992-2010 stacks of data. The building on the lower right of Fig. 10b (photo in Fig. 11a) has a rather homogeneous subsidence rate on its southwest façade. Instead, the building in the upper part of Fig. 10b (photo in Fig. 11b) shows higher velocities in its northern edge, facing the center of the paleo-river valley, now

corresponding to the trace of Viale Giustiniano Imperatore, thus confirming its evident rigid rotation. On the contrary, the close building demolished in 2005 (Fig. 10) was affected by internal strain of its frame. So far, expensive stabilizing works have been carried out for a number of buildings and only one building has been demolished, but a similar fate was proposed for other constructions within a plan, now suspended, of general reorganization of the urban texture.

More recent constructions, which evidently have better planned foundations, are characterized by detached, rapidly sinking sidewalks, which testify to the still active compaction of Holocene deposits and artificial fill (Fig. 11c).

A similar pattern of subsidence and differential settling characterizes other areas built on the Holocene valleys of the Tiber tributaries, like the Almone-Caffarella creek or the Vallerano creek.

4.1.2 Fosso di Vallerano area

Moving south, in the Torrino district, the thick alluvial deposits of Tiber and its tributary Fosso di Vallerano undergo the same process of consolidation, so that here also some buildings suffer damage: it is the case of a nursery school resting on bored piles 10 m deep, placed on the left side of the Fosso di Vallerano (Fig. 12). Soon after its construction, because of differential subsidence, the partition walls cracked and, over time, separated in places from the frame, made of reinforced concrete beams and columns. The stratigraphy underneath the school consists of: 0.0–8.4 m of fill material and alluvial deposits (silt and sand) with an angle of repose of 39° and a cohesion of 0 kPa; 8.40–21.50 m recent alluvial deposits (clayey sandy silt) with an angle of repose of 29° and a cohesion of 22 kPa; 21.50–30 m recent alluvial deposits (silty clay with interbedded sand levels) with an angle of repose of 26° and a cohesion of 0 kPa. As a whole, the bearing capacity determined by a standard penetration test (SPT), dynamic penetration and geotechnical laboratory tests is rather low. The subsidence of the school has been measured by PSInSAR data: a PS on the roof (Fig. 12) shows a mean LOS velocity of −10.65 mm/year (−11.57 mm/year in the vertical direction) in the years 1992–2000. The lowering

Figure 11
a Building near the southern flank of the Grotta Perfetta paleovalley evidently rotated toward the valley edge (location in Fig. 10), suggesting that the model of a basin of soft sediments thickening toward the center-valley may have local deviations with important influence on building stability. **b** Building rotated toward the center of the Grotta Perfetta paleovalley. In both building blocks, the rotation appears to be mostly rigid, without relevant strain of the concrete frame; **c** Detachment and sinking of the sidewalk around deeply rooted buildings along *Viale Giustiniano Imperatore*, caused by continuing compaction of the recent deposits, which can be over 1 cm/year. The same is not observed for the buildings with shallower foundations, which appear to follow the fate of the surrounding subsiding ground

trend has been confirmed by digital monitoring carried out during 2007 and 2008 in several parts of the supporting structure of the school. Measurements have shown sags of 6–7 mm during 2007, and differential settlements of the structure between 2.4 and 7 mm in 2008 (CALZONA 2008).

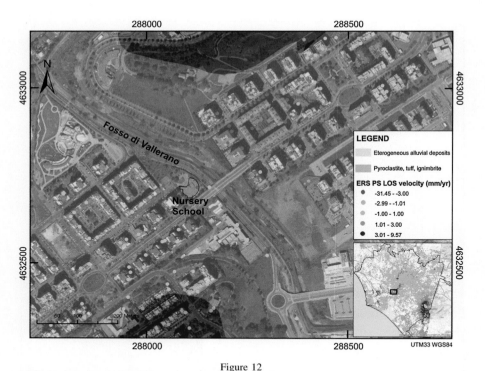

Figure 12
The PS on the nursery school located on the left side of the *Fosso di Vallerano* shows mean velocity values of −10.65 mm/year in the 1992–2000 period. The school is located on recent alluvial sediments of the Tiber and *Fosso di Vallerano*. Location in *inset*

4.1.3 The Tiber delta area and the International Airport Leonardo da Vinci

Approaching the Tiber delta, the velocities generally increase, exceeding −10 mm/year, presumably due to the thickness of compressible sediments in marsh zones, some of which have undergone land reclamation works since the Roman age (GIRAUDI 2011). Within the delta, the most rapidly sinking zones (up to −25 mm/year) correspond to the *Le Pagliete, Maccarese* and *Ostia* palustrine areas. The historical map of Fig. 13 (CINGOLANI 1692) shows the *Levante* and *Ponente* ponds (corresponding to *Ostia* and *Maccarese* marshes) before the last land reclamation, started in 1884 and completed in 1934.

The reclaimed lands are characterized by palustrine and lacustrine sediments, with interbedded alluvial and coastal back ridge deposits. These sediments are highly susceptible to compaction, being made prevalently of silt and organic clays, with frequent peat layers. After drainage of the marshes, the sediments started to consolidate and this process is still active. Although the subsidence in this

zone is mainly a natural process, anthropogenic activities (e.g., water abstraction, increase of loading by buildings, roads and other structures) locally contribute to amplify this phenomenon. Indeed, several buildings are affected by differential settlements and, consequently, by structural damage. Mat foundations of the buildings in Fig. 14 (about 10 years old) rest on low bearing-capacity sediments approximately 7 m below ground level. In December 2013, a check of the verticality of these buildings, carried out by means of a robotic total station and an integrated laser rangefinder, detected a deviation from the plumb line up to 14 cm at the base of their cornice, while the PSInSAR data (1992–2005) show negative LOS velocities up to −30 mm/year.

The rapid urban modification occurring in the Tiber delta, which has become the main sprawling area of *Roma*, with drastic land use changes from mostly farming to housing and infrastructures, has sensibly also risen the risk from flooding, as proved by recent events like that of February 2014 (REGIONE LAZIO 2014).

Subsidence can cause damage to transport infrastructures, requiring frequent reparation works.

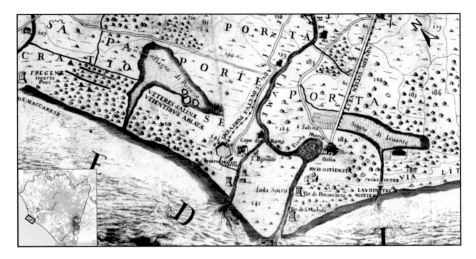

Figure 13

Historical map of the Tiber delta with the location of the *Levante* and *Ponente* Ponds (corresponding to *Ostia* and *Maccarese* marshes) before land reclamation (modified, from CINGOLANI 1692). The *red dashed line* represents the location of the runway #3 of the Airport *Leonardo da Vinci* (see Fig. 15). Location in *inset*

Particularly interesting is the differential ground lowering across runway #3 evidenced by PSs inside the *Roma* International Airport *Leonardo da Vinci* (Fig. 15). A step grows following a dramatic and abrupt change of compaction rate evidently due to the stratigraphic variability of the subsoil along the runway (Fig. 15b, from BELLOTTI *et al.* 1989, redrawn and modified). In particular, high compressible dark organic peat clays (corresponding to the northwestern edge of the *Stagno di Ponente*; Fig. 13) are heteropic toward the NNW with much less compressible sandy deposits.

The Holocene sand dune deposits along the coastal delta Tiber system, whose thickness can reach 10 m, locally overlie fine, alluvial and marsh deposits, contributing to their compaction that shows a rate generally on the order of some mm/year.

4.1.4 Subsiding areas in the uplifting Alban Hills

The *Roma*'s GSL encloses also the western half of the Alban Hills volcanic complex, which is undergoing a general uplift, interpreted as a post-caldera volcanic inflation (AMATO and CHIARABBA 1995; CHIARABBA *et al.* 1997; MARRA *et al.* 2003), well evident in the PS data (Figs. 5, 6 and 16). According to the Geolitho-logical Map (AMANTI *et al.* 2007), about 86 % of the area is covered by volcanic deposits: mainly pyroclastic flows and ignimbrites, but also basic lavas and subordinately scoria and pumices. The uplift is more intense in the area of most recent volcanic activity that, in recent years, has been carefully monitored: according to SALVI *et al.* (2004), PS data for the period 1993–2000 show average surface deformation rates up to 2.6 mm/year in the satellite LOS (close to Ariccia and Albano villages). Levelling data for the period 1950–1997 provide quite larger values (up to 4.5 mm/year), confirmed by GPS monitoring (3.3–6.0 mm/year) carried out for about three years at the Rocca di Papa station (ANZIDEI *et al.* 1998; RIGUZZI *et al.* 2009). The 1950–1993 levelling surveys along the Appian Way detected a remarkable uplift (about 30 cm), not confirmed in the subsequent period until 2002; conversely, a small uplift was measured from 2002 to 2006, indicating a pulsating inflation behaviour (ANZIDEI *et al.* 2009).

Despite the general uplift trend, small sectors of the volcanic Alban Hills show moderate to fast subsidence. Two evidently subsiding areas are located near the northern flank of the intracalderic Faete volcanic edifice (Fig. 16). They were already recognized by SALVI *et al.* (2004), who attributed the subsidence to sediment compaction caused by the dramatic lowering of the water table induced by uncontrolled water abstraction.

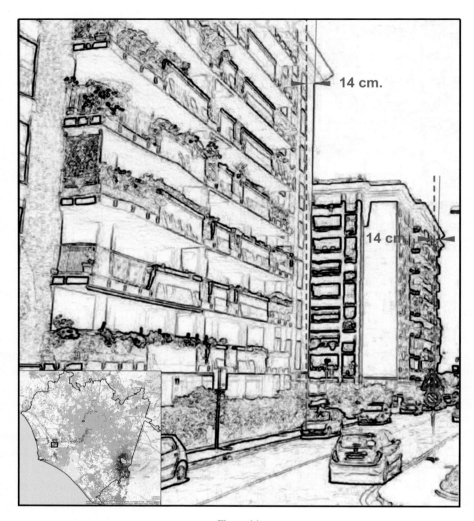

Figure 14
Optical monitoring (December 2013) of the ~10-year-old buildings detected a deviation from their verticality up to 14 cm at the base of their cornice. Location in *inset*

The main subsiding area (Area A in Fig. 16) is located north of *Rocca di Papa* village, along the SR 218—*Frascati* road and, particularly, along the SP 83B-*Barozze* road, where velocities are above — 3 mm/year, without accounting for the general uplift rate of the whole area. Field surveys have not found any evidence of a gravity phenomena, i.e., no fractures or remarkable damage have been identified in the buildings, nor on the ground floor. Aerial photographs and comparison with historical topographic maps show that the district is located in a watershed area (named *Valle Scura*, Dark Valley). At present, the valley incision can be identified in the uphill sector but

not in the downhill area, where the stream has been likely forced into a culvert and the sector artificially filled. Moreover, *Valle Scura* was already interested by the construction of a funicular (Fig. 16) that started operating in 1907. This construction resulted in rehashes of local volcanic deposits and creation of artificial fill. In the subsiding areas, the stratigraphy is characterized by eluvio-colluvial deposits and locally by manmade landfills; due to its concave morphology, much water flows in the area, as also evidenced by the name *Grotte dell'Acqua* (Water Caves) in the vicinity. Moreover, the available geological maps and stratigraphic data (DE RITA *et al.* 1995, 1988; GIORDANO

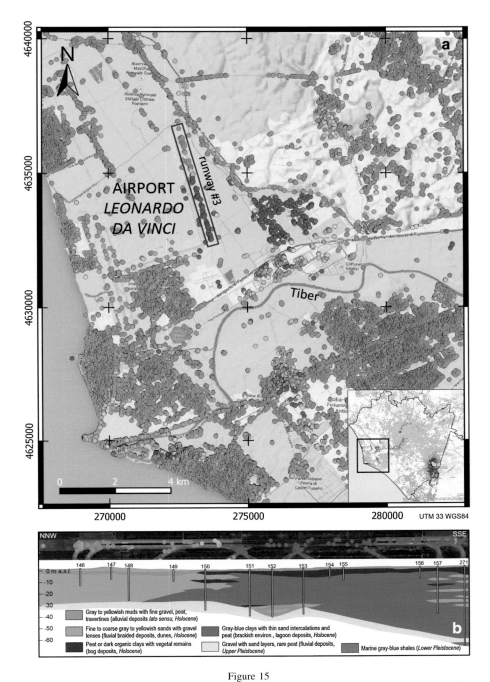

Figure 15
Airport *Leonardo da Vinci* and Tiber delta zone. **a** ERS-1/2 (1992–2000) PSInSAR data. Legend as in Fig. 5. Location in *inset*. **b** Geological profile under runway #3 (modified, from BELLOTTI *et al.* 1989)

et al. 2006) show that the main lowering areas are occupied by the *Campi di Annibale* hydromagmatic unit (Upper Pleistocene p.p.), constituted by altered pyroclastic deposits with volcanic ash, with a thickness ranging from 1 to 5 m.

The other subsiding area is located east of the *Grottaferrata* village (Fig. 16, Area B), in a flat area between the *Mt. Tuscolano* inner sector of the caldera ring and the northern end of the flanks of the main volcanic cone. The area shows the characteristics of a

59

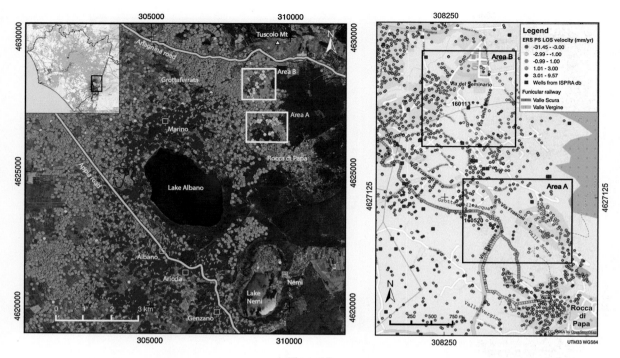

Figure 16
In the generally uplifting Alban Hills, two areas (**a**, **b**) are affected by subsidence. ERS-1/2 (1992–2000) PSInSAR data. Location in *inset*

small flat alluvial plain, where alluvial and eluvio-colluvial deposits are overimposed on the *Campi di Annibale* Formation. The flat morphology, together with the centripetal and partially closed drainage conditions, induce concentration of runoff water. The surface deposits are very soft, highly compressible and rich in water. Aerial photographs show that this area has been interested by diffuse urbanization in the 1988–1998 period, with a new increase of residential and commercial facilities up to 2004 (Fig. 17). Recent, well-founded reinforced concrete-framed constructions do not show evidence of movement. On the contrary, some surface-founded masonry houses show fractured walls. Locally, some streams have been canalized and filled up, in particular in the central sector of the subsiding area.

According to the ISPRA catalogue of boreholes deeper than 30 m (http://sgi.isprambiente.it/geoportal/catalog/content/project/indagini464.page), only a few wells fall in Areas A and B. Therefore, even if some deep wells might have escaped observation, the amount of water abstraction seems to be irrelevant here. Moreover, the concentration of wells is higher in the uplifting zones than in the subsiding

ones. Thus, the observed subsidence can only be attributed to compaction of the very young soft surface deposits, including man-made landfills, recently loaded by widespread urban settlements, in addition to inadequate surface drainage.

The ISPRA database confirms this interpretation: the stratigraphy of borehole 160113 in Area B of Fig. 16 reveals very compressible deposits for about 30 m from the surface, made of 4 m of vegetative soil and 26 m of blackish silty alluvial and detrital deposits (including *pozzolane* and tuffs); borehole 160520 in Area A shows about 10 meters of artificial deposits and detrital soil.

4.2. Sinkhole hazard

Sinkholes are one of the geohazards of interest to *Roma* addressed in the Pangeo project. Areas affected by sinkholes and areas *potentially* affected by sinkholes have been delimited by polygons in the GSL (COMERCI *et al.* 2013), based on the databases reported in Table 1. Most of the sinkholes are induced by present-day and past man activities. In fact, the anthropogenic sinkholes in the *Roma* area

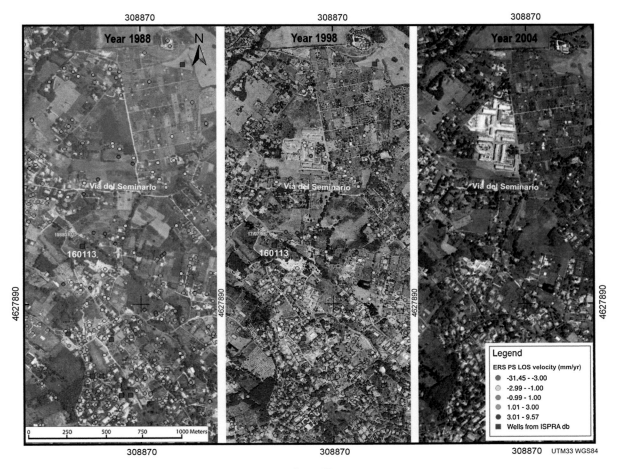

Figure 17
A comparison between 1988, 1998 and 2004 aerial photos reveals the urban sprawl in Area B of Fig. 16 since the '90 s, and the increase of commercial and residential facilities up to 2004. ERS-1/2 (1992–2000) PSInSAR data

are closely related to a widespread warren of underground cavities (quarries, catacombs, buried archaeological remains, hydraulic networks, etc.), heritage of the more than 25 centuries of history. Such a network of tunnels and underground passages, sometimes even still unknown, is locally exposed to progressive erosion of the vaults, giving rise to surface chasms of metric dimensions. An additional cause of cavity and collapse development is the leakage from the underground culverts of hydraulic utilities (aqueducts, sewers and drains).

Over the past ten years, this phenomenon has become more frequent in the city centre, representing a growing danger to the population and infrastructures, with damaged streets and underground utilities (water, electric, telephone and gas mains, etc.). In

Roma there is also a serious menace to the archaeological heritage and its preservation: an example is represented by the collapse that occurred close to the Nero's Golden House in 2010. For this and other areas in the *Roma* historical centre, TAPETE *et al.* (2012b) and ZENI *et al.* (2011) showed the opportunity offered by satellite radar interferometry monitoring to complement periodic inspections and targeted terrestrial surveys.

There are more than 2500 sinkholes mapped in the municipality of *Roma*, mostly located in the urban area with the remaining near its northern boundary. In this scenario, the determination of the actual risk induced by the occurrence of sinkholes is a difficult task. Assessing the susceptibility is easier, defined as the probability that an event of anthropogenic

Reprinted from the journal

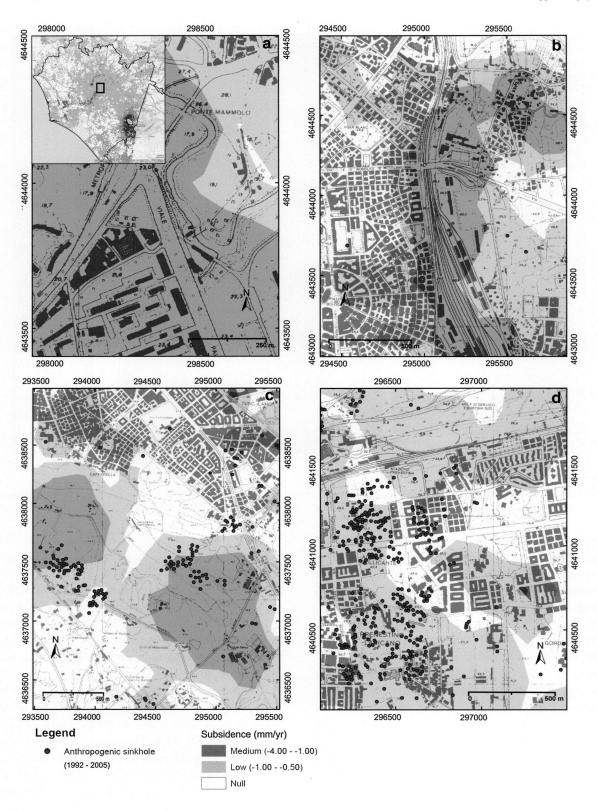

Legend

● Anthropogenic sinkhole
(1992 - 2005)

Subsidence (mm/yr)

◼ Medium (-4.00 - -1.00)
◼ Low (-1.00 - -0.50)
☐ Null

Figure 18

a *Ponte Mammolo* (Tiburtino borough). The subsiding zones are due to the alluvial sediments of the Aniene River. **b** Tiburtino borough. The subsiding area is likely related to compaction of the road embankments (*Circonvallazione tiburtina-nomentana*). **c** *Caffarella—Appio Latino* borough. Locally, a direct correspondence is observed between low to medium subsidence and susceptibility to sinkholes. Some other susceptible areas are not affected by ground sinking. **d** *Prenestino* borough. Despite the large number of sinkholes, subsidence is absent or modest. In the central part of the figure, an area affected by many ground collapses is not undergoing any subsidence. An explanation may be the relatively old age of the ground failures and the subsequent reinforcing works. Location of the areas **a**, **b**, **c** and **d** in *inset*

subsidence occurs in a given place with given geological-morphological characteristics in an infinite time span.

Recent susceptibility studies have targeted some predisposing factors (geological, morphological, hydrological), and the presence of cavities, underground utilities, thicknesses of man-made ground, etc. (CIOTOLI *et al*. 2013).

In order to redefine a forecasting model, the sinking phenomena were compared here with the ground subsidence portrayed by PSInSAR data (the low-lying subsiding areas along the Tiber and its tributaries have been excluded, focusing on the areas with the highest susceptibility to collapse).

The quantitative motion data provided by radar interferometry have proved their great potential in the study of geomorphic processes involving the deformation of ground surfaces (ROTT 2009), and also in cases of sinkholes. Many studies have proved their effectiveness in different environments all around the world: FERRETTI *et al*. (2000b, 2004) detected gradual subsidence before a collapse sinkhole that caused the destruction of several houses in Camaiore (Italy); BAER *et al*. (2002) measured noteworthy subsidence related to sinkholes in the Dead Sea coast area; CLOSSON *et al*. (2005, 2010) measured high subsidence rates preceding a collapse of a salt dike in a Dead Sea zone; PAINE *et al*. (2009) observed huge subsidence rates in a Texas area affected by large sinkholes related to the dissolution of a salt formation induced by oil exploitation boreholes; CASTAÑEDA *et al*. (2009) analyzed the possibilities and limitations of the SBAS technique in the Ebro Valley (Spain); where GUTIÉRREZ *et al*. (2011) delineated successfully the limits of an area with active buried sinkholes.

In our study, the metropolitan area of *Roma*, covered by a satisfactory number of PSs, was classified according to three levels of lowering rate (<-4 mm/year; -4 mm/year to -1 mm/year; -1 to -0.5 mm/year). On this layer, obtained by interpolating PS data through the inverse distance weighted method (the resulting surface is a weighted average of the neighbourhood PSs, whose weights are set as decreasing with distance), were stacked on all the catalogued sinkholes (years 1888–2014). The results highlight areas of point subsidence that either overlap with known collapses, or lack known sinkholes in the surroundings. It was possible to differentiate some typical cases. Where subsidence is not accompanied by known collapses, two cases are distinguished: (a) subsidence probably due to highly compressible alluvial and palustrine sediments (e.g., areas east of *Roma*, likely corresponding to ancient meanders of the Aniene river; Fig. 18a, b lowering due to recent man-made ground, for the realization of road infrastructures (Fig. 18b).

Then, there are places where the ground lowering revealed by PS data closely corresponds to an area susceptible to collapse. This good fit is observed where the network of underground cavities is particularly dense, calling for stabilization works. Figure 18c, d, which report the sinkholes that occurred in 1992–2005, locally show areas affected by low to medium subsidence susceptible to collapse, in the boroughs *Caffarella–Appio Latino* and *Prenestino-Labicano*.

Finally, Fig. 18c, d also show several areas susceptible to collapse that, in the documented time span (1992–2005), do not show ground deformation. In some instances, it might be possible to explain this phenomenon with stabilization works carried out before the observation period. Clearly, it might be also the case that no detectable ongoing deformation is heralding a future collapse.

5. Conclusions

The PanGeo project (www.pangeoproject.eu) has enabled obtaining a thorough picture of the ground motion in the 1992–2005 time interval and the related geohazards affecting the *Roma* territory (properly, the

Roma GSL area). The envisaged service would be aimed at helping government and local authorities, planners and regulators who are concerned with managing development control and risk (CAPES 2012). In fact, PanGeo products provide relevant indications and highlight areas affected by geohazards and where local authorities can focus protection activities. Ground motion detected in the *Roma* GSL by means of PSInSAR data is in good agreement with the results presented by SALVI *et al.* (2004), MANUNTA *et al.* (2008), STRAMONDO *et al.* (2008), FORNARO *et al.* (2010), ZENI *et al.* (2011), TAPETE *et al.* (2012b), ARANGIO *et al.* (2013) and FORNARO *et al.* (2014).

The PSInSAR technique has proved extremely efficient in detecting subsidence (also at the scale of a single building) and uplifting phenomena, whose causing mechanism and evolution has been interpreted based on in situ data: the subsiding zones occupied by the unconsolidated deposits of the Tiber River and its tributaries are clearly defined based on satellite data, as well as the uplift of the Alban Hills volcanic complex. Integration with in situ data has been especially relevant for landslides and sinkholes. Information on landslides, indeed, derives almost entirely from historical databases, being that the satellite dataset made available for the project is less effective at detecting slope movements. In the case of sinkholes, the joint analysis of PSInSAR and in situ data offers the opportunity to identify susceptible areas affected by correlated subsidence, where ground collapse has not yet occurred and where local authorities should first take action to prevent the phenomenon.

Another relevant outcome of the PanGeo project is to enable geohazards to be related to complementary datasets such as land use (e.g., Urban Atlas) or population, facilitating generation of further statistics.

The PanGeo methodology, providing valuable results, has proved effective and is liable to be adopted for future applications of satellite data aimed at monitoring ground instabilities. Certainly, the analysis of up-to-date PSInSAR data (derived from the new European satellites, e.g., COSMO-SkyMed and Sentinel) will effectively portray, with improved precision, the evolution trend of the considered geohazards.

Acknowledgments

This work was performed in the framework of the PanGeo project (http://www.pangeoproject.eu), funded by the European Commission within the 7th Framework Programme under the Global Monitoring for Environment and Security (GMES)—Copernicus initiative, with Grant Agreement No. 262371. The authors acknowledge Tele-Rilevamento Europa (TRE), Milan for processing the ERS-1/2 and ENVISAT imagery with the PSInSAR technique and for their fruitful cooperation. The basemaps have been provided by Esri, DigitalGlobe, GeoEye, i-cubed, the USDA, the USGS, AEX, Getmapping, Aerogrid, IGN, swisstopo, Google Earth and the GIS user community. The authors also thank Prof. R. Calzona for the technical documentation concerning the nursery school, C. Alimonti for geotechnical data, F. Valeri and C. Del Vecchio for their technical support, and two anonymous reviewers for criticisms that strongly helped improve the manuscript.

REFERENCES

AMANTI, M., BATTAGLINI, L., CAMPO, V., CIPOLLINI, C., CONGI, M.P., CONTE, G., DELOGU, D., VENTURA, R., SONETTI, C. (2007), *La carta litologica d'Italia alla scala 1:100.000*, Atti del VI Forum italiano di Scienze della Terra, Geoitalia 2007, Rimini.

AMATO, A., and CHIARABBA, C. (1995), *Recent uplift of the Alban Hills volcano (Italy), evidence for magmatic inflation?*, Geophys. Res. Lett., *22*, 1985–1988.

ANZIDEI, M., BALDI, P., CASULA, G., GALVANI, A., RIGUZZI, F. and ZANUTTA, A. (1998), *Evidence of active crustal deformation in the Colli Albani volcanic area (Central Italy) by GPS surveys*, Journal of Volcanology and Geothermal Res., *80*, 55–65.

ANZIDEI, M., RIGUSSI, F., STRAMONDO, S. (2009), *Current geodetic deformation of the Colli Albani volcano: A review*, in The Colli Albani Volcano, Eds., R. Funiciello and G. Giordano, Special Pubblication of IAVCEI, The Geological Society of London.

ARANGIO, S., CALÒ, F., DI MAURO, M., BONANO, M., MARSELLA, M., MANUNTA, M. (2013), *An application of the SBAS-DInSAR technique for the Assessment of structural damage in the city of Rome*. Structure and Infrastructure Engineering: Maintenance, Management, Life-Cycle Design and Performance, 1–15. doi:10.1080/15732479.2013.833949

ARCHIVIO STORICO CAPITOLINO (2002), *Roma in CD dal XVI al XX secolo nelle mappe e nelle vedute della Biblioteca Romana dell'Archivio Capitolino*. CD-Rom realized by the Soprintendenza ai Beni Librari della Regione Lazio, GAP, Roma.

AUTORITÀ DI BACINO DEL FIUME TEVERE (2004), *Carta geomorfologica dei corridoi fluviali del fiume Tevere e del Fiume Aniene*. Piano stralcio per il tratto metropolitano del Tevere da Castel

Giubileo alla foce. PS5, scala 1:25.000, Roma. http://www.abtevere.it/node/324

BAER, G., SCHATTNER, U., WACHS, D., SANDWELL, D., WDONWINSKI, S., FRYDMAN, S. (2002), *The lowest place on Earth is subsiding : An InSAR Interferometric Synthetic Aperture Radar perspective.* Geological Society of America Bulletin, *114*, 12–23.

BARTOLE, R. (1984), *Tectonic structures of the Latium-Campania shelf (Tyrrhenian sea).* Boll. Ocean. Teor. Appl., *2*, 197–230.

BATESON, L., CUEVAS, M., CROSETTO, M., CIGNA, F., SCHIJF, M., and EVANS, H. (2012), *PanGeo D3.5 Production Manual.* Version 1.1, 25 th July 2012. Available at: http://www.pangeoproject.eu

BELLOTTI, P., CARBONI, M.G., MILLI, S., TORTORA, P. and VALERI, P. (1989), *La piana deltizia del Fiume Tevere: analisi di facies e ipotesi evolutiva dall'ultimo low stand glaciale all'attuale.* Giornale di Geologia serie 3, vol. 51/1.

BELLOTTI, P., CHIOCCI, F. L., MILLI, S., TORTORA, P. and VALERI, P. (1994), *Sequence stratigraphy and depositional setting of the Tiber delta: integration of hy resolution seismics well logs and archeological data.* Journ. Sedimentary Research, *B 64*, 416–432.

BELLOTTI, P., MILLI S., TORTORA, P. and VALERI, P. (1995), *Physical stratigraphy and sedimentology of the late Pleistocene-Holocene Tiber delta depositional sequence.* Sedimentology, *42*, 617-634.

BERARDINO, P., FORNARO, G., LANARI, R. and SANSOSTI, E. (2002). *A new algorithm for surface deformation monitoring based on small baseline differential SAR interferograms.* IEEE Transactions on Geosciences and Remote Sensing, *40*, 2375–2383.

BERSANI, P. and BENCIVENGA, M. (2001), "*Le piene del Tevere a Roma. Dal V secolo a.C. all'anno 2000*". Presidenza del Consiglio dei Ministri, Dipartimento per i Servizi Tecnici Nazionali, Servizio Idrografico e Mareografico Nazionale, pp. 100.

BERTOLETTI, E., CIUFFREDA, M., SUCCHIARELLI, C., CIPOLLONI, C., COMERCI, V., DI MANNA, P., GUERRIERI, L., VITTORI, E., (2013), *Il Progetto Europeo PanGeo: monitoraggio dei movimenti del suolo urbanizzato di Roma Capitale mediante dati satellitari PSI.* Poster session at 14a Conferenza italiana utenti ESRI, Roma 17–18 April 2013, Roma.

BOSCHI, E., GUIDOBONI, E., FERRARI, G., VALENSISE, G. and GASPERINI, P. (editors) (1997), *Catalogo dei Forti Terremoti in Italia dal 461 a.C. al 1990.* ING – SGA Bologna, pp. 644.

BOZZANO, F., ANDREUCCI A., GAETA M. and SALUCCI R. (2000), *A geological model of the buried Tiber River valley beneath the historical centre of Roma*, Bulletin of Engineering Geology and the Environment *59*, 1–21.

BRU, G., HERRERA, G., TOMÁS, R., DURO, J., DE LA VEGA, R., MULAS, J. (2013), *Control of deformation of buildings affected by subsidence using persistent scatterer interferometry.* Structure and infrastructure engineering *9*, 188–200.

CALZONA, R. (2008), *Scuola materna "Amico Ulivo", Relazione e parere sulle condizioni statiche ed abitative*, Municipio Roma XII. Roma, pp. 26 (unpublished report).

CAMPOLUNGHI, M.P., CAPELLI, G., FUNICIELLO, R., LANZINI, M. (2007), *Geotechnical studies for foundation settlement in Holocenic alluvial deposits in the City of Rome (Italy).* Engineering Geology, *89*, 9–35.

CAMPOLUNGHI, P., CAPELLI, G., FUNICIELLO, R., LANZINI, M., MAZZA, R. and CASACCHIA, R. (2008), *Un caso esemplare: la stabilità degli edifici nell'area intorno a Viale Giustiniano Imperatore (Roma, IX Municipio).* In: Funiciello, R., Praturlon, A. & Giordano, G. (eds). La geologia di Roma: dal centro storico alla periferia. Memorie Descrittive della Carta Geologica d'Italia, LXXX, *(2)* 195–219.

CAPES, R. (2012), *PanGeo: monitoring ground instability for local authorities. Window on GMES*, GMES4Regions, Special Issue. SSN 2030–5419

CASAGLI, N., CATANI, F., DEL VENTISETTE, C. and LUZI, G. (2010), *Monitoring, prediction, and early warning using ground-based radar interferometry.* Landslides, *7*, 291–301.

CASTAÑEDA, C., GUTIÉRREZ, F., MANUNTA, M., GALVE, J.P. (2009). *DInSAR measurements of ground deformation by sinkholes, mining subsidence, and landslides, Ebro River, Spain.* Earth Surface Processes and Landforms, *34*, 1562–1574.

CECCHINI, D. (2005), Rifare Città. Gangemi, Roma

CHIARABBA, C., AMATO, A., and DELANEY, P.T. (1997), *Crustal structure, evolution, and volcanic unrest of the Alban Hills, Central Italy*, Bull. Volcanol., *59*, 161–170.

CIGNA, F., DEL VENTISETTE, C., LIGUORI, V., CASAGLI, N. (2011), *Advanced radar-interpretation of InSAR time series for mapping and characterization of geological processes.* Nat. Hazards Earth Syst. Sci., *11*, 865–881.

CIGNA, F., DEL VENTISETTE, C., GIGLI, G., MENNA, F., AGILI, F., LIGUORI, V., CASAGLI, N. (2012a), *Ground instability in the old town of Agrigento (Italy) depicted by on-site investigations and Persistent Scatterers data.* Nat. Hazards Earth Syst. Sci., *12*, 3589–3603.

CIGNA, F., OSMANOĞLU, B., CABRAL-CANO, E., DIXON, T.H., ÁVILA-OLIVERA, J.A., GARDUÑO-MONROY, V.H., DEMETS, C., WDOWINSKI, S. (2012b), *Monitoring land subsidence and its induced geological hazard with Synthetic Aperture Radar Interferometry: a case study in Morelia, Mexico.* Remote Sensing of Environment, *117*, 146–161.

CIGNA, F., BIANCHINI, S., CASAGLI, N. (2013), *How to assess landslide activity and intensity with Persistent Scatterer Interferometry (PSI): the PSI-based matrix approach.* Landslides, *10*(3), 267–283. doi:10.1007/s10346-012-0335-7

CIGNA, F., LASAPONARA, R., MASINI, N., MILILLO, P., TAPETE, D. (2014), *Persistent Scatterer Interferometry Processing of COSMO-SkyMed StripMap HIMAGE Time Series to Depict Deformation of the Historic Centre of Rome, Italy.* Remote Sens., *6*, 12593–12618. doi:10.3390/rs61212593

CINGOLANI, G.B. (1692), *Topografia Geometrica dell'Agro Romano.*

CINTI, F.R., MARRA, F., BOZZANO, F., CARA, F., DI GIULIO, G., and BOSCHI, E. (2008), *Chronostratigraphic study of the Grotta Perfetta alluvial valley in the city of Roma (Italy): investigating possible interaction between sedimentary and tectonic processes.* Annals of Geophysics, *51*, no 5–6, 849–865.

CIOTOLI, G., CORAZZA, A., FINOIA, M.G., NISIO, S., SERAFINI, R., and SUCCHIARELLI, C. (2013), *Sinkholes antropogenici nel territorio di Roma Capitale.* Mem Descr. Carta Geol d'It. *XCIII*, 143–182.

CLOSSON, D., KARAKI, N.A., KLINGER, Y., HUSSEIN, M.J. (2005), *Subsidence and sinkhole hazards assessment in the southern Dead Sea area, Jordan.* Pure and Applied Geophysics, *162*, 221–248.

CLOSSON, D., KARAKI, N.A., MILISAVLJEVIC, N., HALLOT, F., ACHEROY, M. (2010), *Salt dissolution-induced subsidence in the Dead Sea area detected by applying interferometric techniques to ALOS Palsar Synthetic Aperture Radar images.* Geodinamica Acta, *23*, 65–78.

COMERCI, V., CIPOLLONI, C., DI MANNA, P., GUERRIERI, L., VITTORI, E., BERTOLETTI, E., CIUFFREDA M., and SUCCHIARELLI, C., (2013), *Geohazard Description for Roma. PanGeo – Enabling Access to Geological Information in Support of GMES.* Seventh

Framework Programme, Cooperation: Space Call 3, FP7-Space-2010-1, European Commission, Research Executive Agency, pp. 175. www.pangeoproject.eu

CORAZZA, A., MARRA, F. (1995), *Carta dello spessore dei terreni di riporto*. In: R. Funiciello (ed.), La geologia di Roma. Il Centro Storico. Memorie Descrittive della Carta Geologica d'Italia, 50, pp. 550.

CORAZZA, A., LANZINI, M., ROSA, C., SALUCCI, R., (1999), *Caratteri stratigrafici, idrogeologici e geotecnici delle alluvioni tiberine nel settore del centro storico di Roma*. Il Quaternario, *12*(2), 215–235.

COSENTINO, D., CIPOLLARI, P., DI BELLA, L., ESPOSITO, A., FARANDA, C., GIORDANO, G., MATTEI, M., MAZZINI, I., PORRECA, M. and FUNICIELLO, R. (2009), *Tectonics, sea-level changes and palaeoenvironments in the early Pleistocene of Roma (Italy)*. Quaternary Research, *72*, 143–155.

DE RITA, D., FUNICIELLO, R., PAROTTO, M. (1988), *Carta Geologica del Complesso Vulcanico dei Colli Albani (scala 1 : 50.000)*. Progetto Finalizzato Geodinamica, CNR, Roma.

DE RITA, D., MILLI S., ROSA, C., ZARLENGA, F. and CAVINATO G.P. (1994), *Catastrophic eruptions and eustatic cycles: example of Latium Volcanoes*. In: Large esplosive eruptions. International symposium, Roma, 24 ~ 25 May 1993. Atti dei Convegni Lincei, 112, 135–142.

DE RITA, D., FACCENNA, C., FUNICIELLO, R. and ROSA, C. (1995), *Stratigraphy and volcano-tectonics*. In: Trigila (Ed.), The Volcano of the Alban Hills. Tipografia SGS, Roma, pp. 33–71.

DIXON, T.H., AMELUNG, F., FERRETTI, A., NOVALI, F., ROCCA, F., DOKKA, R., et al. (2006), *Space geodesy: subsidence and flooding in New Orleans*. Nature, *441*, 587–588.

EEA (2011), *Mapping Guide for European Urban Atlas*, EEA/EC web page 2010. Available at: http://www.eea.europa.eu/data-and-maps/data/urban-atlas/mapping-guide/urban_atlas_2006_mapping_guide_v2_final.pdf/at_download/file

EZQUERRO, P., HERRERA, G., MARCHAMALO, M., TOMÁS, R., BÉJAR-PIZARRO, M., MARTÍNEZ, R.A. (2014), *Quasi-elastic aquifer deformational behavior: Madrid aquifer case study*. Journal of Hydrology, *519*, 1192–1204.

FACCENNA, C., FUNICIELLO, R., and MARRA, F. (1995), *Inquadramento geologico strutturale dell'area romana*. Servizio Geologico Nazionale, Memorie descrittive della Carta geologica d'Italia, volume L, La geologia di Roma – Il centro storico, 31–47.

FERRETTI, A., PRATI, C. and ROCCA, F. (2000a), *Non-linear Subsidence Rate Estimation Using Permanent Scatterers in Differential SAR Interferometry*. IEEE Trans. on Geoscience and Remote Sensing, *38*, 5, 2202–2212.

FERRETTI, A., FERRUCCI, F., PRATI, C., ROCCA, F. (2000b), *SAR analysis of building collapse by means of the permanent scatterers technique*. Proceedings of the 2000 IEEE International Geoscience and Remote Sensing Symposium, Honolulu, *7*, 3219–3221.

FERRETTI, A., PRATI, C. and ROCCA, F. (2001), *Permanent Scatterers in SAR Interferometry*. IEEE Trans. on Geoscience and Remote Sensing, *39*, 1, 8–20.

FERRETTI, A., BASILICO, M., NOVALI, F., PRATI, C. (2004), *Possibile utilizzo di dati radar satellitari per individuazione e monitoraggio di fenomini di sinkholes*. In: Nisio, S., Panetta, S., Vita, L., (Eds.), Stato dell'arte sullo studio dei fenomeni di sinkholes e ruolo delle amministrazioni statali e locali nel governo del territorio, APAT, Roma, 415–424.

FERRETTI, A., FUMAGALLI, A., NOVALI, F., PRATI, C., ROCCA, F., RUCCI, A. (2011), *A New Algorithm for Processing Interferometric Data-Stacks: SqueeSAR*. IEEE Trans. Geosci. Remote Sens., *49*, 3460–3470.

FORNARO, G., SERAFINO, F., REALE, D. (2010), *4-D SAR Imaging: The Case Study of Rome*. Geoscience and Remote Sensing Letters, IEEE, *7*(2), 236–240. doi:10.1109/LGRS.2009.2032133

FORNARO, G., REALE, D., PAUCIULLO, A., ZHU, X., BAMLER, R. (2012), *SAR Tomography: an advanced tool for spaceborne 4D radar scanning with application to imaging and monitoring of cities and single buildings*. IEEE Geoscience and Remote Sensing Newsletter, Dec. 2012, 10–18.

FORNARO, G., PAUCIULLO, A., REALE, D., VERDE, S. (2013), *SAR Coherence Tomography: A new approach for coherent analysis of urban areas*. Proc. 2013 IEEE IGARSS Conf., Melbourne, Australia, July 21–26, 73–76.

FORNARO, G., PAUCIULLO, A, REALE, D., VERDE, S. (2014), *Multilook SAR Tomography for 3-D Reconstruction and Monitoring of Single Structures Applied to COSMO-SKYMED Data, Selected Topics in Applied Earth Observations and Remote Sensing*, IEEE Journal of, *7*(7), 2776,2785. doi:10.1109/JSTARS.2014.2316323

FORNARO, G., VERDE, S., REALE, D., PAUCIULLO, A. (2015), *CAESAR: An Approach Based on Covariance Matrix Decomposition to Improve Multibaseline-Multitemporal Interferometric SAR Processing*, Geoscience and Remote Sensing, IEEE Transactions on, *53*(4), 2050–2065. doi:10.1109/TGRS.2014.2352853

FUNICIELLO, R., (1995), *La geologia di Roma*. Il Centro Storico (Memorie Descrittive della Carta Geologica d'Italia, 50, pp. 550, Roma 1995).

FUNICIELLO, R., and PAROTTO, M. (2001), *General geological features of the Campagna Romana*. The World of Elephants – International Congress – Roma.

FUNICIELLO, R., GIORDANO, G., DE RITA, D. (2003), *The Albano maar lake (Colli Albani Volcano Italy): recent volcanic activity and evidence of pre-Roman Age catastrophic lahar events*. Journal of Volcanology and Geothermal Research. *123*, 43–61.

FUNICIELLO, R., GIORDANO, G., ADANTI, B., GIAMPAOLO, C., PAROTTO, M. (2004), *Walking through downtown Roma: a discovery tour on the key role of geology in the history and urban development of the city*. Field Trip Guide Book D05. 32nd International Geological Congress. From the Mediterranean Area Toward a Global Geological Renaissance. Geology, Natural Hazards, and Cultural Heritage. Florence-Italy August 20–28, 2004, APAT.

FUNICIELLO, R., CAMPOLUNGHI, M.P., CECILI, A., TESTA O. (2005), *La struttura geologica dell'area romana e il Tevere*. Atti dei Convegni Lincei 218, Ecosistema Roma Roma, 14–16 aprile 2004, Accademia Nazionale dei Lincei, 149–208.

FUNICIELLO R., and GIORDANO, G. (2008a), *La nuova Carta Geologica di Roma: litostratigrafia e organizzazione stratigrafica*. Special volume "La geologia di Roma. Dal centro storico alla periferia", Memorie descrittive della Carta Geologica d'Italia, *80*, 39–85.

FUNICIELLO, R., and GIORDANO, G. (2008b), *Geological Map of Italy 1:50,000*, sheet n. 374 "Roma" and explanatory notes. Servizio Geologico d'Italia. Se.l.c.a. Firenze, Italy.

FUNICIELLO, R., PRATURLON, A., and GIORDANO, G. (2008), *La geologia di Roma: dal centro storico alla periferia*. Parte Seconda. Memorie Descrittive della Carta Geologica d'Italia, LXXX, pp. 313.

FUNICIELLO, R., GIORDANO, G. (editors) (2010), *The Colli Albani volcano*. Special Publications of IAVCEI, 3, 400 pp., The Geological Society, London, UK.

GABRIEL, A.K.; GOLDSTEIN, R.M.; ZEBKER, H.A. (1989), *Mapping small elevation changes over large areas: Differential radar interferometry*. J. Geophys. Res., *94*, 9183–9191.

GIORDANO, G., DE BENEDETTI, A.A., DIANA, A., DIANO, G., GAUDIOSO, F., MARASCO, F., MICELI,M., MOLLO, S., CAS, R.A.F. and FUNICIELLO, R. (2006), *The Colli Albani caldera (Roma, Italy): stratigraphy, structure and petrology*. In: Cas R.A.F. & Giordano G. (eds) Explosive Mafic Volcanism, Journal of Volcanology and Geothermal Research, Spec. Vol., 155, 49–80.

GIORDANO, G. (2008), *I vulcani di Roma: storia eruttiva e pericolosità*. Special volume "La geologia di Roma. Dal centro storico alla periferia", Memorie descrittive della Carta Geologica d'Italia, *80*, 87–95.

GIRAUDI, C. (2011), *The Holocene record of environmental changes in the 'Stagno di Maccarese' marsh (Tiber river delta, central Italy)*. The Holocene, *21(B)*, 1233–1243. doi:10.1177/0959683612455543

GUTIÉRREZ, F., GALVE, J.P., LUCHA, P., CASTAÑEDA, C., BONACHEA, J., GUERRERO, J. (2011), *Integrating geomorphological mapping, trenching, InSAR and GPR for the identification and characterization of sinkholes: A review and application in the mantled evaporite karst of the Ebro Valley (NE Spain)*. Geomorphology, *134*(1–2), 144–156.

HEIKEN, G., FUNICIELLO, R., and DE RITA, D. (2007), *The Seven Hills of Roma: A Geological Tour of the Eternal City*. Princeton University Press, 264 pp., ISBN: 9780691130385.

HERRERA, G., GARCÍA-DAVALILLO, J.C., MULAS, J., COOKSLEY, G., MONSERRAT, O., PANCIOLI, V. (2009a), *Mapping and monitoring geomorphological processes in mountainous areas using PSI data: Central Pyrenees case study*. Nat. Hazards Earth Syst. Sci., *9*, 1587–1598.

HERRERA, G., TOMÁS, R., LÓPEZ-SÁNCHEZ, J.M., DELGADO, J., VICENTE, F., MULAS, J., COOKSLEY, G., SÁNCHEZ, M., DURO, J., ARNAUD, A., BLANCO, P., DUQUE, S., MALLORQUÍ, J.J., VEGA-PANIZO, R., MONSERRAT, O. (2009b), *Validation and comparison of Advanced Differential Interferometry Techniques: Murcia metropolitan area case study*. ISPRS Journal of Photogrammetry and Remote Sensing, *64*, 501–512.

HERRERA, G., FERNÁNDEZ, J.A., TOMÁS, R., COOKSLEY, G., MULAS J. (2009c), *Advanced interpretation of subsidence in Murcia (SE Spain) using A-DInSAR data - modelling and validation*. Natural Hazards and Earth System Sciences 9, 647–661

HERRERA, G., TOMÁS, R., MONELLS, D., CENTOLANZA, G., MALLORQUI, J.J., VICENTE, F., NAVARRO, V. D., LOPEZ-SANCHEZ, J., CANO, M., MULAS, J., SANABRIA, M. (2010), *Analysis of subsidence using TerraSAR-X data: Murcia case study*. Engineering Geology, *116*, 284–295

HERRERA, G., ÁLVAREZ FERNÁNDEZ, M.I., TOMÁS, R., GONZÁLEZ-NICIEZA, C., LOPEZ-SANCHEZ, J. M., ÁLVAREZ VIGIL, A.E. (2012), *Forensic analysis of buildings affected by mining subsidence based on Differential Interferometry (Part III)*. Engineering Failure Analysis 24, 67–76.

HILLEY, G.E., BÜRGMANN, R., FERRETTI, A., NOVALI, F., ROCCA, F. (2004), *Dynamics of slow-moving landslides from permanent scatterer analysis*. Science, *304*(5679), 1952–1955.

HOOPER, A., ZEBKER, H., SEGALL, P., KAMPES, B. (2004), *A new method for measuring deformation on volcanoes and other natural terrains using InSAR persistent scatterers*. Geophys. Res. Lett., *31*. doi:10.1029/2004GL021737

INSPIRE DS-D2.8/III-12 (2013), *INSPIRE Data Specification on Natural Risk Zone* – technical Guidelines, v. 3.0. Published by EC on JRC at: http://inspire.jrc.ec.europa.eu/documents/Data_Specifications/INSPIRE_DataSpecification_NZ_v3.0.pdf

KARNER, D.B., MARRA, F. and RENNE, P.R. (2001), *The history of the Monti Sabatini and Alban Hills volcanoes: Groundwork for assessing volcanic-tectonic hazards for Roma*. Journal of Volcanology and Geothermal Research, *107*(1–3), 185–215.

KOMAC, B., ZORN, M. (2013), *Geohazards, In Enciclopedia of Natural Hazards* (ed. Bobrowsky P.T.) (Springer science Business Media B.V. 2013) p. 387. doi:10.1007/978-1-4020-4399-4

LANARI, R., MORA, O., MANUNTA, M., MALLORQUÌ, J.J., BERARDINO, P. and SANSOSTI, E. (2004), *A small baseline approach for investigating deformations on full resolution differential SAR interferograms*. IEEE Trans. Geosci. Remote Sens., *42*(7), 1377–1386.

LANCIANI, R. (1901), *Forma Urbis Romae*. Scale 1:1000, Regia Accademia dei Lincei, published in parts from 1893 to 1901.

MALINVERNO, A. and RYAN, W. (1986), *Extension in the Tyrrhenian sea and shortening in the apennines as result of arc migration driven opening*. Bollettino di Geofisica Teorica ed Applicata, *28*, 75–156.

MANUNTA, M., MARSELLA, M., ZENI, G., SCIOTTI, M., ATZORI, S. and LANARI, R. (2008). *Two-scale surface deformation analysis using SBAS-DInSAR technique: a case study of the city of Rome, Italy*. Int. J. Remote Sens. 29, 1665–1684. doi:10.1080/01431160701395278

MARRA, F., and ROSA, C. (1995), *Statigrafia e assetto geologico dell'area romana*. Servizio Geologico Nazionale, Memorie descrittive della Carta geologica d'Italia, L, La geologia di Roma – Il centro storico, 49–112.

MARRA, F., CARBONI, M.G., DI BELLA, L., FACCENNA, C., FUNICIELLO, R. and ROSA, C. (1995), *Il substrato Plio - Pleistocenico nell'area romana -* Boll. Soc. Geol. It., *114*, 195–214.

MARRA, F., ROSA, C., DE RITA, D., and FUNICIELLO, R. (1998), *Stratigraphic and tectonic features of the middle Pleistocene sedimentary and volcanic deposits in the area of Roma*. Quaternary International, *47/48*, 51–63.

MARRA, F., FREDA, C., SCARLATO, P., TADDEUCCI, J., KARNER, D. B., RENNE, P. R., GAETA, M., PALLADINO, D. M., TRIGILA, R., and CAVARRETTA, G. (2003), *Post-caldera activity in the Alban Hills volcanic district (Italy): 40Ar/39Ar geochronology and insights into magma evolution*. Bull. Volcanol., *65*(4), 227.

MATTEI, M., FUNICIELLO, R. and PAROTTO, M. (2008), *Roma e contesto geodinamico recente dell'Italia Centrale*. Special volume "La geologia di Roma. Dal centro storico alla periferia", Memorie descrittive della Carta Geologica d'Italia, *80*, 13–24.

MATTEI, M., CONTICELLI, S. and GIORDANO, G. (2010), *The Tyrrhenian margin geological setting: from the Apennine orogeny to the K-rich volcanism*. In Funiciello, R. and Giordano, G. (eds): The Colli Albani Volcano. Special Publications of IAVCEI, The Geological Society, London, UK, *3*, 7–27.

MEISINA, C., ZUCCA, F., FOSSATI, D., CERIANI, M., ALLIEVI, J. (2006), *Ground deformation monitoring by using the Permanent Scatterers Technique: the example of the Oltrepo Pavese (Lombardia, Italy)*. Engineering Geology, *88*, 240–259.

MEISINA, C., ZUCCA, F., NOTTI, D., COLOMBO, A., CUCCHI, A., SAVIO, G., GIANNICO, C., BIANCHI, M. (2008), *Geological interpretation of PSInSAR Data at regional scale*. Sensor, 8, 7469–7492.

MOLIN, D., CASTENETTO, S., DI LORETO, E., GUIDOBONI, E., LIBERI, L., NARCISI, B., PACIELLO, A., RIGUZZI, F., ROSSI, A., TERTULLIANI, A., TRAINA, G. (1995), *Sismicità*. In: La Geologia di Roma, il centro

Reprinted from the journal

storico, Memorie descrittive della Carta Geologica d'Italia, volume L, 323–408.

MORA, O., MALLORQUI, J.J. and BROQUETAS, A. (2003), *Linear and nonlinear terrain deformation maps from a reduced set of interferometric SAR images*. IEEE Trans. Geosci. Remote Sens., *41*, 2243–53.

NOTTI, D., HERRERA, G., BIANCHINI, S., MEISINA, C., GARCÍA-DAVALILLO, J.C., ZUCCA, F. (2014), *A methodology for improving landslide PSI data analysis*. International Journal of Remote Sensing, *35*(6).

PAINE, J.G., BUCKLEY, S., COLLINS, E.W., WILSON, C.R., KRESS, W. (2009), *Assessing sinkhole potential at Wink and Daisetta, Texas using gravimetry and radar interferometry*. 22nd Symposium on the application of geophysics to engineering and environmental problems, Fort Worth, Texas, 480–488.

PAROTTO, M. (2008), *Evoluzione paleogeografica dell'area romana: una breve sintesi*. Special volume "La geologia di Roma. Dal centro storico alla periferia", Memorie descrittive della Carta Geologica d'Italia, *80*, 25–38.

PATACCA, E., SARTORI, R. and SCANDONE, P. (1992), *Tyrrhenian basin and Apenninic arcs: kinematic relations since late Tortonian times*. Mem. Soc. Geol. Italiana, *45*, 425–451.

PRATI, C., FERRETTI, A. and PERISSIN, D. (2010), *Recent advances on surface ground deformation measurement by means of repeated space-borne SAR observations*. J. Geodyn., *49*, 161–170.

RASPINI, F., CIGNA, F., MORETTI, S. (2012), *Multi-temporal mapping of land subsidence at basin scale exploiting Persistent Scatterer Interferometry: case study of Gioia Tauro plain (Italy)*. Journal of Maps, *8*(4), 514–524.

REGIONE LAZIO (2014), *Rapporto di evento del 31 gennaio - 04 febbraio 2014*. Centro funzionale regionale. www.idrografico.roma.it/documenti/RapportiEvento/Anno%202014/01%20-%20Gennaio/Rapporto%20Evento%2031%20gen-04%20feb%2014.pdf

REIGBER, A., MOREIRA, A. (2000), *First demonstration of airborne SAR tomography using multibaseline L-band data*, Geoscience and Remote Sensing, IEEE Transactions on, 38(5), 2142–2152. doi:10.1109/36.868873

RIGHINI, G., PANCIOLI, V., CASAGLI, N. (2012), *Updating landslide inventory maps using Persistent Scatterer Interferometry (PSI)*. Int. J. Remote Sens., *33*, 2068–2096.

RIGUZZI, F., PIETRANTONIO G., DEVOTI R., ATZORI S., and ANZIDEI M. (2009), *Volcanic unrest of the Colli Albani (central Italy) detected by GPS monitoring test*. Earth Planet Inter, *177*(1–2), 79. doi:10.1016/j.pepi.2009.07.012.

ROSEN, P.A., HENSLEY, S., JOUGHIN, I.R., LI, F.K., MADSEN, S.N., RODRIGUEZ, E., GOLDSTEIN, R.M. (2000), *Synthetic aperture radar interferometry*. Proc. IEEE, *88*, 333–382.

ROTT, H. (2009), *Advances in interferometric synthetic aperture radar InSAR in earth system science*. Progress in Physical Geography, *33*, 769–791.

ROVIDA, A., CAMASSI, R., GASPERINI, P. and STUCCHI, M. (2011), *CPTI11, la versione 2011 del Catalogo Parametrico dei Terremoti Italiani*. Milano, Bologna, http://emidius.mi.ingv.it/CPTI

SALVI, S., ATZORI, S., TOLOMEI, C., ALLIEVI, J., FERRETTI, A., ROCCA, F., PRATI, C., STRAMONDO, S., FEULLET, N. (2004), *Inflation rate of the Colli Albani volcanic complex retrieved by the permanent scatterers SAR interferometry technique*. Geophys. Res. Lett., *31*, L12606. doi:10.1029/2004GL020253.

SANABRIA, M.P., GUARDIOLA-ALBERT, C., TOMÁS, R., HERRERA, G., PRIETO, A., SÁNCHEZ, H., TESSITORE, S. (2014), *Subsidence activity maps derived from DInSAR data: Orihuela case study*. Natural Hazards and Earth Science Systems, *14*, 1341–1360.

SCIOTTI, M., FRAIOLI, A., and SOLDÀ, R. (2000), *Il rischio cavità sotterranee nell'area del Comune di Roma*, Università degli Studi di Roma "la Sapienza", Comune di Roma, Roma.

STRAMONDO, S., BOZZANO, F., MARRA, F., WEGMULLER, U., CINTI, F.R., MORO, M. and SAROLI, M. (2008), *Subsidence induced by urbanization in the city of Roma detected by advanced InSAR*. Remote Sensing of Environment *112* (2008) 3160–3172. doi:10.1016/j.rse.2008.03.008.

TAPETE, D., CASAGLI, N., FANTI, R. (2012a), *Radar interferometry for early stage warning of monuments at risk*. In Landslide Science and Practice – Risk Assessment and Mitigation: Special Issue of The Second World Landslide Forum (Rome, Italy, 3–9 Oct. 2011) (Berlin: Springer 2012) vol 6, chapter 4 (Landslides and cultural heritage) pp. 1–6.

TAPETE, D., FANTI, R., CECCHI, R., PETRANGELI, P., CASAGLI, N. (2012b), *Satellite radar interferometry for monitoring and early-stage warning of structural instability in archaeological sites*. Journal of Geophysics and Engineering, *9*(4), 10–25. doi:10.1088/1742-2132/9/4/S10.

TOSI, L., TEATINI, P., STROZZI, T. (2013), *Natural versus anthropogenic subsidence of Venice*. Scientific Reports, *3*:2710. doi:10.1038/srep02710

VENTRIGLIA, U. (2002), *Geologia del territorio del Comune di Roma* (Amministrazione Provinciale di Roma, Roma 2002).

WEGMÜLLER, U., WERNER, C., STROZZI, T. and WIESMANN, A. (2004), *Multi-temporal interferometric point target analysis*. In Smits, P. and Bruzzone, L. (Eds.), Analysis of Multi-temporal remote sensing images. Series in Remote Sensing, World Scientific (ISBN 981-238-915-6), Vol. 3, 136–144.

WERNER, C., WEGMULLER, U., STROZZI, T. and WIESMANN, A. (2003), *Interferometric point target analysis for deformation mapping*. Proceedings of IGARSS '03, Vol. 7, 4362–4364.

ZENI, G., BONANO, M., CASU, F., MANUNTA, M., MANZO, M., MARSELLA, M., PEPE, M. and LANARI R. (2011), *Long-term deformation analysis of historical buildings through the advanced SBAS-DInSAR technique: the case study of the city of Rome, Italy*. J. Geophys. Eng. 8, S1–12. doi:10.1088/1742-2132/8/3/S01.

(Received April 14, 2014, revised February 25, 2015, accepted March 2, 2015, Published online March 31, 2015)

Pure Appl. Geophys. 172 (2015), 3029–3042
© 2014 Springer Basel
DOI 10.1007/s00024-014-0908-6

New results on ground deformation in the Upper Silesian Coal Basin (southern Poland) obtained during the DORIS Project (EU-FP 7)

MAREK GRANICZNY,[1] DAVIDE COLOMBO,[2] ZBIGNIEW KOWALSKI,[1] MARIA PRZYŁUCKA,[1] and ALBIN ZDANOWSKI[1]

Abstract—This paper presents application of satellite interferometric methods (persistent scatterer interferometric synthetic aperture radar (PSInSAR™) and differential interferometric synthetic aperture radar (DInSAR)) for observation of ground deformation in the Upper Silesian Coal Basin (USCB) in Southern Poland. The presented results were obtained during the DORIS project (EC FP 7, Grant Agreement n. 242212, www.doris-project.eu). Several InSAR datasets for this area were analysed. Most of them were processed by Tele-Rilevamento Europa - T.R.E. s.r.l. Italy. Datasets came from different SAR satellites (ERS 1 and 2, Envisat, ALOS- PALSAR and TerraSAR-X) and cover three different SAR bands (L, C and X). They were processed using both InSAR techniques: DInSAR, where deformations are presented as interferometric fringes on the raster image, and PSInSAR, where motion is indentified on irregular set of persistent scatterer (PS) points. Archival data from the C-band European Space Agency satellites ERS and ENVISAT provided information about ground movement since 1992 until 2010 in two separate datasets (1992–2000 and 2003–2010). Two coal mines were selected as examples of ground motion within inactive mining areas: Sosnowiec and Saturn, where mining ceased in 1995 and 1997, respectively. Despite well pumping after closure of the mines, groundwater rose several dozen meters, returning to its natural horizon. Small surface uplift clearly indicated on satellite interferometric data is related to high permeability of the hydrogeological subregion and insufficient water withdrawal from abandoned mines. The older 1992–2000 PSInSAR dataset indicates values of ground motion ranging from –40.0 to 0.0 mm. The newer 2003–2010 dataset shows values ranging from –2.0 to +7.0 mm. This means that during this period of time subsidence was less and uplift greater in comparison to the older dataset. This is even more evident in the time series of randomly selected PS points from both coal mines. The presence of bentonite deposits in the Saturn coal mine can also have influence on the ground surface uplift. Analysis of interferometric L and X-band data in Upper Silesia has enabled observation and monitoring of the underground mining front for several months. It was indicated by the example from the Halemba–Wirek coal mine. Analysis of the TerraSAR–X dataset processed by SqueeSAR algorithms proved to be the most effective for this purpose. X-band persistent scatterer interferometry (PSI) time series can help to indentify small seemingly negligible movements and are successfully supplemented by fringes when displacement becomes significant. Differential interferograms from the L-band dataset detect similar displacement values but, thanks to longer wavelength, are characterized by better coherence, especially in the middle of the subsidence trough. Results on ground deformation proved that ground motion above abandoned mines continues long after their closure. Therefore, existing regulations stating that abandoned mines are considered fully safe five years after mine closure should change. Moreover, it should be emphasised that construction in these areas should be avoided due to existing potential risks.

Key words: InSAR, ground deformations, mining, hydrogeology, monitoring, PSInSAR, upper Silesian Coal Basin.

1. Introduction

The objective of the DORIS project (EC FP 7, Grant Agreement n. 242212, www.doris-project.eu) concerned development of a national downstream service for detection, mapping, monitoring and forecasting of ground deformation at different temporal and spatial scales and in various physiographic and environmental settings using Earth observation (EO) techniques.

In particular, the two main ground deformation phenomena analyzed during the project were those related to ground subsidence and landslides. Therefore, several test sites characterized by different typology, dynamics of ground motion and geological background in five European countries (Italy, Hungary, Poland, Spain and Switzerland) were chosen for the study of ground subsidence and landslide dynamics.

The Polish case study was focused on ground subsidence in the region where subsidence is greatest

[1] Polish Geological Institute, National Research Institute, Rakowiecka 4, 00-975 Warsaw, Poland. E-mail: marek.graniczny@pgi.gov.pl; zbigniew.kowalski@pgi.gov.pl; maria.przylucka@pgi.gov.pl; plalbin.zdanowski@pgi.gov.pl
[2] Tele-Rilevamento Europa-T.R.E. s.r.l, Ripa di Porta Ticinese, 79, 20143 Milan, Italy. E-mail: davide.colombo@treuropa.com

countrywide and is mainly related to mining activities. However, the size of the subsidence is the result of many factors: the progress of mining, geological conditions (including tectonics) and changes of hydrogeological conditions. Interpretation and discussion of the different selected parts of the Upper Silesian Coal Basin (USCB) at three coal mines (Sosnowiec, Saturn and Halemba–Wirek) are the subject of this article.

2. Description of the study area

The area of interest is about 460 km^2 and is located between the cities of Zabrze, Sosnowiec, Myslowice and Katowice in Upper Silesia in Poland (Fig. 1).

The USCB covers an area of about 7,400 km^2 in southern Poland and in the Ostrava-Karvina region in the Czech Republic. The Polish part is about 5,800 km^2. It is the most important coal basin of

Poland and also one of the largest in Europe. Up to 30 percent of the deposit has been exploited by recent mining operations. The basin comprises a thick sequence (up to 8,500 m) of upper carboniferous sediments. The rocks of that sequence include four lithostratigraphic units, the Paralic Series, the Upper Silesian Sandstone Series, the Siltstone Series and the Cracow Sandstone Series. They represent Namurian and Westphalian eras. The thickness of the seams can be up to 6–7 m (VOLKMER 2008).

According to the geological structure and recharge conditions of the Carboniferous aquifer, two hydrogeological subregions are distinguished in the Upper Silesia: the "exposed" northeastern part and the "covered" southwestern part (Fig. 1). In the southwestern subregion carboniferous strata are predominantly mudstones and claystones overlying sandstones. The carboniferous aquifer is covered by thick, impermeable Neogene clays. In the "exposed" subregion sandstones with high permeability and porosity predominate. Carboniferous deposits are

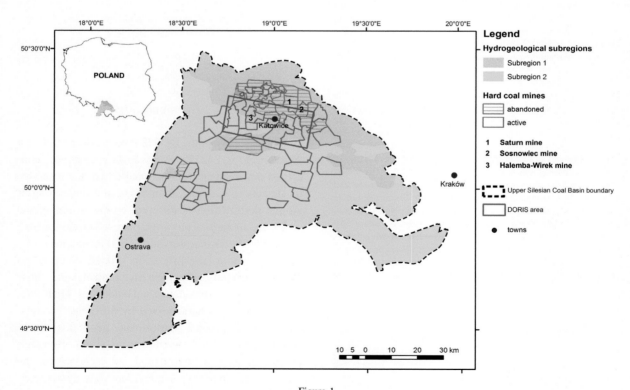

Figure 1
Geographical setting of the study area overlaying the hydrogeological subregions in the Upper Silesian Coal Basin: "exposed" subregion 1 and "covered" subregion 2

characterized by a high percentage of montmorillonite clays and swelling minerals (Razowska-Jaworek *et al.* 2008).

Mining operations in this basin are complicated because of the large scale faulting and folding caused by intense tectonic activity. Coal has been mined in the USCB since the 19th century. At present there are thirty active coal mines in the USCB (estimated 70 million tons/year).

In the study area hazardous ground deformations are caused primarily by the extensive longwall mining operations located mainly in the vicinity of the cities of Katowice, Zabrze and Ruda Slaska. Subsidence in Upper Silesia reaches typical velocities of a few centimeters per month, but there are many areas with subsidence of one centimeter per day. Exploitation of coal deposits in the USCB for more than 200 years has caused complex conditions of stress and deformation that are the cause of dynamic phenomena manifested in the form of rock mass shock. The surface becomes pitted with numerous collapsed cavities or basins with depths that may reach tens of meters. They may remain dry or be filled with water (land surface inundation) depending on the local hydrogeological conditions. Subsidence areas can be endangered during floods because they form a kind of interior basin. Subsidence is particularly dangerous because it causes severe damage to gas and water pipelines, electric cables and to sewer systems. In Upper Silesia it is common to find houses strengthened with iron bars anchored in the walls in order to prevent further damage or collapse, but even such reinforced buildings show cracks and joints on their walls. Collection of systematic information on the ground instability in these parts of the mining basin is very important because of ongoing changes in land use planning.

The USCB is also the most seismically active mining areas in Poland. Seismic observations supported by the Central Mining Institute date back to the 1950s and include an extensive data bank of mine tremors. More than 55,900 mine tremors of energy $E \geq 10^5$ J (local magnitude $M_L \geq 1.5$) occurred from 1974 to 2005 (Stec 2007). Two types of tremors can be distinguished in the USCB. These are the so-called "mining-tectonic" seismicity and "mining" seismicity. Mining-tectonic seismicity results from interaction between mining and tectonic factors. These seismic events appear to be located in tectonically disturbed zones and their sources are visibly more energetic. The second one, mining seismicity, is strongly associated with mining activity and seismic events that occur in the vicinity of active mining excavations. These events are energetically weak and their source mechanism can be of an explosive type. Twenty to fifty percent of the total seismic moment tensor is explosive, practically the same amount as the uniaxial compressional component, and a very small amount is comprised of a shear component (less than 10 percent).

Detailed analysis areas of three coal mines are presented in Fig. 1 (numbers 1–3).

The Saturn and Sosnowiec coal mines (numbers 1 and 2) are abandoned, were ground subsidence and uplift related to changes in hydrological conditions after mine colusre were detected by C-band PSI datasets. The Saturn coal mine was created by connecting three mines (Milowice, Czeladź and Czerwona Gwardia) between 1973 and 1975. Hard coal exploitation at the territory of the coal mine started in 1822. Exploitation was managed with roof–rock collapse systems or hydraulic filling according to geological conditions. From 1965 to 1992 mining of montmorillonite clays (bentonite) as an associated mineral product was conducted parallel to coal mining. Mining stopped on the 31st of December 1995.

The Sosnowiec coal mine (formerly Count Renard) dates back to 1876. Room and pillar exploitation was carried out until 1905, resulting in several rock collapses. Hydraulic filling was introduced subsequently. Mechanized brattices were used for the first time after 1970. Mining stopped on the 31st of December 1997.

The still active Halemba-Wirek coal mine (number 3) is a large mine located in Ruda Śląska. Rapid mining-induced changes in the ground surface were detected here by X and L-band datasets (PSInSAR and DInSAR). Its construction started in 1943 operating as a German factory ("Schaffgotsch"). This was the first coal mine in Poland created after World War II. Nowadays Halemba-Wirek represents one of the largest coal reserves in Poland, having estimated reserves of 120 million tonnes of coal. Annual coal production is around 3.36 million tonnes.

Exploitation is managed with roof–rock collapse systems using hydraulic filling. The mine has also large reserves of geothermal energy rated at 6,260,000 GJ/km^2.

3. Description of the interferometric techniques

DInSAR is an attractive technique for detecting and monitoring ground surface deformations arising from regional scale processes (seismic, volcanic, tectonic) (GABRIEL et al. 1989; MASSONNET and FEIGL 1998). The technique, however, may be ineffective for site-specific evaluations due to coarse resolution, coherence loss (a typical problem for vegetated areas) and atmospheric artefact limitations (WASOWSKI et al. 2002; COLESANTI and WASOWSKI 2006). PSInSAR^{TM}, developed at Politecnico di Milano - POLIMI (FERRETTI et al. 2001, 2006), and similar innovative multi-temporal DInSAR techniques (BERARDINO et al. 2002; WERNER et al. 2003) overcome in part the limitations of DInSAR and extend the applicability of radar interferometry to sub-regional and local-scale geological investigations of slope instability and ground subsidence. PSI analysis allows identification of numerous radar targets (the PS points) where very precise displacement information can be obtained. The main advantages of this technique are a regular re-visit time and a wide-area coverage of satellite radar sensors. PS points usually correspond to buildings and other man-made structures; therefore, this technique is particularly suitable for application in urban/peri-urban environments. Considering these factors the technique can be used as a supplement to GPS or conventional topographic surveying (FERRETTI et al. 2006).

The effectiveness of PSI in monitoring ground deformations was confirmed by numerous examples of practical application in different regions (FERRETTI et al. 2006; HERRERA et al. 2009; RASPINI et al. 2014; www.terrafirma.eu.com).

PS data can assist in:

- Identification and delimitation of areas affected by slow deformations.
- Estimation of surface velocity and acceleration fields with millimetre precision.
- Identification of the source of ground instability by analysing *in situ* and multi-temporal remotely sensed data.

The algorithm aims at identifying and then exploiting PS points only slightly affected by temporal and geometrical decorrelation, where very precise displacement measurements can be carried out (FERRETTI et al. 2000, 2001). Operational use of PSInSAR has originated the request for a more effective monitoring tool over non-urban areas, where the density of measurement points (i.e., PS points) is usually low. These areas, however, are characterized by the presence of distributed scatterers (DSs) affected by both temporal and geometrical decorrelation phenomena, but still exhibiting good coherence levels in some interferograms (FERRETTI et al., 2009a, b, 2011; DE ZAN and ROCCA 2005).

In SqueeSAR (FERRETTI et al., 2009a, b, 2011), recently developed by POLIMI and TRE, a Cherence matrix approach is used to optimally estimate the time series of the optical ray-path values of each DS, preserving, however, the information associated with the point-wise PS (FERRETTI et al. 2011). This new technique is based on an algorithm that selects statistically homogeneous pixels (for DS identification), followed by a maximum likelihood (ML) estimation of a vector of phase values (one vector for each DS) fitting all interferograms (i.e., the CM elements).

The SqueeSAR approach allows one to process jointly and effectively both PS and DS. For each DS the coherence matrix is "squeezed" to pass from $N(N - 1)/2$ interferograms to N "optimum" phase values, to estimate the temporal evolution of range values. The results provided by this processing chain have shown a significant increase (typically by a factor of 10) of measurement points over non-urban areas. Moreover, the time series of deformation are less noisy compared to PSInSAR results for those DS that, by chance, are identified as PS by the standard PSInSAR algorithm.

4. Characteristic of interferometric data

Several InSAR datasets were analysed for this area. Most of them were processed by Tele-

Rilevamento Europa-T.R.E. s.r.l. Italy within the DORIS project. Datasets came from three different SAR bands (C, L and X) and were processed in both InSAR techniques: DInSAR, where deformations are presented as interferometric fringes on the raster image, and PSInSAR, where motion is identified on irregularly-spaced PS points. In particular for C-band data 70 descending ERS satellite images from 17/05/1992 to 20/12/2000 and 31 descending Envisat satellite images from 05/03/2003 to 29/09/2010 were processed. Processing resulted in 71,709 PS points for the ERS dataset and 32,341 PS points for the Envisat dataset. Velocity of motion ranges between −40 and +7 mm per year. Larger movement cannot be indentified due to limitations of processing connected to the wavelength, the non-linear character of changes or the land cover. As such, the central part of the area is not covered by PS points.

Subsequently, the PSI data has been extended for a set of differential interferograms from the L-band satellite. In this study five differential interferograms form ALOS-PALSAR satellite scenes were used. The interferograms acquired for the Terrafirma project (ESRIN/Contract no. 17059/03/I-IW) were processed by GAMMA Remote Sensing and Consulting AG in Switzerland. They cover the period 22/02/2007 to 27/05/2008 and have 45, 92 or 138-day time spans. These data complemented information about very fast movement (up to 30 cm per 45 days) derived from the PSI dataset. A direct correlation was found between the interferometric fringes and the mining zones. Analyses of PS distribution with the location of active mines generally shows that PS points are present at the edges or outside the mine boundaries, possibly indicating slow terrain motions after mining exploitation.

For X-band data 30 TerraSAR-X satellite images from 05/07/2011 to 21/06/2012 were processed using the SqueeSAR approach. In total 824,220 PS points were acquired with velocity values between −337 and +57 mm per year. Shorter wavelength and revisit time resulted in a significantly denser PS dataset with identification of stronger movement values. Nevertheless, there were still some areas not covered by the PS data, due to either very strong motion, non-linear behaviour of the displacement or loss of coherence on vegetated areas. Additionally, 28 subsequent differential interferograms were acquired, successfully complementing the information about subsidence. The interferograms have 11-day time spans, which enables detailed monitoring of surface changes.

The summary of used datasets is presented in Table 1.

5. Potential for SAR data being used for ground movement identification: examples and discussion

Examples of application of this method for mining-induced ground deformation monitoring can be seen in the Terrafirma Atlas that was published in 2009. These examples mainly include the use of C–band data. Fifty four results from thirty countries are presented to illustrate many different user requirements for measuring and monitoring terrain—motion across Europe to which PSI has been applied. Upper Silesia and two other case studies of coal mining areas were included in the Terrafirma Atlas—Stoke on Trent (United Kingdom) and Liege (Belgium). Both areas have an extensive history of coal mining. As a result the areas have experienced terrain motion during the extraction phase and after the mines were abandoned. Zones of uplift and subsidence are observable via the processed data sets. Uplift in Stoke on Trent is attributed to older undermining, probably

Table 1

Summary of the data sets

No.	Satellite	Geometry/track	Band	No. of scenes	Time period	Type of data
1	ERS 1 and 2	Desc/T222	C	70	17/05/1992-20/12/2000	PSInSAR
2	Envisat	Desc/T222	C	31	05/03/2003-29/09/2010	PSInSAR
3	TerraSAR-X	Desc/T108	X	30	05/07/2011-21/06/2012	PSInSAR (SqueeSAR) and DInSAR
4	ALOS-PALSAR	n/a	L	6	22/02/2007-27/05/2008	DInSAR

due to elastic rebound from groundwater recharge. Subsidence is associated with more recent under-mining in areas of compressible alluvial soils and areas underlain by salt (CULSHAW 2009). Similar phenomena are present in Liege where a rising water table has resulted in mine collapse and flooding, and both uplift and subsidence are common (DEVLEESSC-HOUWER and DECLERCQ 2009).

The above-mentioned areas and the USCB are ideal for satellite interferometryused to discern ground deformation. Use of InSAR for ground movement monitoring in the USCB was presented in various studies conducted by Polish researchers (PERSKI 1999, 2010; GRANICZNY *et al.* 2005, 2006a, b, 2007; KRAWCZYK *et al.* 2007; LESNIAK and PORZYCKA 2007, 2009). Extensive material was collected by the Polish Geological Institute—NRI during the Terra-firma project (ESRIN/Contract no. 17059/03/I-IW). The first dataset was processed in mid March 2004 and included PSI results for the Sosnowiec area. Processing covered 54 scenes of the ERS-1 and ERS-2 images registered between 1992 and 2003. Ground vertical movement was identified in the areas of active mining and post-mining areas. The relationship between interferometric data values and major faults, seismic activity and geological conditions, as well as mining activities was analyzed and confirmed. The research has been the subject of many publications (Graniczny 2008a, b, 2009).

5.1. C-band

Archival data from the C-band European Space Agency satellites ERS and ENVISAT provided information about ground movement from 1992 until 2010 in two separate datasets (1992–2000 and 2003–2010). This information is extremely valuable because it gives a unique opportunity to retrieve data describing movement taking place 20 years ago. No other monitoring technique enables such. In situ geodetic measurements, mainly used nowadays for monitoring mining-induced deformation, refer only to the present and cannot be used to identify past displacements if they were not been carried out during the period of concern.

Secondly, PSI datasets illustrate very slow ver-tical movements from +7 mm/year to −40 mm/

year. Such a range of velocity values is limited to 4 cm per year due to specification of the processing that is dependent on the radar wavelength and revisit time. Faster movement, which usually takes place in active mining areas, as well as non linear motion, is not identified. This can be a disadvantage; but on the other hand the C-band PSI dataset enables observation of seemingly harmless movement, not detectable by technical levelling, which can cause destruction of buildings on a long term impact. Also, this particular range of velocities allowed shaping boundaries of mining activity influence for the whole study area. In the central part of the area most of the mines are still active, where ground movement reaches the values of centimetres per day. For these reasons C-band PS points are not expected to be present in this area.

However, since mines are pretty close to each other, when mining occurs different subsidence bowls mix their effects on the surface. In such conditions it is difficult to identify discrete and limited displace-ments and subsidence is measured rather as a complex polygon. In order to set a border for moving areas PS points with a velocity ranging from −2 to +2 mm/yr are selected. This defines the areas where mining subsidence is observed. Within this boundary (in the subsidence bowl) points are characterized by the gradually increasing values of velocity, reaching values of a few negative centimetres (−cm) per year.

After separation of the two different time periods comparison of movement before and after closure of a few of the mines was possible.

As examples, two coal mines were selected: Sosnowiec and Saturn (Fig. 2). The Sosnowiec mine closed December 31, 1997 and the last productive panels were at −10 and −180 m.a.s.l. Nowadays the mine is not accessible and the former shaft

Figure 2 ▶

Example of the C-band PSI dataset over parts of the Sosnowiec and Saturn coal mines, abandoned in 1997 and 1995, respectively. The ERS dataset (**a**) covers 1991 until 2000, before mine closure, whereas the ENVISAT dataset (**b**) covers 2003 to 2010, after mine closure. On the bottom is a time series of randomly selected PS points referring to the areas marked as *1, 2, 3* and *4* on the **a** and **b** pictures on the top. *Red* highlights values relating to the time before closure of the mine, *while blue* indicates values after closing the mine

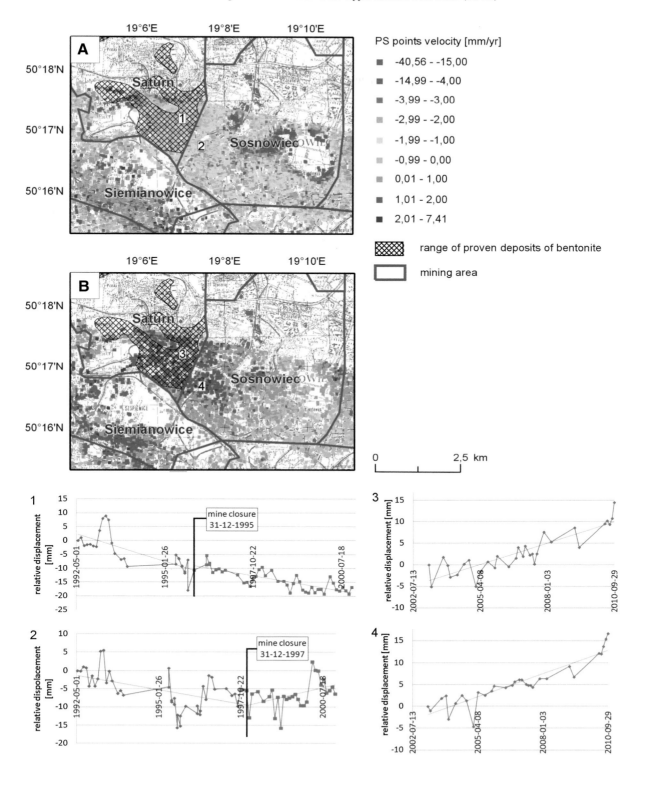

(Szczepan) is used as a dewatering well. While the target water level is +90 m.a.s.l, the water level is currently at level −140 m.a.s.l.

The Saturn coal mine was closed December 31, 1995. During the last years of activity the exploitation was running in two levels: 90 m.a.s.l and −50 m.a.s.l. Water level in the shaft is kept at +38 m.a.s.l. The permissible water level is +69 m.a.s.l.

The older PSInSAR dataset from 1992 to 2000 (before mine closure) indicates values of ground motion from −40.0 to 0.0 mm. The newer dataset from 2003 to 2010 (after mine closure) shows values from −2.0 to +7.0 mm. This means subsidence was less and uplift greater as compared to the older dataset. This is even more evident in the time series of randomly selected PS points from both coal mines (Fig. 2). The reason for changes in the ground

Table 2

Water flow into abandoned coal mines, from 2001 to 2008, with characteristic levels and water volumes in the flooded mines (Czapnik et al. 2009)

Hard coal mine	Water flow into abandoned coal mines [m³/min]								Characteristic levels (m a.s.l.)		Volume of flooded mine workings [mln. m³]
	2001	2002	2003	2004	2005	2006	2007	2008	max	range of flooded mine workings	
Sosnowiec	7.4	7.9	7.1	6.7	6.3	6.8	6.3	6.3	90	(−200)–(−20)	4.50
Saturn	22.6	29.0	23.2	22.2	19.0	20.2	23.5	24.5	69	(−130)–(55)	7.40

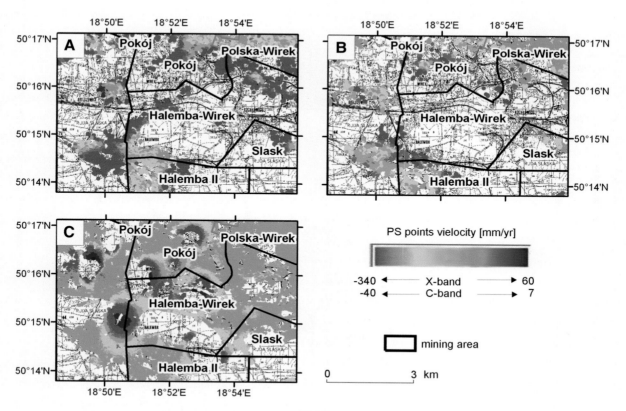

Figure 3

Comparison between different satellite PSI datasets for the Halemba-Wirek mining area. C-band (*top*): **a** ERS and **b** Envisat. X-band (*bottom*): **c** TerraSAR-X

movement may be the hydrogeological conditions after mine closure.

There are three systems of mine dewatering conducted by the Central Department of Mine Dewatering:

- submersible dewatering,
- stationary dewatering and
- complex dewatering combining submersible and stationary dewatering.

Submersible dewatering is applied in the Sosnowiec coal mine. The Saturn coal mine utilizes a complex dewatering system introduced in 2004. Water ilow into the abandoned coal mines is shown in Table 2, as well as ranges of inundated exploited chambers and permissible water levels.

In the hydrogeologically "exposed" northeastern part of the USCB (subregion 1 in Fig. 1) Cenozoic and Mezozoic groundwater bodies are in hydraulic connection with the carboniferous aquifer. Permeable carboniferous sandstones are characterized by low strength and high strain and are sensitive to the influence of water. Despite dewatering of abandoned mines the volume of flooded mine workings reached 7.4 million cubic meters at the Saturn mine and 4.5 million cubic meters in the Sosnowiec mine (Table 2). During the last 10 years the water table rose several dozen meters and returned to its natural horizon. Increased hydraulic pressure caused quick circulation of Quaternary, Triassic and carboniferous groundwaters. Small uplift on the surface clearly indicated on satellite interferometric data is related to elastic rebound from groundwater recharge.

Bentonite deposits can also have an influence on ground surface uplift. These deposits laying between the Upper Silesian Sandstone and the Paralic Series have a thickness up to 2.9 metres (Fig. 2). Montmorillonite clays (bentonite) belong to the group of swelling clays and change volume significantly according to how much water they contain. It is estimated that in extreme conditions bentonite saturated by water can increase volume even eight times.

5.2. X-band

A completely different SAR dataset was processed from the X-band images. It covers July 2011 to June 2012 and consists of 30 images. Processing resulted in a PSI dataset 10 times more dense than the ERS dataset and 25 times more dense than the Envisat dataset. A comparison between the datasets for the Halemba-Wirek mining area can be seen in Fig. 3. The top two images refer to C-band data and the bottom images refer to X-band data. It is clearly visible that the density of the PSI points is much higher for the X-band, and the X-band data also provide greater coverage on vertical ground movement. The colour scale used for velocity is the same (red—subsidence, green—stability, blue—uplift), but the velocity range is very different. X-band PS points cover motion up to -340 mm per year, whereas C-band PS points were only able to detect motion to -40 mm per year. This significant difference and the period of acquisition leads to the conclusion that these two data types can be used for different applications. C-band data appears to be good for indentifying boundaries of areas that were under the influence of mining for an extended period. On the other hand, X-band data give more continuous information on rate and extension of the affected areas and it is suitable both for identifying boundaries of mining activity and for delineating single troughs.

The study of a single subsidence trough can be complemented by differential interferograms. As it is presented in Fig. 4 and Fig. 3c, despite the very wide velocity range, there are still places with a lack of PSI points in the middle of the subsidence troughs; these are areas where very strong ground movements may occur, often in a non-linear fashion. These gaps can be closed with interferometric fringes. The very short revisit time of the TerraSAR-X satellite (11 days) enabled production of 28 subsequent interferograms. Fringes are presented in a colour scale that helps to identify deformation values. A full colour circle (from red to red) means 15 mm (half of the X-band radar wavelength) of ground deformation that happened in 11 days. A second full circle is another 15 mm giving 30 mm, a third 45 mm and so on. In the USCB interferometric fringes appear dependent on mining activity, which take place directly under the study surface. Usually fringes appear in a certain month and then are still visible on a few subsequent interferograms, showing at first very high deformation and then a decreasing trend, finally disappearing

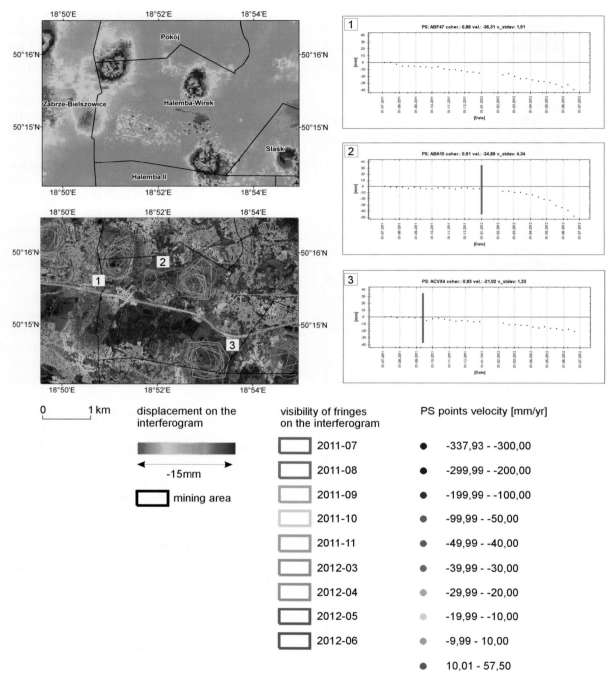

Figure 4
An example of X-band datasets for the Halemba-Wirek coal mining area. A differential interferogram (*top*) is complemented by PSI points (*bottom*). The appearance of fringes on the interferograms from various months is presented by colourful polygons in the bottom picture. The time series on the left refer to random PS points selected near areas 1, 2 and 3. The *red vertical line* indicates the time when subsidence started

after a few months. This kind of relation is presented in Fig. 4 (bottom) by polygons that refer to locations of the fringes. The colour of the polygon means the month in which the fringes were visible. In some places fringes were visible during the whole period of acquisition (e.g., place 1 in Fig. 4), but in other places

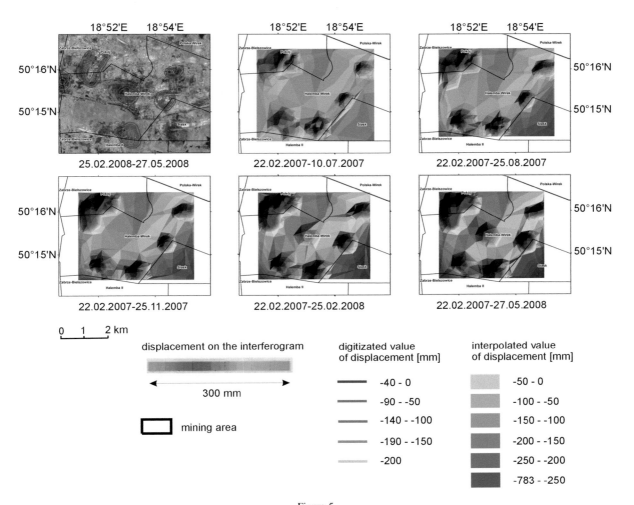

A subsidence trough in the Halemba-Wirek mining area from 22.02.2007 to 27.05.2008. The first picture is a part of an unwrapped interferogram form the ALOS-PALSAR satellite (L-band) with digitalized lines referring to certain values of displacement. The other five pictures show development of a deformation surface based on the surfaces from each of the interferograms via line interpolation. The last picture illustrates total displacement between 22.02.2007 and 27.05.2008

they started to appear in a certain month (e.g., in April 2012 in place 2 or in October 2011 in place 3). A precise study of the ground stability in a certain place can be fulfilled by a PS point time series graph as presented in Fig. 4. A time series of a random point near subsidence trough no. 1 show a subsidence trend for the whole period, whereas on a time series of points near troughs no. 2 and 3 there is stability to a certain point and then the start of downward vertical movement is visible. Usually points show a subsidence trend that began earlier than the one visible on

the interferograms. While it is possibile via PSI data to catch millimetre-sized displacements; each complete fringe on the interferogram refers to a 15 mm movement. Simultaneous and complementary use of these two datasets allows for a better understanding of the phenomena.

5.3. L-band

Similar observations of deformation changes can be seen on the L-band interferograms. This dataset consists of 5 differential interferograms with the following dates of acquisition: 22.02.2007–10.07.2007; 10.07.2007–25.08.2007; 25.08.2007–25.11.2007; 25.11.2007–25.02.2008; 25.02.2008–27.05.2008; the time span of each interferogram is 45, 92 or 138 days. Fringes on the interferograms illustrate subsidence but, in contrast to the X-band interferograms, they were unwrapped. In other words, the colour scale used for deformation characteristics provides information about the value of displacement without needing to identify subsequent fringes.

An example of the possible use of these interferograms is presented in Fig. 5. The first picture shows a part of one of the interferograms from the Halemba-Wirek mining area. Using a colour scale single lines that refer to a certain deformation value can be distinguished and digitized. Interpolation of these lines gives the deformation surface, a displacement that happened during the time span of the interferogam. The deformation surface was obtain for each of the 5 interferograms. These surfaces can be added to each other as presented in the five pictures in Fig. 5. Each subsequent picture presents deformation that happened between the first scene used in the set (22.02.2007) and the end scene of the following interferogram. Finally, the sixth picture shows total displacement during the whole time span (22.02.2007–27.05.2008). In this example maximum subsidence reached a value of 78 cm during 15 months. This kind of analysis can help to describe the evolution of where and when the subsidence trough is formed and the range of deformation between certain periods.

6. Main Conclusions

In regards to methodology of interferometric data and processing:

- DInSAR proved to be an attractive technique for detecting and monitoring ground surface deformations at the active mining areas on a regional scale.
- PSInSAR^{TM} extends applicability of radar interferometry for discerning ground subsidence to a sub-regional and local- scale. PSI analysis enables identification of numerous radar targets (PS points) where very precise displacement information can

be obtained. The main advantages of this technique are a regular re-visit time and wide-area coverage of satellite radar sensors.

- This technique can be used as a supplement to GPS or conventional topographic surveying.
- Different sets of interferometric data collected in the USCB enabled extensive analysis of ground motion. In particular, it was noted that different types of data coming from different satellites are applicable for different ranges of surface deformation.
- The SqueeSAR algorithm, recently developed by POLIMI and TRE, allows one to process jointly and effectively both PS and DS points. The results provided by this processing chain have shown a significant increase of measurement points over non-urban areas (typically by a factor of 10). It was clearly confirmed by our analysis.

In regards to new knowledge of ground motion in the USCB:

- Analysis of interferometric L and X-band data enabled observation and monitoring of the underground mining front in a period of several months; this was indicated in the example from the Halemba–Wirek coal mine. Analysis of the Terra-SAR–X dataset, processed by SqueeSAR algorithms, proved to be most effective for this purpose. The X-band PSI time series can help indentify small seemingly negligible movements and are successfully supplemented by fringes when displacement becomes significant.
- The combined X-band data differential interferograms and PS time series enabled establishing the start of the subsidence to the nearest month, and in some cases even to 11-day precision.
- Differential interferograms of the same area from the L-band dataset detected similar displacement values but, thanks to longer wavelengths, are characterized by better coherence, especially in the middle of subsidence troughs. Finally, the sixth dataset shows total displacement during the whole time span. In this example maximum subsidence reached a value of 78 cm during 15 months. This kind of analysis can describe the evolution of where and when a subsidence trough is formed and the range of deformation between certain periods.

- Interferometric data indicate uplift of the abandoned mining area around the former Sosnowiec and Saturn coal mines. This is primarily related to groundwater recharge, leading to an increase of hydrostatic pressure in the mine aquifer and stress in the overburden. An additional factor could be the presence of saturated bentonite layers. These two factors are most probably responsible for the uplift. The Sosnowiec coal mine was closed in 1997 and the Saturn mine in 1995. Correlation of these dates with the two datasets is highly indicative. In the PSI dataset covering 1992 to 2000 ground motion registered from –40.0 to 0.0 mm and for 2003 to 2010 ground motion varied between –2.0 and 7.0 mm.

- Ground deformation results proved that ground motion above abandoned mines continues long time after their closure. Therefore, existing regulations stating that abandoned mines are considered as fully safe in five years after closure should change. Moreover, it should be emphasised that construction in these areas should be avoided due to the existing risks.

REFERENCES

BERARDINO, P., FORNARO, G., LANARI, R., & SANSOSTI, E. (2002) A new algorithm for surface deformation monitoring based on small baseline differential SAR interferograms. IEEE Transactions on Geoscience and Remote Sensing, 40(11), 2375–2383.

COLESANTI, C. & WASOWSKI, J. (2006) Investigating landslides with satellite Synthetic Aperture Radar (SAR) interferometry. Engineering Geology, 88 (3-4), 173–199.

CULSHAW M. (2009), Stoke-on-Trent, United Kingdom. [In]: The Terrafirma Atlas—The terrain-motion information service for Europe (ed. Capes R., Marsh S.), GMES–ESA, June 2009. TerraFirma project, ESA publication, p. 43.

CZAPNIK, A., JANSON, E., JASIŃSKA, A. (2009) Selected problems with monitoring in monitoring ground abandoned hard coal mines in the Upper Silesian Coal Basin. Biuletyn Państwowego Instytutu Geologicznego, 436, 55–60.

DEVLEESSCHOUWER X., DECLERCQ P. (2009) Liège, Belgium. [In]: The Terrafirma Atlas—The terrain-motion information service for Europe (ed. Capes R., Marsh S.) GMES–ESA, June 2009. TerraFirma project, ESA publication, p. 46–47.

DE ZAN F. and ROCCA F. (2005) Coherent processing of long series of SAR images, Proc. IGARSS 2005, Seoul, pp. 1987–1990.

FERRETTI, A., PRATI C., and ROCCA F. (2000) Nonlinear sub-sidence rate estimation using permanent scatterers in differential SAR interferometry, IEEE Trans. on Geosc. and Rem. Sensing, 38, pp. 2202–2212.

FERRETTI, A., PRATI, C. & ROCCA F. (2001) Permanent Scatterers in SAR Interferometry. IEEE Trans. Geoscience And Remote Sensing, 39(1), 8–20.

FERRETTI, A., PRATI, C., ROCCA, F., & WASOWSKI, J. (2006) Satellite interferometry for monitoring ground deformations in the urban environment. Proc. 10th IAEG Congress, Nottingham, UK (CD-ROM).

FERRETTI A., FUMAGALLI A., NOVALI F., PRATI C., ROCCA F., RUCCI A. (2009a) Exploitation of Distributed Scatterers in Interferometric Data Stacks, Presented at IGARSS 2009 Conf. Cape Town.

FERRETTI A., FUMAGALLI A., NOVALI F., PRATI C., ROCCA F., RUCCI A. (2009b) The Second Generation PSInSAR Approach: Squee-SAR, Presented at Fringe Conf. 2009 Frascati.

FERRETTI A., FUMAGALLI A., NOVALI F., PRATI C., ROCCA F., RUCCI A. (2011) A New Algorithm for Processing Interferometric Data-stacks: SqueeSAR, IEEE Trans. On Geoscience and Remote Sensing, Volume: 49, Issue: 9, p.3460–3470.

GABRIEL A.K., GOLDSTEIN R. M. & ZEBKER H. A. (1989) Mapping Small Elevation Changes over Large Areas: Differential Radar Interferometry. Journal of Geophys., Res.94, N°B7, pp 9183–9191.

GRANICZNY M., (2009) Sosnowiec, Poland. [In]: The Terrafirma Atlas—The terrain-motion information service for Europe (ed. Capes R., Marsh S.), GMES–ESA, June 2009. TerraFirma project, ESA publication, p. 34.

GRANICZNY M., KOWALSKI Z., CZARNOGÓRSKA M. (2005) TerraFirma Project—monitoring of subsidence of the northeastern part of the Upper Silesian Coal Basin; Mass movements hazard in various environments. 20–21 October 2005, Kraków, Poland. Abstracts and Field Trip Guide-Book, Centre of Excellence REA, Polish Geological Institute, 18–19.

GRANICZNY M., KOWALSKI Z., JURECZKA J., CZARNOGÓRSKA M., (2006a) Practical Application of TerraFirma PS-InSAR in Poland. 26th EARSeL Symposium, May 29–June 2, Warsaw, Poland. EARSeL and Institute Geodesy and Cartography, Warsaw: 25-26.

GRANICZNY M., KOWALSKI Z., JURECZKA J., CZARNOGÓRSKA M. (2006b) TerraFirma Project—monitoring of subsidence of northeastern part of the Upper Silesian Coal Basin. Pol. Geol. Inst. Sp. Pap., 20: 59–63.

GRANICZNY M., KOWALSKI Z., LEŚNIAK A., CZARNOGÓRSKA M., PIĄTKOWSKA A. (2007) Analysis of the PSI data from the Upper Silesia—SW Poland. The International Geohazard Week 5–9 November 2007 ESA-ESRIN Frascati Rome, Italy. The International Forum on Satellite EO and Geohazards: 17.

GRANICZNY M., KOWALSKI Z., JURECZKA J., CZARNOGÓRSKA M., PIĄTKOWSKA A. (2008a) Preliminary interpretation of PSI data of the northeastern part of the Upper Silesian Basin (Sosnowiec test site)—TerraFirma project. Pol. Geol. Inst. Sp. Pap., 24: 29–35.

GRANICZNY M., CZARNOGÓRSKA M., KOWALSKI Z., LEŚNIAK A., JURECZKA J. (2008b) Metoda punktowej, długookresowej satelitarnej interferometrii radarowej (PSInSAR) w rozpoznaniu geodynamiki NE części Górnośląskiego Zagłębia Węglowego. Przegląd Geologiczny, 56, 9: 826–835.

HERRERA G., FERNÁNDEZ J.A., TOMÁS R., COOKSLEY G., MULAS J. (2009) Advanced interpretation of subsidence in Murcia (SE Spain) using A-DInSAR data—modelling and validation, Nat. Hazards Earth Syst. Sci., 9, 647–661, doi:10.5194/nhess-9-647-2009, 2009.

KRAWCZYK A., PERSKI Z., HANSSEN R.F. (2007) Application of ASAR Interferometry for Motorway Deformation Monitoring,

Reprinted from the journal

ESA ENVISAT Symposium, Montreux, Switzerland, 23–27 April 2007: 4.

Leśniak A., Porzycka S. & Graniczny M. (2007) Detekcja długookresowych pionowych przemieszczeń gruntu na obszarze terenów górniczych kopalń Zagłębia Dąbrowskiego z zastosowaniem technologii PSInSAR. [In:] Warsztaty Górnicze 2007 z cyklu„Zagrożenia naturalne w górnictwie", Ślesin k. Konina, 4–6 czerwca 2007. WUG, Katowice: 283–295.

Leśniak A. & Porzycka S. (2009) Impact of tectonics on ground deformations caused by mining activity in the north-eastern part of the Upper Silesian Coal Basin. Gosp. Sur. Min., 25: 227–238.

Massonnet D. & Feigl K. (1998) Radar interferometry and its application to changes in the Earth's surface. *Rev. Geophys.*, 36, 441–500.

Perski, Z. (1999) Osiadania terenu GZW pod wpływem eksploatacji podziemnej określane za pomocą satelitarnej interferometrii radarowej (InSAR). Przegląd Geologiczny, 2: 171–174.

Perski Z. (2010) Kompleksowa analiza interferogramów. [In:] Geneza i charakterystyka zagrożenia sejsmicznego w Górnośląskim Zagłębiu Węglowym, Zuberek W.M. & Jochymczyk K. (ed.). Wyd. UŚ, Katowice: 41–45.

Raspini F., Loupasakis C., Rozos D., Adam N., Moretti S. (2014) Ground subsidence phenomena in the Delta municipality region (Northern Greece): Geotechnical modeling and validation with Persistent Scatterer Interferometry, International Journal of Applied Earth Observation and Geoinformation, Volume 28, May 2014, pp 78–89.

Razowska-Jaworek, L., Pluta, I., Chmura, A. (2008) Mine waters and their usage in the Upper Silesia in Poland. Examples from selected regions. Technical University of Ostrava, Faculty of Mining and Geology, Proceedings of the 10th IMWA Congress, 2008, pp. 101–104.

Stec K. (2007) Characteristics of seismic activity of the Upper Silesia Coal Basin in Poland, Geophysical Journal International, 168 (2), pp 757–768.

Tele-Rilevamento Europa - T.R.E. s.r.l. (2013) Monitoring Reservoir Deformation from Space (leaflet), pp 1–4.

Volkmer G. (2008) Coal deposits of Poland, including discussion about the degree of peat consolidation during lignite formation, TU Bergakademie Freiberg, 2008.

Wasowski, J., Refice, A., Bovenga, F., Nutricato, R. & Gostelow P. (2002) On the Applicability of SAR Interferometry Techniques to the Detection of Slope Deformations, Proc. 9th IAEG Congress, Durban, South Africa, 16–20 Sept. 2002, CD ROM.

Werner, C., Wegmüller, U., Strozzi, T., Wiesmann, A. (2003) Interferometric Point Target Analysis for Deformation Mapping. Proc. IEEE International Geoscience Remote Sensing Symposium (IGARSS 2003), vol. 7, pp. 4362–4364. www.terrafirma.eu.com.

(Received February 4, 2014, revised April 24, 2014, accepted July 14, 2014, Published online September 10, 2014)

Pure Appl. Geophys. 172 (2015), 3043–3065
© 2014 The Author(s)
This article is published with open access at Springerlink.com
DOI 10.1007/s00024-014-0839-2

Pure and Applied Geophysics

Multi-Temporal Evaluation of Landslide Movements and Impacts on Buildings in San Fratello (Italy) By Means of C-Band and X-Band PSI Data

Silvia Bianchini,[1] Andrea Ciampalini,[1] Federico Raspini,[1] Federica Bardi,[1] Federico Di Traglia,[1,2]
Sandro Moretti,[1] and Nicola Casagli[1]

Abstract—This work provides a multi-temporal and spatial investigation of landslide effects in the San Fratello area (Messina province within the Sicily region, Italy), by means of C-band and X-band Persistent Scatterer Interferometry (PSI) data, integrated with in situ field checks and a crack pattern survey. The Sicily region is extensively affected by hydrogeological hazards since several landslides regularly involved local areas across time. In particular, intense and catastrophic landslide phenomena have recently occurred in the San Fratello area; the last event took place in February 2010, causing large economic damage. Thus, the need for an accurate ground motions and impacts mapping and monitoring turns out to be significantly effective, in order to better identify active unstable areas and to help proper risk-mitigation measures planning. The combined use of historical and recent C-band satellites and current X-band Synthetic Aperture Radar sensors of a new generation permits spatially and temporally detection of landslide-induced motions on a local scale and to properly provide a complete multi-temporal evaluation of their effects on the area of interest. PSI ground motion rates are cross-compared with local failures and damage of involved buildings, recently recognized by in situ observations. As a result, the analysis of landslide-induced movements over almost 20 years and the validation of radar data with manufactured crack patterns, permits one to finally achieve a complete and reliable assessment in the San Fratello test site.

Key words: Synthetic Aperture Radar, Persistent Scatterer Interferometry, field survey, landslides, San Fratello.

1. Introduction

The occurrence of landslides in populated areas can pose a serious threat to human lives, property and structures. Moreover, where significant cultural heritage is present, the socio-economic losses and damages are stronger because of the higher value of the elements at risk.

The detection of active ground movements on unstable slopes and landslide-prone areas can greatly benefit from advanced remote sensing techniques, i.e,. Persistent Scatterer Interferometry (PSI), thanks to their non-invasiveness, availability and high precision (Ferretti *et al.* 2001). Furthermore, radar satellite data analysis and traditional geomorphological tools, like field surveys and in situ observations are complementary for the mapping and monitoring of the impacts of such natural phenomena on buildings and manufactures of affected areas.

Persistent Scatterer Interferometry throughout the use of medium resolution Synthetic Aperture Radar (SAR) data in C-band (e.g., from ERS/ENVISAT satellites) has been demonstrated to be a valuable tool for back-monitoring slow-moving landslides, with good accuracy (up to 1 mm/year) and maximum detectable movement of about 15–20 cm/year (Hanssen 2005; Ferretti *et al.* 2005; Adam *et al.* 2008; Cascini *et al.* 2010; Cigna *et al.* 2013).

The launch of new SAR sensors that operate at 3 cm wavelength in X-band, i.e., TerraSAR-X and COSMO-SkyMed, with higher spatial resolution and reduced revisiting time (4–16 days) compared to the previous C-band satellites, has enhanced PSI capability for landslides detection and monitoring, allowing the identification of more recent and faster ground movements affecting small areas with improved precision. X-band SAR images make the number of retrieved PS targets higher by a factor of approximately 100–200, compared to medium resolution data (Cuevas *et al.* 2011). Most of the PS

[1] Department of Earth Sciences, University of Firenze, Via G. La Pira 4, 50121 Florence, Italy. E-mail: silvia.bianchini@unifi.it

[2] Department of Earth Sciences, University of Pisa, Via Santa Maria 53, 50126 Pisa, Italy.

targets show up on housetops and especially on facades and roofs of buildings, enabling a site-specific investigation. Therefore, the use of X-band data significantly improves the level of detail of the analysis, since small structures now act as stable scatterers and PS from different surfaces can be separated due to the very high resolution of up to 1 m in azimuth and range direction (Roth et al. 2003; Ge et al. 2010; Gernhardt et al. 2010; Notti et al. 2010).

C-band satellites provide the availability of long historical archives of motion rates and time series, covering wide areas at a relatively low cost and medium spatial resolution. On the other hand, X-band data, with higher spatial and temporal resolution, allow for a more detailed investigation even at the scale of a single building movement, in a recent and shorter span of time (e.g., some months) (Crosetto et al. 2010; Tomás 2010; Bovenga et al. 2012; Bru et al. 2013).

Radar data can provide an initial and non-invasive evaluation of most critical unstable areas, to be performed "at desk", prior to in situ survey. Thus, PSI data give a preliminary and rapid discrimination of the most unstable slopes over wide areas and, consequently, need to be integrated with additional geo-information and auxiliary data to obtain a more robust interpretation at a local scale. A detailed structural damage analysis of several buildings is required for cross-comparing crack pattern survey with PSI motion rates, for achieving a complete investigation (Herrera et al. 2010).

The Sicily region of Italy is extensively affected by hydrogeological hazards, and several landslides occurred at localized areas across time, causing casualties and large economic damage (Ardizzone et al. 2012; Ciampalini et al. 2012; Cigna et al. 2012). In particular, intense and catastrophic landslides have recently occurred in the San Fratello area. The recent activity coupled with the historical significance of many of the affected structures, field observations of damage, and the availability of SAR data make this an ideal field site.

In this paper the impacts of landslides in San Fratello are investigated from 1992 to 2012 by combining the available C-band and X-band SAR data along with a field survey of structural damages.

Multi-temporal estimation of radar velocities and related impacts on cultural and social heritage lead to an assessment of ground movements and landslide damage occurring within San Fratello over 20 years.

2. Study Area

2.1. Geographical and Geological Setting

San Fratello village is located in the NE sector of the Sicily Region (Southern Italy), within Messina province, at 640 m a.s.l. on the Nebrodi Mountains, which, together with the Peloritani Mountains, represent part of the Apenninic-Maghrebian orogenic chain (Cubito et al. 2005) (Fig. 1).

This area is made up of imbricate sheets of Mesozoic–Tertiary rocks, made of the lowermost autochthonous African foreland units, overlapped by the Appenninic-Maghrebian sequences (Corrado et al. 2009). These Appenninic units are tectonically overthrusted by allochthonous Kabilo-Calabrian Units, which represent different tectonic assemblages derived from the European continental margin (according to Ogniben 1969; Knott 1987; Dietrich 1988) or, according to an opposite interpretation, to an Eo-Alpine chain (Austroalpine sector) piled up toward the European foreland (Amodio-Morelli et al. 1976). Overall, the tectonic nappes are E–SE verging and show a total thickness of about 15 km.

The rocks outcropping in the area consist of a sequence of terrigenous to calcareous sedimentary sequences belonging to the different already mentioned paleogeographic domains (Fig. 1). The western and southern part of the study area are mostly characterized by terrigenous terrains, since the lower Cretaceous clayey sequences—called the Argille Scagliose Unit—extensively crop out (Appeninic-Maghrebid Units). In the northern portion of the area, the top units (Kabilo-Calabride Units) made of predominantly carbonate complexes outcrop, represented by Liassic limestone platform sequences, overlapped by a terrigenous Late Eocene–Oligocene Flysch (Frazzanò Flysch). The uppermost Cretaceous pelagic dolostones and limestones close the tectonic sequence (San Marco D'Alunzio Unit), outcropping towards N–NW San Fratello village (Nigro and Sulli 1995; Lavecchia et al. 2007).

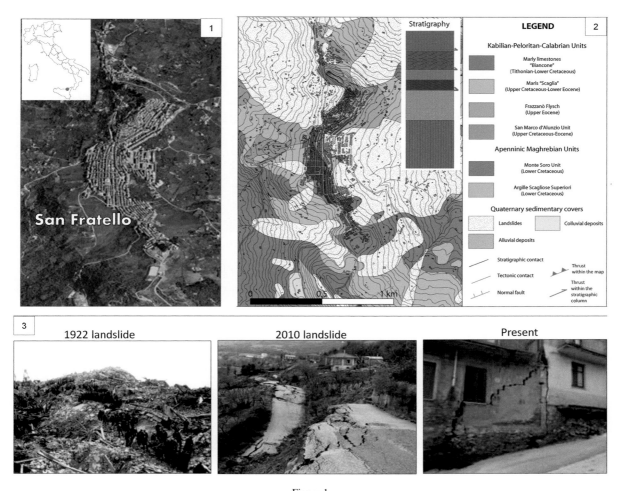

Figure 1

Study area: *1* Geographical location of San Fratello village (Messina province) in South Italy; *2* geological map and stratigraphic sequence; *3* photos referred to the most recent recorded landslides and present scenario

Throughout the inhabited area, an aquifer is also present, at a depth of between 0.5 and 2.5 m from the surface ground level (D.R.P.C. 2010).

From a geomorphological point of view, the test site is strongly influenced by the geo-structural conditions and the recent tectonic activity. The landscape is typical of recently uplifted areas: steep slopes, narrow valleys, high topographical gradient and remarkable relief energy are the most impressive geomorphological features of the study area. Moreover, all the geological units are highly tectonized, being these clays highly fissured and the stone-like lithotypes extensively fractured.

3. Landslides Occurrence

Messina province is prone to landslide hazard, mainly due to the steep topography, the nature of the lithotypes, mainly consisting of flysch units with tectonized silt–clay levels, and the occurrence of intense and seasonally high rainfall events (MONDINI *et al.* 2011; DEL VENTISETTE *et al.* 2012; RASPINI *et al.* 2013).

The main landslide types can be prevalently classified as debris flows, complex slides (VARNES 1978) and shallow and deep-seated landslides.

San Fratello has been chronically affected by landslides (Fig. 1). A severe landslide event dates

Figure 2
Boundaries and main directions of the three most important and recorded landslides in San Fratello, occurring in 1754, 1922, 2010. Location and photos of the main cultural sites of interest: *1* sanctuary of the three Saints Alfio, Filadelfio and Cirino on the Old Mountain; *2* Maria St. delle Grazie Church; *3* rocky massif Roccaforte; *4* St. Antonio Abate Church; *5* St. Nicolò Old Cathedral; *6* complex of St. Maria Assunta, former convent of S.S. Maria di Gesù and library; *7* St. Crocifisso Church; *8* St. Nicolò New Cathedral (now demolished); *9* St. Benedetto Il Moro Church; *10* Apollonia Archeological site; *11* ruins of St. Filadelfio Castle; *12* St. Nicolò Arcway *13* Stesicorea Arcway; *14* Historical Mammana palace; *15* Historical Stairway of Vittorio Veneto knights

back to 1754 and almost completely destroyed the village. The most recent phenomena are recorded in 1922 and 2010 (Fig. 2). On 8 January 1922, a landslide occurred in the northwestern sector of San Fratello, causing hundreds of deaths; about ten thousand people were evacuated and another village (called Acquedolci) was built along the coast, as

ordered by a Royal Decree (FARANDA 2010). However, San Fratello was re-populated again across time and, more recently, on 14 February 2010, another wide landslide, triggered by intense rainfall and extended up about 1 km^2, developed on the opposite southern–eastern slope, causing huge damages to the roads and structures (Fig. 2). This landslide affected the eastern urban districts (i.e., the Stazzone and Riana districts) and the first boundary of damaged area was released on 22February 2010 (Fig. 2). About 2,000 inhabitants were initially evacuated, approximately 300 houses were slightly damaged and 50 needed to be demolished (D.R.P.C. 2010).

This latest landslide affected the whole E-facing slope, consisting of roto-translational slide and flow, and involving the surface debris cover, which is about 10 m thick and mainly made of wet and fissured clayey lithotypes. As a result, the phenomenon has been caused by predisposing variables that deal with soil and rock geo-mechanical properties, and with the geostructural and hydrogeological setting of the area. The triggering factors may have been the intense rainfall that has increased the static water table within the shallow aquifer, determining soil saturation processes in the clays and causing mass movements.

Ground movements kept on being active up to nowadays, and, thus, the instability scenario is still very critical in San Fratello area.

4. Cultural Heritage

San Fratello is an old village, characterized by several sites of cultural-artistic interest that have been affected by the long-lasting catastrophic natural phenomena (Fig. 2). The remains of the very first inhabited territory of San Fratello, dating back to the III century B.C., are located uphill on the Old Mountain (718 m a.s.l.), northward of the present town (Fig. 2, point 1), where the ancient Norman Sanctuary of the three Saints, built up in the XII century, is also located (Fig. 2, point 1). Since the Norman age, San Fratello village expanded near the rocky massif called Roccaforte (Fig. 2, points 3 and 11) and developed until the Middle Age, when many churches and religious sites were built across time. Thus, on the one hand ,San Fratello was a rural village, mainly inhabited by farmers and artisans, but on the other hand, about a hundred households were among the richest and most powerful of Sicily region and undertook a struggle for possessions and interests even in building and embellishing the churches of the town: some examples are the St. Crocifisso Church characterized by an octagonal medieval shape plant

(Fig. 2, point 7), the St. Maria delle Grazie Church dating back to the XVIII century (Fig. 2, point 2), the St. Benedetto il Moro Church (Fig. 2, point 9) and the St. Antonio Abate Church (Fig. 2, point 4).

Many cultural sites of San Fratello have been destroyed by landslides and re-built again across time. The landslide that occurred on 8th January 1922 destroyed most of the town (about two-thirds of San Fratello village). The Maria St. Assunta mother-church, built in the XIII century in the western portion of San Fratello, was completely destroyed by the phenomenon. The newly built Maria St. Assunta church, together with the former convent of Santissima Maria di Gesù and the library (Fig. 2, point 6), can be regarded today as the center of religion, culture and art of San Fratello.

The Old St. Nicolò Cathedral (Fig. 2, point 5), dating back to the XVI century, has been severely damaged by the 1922 landslide and nowadays the only remaining portions are the right sector and the lower part of the bell tower. In the 50s a new St. Nicolò Cathedral was built in the modern Stazzone district (Fig. 2, point 8). Unfortunately also this church has been affected by the 14 February 2010 landslide and has recently been demolished, in February 2013, due to the irreversible and non-repairable damages.

The majority of the historical old houses of San Fratello, the oldest ones dating back to the Norman settlement, have remained uninjured after the natural disasters up to nowadays (Fig. 2, points 12–15). Hence, the damage assessment and conservation strategies for the cultural heritage are strongly recommended and can be addressed to the most important sites of interest of the village.

5. Methodology

The approach of this work consists in a multi-temporal and spatial investigation of landslide effects by means of PSI technique and field survey (Fig. 3). Firstly, we analyzed the available PSI data that highlighted ground motions in historical, recent and current time intervals. A down slope projection of Line-of-Sight (LOS) velocities was performed, following the procedure of COLESANTI and WASOWSKI

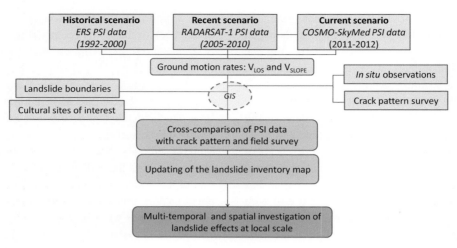

Figure 3
Methodology flow chart

(2006), in order to homogenize all the PSs in the same direction of the maximum local slope, and to obtain vectors of displacement that account more specifically for topographic features of landslides phenomena.

Then, radar movement rates were cross-validated and compared with local failures and with the in-situ observations on historical buildings carried out in the spring of 2010 and in the winter of 2012–2013. The radar mapping and subsequent zoning of the unstable urban sectors of San Fratello were performed by focusing on individual targets, i.e., buildings within the built-up area. In particular, PSI data, local cadastre, the distribution of churches and sites of cultural interest, and pre-existing landslide boundaries were compared and over-layered in a Geographical Information System environment (Fig. 3).

Radar-interpretation and photo-interpretation procedures (e.g., FARINA et al. 2006; BIANCHINI et al. 2012; CIGNA et al. 2013; RIGHINI et al. 2012) combined with a field survey allowed the mapping and characterization of landslides. In particular, on the one hand, the recognition of features related to topographic surface movements and the typology classification were mainly based on visual interpretation of orthophotos; on the other hand, the evaluation of the velocity and state of activity of phenomena took advantage from multi-interferometry-based information. For newly detecting

and enlarging phenomena, the two most recent available datasets (i.e., RADARSAT-1 and COSMO-SkyMed data) were employed when improving the landslide inventory map of the study area.

Only "very slow" and "extremely slow" phenomena (velocity <16 mm/year and 16 mm/year ≤velocity <1.6 m/year, respectively, according to CRUDEN and VARNES 1996) can be detected by PSI data due to the satellite technical acquisition parameters (i.e., signal wavelength and revisiting time; CANUTI et al. 2004). Moreover, N–S oriented ground movements are not or only partially illuminated by InSAR sensors, due to the intrinsical acquisition parameters of the satellites that move along N–S orbits with a right-side looking system.

Overall, the outcomes of the work leads to an accurate mapping and monitoring of ground motions and impacts at a local scale, permitting proposing new boundaries of the landslide-affected areas in San Fratello village, and to update the landslide inventory map of the site (Fig. 3).

6. InSAR Processing and Data

Available radar data used in this work consist of SAR images acquired in historical (1992–2001), recent (2005–2010) and current (2011–2012) time intervals. In particular, in C-band (5.6 cm wavelength), 104 SAR images were acquired by ERS 1/2

Table 1

Main acquisition characteristics of the used SAR datasets and PSI velocity values before and after downslope projection of LOS values

Satellite	ERS 1/2	RADARSAT-1	COSMO-SKYMED
Microwave band	C	C	X
Acquisition mode	Ascending and descending	Ascending and descending	Descending
Incidence angle (°)	23	34	26
Track angle (°)	348	349	185
	192	191	
Repeat cycle (days)	35	24	4
Cell resolution in azimuth (m) and range (m)	4×20	4×10	3×3
Critical baseline (m)	1,286	1,825	5,728
Number of SAR images	34 Ascending	46 Ascending	32
	70 Descending	47 Descending	
Temporal span	1992–2001	2005–2010	2011–2012
Acquisition dates interval	Ascending	Ascending	16/05/2011–02/05/2012
	11/09/1992–05/06/2001	30/12/2005–26/01/2010	
	Descending	Descending	
	01/05/1992–08/01/200	31/01/2005–03/02/2010	
Processing technique	SqueeSAR^{TM}	SqueeSAR^{TM}	SqueeSAR^{TM}
PS density (PS/km^2)	16	112	400
V_{LOS} PSI data velocity range (mm/yr)	Ascending (−9.5, +7.2)	Ascending (−46.8, +19.8)	(−56.4, +31.8)
	Descending (−26.8, +8.6)	Descending (−26.3, +20.5)	
Correction factor (*C*)	Ascending (−0.33, +0.96)	Ascending (−0.54, +0.99)	(−0.45, +1.00)
	Descending (−0.41, +0.96)	Descending (−0.52, +1.00)	
Maximum V_{SLOPE} values (V_{LOS}/C)	Ascending (−33.4, +6.8)	Ascending (−142.1, +146.6)	(−119.8, +111.3)
	Descending (−84.5, +50.0)	Descending (−71.5, +21.0)	
V_{SLOPE} PSI data velocity range (mm/yr)	Ascending (−33.4, +0.0)	Ascending (−142.1, +0.0)	(−119.8, +0.0)
	Descending (−84.5, +0.0)	Descending (−71.5, +0.0)	

satellites in the period 1992–2001 in ascending (34 scenes) and descending orbit (70 scenes), and 93 SAR images were acquired by RADARSAT-1 satellite, in ascending (46 images) and descending (47 images) modes, in the spanning time 2005–2010. Moreover, 32 SAR scenes were collected by COSMO-SkyMed satellite in X-band (3.1 cm wavelength) in descending geometry, in a 1 year-long period from May 16th, 2011 to May 2, 2012 (Table 1).

All the SAR images were processed through the SqueeSAR^{TM} algorithm (FERRETTI *et al.* 2011) to obtain PSI data. The SqueeSAR^{TM} is a new multi-temporal interferometric processing technique, being an advance on the PSInSAR^{TM} algorithm (FERRETTI *et al.* 2011), which permits measurement of ground displacements by means of traditional Permanent Scatterers (PS) like buildings, rock and debris, as well as from Distributed Scatterers (DS). DS are homogeneous areas spread over a group of pixels in a SAR image such as rangeland, pasture, shrubs and bare soil. These targets do not produce the same high signal-to-noise ratios of PS, but are, nonetheless, distinguishable from the background noise and their reflected radar signals are less strong, but statistically consistent. The SqueeSAR^{TM} algorithm was developed to process the signals reflected from these low-reflectivity homogeneous areas, but it also incorporates PSInSAR^{TM}; hence, no information is lost and movement measurement accuracy is improved (FERRETTI *et al.* 2011). As a result, the SqueeSAR^{TM} algorithm extracts geophysical parameters not only from point-wise deterministic objects (i.e., PS), but also from DS. PS and DS are jointly processed taking into account their different statistical behavior. The coherence matrix associated with each DS is properly "squeezed" to provide a vector of optimum (wrapped) phase values (FERRETTI *et al.* 2011). The SqueeSAR^{TM} technique allows an increase of density of the point targets that register ground motion, especially in non-urban areas, as sparse vegetation landscapes (MEISINA *et al.* 2013; RASPINI *et al.* 2013; BELLOTTI *et al.* 2014).

ERS Descending (1992 – 2001)

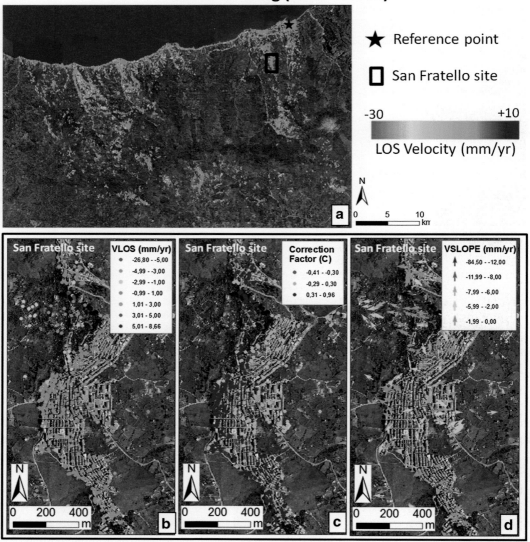

Figure 4
ERS PSI dataset in descending geometry within the study area: **a** PS LOS data distribution and reference point location; **b** V_{LOS} values on the San Fratello village; **c** correction factor (C) distribution values; **d** V_{SLOPE} values on the San Fratello village

Persistent Scatterer Interferometry datasets for each satellite used on the San Fratello village in descending geometry are shown in Figs. 4, 5 and 6

Since satellite systems measure velocities just along their LOS, only the component of motion that is parallel to the LOS direction is measured. A projection of the LOS displacement measures along the most probable direction of movement can be performed. Thus, assuming a simple translational movement parallel to the slope, in this work LOS velocity of each available PS point (V_{LOS}) was projected along the direction of the maximum slope (V_{SLOPE}), in order to account more specifically for topographic and geomorphological slope conditions within a local-scale landslides analysis. Moreover, this conversion permits comparing landslide velocities with different slope orientations, resolving the satellite acquisition orbit differences and allowing a more feasible interpretation.

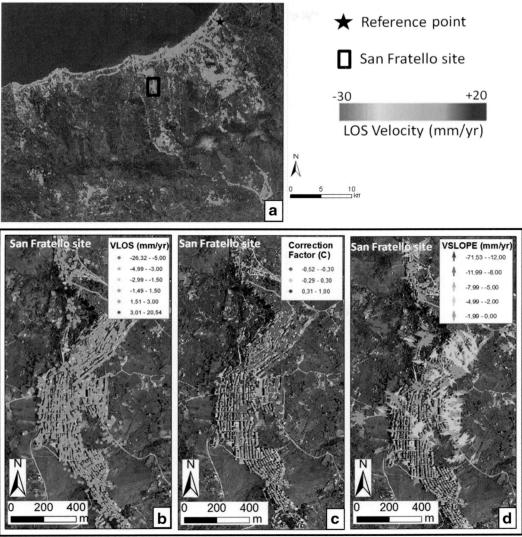

Figure 5
RADARSAT-1 PSI dataset in descending geometry within the study area: **a** PS LOS data distribution and reference point location; **b** V_{LOS} values on the San Fratello village; **c** Correction factor (C) distribution values; **d** V_{SLOPE} values on the San Fratello village

Following the procedure initially proposed by COLESANTI and WASOWSKI (2006), and then success-fully applied in several scientific works (CIGNA *et al.* 2013; GRIEF and VLCKO 2012; BIANCHINI *et al.* 2013; HERRERA *et al.* 2013), a correction factor (C) was applied to each LOS measurement, in order to determine the "real" V_{SLOPE} velocity (intended as not the one measured in the LOS direction, but the one occurring in the landslide direction), taking into account satellite-dependant parameters, i.e.,

incidence angle and track angle, as well as topo-graphic parameters, i.e., terrain slope and orientation, The track angle and the incidence angle are provided within the processed SAR images for each satellite (Table 1). The slope and aspect of the area of interest are derived from a DEM with 20 m cell resolution.

The C correction factor represents the fraction of movement that can be registered by the SAR sensor, ranging from a negative value up to 1. It depends on the angle between the steepest slope and the LOS

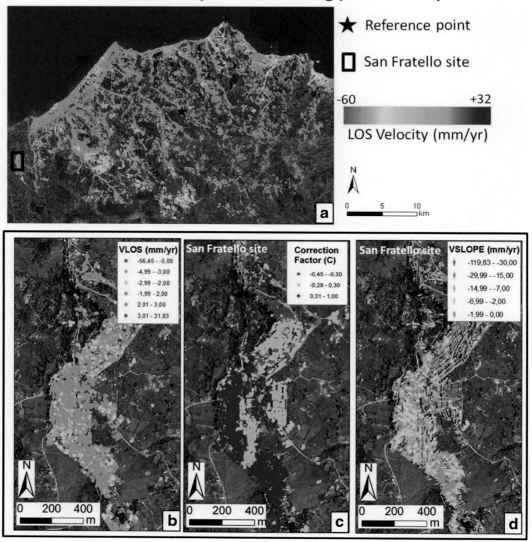

Figure 6
COSMO-SkyMed PSI dataset in descending geometry within the study area: **a** PS LOS data distribution and reference point location; **b** V_{LOS} values on the San Fratello village; **c** correction factor (C) distribution values; **d** V_{SLOPE} values on the San Fratello village

direction, being close to 0 when this angle is almost 90°. The factor C shows negative values when the movement is registered with reverse direction. The V_{SLOPE} values are obtained through the V_{LOS}/C ratio (BIANCHINI *et al.* 2013) (Figs. 4, 5, 6). When the C value is close to 0, then the V_{SLOPE} rate tends to infinity. In order to reduce any exaggeration of the downslope projection when C tends towards 0, we set $C = -0.3$ when $-0.3 < C < 0$ and $C = 0.3$ when $0 < C < 0.3$, according to previously tested

procedures (BIANCHINI *et al.* 2013; HERRERA *et al.* 2013). The maximum V_{SLOPE} values obtained after the downslope projection calculations, as well as the PS velocity distribution features are included in Table 1.

It is worthwhile to highlight that PS rates on flat areas (slope gradient lower than 5°) were not projected downslope and that positive V_{SLOPE} values, which would represent uphill movement, were discarded, following the approach of BIANCHINI *et al.*

(2013) and HERRERA et al. (2013). This is because landslide occurrence on almost flat areas is very rare and positive V_{SLOPE} values indicate that a landslide is going up the slope. Although positive movements may be present at the toe of landslides where vertical displacements can occur, the horizontal vector of the movement should remain oriented downhill (HERRERA et al. 2013).

V_{LOS} and V_{SLOPE} measurements of each satellite employed within the analysis on San Fratello village, as well as the C factor values, are shown in Figs. 4, 5 and 6, As the coefficient C represents the percentage of real motion measured by the satellite, the projectability map showing C distribution reveals the amount of velocity along the local slope seen for each PS and can be considered as a quantitative evaluation of the projection procedure of V_{LOS} along the local slope.

The pre-existing available inventory map of the study area is the Piano Assetto Idrogeologico (PAI, Hydrogeological Setting Plan), which is dated up to 2012 and includes landslide phenomena classified according to the type and the state of activity. The most representative typologies are slow-moving complex, translational slides, and earth slips. Landslides are classified as active, dormant, inactive (including relict and abandoned phenomena) and stabilized, according to a simplified version of CRUDEN and VARNES (1996) classification.

In situ observations and crack pattern survey were performed in two different time periods: just after the 2010 landslide, and during the period November 2012–January 2013.

Field surveys focused on both ground surface cracks and building cracks. Fractures on buildings were qualitatively classified with respect to orientation (vertical, oriented or horizontal) and typology, according to ALEXANDER (1989) (Fig. 7a, b). Although building cracks differently develop according to building material and foundation typology, from the features and distribution of the crack system observed on the facades, the main cause and movement direction can be supposed (HARP 1998). Some examples are shown in Fig. 7c. Many extension cracks with a unique dip direction on a facade reveal a failure due to a differential foundation settlement induced by a translational movement (Fig. 7c1). Cracks showing up as an arc on a facade can be induced by a sinking motion of the foundation (Fig. 7c2). Vertical extensional cracks are usually caused by a translational mass movement and show up orthogonally to the main tensile stress direction (Fig. 7c3) (DI ROMOLO 2008).

7. Cross-Validation Between PSI Data and the Field Survey

Radar-interpretation combined with photo-interpretation analysis (FARINA et al. 2006) permitted one to successfully update the pre-existing inventory map of the whole area around San Fratello village, extended up about 25 km² (ADB 2012). In particular, this procedure allowed detecting some new potentially hazardous areas and enlarging the boundaries of most of the already mapped phenomena (BIANCHINI et al. 2014) (Fig. 8).

At a more detailed and local scale, PSI analysis compared with the in situ survey was exploited over the urban fabric of San Fratello village. Historical, recent and current PS V_{LOS} and V_{SLOPE} ground motion rates were analyzed for instability detection over the most significant areas of San Fratello, primarily considering the distribution of the sites of cultural interest of the village and the boundaries of the two most recent catastrophic landslides occurring in 1922 and 2010.

The soil crack pattern mapping within San Fratello built-up area was performed just after the 2010 landslide. Moreover, further and more recent site-specific field checks and building cracks surveys were carried out in November 2012 and January 2013, in order to validate PSI-based impact assessment performed at a desk, prior to in situ investigations.

At the northern entrance of San Fratello, where the Church of St. Maria delle Grazie is located on a raised position close to the road running along the slope crest (Fig. 9), the spatial distribution of V_{LOS} and V_{SLOPE} displacements in the historical time period (ERS descending data in 1992–2000) seems to show a quite relative stability over the buildings and infrastructure (i.e., yearly motion rate not exceeding ±2.0 mm/year), with no PS identified over the east-facing slope. Nevertheless, PS RADARSAT-1 data acquired in the recent period (2005–2010) reveal a significant ground instability of the area, with velocity rates up to −15 mm/year along the slope. Field checks also allowed the recognition of indicators of landslide movements and soil creep over the upper part of the slope, such as the loss of verticality of the lights poles and vineyards (Fig. 9). Furthermore, a severe crack pattern affects the concrete structures and retaining walls located along the scarp. The presence of extension fractures, with wide aperture up to 2 cm (Fig. 9), confirms the slow-moving slide displacement downslope, permitting one to locally update the pre-existing landslide inventory accordingly (Fig. 9).

High motion rates were detected in historical, recent and current time periods, in and close to the two main landslides occurring on opposite slopes in

Figure 8 ▶

Updating of the landslide inventory map within the whole study area around San Fratello village: **a** Pre-existing landslide inventory map from PAI (Piano di Assetto Idrogeologico–Hydrogeological Setting Plan) referred to 2012 (ADB 2012). Phenomena are classified according to typology and state of activity; **b** Improved landslide inventory map updated to 2012 by means of photo-interpretation and radar-interpretation. Phenomena are classified according to typology and comparison with a pre-existing inventory map

1922 and 2010, affecting several urban sectors of San Fratello.

In particular, in the north-eastern portion of the village, both historical and recent PS data allow detecting the persistence of a suspicious W-directed displacement pattern over an enclosed sector within the area of the 1922 landslide, north-westward of the old districts of the village (Fig. 10). The yearly

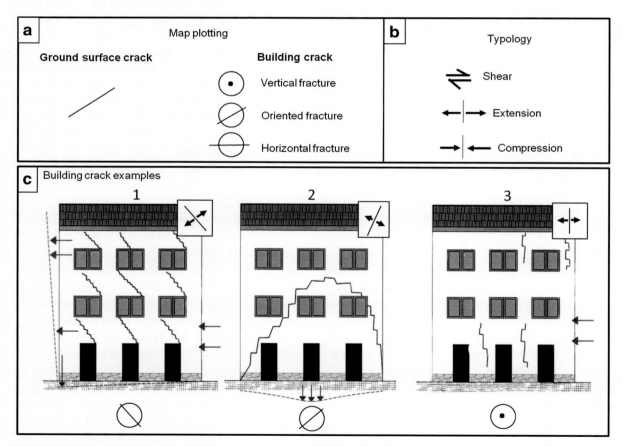

Figure 7

Symbology for ground surface and buildings cracks: **a** symbology for plotting cracks on the map; **b** typology of cracks (from ALEXANDER 1989); **c** building crack examples (modified from: http://www.controllofessure-mg.it/): *1* oriented extensional cracks showing a unique orientation and dip direction; *2* extensional cracks distributed as "an arc" on the building façade; *3* vertical tension cracks

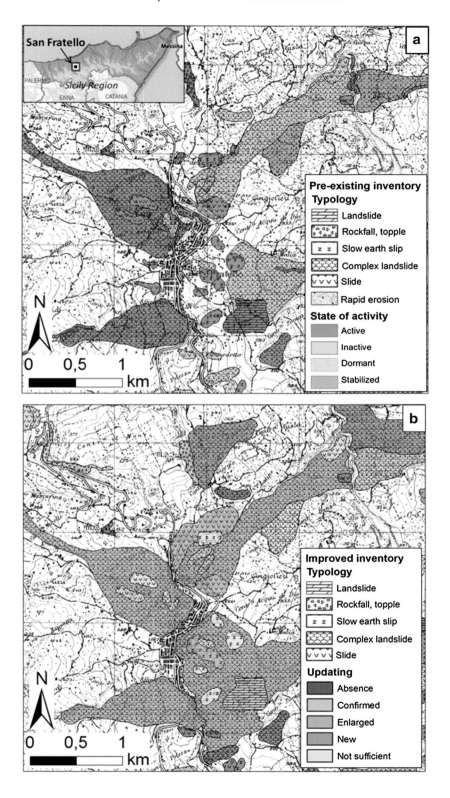

Figure 9
Northern portion of San Fratello village: *1* PSI ERS (V_{LOS} and V_{SLOPE}) displacement map in the historical time interval, 1992–2001 *2* PSI RADARSAT-1 (V_{LOS} and V_{SLOPE}) displacement map in the recent time interval, 2005–2010. Photos are explained within the text

motion rates reach values of −6 mm/year measured along the LOS (V_{LOS}) and −11 mm/year along the local slope (V_{SLOPE}) in the historical period (ERS data), and values ranging from −5 to −14 mm/year and from −8 to −19 mm/year, respectively, in the LOS and local slope directions, during the recent acquisition time (RADARSAT-1 data). These rates indicate that the area has been continuously unstable in the last 20 years, and allow confirming the boundary and state of activity of this landslide-affected sector of the slope. Persistent and comparable high ground motion rates from 1992 up to 2012 are also observable along the mapped 1922 landslide boundary, and needed to be taken into account, due to

the near presence of some cultural sites of interest (the Roccaforte and the St. Filadelfio Castle ruins) (Fig. 10).

Close to the eastern boundary of the 1922 landslide, unexpected high displacement rates were detected (Fig. 11), especially by COSMO-SkyMed data, around the Old Cathedral of St. Nicolò, which was severely damaged by the 1922 landslide. Ground surface cracks and building fractures near the Old Cathedral of St. Nicolò were surveyed in November 2013. Radar-detected movements retrieved by COSMO-SkyMed show good correlation with the distribution and the opening of the cracks along the pavement, which are located orthogonally to the

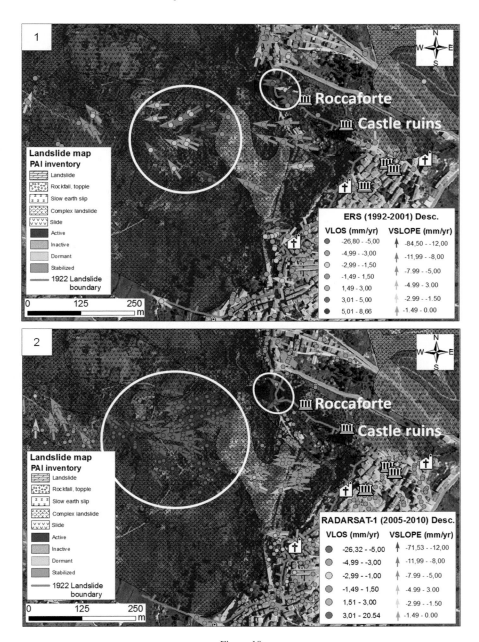

Figure 10
Area within the 1922 landslide on the W-facing slope in San Fratello village: *1* PSI ERS displacement map (V_{LOS} and V_{SLOPE}) in the historical time interval, 1992–2001; *2* PSI RADARSAT-1 (V_{LOS} and V_{SLOPE}) displacement map in the recent time interval, 2005–2010

ground motion direction, and along wall surfaces of the old districts. The vertical extension cracks observed on the walls reveal the subhorizontal vector of displacement. Overall, the instability affects the Old Cathedral of St. Nicolò and most of the surrounding civil buildings along the scarp, as well as the apparently stable Church of St. Crocifisso, with

V_{SLOPE} displacement up to −30 mm/year estimated in the most recent time interval, 2011–2012.

It is worth noting that for the Church of St. Crocifisso, the field checks completed the PS-based mapping of ground deformation. Although the radar processing did not provide a sufficiently dense set of PS over the monument neither in X-band, the in situ

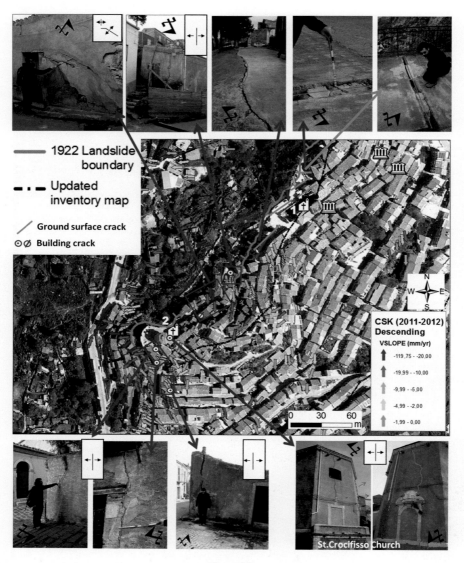

Figure 11
Displacement map of the area in and close to the crown of the 1922 landslide, representing the oldest reaming part of San Fratello village: V_{SLOPE} COSMO-SkyMed PSI data acquired in 2011–2012 overlapped on a recent (2011) orthophoto. *Above* and *below* the map: photos showing some of the pavement and building cracks surveyed in November 2012–January 2013

inspections permitted one to discover and survey not only surface cracks, but also a sort of bulging of the external walls (Fig. 11), as a further indicator of the structural instability currently affecting the church, due to the general instability of the terrace on which it is built.

As a result, radar data validated with in situ data led us to assign a high level of criticality to the entire sector of the old districts along the scarp, and to update the boundary of the damaged area of the 1922

phenomenon, including further buildings and monuments within the crown of the ancient landslide. The proposed newly mapped boundary is shown in Fig. 11. Moreover, this area will need careful monitoring, since it is the oldest remaining part of San Fratello village.

Regarding the opposite slope and the area in and close to the 2010 landslide, PS data and field validation survey allowed the landslide boundary to be updated, thereby including further urban districts

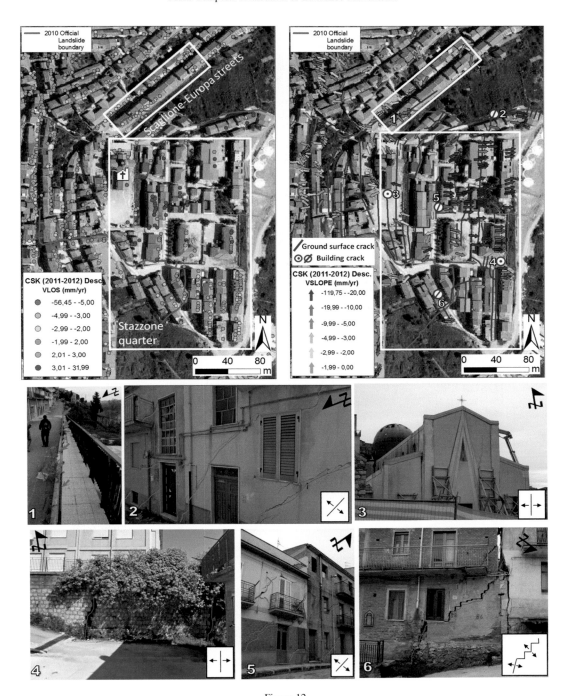

Figure 12

Displacement map of the area in and close to 2010 landslide, focusing on Scaglione-Europa streets area and Stazzone quarter: **a** V_{LOS} COSMO-SkyMed PSI data; **b** V_{SLOPE} COSMO-SkyMed PSI data and cracks plotting: pavement cracks mapped in March 2010 and location of the main building cracks surveyed in November 2012–January 2013. Below the map: *1* pavement crack in Scaglione street (November 2012); *2* oriented extensional cracks on buildings in Fontana Nuova street (November 2012); *3* St. Nicolò New Cathedral (photo taken in November 2012), now demolished; *4* severely damaged wall in Generale Artale street (March 2010); *5* oriented extensional cracks on buildings in Pirandello street; *6* extensional cracks in Stazzoni street (November 2012)

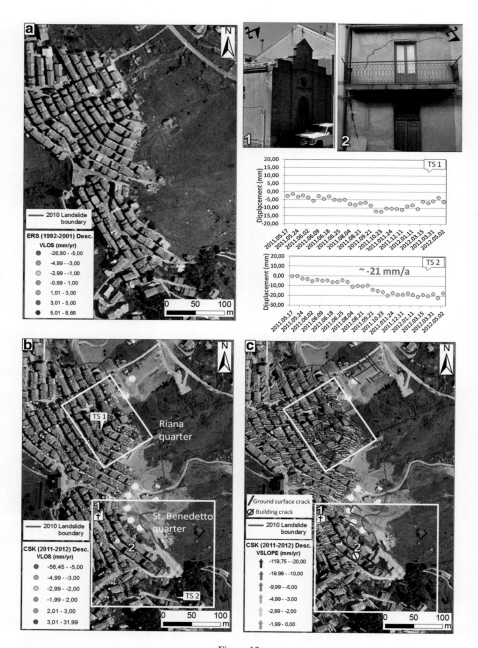

Figure 13
Displacement map of the area in and close to 2010 landslide, focusing on Riana and St. Benedetto districts: **a** V_{LOS} ERS PSI data overlapped on a historical (2000) orthophoto (Volo Italia 2000) and time series of two selected PS targets; **b** V_{LOS} COSMO-SkyMed PSI data overlapped on a recent (2011) orthophoto; **c** V_{LOS} COSMO-SkyMed PSI data and cracks plotting, overlapped on a recent (2011) orthophoto. Photos: *1* St. Benedetto Il Moro Church; *2* Example of damages on a building in Roma street

previously not considered as being critical (Figs. 12, 13). COSMO-SkyMed (2011–2012) PS data in descending geometry show up on housetops and facades of buildings, enabling a highly detailed investigation and precisely detecting the buildings

affected by ground deformation. In particular, four main areas located close to the official zoning of the 2010 landslide were definitely recognized as affected by ground motions instability in the last years of monitoring (Figs. 12, 13). A strong spatial correlation between the identified COSMO-SkyMed PS targets and the surveyed cracks and damages was found.

Northward of the 2010 landslide boundary, some PS show high motion rates, confirmed by the damages observed in situ along Scaglione and Europa streets. In particular, the presence of tension cracks along the pavement may indicate the occurrence of a surface ground deformation (D.R.P.C. 2010) (Fig. 12).

Another unstable area is the Stazzone district, located on the upper part of the 2010 landslide. The N–S-oriented cracks on the ground, mapped after the 2010 landslide, derive from tensile stresses within the translational landslide characterized by a main planar motion component (D.R.P.C. 2010). Severe crack patterns were found on the buildings of the whole quarter during the recent field survey, revealing that soil moved away from foundations. The New Cathedral of St. Nicolò was demolished in February 2013 due to non-repairable damages (Fig. 12).

PS data and field checks in the Riana district, located just out of the mapped landslide boundary, also reveal high motion rates up to −12 mm/year along the satellite LOS and up to −15 mm/year along the steepest slope, during the acquisition period 2011–2012 (Fig. 13).

In the southern St. Benedetto district, ground deformations are the highest (up to −21 mm/year in 2011–2012) and visible only through the current data measured by the satellite COSMO-SkyMed, while, in the previous periods, ERS and RADARSAT data do not reveal any movements (Fig. 13). Therefore, buildings in this part of the village started being affected by instability since 2011, as confirmed by the inhabitants and from the in situ survey carried out in November 2012.

In conclusion, thanks to the cross-comparison of local failures and displacement features of single edifices observed in situ with radar data, an updating of the pre-existing landslide inventory map of the whole study area was performed. In particular, within the San Fratello urban area, radar data validated by localized field checks allowed a new zoning of the western boundary of the 2010 landslide with respect to the official mapping released on 22 February 2010, enlarging it to include urban areas previously not considered as being unstable.

If the above result was reasonably expected in light of the state of activity and actual extent of the recent 2010 landslide, the satellite data evidenced critical conditions retrieved also for the historical districts of San Fratello along the scarp of the old 1922 landslide, proposing a new mapping also for the eastern boundary of this mapped phenomenon.

8. Discussion

The multi-temporal comparison of PS motion velocities of all the available datasets (i.e., ERS-1/2 1992–2001; RADARSAT-1 2005–2010; COSMO-SkyMed 2011–2012) highlighted persistent ground deformation in San Fratello village, especially over the old districts around the crown of the 1922 landslide and along the western boundary of the 2010 landslide. In particular, COSMO-SkyMed data definitely proved the current critical condition of this sector of San Fratello, thereby increasing the alert level for an area otherwise classified as stable.

Ground motion within the area is assumed to be characterized by a steady-state nature, given the linear deformation model assumption made in the PSI processing approach (CROSETTO et al. 2010; HOOPER 2006). A linear regression model is fitted to the data within the processing technique, and, thus, PSI time series show a linear deformation trend in time (Fig. 13). As a result, the non-linear nature of the deformation cannot be analyzed by the PSInSAR™ and SqueeSAR™ (HOOPER 2006). However, in this case, the non-steady and episodic motion within the landslide is assumed to be actually small in magnitude, just determined by intense rainfall that cause acceleration and groundwater level change, because plastic deformations were recognized along the whole slope, being independent from the recently occurred catastrophic movements (D.R.P.C. 2010).

The affordability of X-band data rely on their potential for accurate deformation measurements that benefits from high spatial and temporal resolution. Overall, the use of X-band SAR sensors, i.e., COSMO-SkyMed, with high spatial resolution up to 1 meter (ROTH et al. 2003; GERNHARDT et al. 2010) and short revisiting time improves PSI capability for ground motions detection (BIANCHINI *et al.* 2013; HERRERA *et al.* 2010; NOTTI *et al.* 2010).

X-band radar data turn out to be particularly suited for local detection of landslide processes at small scales, especially in urbanized areas, since the great density of PS point targets in X-band permits to better understand and accurately describe deformation phenomena, entering at the level of the site-specific ground motions investigation. In the San Fratello test site, COSMO-SkyMed data show a PS density 40 times higher than the one of the medium resolution satellite (i.e., ERS) (Table 1). These advantages improve the level of detail of the analysis and allow studying highly localized surface displacements and their dynamic evolution patterns. Although temporal decorrelation is more problematic at X-band compared to longer wavelengths like C-band, the high-bandwidth data acquired by COSMO-SkyMed sensor permit more PSI targets to be identified, and so higher deformation gradients can be detected compared to C-band satellites (GE *et al.* 2010). This is due to the shorter monitoring period of only 12 months (05/16/2011–02/05/2012) as well as to the shorter temporal sampling (up to 4 days), allowing the retrieving of more coherent pixels that show displacements. Some of the PS targets identified by X-band sensors would not be detected over a longer monitoring period or with a worst temporal resolution because the coherence would be lost.

The use of the new SqueeSAR^{TM} algorithm (FERRETTI *et al.* 2011), exploiting both 'point-wise' PS and 'spatially distributed scatterers' (DS), increases the PS targets retrieval also in not-densely urbanized areas and produces improvements in the quality of the displacement time series (MEISINA *et al.* 2013; RASPINI *et al.* 2013). PSI technique is not a stand-alone technique, and it must be considered as a complementary support to the analyses of the areas affected by slope instability. Ground movement evidence obtained from radar data need always to be validated and compared as much as possible with other kinds of techniques and auxiliary information, in order to achieve a reliable investigation (FARINA *et al.* 2006; RIGHINI *et al.* 2012). Thus, the combined use of radar data with traditional geomorphological tools like photo-interpretation, field surveys and in situ campaigns give useful effort for the mapping

and monitoring of the impacts of landslide phenomena on buildings and manufactures of the investigated sites (HERRERA *et al.* 2010; PARCHARIDIS *et al.* 2010; CIGNA *et al.* 2011; TAPETE *et al.* 2012; FRATTINI *et al.* 2013).

In this work, PSI radar-interpreted data were successfully cross-compared with a field survey, which includes observations on landslide-induced damages and crack pattern survey of urban structures. A good agreement between the satellite ground motion evidences and the ground truth was found in the San Fratello village.

In order to compare more specifically InSAR results with local failures and building damages, PSI data measured along the satellite LOS were projected downslope along the maximum steepest slope, following the approach presented and applied in previous works by the scientific community (COLESANTI and WASOWSKI 2006; CASCINI *et al.* 2009; CIGNA *et al.* 2012, 2013; GRIEF and VLCKO, 2012; BIANCHINI *et al.* 2013; HERRERA *et al.* 2013).

The so-called V_{SLOPE} values obtained through the projection of LOS displacement values (V_{LOS}) allow a more intelligible interpretation of radar data with respect to the local morphology (CIGNA *et al.* 2013; BIANCHINI *et al.* 2013). Several limitations need to be accounted for when projecting the velocity along the slope. The V_{LOS} projection is only valid when the landslide movement is parallel to the slope, thus, this requirement typically holds for planar slides with slow flow but not for rotational landslides, which generally possess a vertical movement at the crown and horizontal movement at the toe. The landslide movements in San Fratello are mainly ascribed as translational slides associated with localized flows (D.R.P.C. 2010; ADB 2012), so they are compatible with the assumed limitations of the projection procedure. Moreover, V_{LOS} values may vary due to vertical displacements, i.e., soil consolidation, which should not be projected, and, if the V_{LOS} data are noisy, the projection will amplify any errors. In the San Fratello test site, in order to reduce these problems within the PSI analysis, the V_{SLOPE} positive values were discarded as indicating uphill motion and only PS velocities over 5° slope were projected since movement on flat areas (slope <5°) would be related to other causes. As shown in Figs. 4, 5 and 6, for the same area of interest and the same satellite, the value of the correction factor C can vary strongly with the irregularity of the slope. The real direction of motion is most likely more uniform. For this reason, when calculating the C factor for performing the downslope projection, it is better to use a DEM with a lower resolution or resample a detailed DEM to smooth out these variations. In the San Fratello test site, a DEM with 20 m resolution was used to perform the analysis.

The outcomes of this work are a valuable proof of the capability of PSI-based approach to selectively detect areas actually unstable especially if combined with in situ data, useful for cultural heritage applications. As a result, the implementation of PSI data in this work expands the current applications of PSI to landslides monitoring, by suggesting review of the pre-existing landslide inventory and consequently by demonstrating it to be potentially for supporting strategies of land planning and activities of built heritage management.

9. Conclusions

In this paper a multi-temporal landslide effects investigation was performed in San Fratello (Italy) chronically affected by landslide phenomena, in order to evaluate the ground motion impacts on the village, especially on the cultural heritage of the site. The analysis combines InSAR data, which maps the active slide movements over 20 years, with a reconnaissance of damage observed in built structures.

The most recent landslide phenomena in the San Fratello site occurred in 1922 and in February 2010, on the two opposite slopes on which the village is built.

The combined use of historical C-band from 1992 and new generation X-band SAR data up to 2012 permitted one to spatially and temporally detect and monitor ground deformations at a local scale, achieving a comprehensive detailed multi-temporal and spatial investigation of the village.

The InSAR results come from a dual persistent and distributed scatterer approach by means of SqueeSAR^{TM} processing technique.

Persistent Scatterer Interferometry data were analyzed and combined with other available

information on the test site, such as orthophotos and field survey observations carried out in March 2010 just after the 2010 phenomenon, and more recently in November 2012 and January 2013.

The obtained results allow improving the landslide inventory map of the area, properly providing an updated assessment of the instability on the study area. In particular, the outcomes of the work lead to proposed new boundaries of the landslide-affected areas in San Fratello village, not only for the 2010 landslide, where PSI data revealed four main unstable areas otherwise classified as stable, but also for the 1922 landslide, where a suspicious ground motion pattern was recognized by means of historical and recent PSI datasets up to 2010.

The cross-comparison of PSI data with local failures and damage of single edifices observed in situ allowed a validation of radar data, to finally achieve a complete analysis that can be particularly useful for strategies of cultural heritage and building management.

Acknowledgments

This work has been carried out in the framework of the DORIS project funded by the EC-GMES-FP7 initiative (Grant Agreement No. 242212). Persistent Scatterer Interferometry data were processed by Tele-Rilevamento Europa and were available within the DORIS project. Federico Di Traglia is supported by a post-doc fellowship founded by the Regione Toscana (UNIPI-FSE) under the project RADSAFE (UNIPI-4) in the framework of the research agreement between DST-UNIPI, DST-UNIFI and Ellegi s.r.l.—LiSALab. The authors would like to thank the Italian Civil Protection for field survey data collected after the 2010 landslide. Information on the cultural and historical heritage of San Fratello mostly derive from the websites of San Fratello village: http://sottolapietra.blogspot.it/ and http://www.san-fratello.com/.

REFERENCES

ADAM, N., EINEDER, M., YAGUE-MARTINEZ, N., BAMLER, R., High resolution interferometric stacking with TerraSAR-X, In *Proc. of the Geoscience and Remote Sensing Symposium, IGARSS* (Boston, MA, Jul. 7–11, 2008) pp. II-117–II-120.

AdB Regione Sicilia (2012), *PAI – Piano Stralcio di Bacino per l'Assetto Idrogeologico*, http://www.sitr.regione.sicilia.it/pai.

ALEXANDER, D. (1989), *Urban landslides. An International review of geographical work in the natural and environmental sciences*, Progress in Phys Geography *13*, 157–191.

AMODIO-MORELLI, L., BONARDI, G., COLONNA, V., DIETRICH, D., GIUNTA, G., IPPOLITO, F., LIGUORI, V., LORENZONI, F., PAGLIONICO, A., PERRONE, V., PICCARRETA, G., RUSSO, M., SCANDONE, P., ZANETTIN-LORENZONI, E, ZUPPETTA, A. (1976), *L'Arco Calabro-Peloritano nell'orogene appenninico-maghrebide*, Mem Soc Geol It *17*, 1–60.

ARDIZZONE, F., BASILE, G., CARDINALI, M., CASAGLI, N., DEL CONTE, S., DEL VENTISETTE, C., FIORUCCI, F., GARFAGNOLI, F., GIGLI, G., GUZZETTI, F., IOVINE, G., MONDINI, A. C., MORETTI, S., PANEBIANCO, M., RASPINI, F., REICHENBACH, P., ROSSI, M., TANTERI, L., TERRANOVA, O. (2012), *Landslide inventory map for the Briga and the Giampilieri catchments, NE Sicily, Italy*, J Maps, doi:10.1080/17445647.2012.694271.

BELLOTTI, F., BIANCHI, M., COLOMBO, D., FERRETTI, A., TAMBURINI, A. (2014), Advanced InSAR Techniques to Support Landslide Monitoring. In *Mathematics of Planet Earth, Lecture Notes in Earth System Sciences 2014*, (ed. Springer Berlin Heidelberg), 287–290. doi:10.1007/978-3-642-32408-6_64.

BIANCHINI, S, CIGNA, F, RIGHINI, G, PROIETTI, C, CASAGLI, N (2012), *Landslide HotSpot Mapping by means of Persistent Scatterer Interferometry*, Environ Earth Sci *67*(4), 1155–1172.

BIANCHINI, S., HERRERA, G., NOTTI, D., MATEOS, R.M., GARCIA, I., MORA, O., MORETTI, S. (2013), *Landslide activity maps generation by means of Persistent Scatterer Interferometry*, Remote Sens *5*(12), 6198–6222.

BIANCHINI, S., TAPETE, D., CIAMPALINI, A., DI TRAGLIA, F., DEL VENTISETTE, C., MORETTI, S., CASAGLI, N. (2014), Multi-Temporal Evaluation of Landslide-Induced Movements and Damage Assessment in San Fratello (Italy) by Means of C- and X-Band PSI Data. In *Mathematics of Planet Earth, Lecture Notes in Earth System Sciences*, (ed. Springer Berlin Heidelberg), 257–261.

BOVENGA, F., WASOWSKI, J., NITTI, D.O., NUTRICATO, R., CHIARADIA, M.T. (2012), *Using COSMO-SkyMed X-band and ENVISAT C-band SAR interferometry for landslides analysis*, Remote Sens Environ, *119*, 272–285.

BRU, G., HERRERA, G., TOMÁS, R., DURO, J., DE LA VEGA, R., MULAS, J. (2013), *Control of deformation of buildings affected by subsidence using persistent scatterer interferometry*, Struct Infrastruct Eng *9*, 188–200.

CANUTI, P., CASAGLI, N., ERMINI, L., FANTI, R., FARINA, P. (2004), *Landslide activity as a geoindicator in Italy: significance and new perspectives from remote sensing*, Environ Geol *45*(7), 907–919.

CASCINI, L., FORNARO, G., PEDUTO, D. (2009), *Analysis at medium scale of low-resolution DInSAR data in slow-moving landslide affected areas*, J Photogramm Remote Sens *64*(6), 598–611.

CASCINI, L., FORNARO, G., PEDUTO, D. (2010), *Advanced low and full resolution DInSAR map generation for slow moving landslide analysis at different scales*, Eng Geol *112*, 29–42.

CIAMPALINI, A., CIGNA, F., DEL VENTISETTE, C., MORETTI, S., LIGUORI, V., CASAGLI, N. (2012), *Integrated geomorphological mapping in the north-western sector of Agrigento (Italy)*, J Maps *8*(2), 136–145.

CIGNA, F., DEL VENTISETTE, C., LIGUORI, V., CASAGLI, N. (2011), *Advanced radar-interpretation of InSAR time series for mapping and characterization of geological processes*, Nat Haz Earth Sys Sci *11*, 865–881.

CIGNA, F., DEL VENTISETTE, C., GIGLI, G., MENNA, F., AGILI, F., LIGUORI, V., CASAGLI, N. (2012), *Ground instability in the old town of Agrigento (Italy) depicted by on-site investigations and Persistent Scatterers data*, Nat Haz Earth Sys Sci *12*, 3589–3603.

CIGNA, F., BIANCHINI, S., CASAGLI, N. (2013), *How to assess landslide activity and intensity with Persistent Scatterer Interferometry (PSI): the PSI-based matrix approach*, Landslides *10*(3), 267–283.

COLESANTI, C. and WASOWSKI, J. (2006), *Investigating landslides with space-borne Synthetic Aperture Radar (SAR) Interferometry*, Eng Geol *88*, 173–199.

CORRADO, S., ALDEGA, L., BALESTRIERI, M. L., MANISCALCO, R., GRASSO, M. (2009), *Structural evolution of the sedimentary accretionary wedge of the alpine system in Eastern Sicily: Thermal and thermochronological constraints*, Geol Soc Am Bull *121*(11–12), 1475–1490.

CROSETTO, M., MONSERRAT, O., IGLESIAS, R., & CRIPPA, B. (2010), *Persistent Scatterer Interferometry: potential, limits and initial C- and X-band comparison*, Photogramm Eng Remote Sens *76*(9), 1061–1069.

CRUDEN, D.M., VARNES, D.J. (1996), Landslide types and processes. In, *Landslides: Investigation an Mitigation: Sp. Rep.* 247 eds. Turner, A.K. & Schuster, R.L.), Transportation Research Board, National research Council, National Academy Press, Washington, DC, 36–75.

CUBITO, A., FERRARA, V., PAPPALARDO, G. (2005), *Landslide hazard in the Nebrodi Mountains (Northeastern Sicily)*, Geomorphology *66*, (1–4), 359–372.

CUEVAS, M., CROSETTO, M., MONSERRAT, O., Monitoring urban deformation phenomena using satellite images, In *9th International Geomatic Week* (Barcelona, Spain, 15–17 March 2011).

DEL VENTISETTE, C., GARFAGNOLI, F., CIAMPALINI, A., BATTISTINI, A., GIGLI, G., MORETTI, S., CASAGLI, N. (2012), *An integrated approach to the study of catastrophic debris-flows: geological hazard and human influence*, Nat Haz Earth Sys Sci *12*, 2907–2922.

DIETRICH, D. (1988), *Sense of overthrust shear in the Alpine nappes of Calabria (Southern Italy)*, J Struct Geol *10*(4), 373–381.

DI ROMOLO, F., *Lesioni degli edifici. Applicazioni di geotecnica e geofisica nell'analisi dei cedimenti delle fondazioni* (Hoepli ed., Italy 2008).

D.R.P.C. - Dipartimento Regionale Protezione Civile - (2010), *La frana di san Fratello (ME) del 14 febbraio 2010 - Relazione geologica e rapporto di sintesi sulle indagini geognostiche* (Palermo, 2010).

FARANDA, P., *Città - giardino: il piano di Acquedolci. Storia e urbanistica di una città fondata in era fascista (1922–1932)* (Qanat, Palermo, Italy 2010).

FARINA, P., COLOMBO, D., FUMAGALLI, A., MARKS, F., & MORETTI, S. (2006), *Permanent Scatterers for landslide investigations: outcomes from the ESA-SLAM project*, Eng. Geol. *88*, 200–217.

FERRETTI, A., PRATI, C., ROCCA, F. (2001), *Permanent Scatterers in SAR Interferometry*, IEEE Trans Geosci Remote Sens *39*, 1, 8–20.

FERRETTI, A., PRATI, C., ROCCA, F., CASAGLI, N., FARINA, P., YOUNG, B., Permanent Scatterers technology: a powerful state of the art tool for historic and future monitoring of landslides and other terrain instability phenomena, In *Proc. of 2005 International Conference on Landslide Risk Management,* (Vancouver, Canada, 2005).

FERRETTI, A., FUMAGALLI, A., NOVALI, F., PRATI, C., ROCCA, F., RUCCI, A. (2011), *A new Algorithm for Processing Interferometric Data-Stacks: SqueeSAR,* IEEE Transaction on Geo Sci Remote Sens 49(9), 3460–3470.

FRATTINI, P, CROSTA, G.B., ALLIEVI, J. (2013), *Damage to Buildings in Large Slope Rock Instabilities Monitored with the PSInSAR^{TM} Technique,* Remote Sens, Special issue: Remote Sensing for Landslides Investigation: From Research into Practice 5(10), 4753–4773.

GE D., WANG Y., ZHANG L., GUO X., and XIA Y., Mapping urban subsidence with TerraSAR-X data by PSI analysis, In *Proc. of Geoscience and Remote Sensing Symposium (IGARSS), 2010 IEEE International,* (Honolulu, Hawaii, USA, 2010) pp. 3323–3326.

GERNHARDT, S., ADAM, N., EINEDER, M., BAMLER, R. (2010), *Potential of very high resolution SAR for Persistent Scatterer Interferometry in urban areas,* Ann. GIS 16(2), pp. 103–111.

GRIEF, V., VLCKO, J. (2012), *Monitoring of post-failure landslide deformation by the PS-InSAR technique at Lubietova in Central Slovakia.* Environ Earth Sci 66(6), 1585–1595.

HARP E.L. (1998), Origin of fractures triggered by the earthquake in the Summit Ridge and Skyland Ridge areas and their relation to landslides, In *The Loma Prieta, California, Earthquake of October 17, 1989–Landslides,* (ed. Keefer D.K), U.S. Geological survey professional paper, pp. 1551-C.

HANSSEN, R.F. (2005), *Satellite radar interferometry for deformation monitoring: a priori assessment of feasibility and accuracy,* Int J Appl Earth Obs 6, 253–260.

HERRERA, G., TOMÁS, R., MONELLS, D., CENTOLANZA G., MALLORQUI J. J., VICENTE, F., NAVARRO, V. D., LOPEZ-SANCHEZ, J. M., CANO, M., MULAS, J., SANABRIA, M. (2010), *Analysis of subsidence using TerraSAR-X data: Murcia case study,* Eng. Geol 116, 284–295.

HERRERA, G., GUTIÉRREZ, F., GARCÍ-DAVALILLO, J.C., GUERRERO, J., GALVE, J.P., FERNÁNDEZ-MORODO, J.A., and COOKSLEY, G. (2013), *Multi-sensor advanced DInSAR monitoring of very slow landslides: the Tena valley case study (central Spanish Pyrenees),* Remote Sens Environ 128, 31–43.

HOOPER, A. (2006), *Persistent scatterer radar interferometry for crustal deformation studies and modeling of volcanic deformation,* Ph.D. thesis, Stanford University.

KNOTT, S.D. (1987), *The Liguride Complex of Southern Italy-a Cretaceous to Paleogene accretionary wedge,* Tectonophysics 142, 217–226.

LAVECCHIA, G., FERRARINI, F., DE NARDIS, R., VISINI, F., BARBANO, M.S. (2007), *Active thrusting as possible seismogenic source in Sicily (Southern Italy): Some insights from integrated structural-kinematic and seismological data,* Tectonophysics 445, 145–167.

MEISINA, C., NOTTI, D., ZUCCA, F., CERIANI, M., COLOMBO, A., POGGI, F., ROCCATI, A., ZACCONE, A., The use of PSInSAR^{TM} and SqueeSAR^{TM} techniques for updating landslide inventories, In Landslide Science and Practice, In *Proc. of The Second World Landslide Forum, Volume 1: Landslide Inventory and Susceptibility and Hazard Zoning* (eds. Margottini, C., Canuti, C., Sassa, K.) (Roma, Italy, 2013) pp. 81–88.

MONDINI, A.C., GUZZETTI, F., REICHENBACH, P., ROSSI, M., CARDINALI, M., ARDIZZONE, F. (2011), *Semi-automatic recognition and mapping of rainfall induced shallow landslides using optical satellite images,* Remote Sens Environ 115(7), 1743–1757.

NIGRO, F. AND SULLI, A. (1995), *Plio-Pleistocene extensional tectonics in the Western Peloritani area and its offshore (northeastern Sicily),* Tectonophysics 252, 295–305.

NOTTI, D., DAVALILLO, J.C., HERRERA, G., MORA, O., (2010), *Assessment of the performance of X-band satellite radar data for landslide mapping and monitoring: Upper Tena Valley case study,* Nat Haz Earth Sys Sci 10, 1865–1875.

OGNIBEN, L., (1969), *Nota illustrativa dello schema geologico della Sicilia nord-orientale,* Riv Min Sicil 11, n. 64–65, 183–212.

PARCHARIDIS, I., FOUMELIS, M., PAVLOPOULOS, K., KOURKOULI, P.,Ground deformation monitoring in cultural heritage areas by time series SAR interferometry: the case of ancient Olympia site (Western Greece), In *European Space Agency publications, Fringe Conference, ESA* (Noordwijk, Holland, 2010).

RASPINI, F., MORETTI., S, CASAGLI, N., Landslide Mapping Using SqueeSAR Data: Giampilieri (Italy) Case Study, In *Landslide Science and Practice, Proc. of The Second World Landslide Forum, Volume 1: Landslide Inventory and Susceptibility and Hazard Zoning* (eds. Margottini, C., Canuti, C., Sassa, K.) (Roma, Italy 2013) pp. 147–154.

RIGHINI, G., PANCIOLI, V., CASAGLI, N., (2012), *Updating landslide inventory maps using Persistent Scatterer Interferometry(PSI),* Int J Remote Sens 33(7), 2068–2096.

ROTH, A.,TerraSAR-X: A new perspective for scientific use of high resolution spaceborne SAR data, In Proc. Of the 2nd GRSS/ISPRS Joint workshop on remote sensing and data fusion on urban areas, (Berlin, Germany, 2003), pp. 4–7.

TAPETE, D., FANTI, R., CECCHI, R., PETRANGELI, P., CASAGLI, N. (2012), *Satellite radar interferometry for monitoring and early-stage warning of structural instability in archaeological sites,* J Geophys Eng 9, S10–S25.

TOMÁS, R., HERRERA, G., LOPEZ-SANCHEZ, J.M., VICENTE, F., CUENCA, A., MALLORQUÍ J.J. (2010), *Study of the land subsidence in the Orihuela city (SE Spain) using PSI data: distribution, evolution and correlation with conditioning and triggering factors,* Eng Geol 115, 105–121.

VARNES, D. J., Slope movements, type and processes, In *Landslide analysis and control. Transportation Research Board, National Academy of Sciences,* (eds Schuster, R. L., Krizek, R. J.) (Washington, D.C., 1978), Special report 176, pp. 11–33.

(Received December 20, 2013, revised February 24, 2014, accepted March 17, 2014, Published online April 2, 2014)

Pure Appl. Geophys. 172 (2015), 3067–3080
© 2014 Springer Basel
DOI 10.1007/s00024-014-1008-3

Landslide Kinematical Analysis through Inverse Numerical Modelling and Differential SAR Interferometry

R. Castaldo,[1] P. Tizzani,[1] P. Lollino,[3] F. Calò,[1] F. Ardizzone,[2] R. Lanari,[1]
F. Guzzetti,[2] and M. Manunta[1]

Abstract—The aim of this paper is to propose a methodology to perform inverse numerical modelling of slow landslides that combines the potentialities of both numerical approaches and well-known remote-sensing satellite techniques. In particular, through an optimization procedure based on a genetic algorithm, we minimize, with respect to a proper penalty function, the difference between the modelled displacement field and differential synthetic aperture radar interferometry (DInSAR) deformation time series. The proposed methodology allows us to automatically search for the physical parameters that characterize the landslide behaviour. To validate the presented approach, we focus our analysis on the slow Ivancich landslide (Assisi, central Italy). The kinematical evolution of the unstable slope is investigated via long-term DInSAR analysis, by exploiting about 20 years of ERS-1/2 and ENVISAT satellite acquisitions. The landslide is driven by the presence of a shear band, whose behaviour is simulated through a two-dimensional time-dependent finite element model, in two different physical scenarios, i.e. Newtonian viscous flow and a deviatoric creep model. Comparison between the model results and DInSAR measurements reveals that the deviatoric creep model is more suitable to describe the kinematical evolution of the landslide. This finding is also confirmed by comparing the model results with the available independent inclinometer measurements. Our analysis emphasizes that integration of different data, within inverse numerical models, allows deep investigation of the kinematical behaviour of slow active landslides and discrimination of the driving forces that govern their deformation processes.

Key words: Numerical modelling, optimization procedure, differential SAR interferometry, slow landslide.

1. Introduction

Management of risk associated with slow-moving active landslides requires analysis of the kinematical evolution of the unstable mass in terms of displacements, velocity or acceleration over time (Alonso 2012; Ledesma *et al.* 2009), more than the assessment of the stability conditions of the landslide mass, which instead refer to static conditions, such as those corresponding to the triggering or reactivation stages. Indeed, study of the kinematical trend of a landslide represents an effective way for predicting, and therefore mitigating, the eventual damage caused by landslide movements to buildings and infrastructures.

In this framework, numerical modelling allows simulation of the stress–strain state of a landslide process during both the triggering and propagation stages, provided that a suitable modelling approach is pursued and the factors controlling the landslide evolution are correctly implemented in the model, along with the geological and geotechnical information available for the examined slope (Troncone 2005; Lollino *et al.* 2011).

However, when the objective of the numerical analysis is time-dependent simulation of the kinematical evolution of complex phenomena, such as the case of mass movements, a critical role is represented by the selection of a proper physical approach, suitable to correctly describe the investigated phenomenon. In this context, several works have focussed on studying and predicting the time-dependent law of variation of the displacement pattern of slow active landslides (Vulliet and Hutter 1988; Ledesma *et al.* 2009; Crosta *et al.* 2012). In particular, soil viscosity or changes in the slope boundary conditions play a remarkable role as leading factors controlling the landslide dynamics, and should be properly accounted for in the simulation of the landslide evolution. For slopes characterized by deformation patterns that are weakly affected by

---
[1] CNR IREA, via Diocleziano 328, 80127 Naples, Italy. E-mail: castaldo.r@irea.cnr.it
[2] CNR IRPI, via della Madonna Alta 126, 06128 Perugia, Italy.
[3] CNR IRPI, via Amendola 122 I, 70126 Bari, Italy.

Reprinted from the journal

changes in the hydraulic boundary conditions, namely characterized by a steady-state pore water pressure regime, adequate viscous-type sliding laws should be assumed to simulate the slope displacement trends of the involved soils (VULLIET and HUTTER 1988, LEROUEIL 2001; PASTOR *et al.* 2002). In this case, several constitutive approaches are available in the literature to reproduce creep displacement patterns in time-dependent numerical models (PASTOR *et al.* 2002), ranging from simulation of soils in terms of Newtonian or non-Newtonian fluids, to the assumption of visco-plastic behaviour of the soil according to a Bingham fluid (PERZYNA 1966; VULLIET and HUTTER 1988; CROSTA *et al.* 2012) or more advanced visco-plastic fluid approaches (CHEN and LING 1996). In particular, following earlier experimental results from BAGNOLD (1954) and more recent work from HUNT *et al.* (2002), the behaviour of a dispersion of granular particles in a viscous fluid at low strain rates, e.g. in the macro-viscous region where viscosity effects are dominant, can be effectively reproduced by means of a Newtonian fluid model. This approach accounts for a linear relationship between the shear stress and strain rate of the material according to a viscosity parameter that depends on the solid particle concentration. Regarding the behaviour of mudflows and flow-slides, PASTOR *et al.* (2002) state that, after the landslide initiation stage, the soil material under shear behaviour can be considered as subjected to fluidification, and as a consequence, the behaviour of fluidised mixtures of soil and water can be described by rheological laws relating total shear stresses to strain rates. The same authors also highlight that such a total stress approach is mostly valid in extreme cases where the permeability of the soil is very low and pore pressures do not change significantly during the sliding process.

Despite such advancements in the field, the search for proper values of the parameters associated with the physical approach adopted to simulate landslide kinematical evolution remains a critical stage due to the lack of methodologies and procedures capable of efficiently calibrating forward numerical models. An efficient way to overcome this limitation is exploitation of inverse numerical optimization algorithms aimed at estimating the correct values of kinematical

parameters. The use of inverse numerical modelling approaches has already been proposed in other geoscience fields, such as seismology and volcanology (TIZZANI *et al.* 2010, 2013), where they are applied to extract a synoptic view of the investigated natural events.

In this work, we propose a new methodology to properly perform inverse numerical modelling of landslides. The developed procedure is based on the integration of geological, geomorphological, geotechnical and geodetic information. In particular, we exploit satellite differential synthetic aperture radar interferometry (DInSAR) techniques that are proving to be a valuable tool for long-term deformation monitoring of slow-moving landslides and can provide useful information for predicting landslide kinematical evolution, assuming that the slope boundary conditions do not change over time (HILLEY *et al.* 2004; FARINA *et al.* 2006; GUZZETTI *et al.* 2009; CASCINI *et al.* 2009, 2010; CALÒ *et al.* 2012). In our methodology, DInSAR deformation time series are effectively used to calibrate the numerical model through an optimization procedure that minimizes, with respect to a proper cost function, the difference between the modelled displacement field and the DInSAR measurements. A preliminary example of the application of inverse numerical modelling methods to landslide scenarios has already been proposed in CALÒ *et al.* (2014), developed by exploiting mean deformation velocity DInSAR maps. In this work, we present a detailed analysis of the proposed methodology, based on the exploitation of long-term deformation time series.

As a representative case study, maintaining continuity with previous work, the Ivancich landslide (Assisi, central Italy) is considered. This landslide affects a slope formed of pelitic sandstone, and is characterized by a sliding surface at depths ranging from 15 to 60 m from the ground surface. The landslide displacement rate is around 1 cm/year at maximum, and the corresponding trends are observed to be quite independent of water table fluctuations or pore water pressure variations at the depth of the sliding surface. It appears that the landslide process seems to be controlled more by viscous factors than by changes in the hydraulic boundary conditions. Accordingly, the soil flowing within the thin shear zone of the Ivancich landslide is simulated through two different physical approaches, i.e. Newtonian fluid dynamics and a deviatoric creep model. We perform two-dimensional finite-element (FE) analysis aimed at calculating the kinematics of the landslide mass in the recent evolution stage, by investigating the behaviour of the soil characterizing the thin shear band. The numerical results are calibrated through DInSAR deformation time series covering nearly 20 years (1992–2010) and then are validated by means of independent inclinometer measurements performed in four boreholes located along the slope.

Our results clearly demonstrate that integration of advanced numerical modelling approaches and satellite remote sensing monitoring techniques results in a tool suitable for study of the behaviour of the Ivancich landslide and, more generally, analysis of the kinematical evolution of slow landslides characterized by similar geological and mechanical features.

2. Numerical Analysis and Optimization Procedure

Ground deformations are the expression of near-surface and/or deep-seated geological processes. In the earth science context, interpretation of deformation measurements can be effectively performed by setting up inverse problems to constrain the nature of the causative factors and the values of representative parameters. Numerical methods based on inverse analysis exploit the problem's solution space by iterating a large set of forward mathematical models, which instead can be affected by uncertainty as regards specific parameters or problems of oversimplification in model construction.

In this perspective, inverse models based on the FE method are a suitable tool to fill the gap between the accuracy achieved in the field of ground deformation observations and the models used for the corresponding interpretation. For this reason, we propose a numerical approach based on inverse FE models that are constrained by field and DInSAR-based monitoring data as an alternative to standard forward FE models (Tizzani et al. 2010, 2013). In particular, we combine the benefits of the numerical approach with a Monte Carlo optimization procedure referred to as a genetic algorithm (GA) (Gill et al.

1981; Bingul et al. 2000), to analyse and interpret ground deformations measured in active landslide areas. The GA optimization approach derives from the theory of biological evolution. By analogy, the algorithm starts with an initial set of models (population) randomly generated by imposing as input data to the procedure parameter values ranging within appropriate intervals. Within this population, the best solution, defined as the set of parameter values providing the best fit between the numerical solution and the landslide behaviour as observed by available monitoring data, is found through minimization of a defined cost function. Numerical operators for mutation and chromosome crossover (recombination) act on the best individuals, resulting in the breeding of a new population of "evolved" individuals; i.e. only models that survived the preceding selection may reproduce and proceed to the next step (generation). According to Manconi et al. (2009), the procedure is thus iterated until reaching a previously assigned maximum number of generations. The optimization procedure performed in the present analysis is based on minimization of the discrepancy between DInSAR data and the results of the FE model in terms of the calculated displacement rate. In our study, the cost function is based on the root-mean-square error (RMSE) between the modelled and observed data.

To investigate the kinematical evolution of a landslide, various physical approaches can be considered according to the characteristics of the analysed phenomenon. In our case, we focussed on two physical approaches, i.e. the Newtonian and deviatoric creep models, which are suitable to simulate the kinematical trend of the Ivancich landslide, selected as a test site for the proposed methodology. In particular, the former approach consists in the application, within a fluid dynamics context (i.e. solving for the velocity using the Navier–Stokes equations), of a Newtonian viscosity model for which the stress distribution and velocity field are governed by the dynamic viscosity. In the latter approach, the temporal evolution of the ground deformation field is assumed to be governed by the soil creep rate distribution, and accordingly, the role of the secondary creep behaviour in a structural mechanical context (i.e. solving for the displacements using the Navier equations) is investigated. As a tool to integrate all the available data and solve the considered equations in both physical approaches, we use the COMSOL Multiphysics finite-element modeling code (http://www.comsol.com/).

2.1. Fluid Dynamics Model

For active landslides characterized by a quasi-linear displacement trend, the kinematical evolution can be described through a numerical model based on the approximation of the material behaviour as a Newtonian fluid characterized by a viscosity constant over time. Accordingly, we can assume a steady-state viscous flow (Newtonian fluid) solved through the incompressible Navier–Stokes differential equations:

$$-\nabla \cdot \eta\left(\nabla \mathbf{u} + (\nabla \mathbf{u})^T\right) + \rho(\mathbf{u} \cdot \nabla)\mathbf{u} + \nabla p = \mathbf{F}$$
$$\nabla \cdot \mathbf{u} = 0, \tag{1}$$

where $\mathbf{u}$ (m/s) is the deformation velocity vector, $\mathbf{F}$ (Pa/m) is the body force term, ρ (kg/m^3) is the density, p (Pa) is the pressure, and η is the dynamic viscosity (Pa s) (hereafter referred to as viscosity).

The viscosity distribution can be evaluated through an advanced procedure implementing non-linear optimization of the FE model with respect to the available displacement measurements. In our approach, we searched for the best-fit viscosity model explaining the observed DInSAR displacement rates. In particular, the model deformation velocities are calculated at the topographic surface, projected along the satellite line of sight (LOS), and compared with the DInSAR measurements. The best-fit viscosity model is finally selected by considering the minimum (RMSE) as the cost function.

2.2. Creep Model

When dealing with steady-state kinematical trends, as an alternative to the Newtonian fluid model, a deviatoric creep model characterized by a creep rate depending on the deviatoric component of the stress state can be chosen to simulate the behaviour of soil. This model is suitable to simulate soil material undergoing secondary creep (steady-state creep), with the creep rate almost constant over

time and depending on the current stress level of the soil. In the inverse analysis, the creep rate can be adopted as the unknown parameter to be defined through the optimization procedure. In this physical scenario, the creep strain rate (ε_c) is calculated by solving the following equation:

$$\frac{d\varepsilon_c}{dt} = F_{cr}n^D, \tag{2}$$

where $\mathbf{n}^D$ is the deviatoric component of the stress tensor, and F_{cr} (1/s) represents the creep rate. The deviatoric tensor $\mathbf{n}^D$ is defined as follows:

$$n^D = \frac{3}{2} \frac{(\sigma_{11} - \sigma_{33})}{\begin{vmatrix} \sigma_{11}\sigma_{12}\sigma_{13} \\ \sigma_{21}\sigma_{22}\sigma_{23} \\ \sigma_{31}\sigma_{32}\sigma_{33} \end{vmatrix}}, \tag{3}$$

where σ_{11}, σ_{22} and σ_{33}, respectively, represent the maximum, intermediate and minimum principal stresses of the soil and σ_{ij} $_{(i \neq j)}$ is the shear stress in the i–j plane. The creep rate F_{cr} normally depends on the second deviatoric invariant of the stress (in addition to the temperature and other material parameters), and the effective creep strain rate equals the absolute value of F_{cr}:

$$\frac{d\varepsilon_{cc}}{dt} = |F_{cr}|. \tag{4}$$

As for the fluid dynamics model case, the creep rate distribution is evaluated through an advanced procedure allowing for the application of non-linear optimization algorithms to FE models. In particular we search for the best-fit creep rate model with respect to the observed DInSAR deformation measurements.

3. Case Study: the Ivancich Landslide

For our study, we focussed on the Ivancich area, a neighbourhood located in the eastern part of the historical town of Assisi (central Italy), and affected by an active slow-moving landslide since its first urbanization (ANTONINI et al. 2002). Damage caused by slow movement of the landslide to buildings and infrastructure has led local authorities to carry out several geological and geotechnical investigations aimed at implementing effective remedial work and mitigation strategies (Felicioni *et al.* 1996; Angeli and Pontoni 2000, 2011). As a result, the area has been deeply investigated in the last 20 years and extensively monitored through in situ inclinometers and piezometers (Pontoni 1999, 2011; Fastellini *et al.* 2011). Geomorphological and topographic surveys revealed that the phenomenon is an old translational slide, with a rotational component in the source area, moving along a well-defined shear band (Cruden and Varnes 1996). In particular, the mass movement involves a debris deposit, from 15 to 60 m in thickness, overlaying a bedrock that consists of a pelitic sandstone unit and layered limestone (Fig. 1a, b) (Servizio Geologico Italiano 1980; Canuti *et al.* 1986; Angeli and Pontoni 2000; Cardinali *et al.* 2001).

3.1. Ground-Based Data

Several campaigns of sub-surface investigations, consisting of inclinometer and piezometer measurements, have been carried out in the Ivancich landslide area, aimed at defining the geometry of the sliding mass and acquiring data on sub-surface displacements and the groundwater regime. In particular, displacement–depth profiles collected between 1998 and 2009 in four inclinometer boreholes approximately located along the central longitudinal section of the landslide body (Fig. 1a, b) pointed out the existence of a shear zone characterized by thickness of less than 2 m, above which the landslide material moves nearly as a rigid body on the stable bedrock (Fig. 1c) (Angeli and Pontoni 2000; Pontoni 2011). The higher cumulated displacements at the ground surface were recorded in the middle portion of the unstable slope, where two inclinometers, namely inclinometers 113 and 117 in Fig. 1a, respectively detected about 7.5 cm between December 1998 and December 2005 and about 6 cm between December 1998 and July 2004 (Fig. 1c). Therefore, these data highlight the major activity of the mid-slope sector, which has been confirmed by DInSAR data (Sect. 3.2).

Regarding the groundwater regime, unpublished piezometer data acquired in the landslide deposit reveal that the pore water pressure affects the landslide kinematics quite moderately. In particular,

the ground water surface was measured to be, in general, only a few metres above the shear band. This results in very low piezometric heights, when compared with the total stress levels. Also, the piezometric surface was observed to be approximately constant over time, with limited seasonal fluctuations. We can consider this a further indication of the limited influence of the rainfall pattern and of the slope groundwater regime on the landslide kinematics, which is characterized by displacement rates that are approximately constant in time over long periods (Calò *et al.* 2014).

3.2. DInSAR Data

For our purposes we benefited from the capability of the multi-sensor Small BAseline Subset (SBAS) approach (Bonano *et al.* 2012) to generate very long deformation time series exploitable for landslide back-analyses. In particular, such a technique, by jointly processing data acquired by geometrically compatible SAR sensors, as in the case of ERS-1/2 and ENVISAT acquisitions, allows production of deformation time series spanning a period of almost 20 years, with accuracy of about 5 mm (Calò *et al.* 2014).

In this work, we exploit the SBAS results, presented in Calò *et al.* (2014), achieved by processing a SAR dataset of 130 ERS-1/2 and ENVISAT images, acquired from 21 April 1992 to 12 November 2010 over central Umbria (Italy). Application of the SBAS DInSAR technique provided a ground deformation velocity map, and associated time series of displacements [measured along the satellite line of sight (LOS)], for the whole study area. The performed analysis allowed detection of a large number of SAR measurement points over the Ivancich landslide (Fig. 2a). For our back-analysis, we selected six SAR pixels located within a distance of 25 m from the landslide longitudinal section S–S′ (Fig. 2b, c).

4. Model Setup

To investigate the kinematical evolution of the Ivancich landslide, we carried out two-dimensional (2D) time-dependent FE modelling of the active

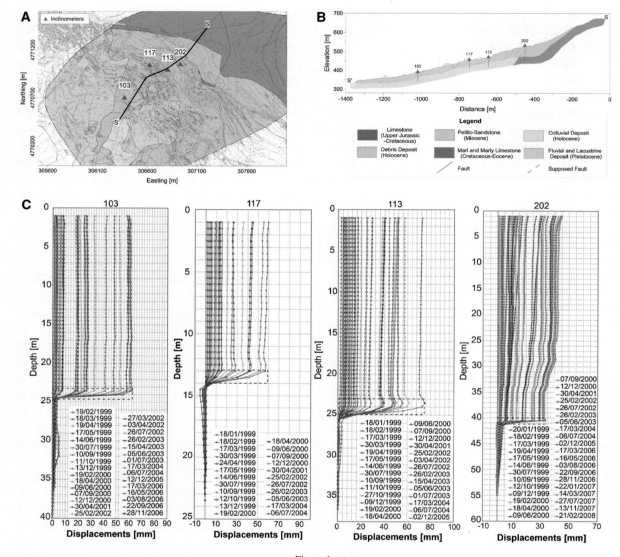

Figure 1

a Geological map of the study area superimposed on a topographic map (modified from Angeli and Pontoni 1999). The *black line* represents the trace of the geological section S–S'. *Red triangles* show locations of inclinometers. **b** Geological section S–S', and location of inclinometers (*red triangles*). **c** Displacement–depth profiles for inclinometers 103, 117, 113 and 202. The *dotted rectangle* represents the depth and thickness of the shear band

ground deformation field using the fluid dynamics and deviatoric creep physical models described above.

We defined the mesh domain of the FE model by exploiting the available geological and geotechnical data obtained from previous investigation campaigns and sub-surface monitoring surveys (Angeli and Pontoni 2000). Based on such information, we set up the 2D model of the whole slope representing the S–

S' longitudinal section of Fig. 2b. In particular, as shown in Fig. 3a, we subdivided the discretization mesh into five geo-mechanical units (see Table 1 for physical properties): (1) a limestone bedrock, in the upper part of the slope, (2) a pelitic sandstone bedrock, in the central and lower portions of the slope, (3) the landslide deposit, formed of unsorted debris, (4) a colluvial deposit, in the landslide toe area, and (5) a shear zone, characterized by thickness of less

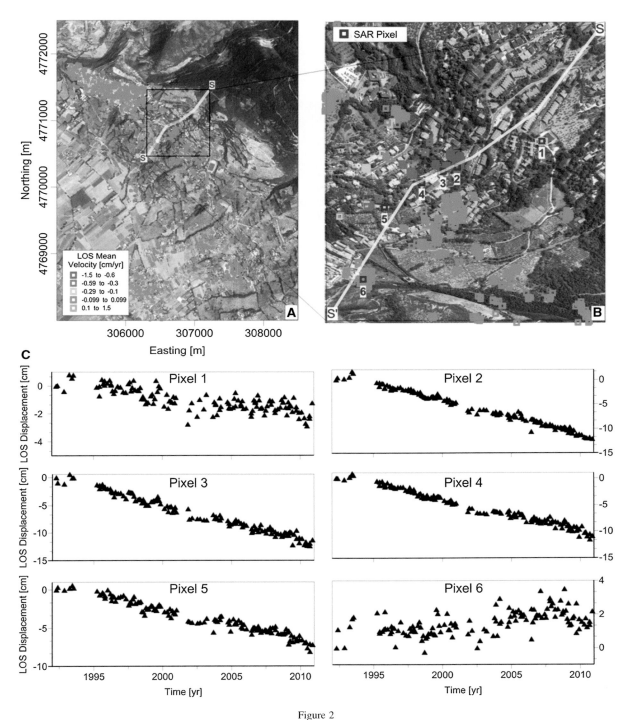

Figure 2

a Ground deformation velocity map obtained through SBAS DInSAR processing of descending ERS-1/2 and ENVISAT data, for Assisi (central Italy). *Inset* shows a zoom over the Ivancich landslide. **b** Zoom over the Ivancich landslide, and location of the six SAR pixels selected for analysis (*blue squares*). *White line* shows the geological section S–S′. **c** Deformation time series of the SAR pixels reported in **b**. Displacements measured along the satellite line of sight (LOS)

113   Reprinted from the journal

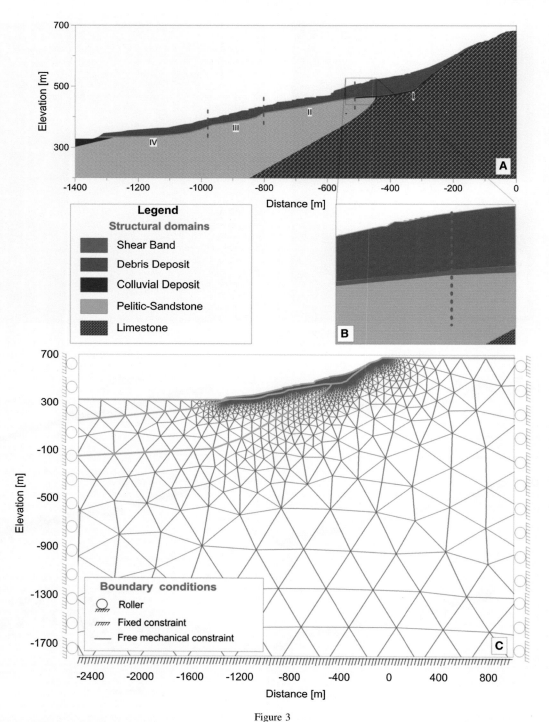

Figure 3

a Simplified 2D model geometry with structural domains. *Inset* **b** shows the shear band domain. **c** Boundary conditions and model geometry superimposed on the triangular FE mesh

Table 1

Physical properties used for different geo-mechanical units

Input parameters

Rock material	Density (kg/m^3)	Young's modulus (MPa)	Poisson ratio (–)	Viscosity (Pa s)
Limestone bedrock	2,200	8×10^3	0.28	1×10^{20}
Pelitic sandstone bedrock	1,850	7×10^3	0.26	1×10^{19}
Debris deposit	1,600	1×10	0.24	1×10^{18}
Colluvial deposit	1,700	6×10	0.24	1×10^{18}
Shear band	1,600	1×10	0.23	–

Table 2

Physical parameter bounds used for the four shear band sectors within the optimization procedure and estimated values

Output parameters

	Lower bound	Upper bound	Estimated value
Viscosity (Pa s)			
η_1	1×10^{14}	5×10^{16}	7×10^{15}
η_2	1×10^{13}	5×10^{15}	1.5×10^{14}
η_3	1×10^{13}	5×10^{15}	1×10^{14}
η_4	1×10^{14}	5×10^{16}	2.2×10^{15}
Creep rate (1/year)			
F_{cr1}	1×10^{-3}	5×10^{-5}	9×10^{-4}
F_{cr2}	1×10^{-2}	5×10^{-4}	8×10^{-3}
F_{cr3}	1×10^{-2}	5×10^{-4}	7×10^{-3}
F_{cr4}	1×10^{-3}	5×10^{-5}	3.5×10^{-4}

than 2 m, with depth ranging between 15 and 60 m from the ground surface (PONTONI 1999, 2000 and ANGELI). According to CALÒ et al. (2014), the shear band domain was divided into four homogeneous sectors in order to comply with the evidence of corresponding inner landslide bodies, characterized by different slope angles, crests and toes, as detected by the available geological and geomorphological information. This evidence is also supported by the analysis of the DInSAR measurements, proving that the identified sectors are characterized by different kinematical rates. It is worth noting that this shear band domain subdivision has no significant impact on

the general applicability of our approach, and a finer discretization, although involving an increase of the computational load, can be easily applied in more complex scenarios.

Note that, in the fluid dynamics scenario, for the geo-mechanical units neighboring the shear surface, the physical model input parameters (Table 1) include significantly high viscosity values (RENZHI-GLOV and PAVLISHCHEVA 1970) to simulate the much lower shear rates; this approach is in agreement with the available inclinometric data. On the other hand, for the deviatoric creep model analysis, we associate the creep rate parameters only with the shear band zone, to simulate its physical behaviour.

Regarding the boundary conditions, the upper part of the model, which represents the ground surface, is considered as unconstrained, whereas the bottom side is fixed in both the vertical and horizontal directions; the vertical side boundaries of the model are instead characterized by null horizontal displacement to make edge effects negligible. The inner domain is characterized by continuity between the different geological units.

Subsequently, we defined the mesh domain, discretized into about 22,000 triangular elements characterized by maximum and minimum sides of 400 and 5 m, respectively, and we applied mesh refinement along the shear band domain (Fig. 3b). The generated mesh was then validated through several resolution tests (ZIENKIEWICZ and TAYLOR 1988; ZHANG and ZHU 1998), which indicated that the use of a finer mesh would affect the results by less than 2 %.

Similarly to GRIFFITHS and LANE (1999) and TIZ-ZANI et al. (2013), we performed our modelling in two stages, i.e. a first step of gravity loading aimed at defining a stress field representative of the current stress state of the slope, and a second step of landslide kinematics simulation.

During the first stage (gravity loading), the stress state of the slope is defined by considering the slope as subjected to the soil gravitational loads and assuming elastic soil behaviour; at this level, only the generated stress field is considered, while the nodal displacements are kept equal to zero.

Reprinted from the journal

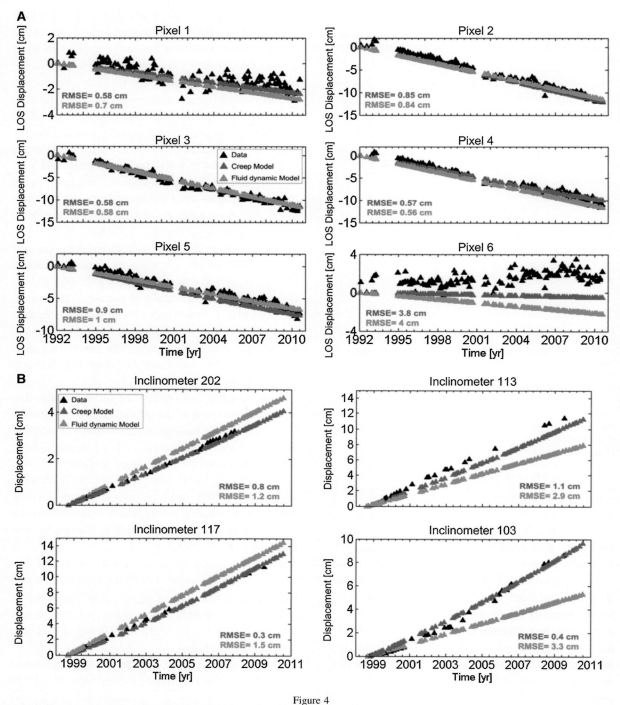

Figure 4
a Comparison between time series of the six selected SAR pixels (*black triangles*), fluid dynamics model (*green triangles*) and creep model (*red triangles*) results. **b** Comparison between time series of four inclinometers (*black triangles*), fluid dynamics model (*green triangles*) and creep model (*red triangles*) results. The RMSE values are also reported

In the second stage (landslide process), the previously calculated stresses are applied to the whole domain, and the two physical approaches, i.e. the fluid dynamics and deviatoric creep models, were alternatively considered to simulate the kinematical trend of the Ivancich landslide during the 1992–2010 time span. Both analyses implicitly assume that the landslide body, delimited by the shear band, is unstable or in a marginal stability condition; this condition has been specifically verified by means of a limit equilibrium calculation that provided a stability factor of the landslide body close to unity. It is worthwhile stressing that, for both approaches, soil is assumed to behave as a single-phase material. This assumption is supported by the available open-pipe piezometric measurements, locally indicating moderate variations of the water level within the landslide debris, and by electric piezometer measurements showing very low variations of the pore water pressure at the depth of the shear band (Sect. 3.1). Accordingly, we can assume that the pore water pressure regime plays a minor role in the landslide evolution process.

5. Results

The optimization procedure, based on a GA, was performed by exploiting the long-term SBAS DInSAR results. In particular, we considered the deformation time series of six SAR pixels located within a distance of 25 m from the landslide longitudinal section S–S' (Fig. 2b). The GA starts by randomly generating ten models, each characterized by viscosity/creep rate values within the lower and upper bounds reported in Table 2. The model providing the best agreement, in terms of the RMSE criterion, with respect to the DInSAR measurements is selected to define the evolution line. Each subsequent generation is then created by considering a narrower range of viscosity/creep rate values centred on the value found in the previous step. This procedure is iterated until 30 generations, for a total number of 300 forward models. Among these, the viscosity/creep rate (Table 2) corresponding to the model that best fits the DInSAR data is chosen as the final parameter value characterizing the soil in the investigated physical process.

The dynamic viscosity values obtained through the optimization procedure are comparable to those obtained in other landslide case studies reported in literature (DI MAIO et al. 2013; CONTE and TRONCONE 2011; TER-STEPANIAN 1975).

Quantitative comparison between the model results, achieved through the above-described optimization procedure, and the DInSAR time series is shown in Fig. 4a for both physical approaches. The figure shows generally good agreement between the DInSAR monitoring data and the model results, except for pixel 6 (Fig. 4a). The discrepancy observed for pixel 6 is supposed to be related to the reduction in LOS distance of the ground surface at the toe of the slope, which is detected by the DInSAR data but is not simulated by the model.

Furthermore, to validate the performed numerical modelling, we exploited the available inclinometer data acquired from 1998 to 2009 in four boreholes located within the landslide deposit (Fig. 1), and compared the model results with such independent in situ measurements (Fig. 4b).

In terms of RMSE, the deviatoric creep model seems to be more suitable, compared with the fluid dynamics model, to describe the kinematical temporal evolution of the Ivancich landslide. This finding is evident by comparing the two modelled time series with the inclinometer ones, particularly in the case of boreholes 103 and 113 (Fig. 4b), for which a slight curvature of the displacement trend is observed. Concerning these two inclinometers, the deviatoric creep approach is capable of simulating the non-linear increase of the displacements over time.

For this physical approach, the modelled distribution of shear stress (τ_{xy}) with respect to lithostatic conditions and of the cumulative displacement over time along the landslide body are reported in Fig. 5a and b, respectively. In particular, Fig. 5b shows that higher displacement rates characterize the central portions of the landslide, whereas significantly lower rates are found in the upper and lower portions of the slope. As a consequence, an increase of of shear stress distribution in the central region of the shear zone is observed (Fig. 5a).

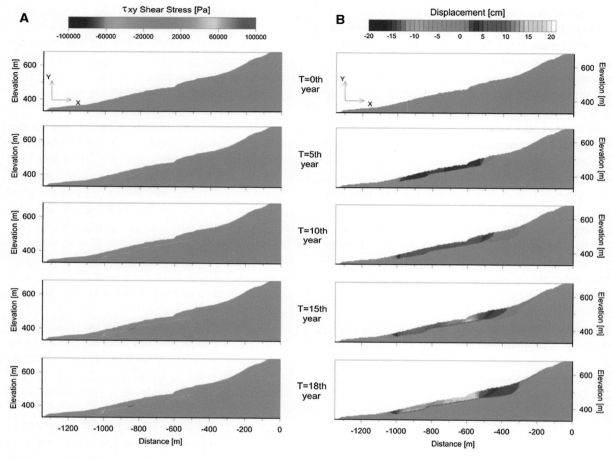

Figure 5
a 2D shear stress and **b** 2D displacement field over time. Both results correspond to the creep numerical simulation

6. Conclusions

In this work, we developed a new procedure to perform inverse numerical modelling of natural phenomena by exploiting DInSAR deformation time series. In particular, we extend the procedure, already proposed in other geoscience contexts such as seismology and volcanology, to investigation of landslide processes.

The proposed methodology is based on integration of the available a priori information, such as geological, geotechnical and geodetic data, into a finite-element environment to build several forward models. By using a Monte Carlo-like optimization procedure, specifically a genetic algorithm, we automatically search for the physical parameter values minimizing the difference between the observed DInSAR deformation time series and modelled displacement field.

The Ivancich landslide (Assisi, central Italy), which is an ancient slow active landslide phenomenon, is considered as a case study. In particular, the kinematical evolution of the unstable mass was analysed by considering a two-dimensional time-dependent FE model using two different physical scenarios, i.e. Newtonian viscous flow and a deviatoric creep model. Quantitative comparison between the results of the two models reveals that, in terms of RMSE (Fig. 4a), the deviatoric creep model is more suitable to describe the temporal evolution of the considered landslide process. This finding is clearly highlighted by comparing the modelled time series with the information derived by inclinometric measurements (Fig. 4b).

Furthermore, based on the optimized creep rate values (Table 2), the highest displacement rates are mainly concentrated in the central regions of the slope, where the minimum values of F_{cr} (1/year) and consequently increase of shear stress are found. Accordingly, the central part of the Ivancich landslide body is supposed to have been active in the last 10,000 years, during the Holocene period. Moreover, analysis of the cumulative displacements points out the presence of an extensional region located near the landslide crown area and a compressive region within the lower area of the landslide body (see $T = $ 18th year in Fig. 5b). In this context, the tensile stress region is more enhanced than the compressive one.

Finally, we stress that integration of data derived from geological surveys and remote-sensing and ground-based monitoring measurements within inverse numerical models, through the implementation of optimization procedures, represents a suitable automatic tool to: (i) deeply investigate the kinematical behaviour of slow active landslides in different geological and geomorphological scenarios, and (ii) discriminate the driving forces governing the physical process responsible for the kinematical evolution of the observed phenomenon.

Acknowledgments

Work conducted in the framework of DORIS (Contract no. 242212) and LAMPRE (Contract no. 312384) EC FP7 projects. F. Calò and R. Castaldo were supported by grants of DORIS and LAMPRE projects.

REFERENCES

ALONSO E.E., (2012). Deformation analysis of landslides: progressive failure, rate effects and thermal interactions. Landslides and Engineered Slopes: Protecting society through improved understanding. Eberhardt et al. (eds). Taylor & Francis, London, 175–214.

ANGELI M. G., PONTONI, F., (2000). The innovative use of a large diameter microtunneling technique for the deep drainage of a great landslide in an inhabited area: the case of Assisi (Italy). Landslides in research, theory and practice. Thomas Telford, London, 1666–1672.

ANGELI M. G., PONTONI, F., (1999). Relazione geologica (Technical Report), 45 pp. (in Italian).

ANTONINI G., ARDIZZONE, F., CACCIANO, M., CARDINALI, M., CASTELLANI, M., GALLI, M., GUZZETTI, F., REICHENBACH, P., SALVATI, P., (2002). Rapporto Conclusivo Protocollo d'Intesa fra la Regione dell'Umbria, Direzione Politiche Territoriali Ambiente e Infrastrutture, ed il CNR-IRPI di Perugia per l'acquisizione di nuove informazioni sui fenomeni franosi nella regione dell'Umbria, la realizzazione di una nuova carta inventario dei movimenti franosi e dei siti colpiti da dissesto, l'individuazione e la perimetrazione delle aree a rischio da frana di particolare rilevanza, e l'aggiornamento delle stime sull'incidenza dei fenomeni di dissesto sul tessuto insediativo, infrastrutturale e produttivo regionale. Unpublished Project Report, May 2002, 140 pp (in Italian).

BAGNOLD et al. (1954). Experiments on a Gravity-Free Dispersion of Large Solid Spheres in a Newtonian Fluid under Shear, Proc. R. Soc. Lond. A August 6, 1954 225 1160 49–63.

BINGUL Z., SEKMEN A., ZEIN-SABATTO S., (2000). Evolutionary approach to multi-objective problems using adaptive genetic algorithms, Systems, Man, and Cybernetics, 2000 IEEE International Conference on, pp. 1923–1927, vol. 3.

BONANO M., MANUNTA M., MARSELLA M., LANARI R., (2012). Long Term ERS/ENVISAT Deformation Time-Series Generation at Full Spatial Resolution via the Extended SBAS Technique, Int. J. Remote Sens., 33, 15, pp. 4756–4783, Feb. 2012, doi:10.1080/01431161.2011.638340.

CALÒ F., CALCATERRA, D., IODICE, A., PARISE, M., RAMONDINI, M., (2012). Assessing the activity of a large landslide in southern Italy by ground-monitoring and SAR interferometric techniques. International Journal of Remote Sensing, 33:11, 3512–3530.

CALÒ F., ARDIZZONE F., CASTALDO R., LOLLINO P., TIZZANI P., GUZZETTI F., LANARI R., ANGELI M-C., PONTONI F., MANUNTA M., (2014). Enhanced landslide investigations through advanced DInSAR techniques: The Ivancich case study, Assisi, Italy. Remote Sensing of Environment, 142, pp. 69–82.

CANUTI P., MARCUCCI, E., TRASTULLI, S., VENTURA, P., VINCENTI, G., (1986). Studi per la stabilizzazione della frana di Assisi. National Geotechnical Congress, Bologna, 14–16 May 1986, Vol. 1, 165–174.

CARDINALI M., ANTONINI G., REICHENBACH P., GUZZETTI, F., (2001). Photo-geological and landslide inventory map for the Upper Tiber River basin. CNR, Gruppo Nazionale per la Difesa dalle Catastrofi Idrogeologiche, Publication n. 2154, scale 1:100,000.

CASCINI L., FORNARO G., PEDUTO D., (2009). Analysis at medium scale of low-resolution DInSAR data in slow-moving landslide-affected areas. ISPRS Journal of Photogrammetry and Remote Sensing, 64, 598–611, doi:10.1016/j.isprsjprs.2009.05.003.

CASCINI L., FORNARO G., PEDUTO D., (2010). Advanced low- and full-resolution DInSAR map generation for slow-moving landslide analysis at different scales. Engineering Geology, 112 (1–4), 29–42, doi:10.1016/j.enggeo.2010.01.003.

CHEN C.L. and LING C.H., (1996). Granular-flow rheology: role of shear-rate number in transition regime. J. Eng. Mech. ASCE, 122, No. 5, 469–481.

CONTE E. and A. TRONCONE (2011). Analytical method for predicting the mobility of slow-moving landslides owing to groundwater fluctuations. J. Geotech. Geoenv. Eng. ASCE, 777–784.

CROSTA G.B., CASTELLANZA R., FRATTINI, P., BROCCOLATO M., BERTOLO, D., CANCELLI P., TAMBURINI A., (2012). Comprehensive understanding of a rapid moving rockslide: the Mt de la Saxe landslide. MIR 2012, Nuovi metodi di indagine monitoraggio e modellazione degli ammassi rocciosi, Barla, G. Ed., Torino, 21–22 novembre 2012, 20 pp.

CRUDEN D.M., VARNES D.J., (1996). Landslide types and processes. In: Turner, A.K., Schuster, R.L. (eds.) 1996 Landslides, Investigation and Mitigation, Transportation Research Board Special Report 247, Washington, D.C., pp. 36–75.

DI MAIO C., VASSALLO R., VALLARIO M. (2013). Plastic and viscous shear displacements of a deep and very slow landslide in stiff clay formation. Engineering Geology, 162, 53–66.

FARINA P., COLOMBO D., FUMAGALLI A., MARKS F., MORETTI, S., (2006). *Permanent scatters for landslide investigations: outcomes from the ESA-SLAM project.* Engineering Geology, *88*, 200–217.

FASTELLINI G., RADICIONI F., STOPPINI A., (2011). *The Assisi landslide monitoring: a multi-year activity based on geomatic techniques*, Applied Geomatics, 3(2), 91–100, doi:10.1007/s12518-010-0042-9.

FELICIONI G., MARTINI E., RIBALDI C., (1996). Studio dei Centri Abitati Instabili in Umbria. Rubettino Publisher, 418 pp (in Italian).

GILL PH., MURRY W., WRIGHT M., (1981), Practical Optimization. Academic.

GRIFFITHS D. V., LANE P. A., (1999). *Slope stability analysis by finite elements*. Geotechnique *49*, No. 3, 387–403.

GUZZETTI F., MANUNTA M., ARDIZZONE F., PEPE A., CARDINALI M., ZENI G., REICHENBACH P., LANARI R., (2009). *Analysis of ground deformation detected using the SBAS-DInSAR technique in Umbria, central Italy.* Pure and Applied Geophysics *166*, 1425–1459, doi:10.1007/s00024-009-0491-4.

HILLEY G., BÜRGMANN R., FERRETTI A., NOVALI F., ROCCA F., (2004). *Dynamics of slow-moving landslides from permanent scatterer analysis*. Science, 304, 1952–1955, doi:10.1126/science.1098821.

HUNT M.L., ZENIT, R., CAMPBELL C.S, BRENNEN C.E., (2002). *Revisiting the 1954 suspension experiments of R. A. Bagnold.* J. Fluid Mech. 452, 1–24.

LEDESMA A., COROMINAS J., GONZALES DA., FERRARI A., (2009). Modelling slow moving landslide controlled by rainfall. In Picarelli L., Tommasi P., Urciuoli G., Versace P. (eds) Proceedings of the 1st Italian Workshop on Landslides, rainfall-induced Landslides: mechanisms, monitoring techniques and nowcasting models for early warning systems, Naples, 8–10 June 2009, vol. *1*, pp. 196–205.

LEROUEIL S., (2001). *Natural slopes and cuts: movement and failure mechanisms.* Géotechnique, Volume *51*, Issue 3, pages 197–243.

LOLLINO, P., SANTALOIA F., AMOROSI A., AND COTECCHIA F. (2011). *Delayed failure of quarry slopes in stiff clays: The case of the Lucera landslide.* Géotechnique, *61*(10), 861–874.

MANCONI A., TIZZANI P., ZENI G., PEPE S. AND SOLARO G., (2009). Simulated Annealing and Genetic Algorithm Optimization using COMSOL Multiphysics: Applications to the Analysis of Ground Deformation in Active Volcanic Areas. Excerpt from the Proceedings of the COMSOL Conference.

PASTOR M., QUECEDO M., FERNANDEZ-MERODO J.A., HERREROS M.I., GONZALEZ E., MIRA P., (2002). *Modelling tailing dams and mine waste dumps failures.* Geotechnique *52* (8): 579–591.

PERZYNA P., (1966). Fundamental Problems in viscoplasticity. Rec. Adv. Appl. Mech. 9, 243–377. Academic, New York.

PONTONI, F., (1999). Unpublished Technical Report, 45 pp (in Italian).

PONTONI, F., (2011). Geoequipe Studio Tecnico Associato Geologia—Ingegneria. Unpublished Technical Report, 4 pp. (in Italian).

RENZHIGLOV N. F. and T. V. PAVLISHCHEVA, (1970). On the viscosity of rocks. Soviet Mining September–October, 1970, Volume 6, Issue 5, pp 582–585.

SERVIZIO GEOLOGICO ITALIANO, (1980). Carta Geologica dell'Umbria. Map at 1:250,000 scale (in Italian).

TER-STEPANIAN G. (1975). *Creep of a clay during shear and its rheological model.* Géotechnique, *25* (2), 299–320.

TIZZANI P., MANCONI A., ZENI G., PEPE A., MANZO M., CAMACHO A., AND J. FERNÁNDEZ, (2010). *Long-term versus short-term deformation processes at Tenerife (Canary Islands)*, J. Geophys. Res., *115*, B12412, doi:10.1029/2010JB007735.

TIZZANI P., CASTALDO R., SOLARO G., PEPE S., BONANO M., CASU F., MANUNTA M., MANZO M., PEPE A., SAMSONOV S., LANARI R., SANSOSTI E., (2013). *New insights into the 2012 Emilia (Italy) seismic sequence through advanced numerical modeling of ground deformation InSAR measurements*, Geophysical Research Letters, Volume *40*, Issue 10, pages 1971–1977.

TRONCONE A., (2005). *Numerical analysis of a landslide in soils with strain-softening behaviour.* Géotechnique, *55*(8), 585–596.

VULLIET L., HUTTER K. (1988). *Viscous-type sliding laws for landslides.* Canadian Geotechnical Journal, 25, 467–477.

ZHANG Z., ZHU J.Z., (1998). *Analysis of the superconvergent patch recovery technique and a posteriori error estimator in the finite element method (II)*. Comput. Methods Appl. Mech. Eng. *163*, 159–170.

ZIENKIEWICZ O.C., TAYLOR R.L., (1988). The Finite Element Method: Basic Formulation and Linear Problems, Volume *1*. McGraw-Hill.

(Received March 10, 2014, revised November 22, 2014, accepted December 5, 2014, Published online December 23, 2014)

Pure Appl. Geophys. 172 (2015), 3081–3105
© 2015 Springer Basel
DOI 10.1007/s00024-015-1071-4

❙Pure and Applied Geophysics

A User-Oriented Methodology for DInSAR Time Series Analysis and Interpretation: Landslides and Subsidence Case Studies

Davide Notti,[1,4] Fabiana Calò,[2] Francesca Cigna,[3] Michele Manunta,[2] Gerardo Herrera,[4,7,8] Matteo Berti,[5] Claudia Meisina,[1] Deodato Tapete,[6] and Francesco Zucca[1]

Abstract—Recent advances in multi-temporal Differential Synthetic Aperture Radar (SAR) Interferometry (DInSAR) have greatly improved our capability to monitor geological processes. Ground motion studies using DInSAR require both the availability of good quality input data and rigorous approaches to exploit the retrieved Time Series (TS) at their full potential. In this work we present a methodology for DInSAR TS analysis, with particular focus on landslides and subsidence phenomena. The proposed methodology consists of three main steps: (1) pre-processing, i.e., assessment of a SAR Dataset Quality Index (SDQI) (2) post-processing, i.e., application of empirical/stochastic methods to improve the TS quality, and (3) trend analysis, i.e., comparative implementation of methodologies for automatic TS analysis. Tests were carried out on TS datasets retrieved from processing of SAR imagery acquired by different radar sensors (i.e., ERS-1/2 SAR, RADARSAT-1, ENVISAT ASAR, ALOS PALSAR, TerraSAR-X, COSMO-SkyMed) using advanced DInSAR techniques (i.e., SqueeSAR™, PSInSAR™, SPN and SBAS). The obtained values of SDQI are discussed against the technical parameters of each data stack (e.g., radar band, number of SAR scenes, temporal coverage, revisiting time), the retrieved coverage of the DInSAR results, and the constraints related to the characterization of the investigated geological processes. Empirical and stochastic approaches were used to demonstrate how the quality of the TS can be improved after the SAR processing, and examples are discussed to mitigate phase unwrapping errors, and remove regional trends, noise and anomalies. Performance assessment of recently developed methods of trend analysis (i.e., PS-Time, Deviation Index and velocity TS) was conducted on two selected study areas in Northern Italy affected by land subsidence and landslides. Results show that the automatic detection of motion trends enhances the interpretation of DInSAR data, since it provides an objective picture of the deformation behaviour recorded through TS and therefore contributes to the understanding of the on-going geological processes.

Key words: Persistent scatterers, small baseline subset, SAR interferometry, time series analysis, quality assessment, subsidence, landslides.

Abbreviations

SDQI	SAR dataset quality index
TS	Time series
NI	Number of images Index
TI	Time Index
MTBI	Mean temporal baseline index
MSBI	Mean spatial baseline index
SRI	Spatial resolution index

[1] Earth and Environmental Science Department, University of Pavia, Via Ferrata, 1, 27100 Pavia, Italy. E-mail: davide.notti@unipv.it

[2] CNR, IREA, via Diocleziano 328, 80124 Naples, Italy. E-mail: calo.f@irea.cnr.it; manunta.m@irea.cnr.it

[3] British Geological Survey (BGS), Natural Environment Research Council (NERC), Nicker Hill, Keyworth, NG12 5GG Nottinghamshire, UK. E-mail: fcigna@bgs.ac.uk

[4] Geohazards InSAR Laboratory and Modeling Group (InSARlab), Geoscience Research Department, Geological Survey of Spain (IGME), Alenza 1, E-28003 Madrid, Spain. E-mail: g.herrera@igme.es

[5] Dipartimento di Scienze Biologiche, Geologiche e Ambientali, Università di Bologna, Via Zamboni 67, 40126 Bologna, Italy. E-mail: matteo.berti@unibo.it

[6] Department of Geography, Institute of Hazard, Risk and Resilience (IHRR), Durham University, Durham, UK. E-mail: deodato.tapete@durham.ac.uk

[7] Unidad Asociada de investigación IGME-UA de movimientos del terreno mediante interferometría radar (UNI-RAD), Universidad de Alicante, P.O. Box 99, 03080 Alicante, Spain.

[8] Earth Observation and Geohazards Expert Group (EOEG), EuroGeoSurveys, The Geological Surveys of Europe, 36-38, Rue Joseph II, 1000 Brussels, Belgium.

1. Introduction

The spatial and temporal monitoring of geological processes, such as subsidence, swelling/shrinkage of soils, or landslides, is crucial to understand their mechanisms, and therefore activate procedures of early warning and deploy suitable risk mitigation measures. In this respect, among the remote sensing technologies used for landslide hazard assessment (METTERNICHT *et al.* 2005), space-borne synthetic aperture radar (SAR) provides data from low to very high spatial resolution (smaller than one metre), day and night and all weather condition images with archives that, dependently on the space mission, can offer a high degree of coverage over Earth's surface. These unique imaging capabilities, thanks to recent advances in sensor technology and processing algorithms, boosted the use of SAR data for a wide range of geosciences and environmental applications. Among these, mapping and monitoring of natural hazards have greatly benefited in the last decade from the development of advanced multi-temporal Differential SAR Interferometry (DInSAR) techniques (*e.g.* PSInSAR™ by FERRETTI *et al.* 2001; SqueeSAR™ by FERRETTI *et al.* 2011; SPN by ARNAUD *et al.* 2003, SBAS by BERARDINO *et al.* 2002; CPT by BLANCO-

SANCHEZ *et al.* 2008; MT-UnSAR by HOOPER 2008, ISBAS by SOWTER *et al.* 2013; PSIG - Cousin PSs (CPSs) by DEVANTHÉRY *et al.* 2014). In particular, such Persistent Scatterer Interferometry (PSI) techniques were used for updating landslide inventories (*e.g.* COLESANTI AND WASOWSKY, 2006; FARINA *et al.* 2006; MEISINA *et al.* 2008; CHEN *et al.* 2010; BIANCHINI *et al.* 2012; Bovenga *et al.* 2012), as well as for detecting and mapping land subsidence (e.g. SOUSA *et al.* 2010; TOMAS et al. 2010, RASPINI *et al.* 2014; TEATINI *et al.* 2012).

So far, most deformation studies using DInSAR data were focused on the spatial analysis of ground movements using mainly the average rates of the displacements. Only in recent years, thanks to the improvement in processing techniques and the possibility to infer non-linear ground motions, it was also possible to take advantage of the capability of

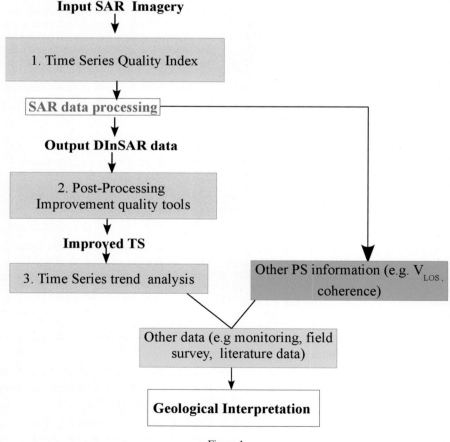

Figure 1
Conceptual flow-chart of DInSAR time series analysis

DInSAR techniques in describing the long-term evolution of natural processes (CALÒ *et al.* 2014). Furthermore, novel works were conducted in order to identify different trends in Time Series (TS) and to detect different phases of temporal evolution of a natural process: acceleration and deceleration, seasonality, sudden change in motion, trend inversion (MILONE and SCEPI, 2011; CIGNA *et al.* 2012; BERTI *et al.* 2013).

Aim of this work is to extract the best out of the information contained in DInSAR data, in order to properly characterize the temporal evolution of ground displacements and improve the interpretation of geological processes. For the purpose, we propose a methodology for the analysis, and improvement, of TS produced by PSI processing techniques, with particular focus on landslides and subsidence phenomena. Our procedure, addressed to DInSAR users interested to fully exploit the potential of TS without working on the development of DInSAR processing chains, is composed by three main steps (see flow-chart in Fig. 1):

1. Pre-processing: evaluation of a SAR Dataset Quality Index (SDQI) that describes the expected quality of DInSAR products prior to the SAR image processing.
2. Post-processing: application of empirical/stochastic methods to improve the quality of already processed TS data.
3. Trend analysis: application and comparison of different approaches of TS trend analysis.

We applied this methodology to SAR data collected by different SAR sensors over different test areas, and processed with different DInSAR techniques as summarised in Table 1. We purposely selected these heterogeneous datasets to prove the effectiveness of the methodology in a wide range of application contexts.

2. Pre-Processing: SAR Dataset Quality Index (SDQI)

2.1. SDQI: Input Parameters and Evaluation

Products of advanced DInSAR techniques, i.e., mean deformation velocity maps and associated TS, can be effectively exploited for hazard and risk assessment, providing valuable information on the spatial and temporal evolution of deformation phenomena such us landslides and ground subsidence. This information can be complemented with ground-based monitoring data (e.g. inclinometers, extensometers, piezometers, and GPS) to deeply understand the long-term kinematical behaviour of geological phenomena, and to investigate the possible triggering factors of the detected ground motion. In this context, it can be very useful for end-users, such as surveyors, geologists, geoscientists, or land use planners, to assess the suitability of deformation TS for this kind of analysis.

To this purpose, we propose an innovative SAR Dataset Quality Index that provides a quantitative assessment of the impact that "decorrelation"

Table 1

Summary of DInSAR datasets used in this work

ID	Satellite	Band	Study area	Area (km²)	Type of processing	Temporal SPAN	No of images	Revisiting time (day)
1	ERS-1/2	C	NW Italy (5 datasets)	>5000	PSInSAR™	1995–2001	60/80	35
2	RADARSAT-1	C	NW Italy (5 datasets)	>5000	SqueeSAR™	2003–2010	80/100	24
3	ERS-1/2	C	Valle de Tena, Spain	33	SPN	1995–2001	29	35
4	ENVISAT ASAR	C	Valle de Tena, Spain	33	SPN	2003–2007	37	35
5	ALOS PALSAR	L	Valle de Tena, Spain	33	SPN	2006–2010	12	46
6	TerraSAR-X	X	Valle de Tena, Spain	33	SPN	05/2008–10/2008	11	11
7	ERS-1/2, ENVISAT	C	Umbria, Central Italy	1200	SBAS	1992–2010	91/37	35
8	COSMO-SkyMed	X	Umbria, Central Italy	300	SBAS	12/2009–02/2012	39	4/16
9	TerraSAR-X	X	Umbria, Central Italy	216	SBAS	07/2011–01/2013	38	11
10	ENVISAT	C	Daunia Apennines, Southern Italy	2200	SBAS	2003–2010	37	35
11	TerraSAR-X	X	Daunia Apennines, Southern Italy	1500	SBAS	01/2010–07/2012	24	11

phenomena, size of the dataset, and sensor wavelength exert on the performance of the multi-temporal DInSAR processing and the generation of ground motion TS. This index refers to the capability of the available SAR datasets to be effectively used to generate deformation velocity maps and associated TS, with a density of measure points suitable to carry out a comprehensive investigation of the study area.

SDQI can be easily implemented by users not necessarily expert in SAR processing, to assess the expected quality of the DInSAR results against the coverage or number of detectable measurement points.

Furthermore, when several SAR datasets are available for the same study area, the SDQI can be used to effectively select the most suitable dataset for the analysis of the investigated deformation phenomenon. In this way, the SDQI supports the design of the DInSAR monitoring activities.

The SDQI is calculated as weighted average of five indexes that account for the main parameters impacting on the number of coherent points detectable through a DInSAR processing:

Table 2

Weights assigned to each parameter of SAR Dataset Quality Index (SDQI)

Parameters	Weight	Weight value
Number of images (NI)	W_{NI}	2
Mean temporal baseline index (MTBI)	W_{MTBI}	2
Time Index (TI)	W_{TI}	1
Mean spatial baseline index (MSBI)	W_{SBI}	1
Spatial resolution index (SRI)	W_{SRI}	1

Each index, below described in detail, is dimensionless, ranges between 0 and 1 and is properly weighted according to the values reported in Table 2. These weights are defined based on our experience related to analyses of phenomena with maximum displacements of the order of few centimetres per year. However, in case of geological processes characterized by high deformation rates (i.e., exceeding 5 centimetres per year), the weights can be easily adapted, in order to make the proposed methodology flexible with respect to the investigated processes, without modifying the calculation of the single

$$SDQI = \frac{(NI \times w_{NI} + TI \times w_{TI} + MTBI \times w_{MTBI} + MSBI \times w_{MSBI} + SRI \times w_{SRI})}{w_{NI} + w_{TI} + w_{MTBI} + w_{MSBI} + w_{SRI}} \qquad (1)$$

Where:

- *NI* (Number of images Index) is associated to the number of acquisitions belonging to the selected SAR dataset;
- *TI* (Time Index) refers to the length of the whole time period spanned by the SAR dataset;
- *MTBI* (Mean Temporal Baseline Index) is associated to the average time interval between consecutive scenes;
- *MSBI* (Mean Spatial Baseline Index) refers to the average spatial baseline of the interferometric pairs;
- *SRI* (Spatial Resolution Index) is associated to the ground range resolution of the SAR scenes;
- and w_{NI}, w_{TI}, w_{MTBI}, w_{MSBI}, w_{SRI} are the respective weights.

indexes. In such a case, the MTBI weight should be properly increased to 3 or 4, in order to make more significant the role of the temporal baseline with respect to other parameters. At the same time if we are not interested in the long-term behaviour of the deformation process, e.g., when we a priori know it is characterized by quasi-linear deformation trend, we can opportunely decrease the weight of Time Index. Moreover, if not all the parameters can be estimated, the SDQI assessment can be performed by setting equal to zero the weights corresponding to the unavailable parameters. For instance the Mean Baseline Spatial Index is not always accessible before processing the SAR dataset; therefore, in such a case, the SDQI could be based on 4 parameters and its reliability will not be affected.

Table 3

SAR Dataset Quality Index (SDQI)

Value	SDQI	Condition
Very low	≤0.25	Low confidence in TS and V_{LOS}
Low	0.25–0.45	Use of average V_{LOS} only
Medium	0.45–0.65	TS are likely to be noisy
High	0.65–0.75	High confidence in TS and V_{LOS}
Very high	>0.75	TS are more likely to fit the expected trend

Table 4

Number of images Index (NI)

Number of images	NI
<10	0
10–20	0.25
20–30	0.5
30–40	0.75
>40	1

Table 5

Time Index (TI)

Temporal span (years)	TI
<0.5	0
0.5–1	0.25
1–2	0.5
2–5	0.75
5–8	1
8–12	0.75
>12	0.5

Table 6

Mean Temporal Baseline Index (MTBI)

Mean temporal baseline			MTBI
L Band (Days)	C Band (Days)	X Band (Days)	
>360	>120	>60	0
180–360	90–120	45–60	0.25
90–180	60–90	30–45	0.5
60–90	20–60	15–30	0.75
<60	<20	<15	1

The estimated SDQI value can be converted in a qualitative evaluation (see Table 3), via the identification of the following SDQI classes: Very Low (if SDQI ≤ 0.25), Low (if 0.25 < SDQI ≤ 0.45), Medium (if 0.45 < SDQI ≤ 0.65), High (if 0.65 < SDQI ≤ 0.75), and Very High (if SDQI > 0.75).

1. Number of Images Index (NI): This index (Table 4) assesses the impact of the number of available SAR images on the achievable coherent point density. Multi-temporal DInSAR processing produces more spatially dense and accurate results when applied to large datasets (Casu *et al.* 2006). Accordingly, this index increases with the number of acquisitions belonging to the SAR dataset. It is worth noting that a minimum number of SAR acquisitions, about 20 (e.g., Marinkovic *et al.* 2005; Crosetto and Cuevas 2011), opportunely distributed over time, should be considered to improve removal of atmospheric phase components from the stack of interferograms and generate reliable ground motion series. We assume that when the number of scenes composing the SAR dataset is lower than 20, this index is lower than 0.5, and in this case DInSAR processing could be strongly affected by noise effects independently from the other indexes.

2. Time Index (TI): This index (Table 5) takes into account the time interval covered by the SAR dataset. In particular, in order to properly estimate and filter out seasonal trends (mostly due to atmospheric contributions), a time period sufficiently long should be considered. In addition, long deformation TS allow a better characterization of the temporal behaviour of the geological processes. For instance, to analyse swelling/shrinkage phenomena or landslides, TS should cover more than one seasonal cycle (>1 year), and ideally be extended to several years to include both dry and wet years. However, long time period DInSAR analyses are affected by temporal "decorrelation" phenomena that reduce the number of coherent points (Bonano *et al.* 2012). Accordingly, TI takes into account these two opposite effects, reaching the maximum value for 5–8 year long analyses and decreasing for longer periods of investigation.

3. Mean Temporal Baseline Index (MTBI): This index (Table 6) quantitatively assesses the impact of the mean temporal baseline (MTB) between consecutive SAR acquisitions, according to the following equation:

$$\text{MTB} = \frac{\sum_{i=1}^{N-1}(T_i - T_{i-1})}{N-1} = \frac{T}{N-1} \qquad (2)$$

where

- N is the number of available SAR images acquired at epochs,
- T is the whole temporal interval of the analysis, expressed in days.

MTBI is based on the average temporal baseline, in order to take in account the effect of very long temporal baselines that can occur in a very irregular temporal sampling.

MTBI considers that the effect of temporal "decorrelation" on different bands (L-, C-, and X-band) increases with decreasing wavelength (Rocca 2007). Moreover, since larger wavelengths allow larger deformations to be estimated, lower values of the mean temporal baseline should be considered with C- and X-band SAR datasets with respect to L-band ones in order to limit phase unwrapping errors.

4. Mean Spatial Baseline Index (MSBI): This index (Table 7) takes into account the geometrical "decorrelation" effect, which depends on the

critical perpendicular baseline computed according to the following equation (Franceschetti and Lanari 1999):

$$\text{Bperp}_{\text{critical}} = \frac{\lambda r'}{2\Delta r} \qquad (3)$$

where:

- λ is the sensor wavelength;
- r' is the sensor-target distance;
- and Δr is the ground range resolution.

MSBI values (Table 7) were set on the basis of the critical spatial baseline for each band, considering the C-band ERS-1/2 case as a reference (ground range resolution of 20 m and a sensor-target distance of 800 km). In addition, the worst case (MSBI equal to 0) for the C-band has been defined as about half of its $\text{Bperp}_{\text{critical}}$ value (cf. Eq. 3). The L- and X-band values have been retrieved by scaling the C-band spatial baseline values with respect to wavelengths. It is worth noting that the impact of the ground range resolution on the critical baseline will be considered within the SRI described below.

By considering N SAR images, characterized by baseline values computed with respect to an acquisition chosen as reference, the mean spatial baseline (MSB) is calculated according to the following equation:

$$\text{MSB} = \frac{B_{\max} - B_{\min}}{N-1} = \frac{B}{N-1} \qquad (4)$$

where $B_{\max}$, $B_{\min}$ and B represent the maximum perpendicular baseline, minimum perpendicular baseline, and the orbital tube of the SAR dataset, respectively.

5. Spatial Resolution Index (SRI): This index (Table 8) accounts for the impact of the ground range resolution on the $\text{Bperp}_{\text{critical}}$ value (cf. Eq. 3). SAR datasets characterized by high to very

Table 7

Mean Spatial Baseline Index (MSBI)

Mean spatial baseline			MSBI
L-band (m)	C-band (m)	X-band (m)	
>1500	>500	>300	0
1250–1500	400–500	250–300	0.25
1000–1250	300–400	200–250	0.5
750–1000	250–300	150–200	0.75
<750	<250	<150	1

Table 8

Spatial Resolution Index (SRI)

Resolution	Satellite	Ground range resolution (m)	SRI
Medium	ERS-1/2, ENVISAT, Standard Beam RADARSAT-1/2	>20	0.25
Medium–high	Fine Beam RADARSAT-1/2, ALOS-1	7–15	0.5
High	StripMap TerraSAR-X and COSMO-SkyMed	3–7	0.75
Very high	Spotlight TerraSAR-X and COSMO-SkyMed	<3	1

high ground range resolution (see Table 8) ease the detection of a larger number of measurement points, so that very spatially dense deformation maps can be produced (CALÒ et al. 2014).

2.2. Application and Performance Assessment

We applied the proposed quality index SDQI to SAR datasets acquired by different satellites, at various frequency bands and spatial resolutions, over different study areas (Table 1 resumes all the parameters). Table 9 summarizes the computed SDQI indexes for the considered datasets. To ease the comparison, the SAR datasets are grouped by study area and processing technique. The average density of PS is used as parameter to assess the performance of SDQI. It's worth pointing out that radar target density depends also on characteristics of study area, e.g., land cover and topography (Table 9).

The analysed data present usually good quality but in some cases (e.g. Tena Valley datasets) the low

SDQI values correspond to noisy TS (e.g. ALOS dataset Fig. 2c) or too short temporal spans (e.g. TerraSAR-X dataset Fig. 2d).

1. Tena Valley Spain. This area covers a medium steep mountainous area in Pyrenees. The main PS targets are represented by talus debris and some spare villages. Landslides are the main land surface processes that affect the area (HERRERA et al. 2013). All the datasets (ERS 1995–2000; EVISAT 2003–2007; ALOS 2006–2010; TerraSAR-X may—October 2008) were processed with SPN algorithm. In the case of ALOS dataset, the low amount of SAR images and the presence of 4–5 image gap (>200 days) makes medium the overall quality of this dataset (SDQI = 0.54, 250 PS/km^2), decreasing the level of confidence in using these TS for landslide hazard assessment. In the case of TerraSAR-X, even if the temporal baseline and the spatial resolution are good (11 images and revisiting time of 11 days), the observation period is too short (5/6 months),

Table 9

SDQI calculation for selected case history datasets compared with the PS density (*) The data of MSBI was assumed equal to 1.0 on the base of PSI data producer indication

Weight		2	2	1	1	1		Relative quality	PS/km^2
Dataset	Processing	NI	MTBI	TI	MSBI	SRI	SDQI		
RSAT NW Italy	SqueeSAR™	1.00	0.75	1.00	1.00	0.25	0.82	Very high	50–200
ERS NW Italy	PSInSAR™	1.00	0.75	1.00	1.00	0.25	0.82	Very high	30–100
ENVISAT Tena Valley	SPN	0.75	0.50	1.00	1.00 (*)	0.25	0.68	High	15
ERS Tena Valley	SPN	0.50	0.50	1.00	1.00 (*)	0.25	0.61	Medium	5
TSX Tena Valley	SPN	0.25	0.75	0.00	1.00 (*)	0.75	0.54	Medium	700
ALOS Tena Valley	SPN	0.25	0.50	0.75	1.00 (*)	0.50	0.54	Medium	250
ERS-ENVISAT Umbria	SBAS	1.00	0.75	0.50	1.00	0.25	0.75	High	110
CSK Umbria	SBAS	0.75	0.75	0.75	1.00	0.75	0.78	Very high	1700
TSX Umbria	SBAS	0.75	1	0.5	1.00	0.75	0.82	Very high	3600
ENVISAT Daunia	SBAS	0.75	0.5	1	1.00	0.25	0.67	High	350
TSX Daunia	SBAS	0.5	0.5	0.75	1.00	0.75	0.64	Medium	330

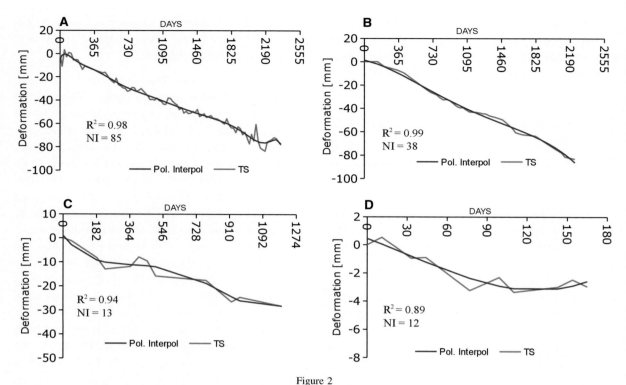

Figure 2
An example of 4 Time series with similar rate of velocities (−10/−20 mm/year) of 4 datasets with different SDQI. NI is the number of images **a** RADARSAT descending dataset NW Italy, 85 images (temporal span 2003–2009), processed with SqueeSAR, SDQI = 0.81; **b** ENVISAT descending datasets, Tena Valley, 38 images (temporal span 2001–2008) processed with SPN, SPN = 0.68; **c** ALOS ascending dataset, Tena Valley, 13 images (temporal span 2006–2010), SDQI = 0.54; TSX descending dataset, Tena Valley, 12 images (temporal span May–October 2008), SDQI = 0.54. It is possible to appreciate the best fit of TS with high SDQI to their polynomial interpolation

decreasing the dataset quality (SDQI = 0.54, 700 PS/km²). ERS and ENVISAT (Fig. 2b) datasets show better TS with medium (SDQI = 0.61, 5 PS/km²) and high quality (SDQI = 0.68; 15 PS/km²), respectively.

2. The NW Italy is covered by many datasets that presents almost the same characteristics in terms of number of images used, satellite and type of SAR processing (Table 1), as a result the SDQI is the same. Once we exclude the 1994 year gap from ERS-1/2 dataset processed by means of PSInSAR™ the SDQI is very good (0.82), which is the same as for the RADARSAT-1 processed by means of SqueeSAR™: this dataset covers the period 2003–2010 (Fig. 2a). The SAR point density range from 30 to 100 (PS/km²) for ERS and from 50 to 200 PS-DS/km² for Radarsat.

3. Umbria region, Central Italy. The landscape shows a hilly or mountainous morphology, with open valleys and large intra mountain basins. The area is widely affected by landslides and subsidence phenomena, with relevant impacts on urban areas and infrastructures. SAR Datasets (ERS-ENVISAT 1992–2010, COSMO-SkyMed 2009–2012, and TerraSAR-X 2011–2013) were processed by applying the SBAS-DInSAR approach. The data provided high quality TS for all datasets: SDQI = 0.75, 110 Points/km² for ERS-ENVISAT, SDQI = 0.78, 1700 Points/km² for COSMO-SkyMed and SDQI = 0.82, 3600 Points/km² for TerraSAR-X.

4. Daunia Apennines (southern Italy). The study area is characterized by gentle hills and low mountains, only locally exceeding 1000 m above sea level. Complex mass movements widely occur in the

region, heavily damaging villages and infrastructures. SAR datasets (ENVISAT 2003–2010, and TerraSAR-X January 2010–June 2012) were processed by applying the SBAS-DInSAR approach. The data provided high quality TS for ENVISAT (SDQI = 0.67, 330 Points/km^2), and medium quality TS for TerraSAR-X (SDQI = 0.64, 350 Points/km^2).

3. Post-Processing: Time Series Improvement

The post-processing analysis can be carried out to improve the quality of DInSAR data and correct possible errors. In literature it is possible to find works on TS error estimation working on process chain (HANSSEN, 2001). For instance GONZÁLEZ and FERNÁNDEZ (2011) provide a model based on Monte Carlo methodology to asses the error in the time series during the processing steps related to atmosphere and develop a model to fit a non linear deformation. In this paper we provide a wider set of post-processing approaches that can be used also by SAR final user and oriented by ground truth observation. If the analysis of the time series does not show any particular noisy data, TS can be

directly analysed as described in Sect. 3 (i.e. trend analysis).

3.1. Removing Noise and Regional Trends

The analysed DInSAR TS may be affected by trends or anomalous estimates not related to real ground motion. These trends are quite easy to detect because they usually affect the whole dataset. They can be due to different causes: errors in data processing (e.g., atmospheric phase screen);

- uncompensated orbital phase ramps
- regional trends related to long-term geological process (e.g., tectonic creep) onto which the local geological processes of interest (e.g., landslides, subsidence) overlap;
- presence of land movements at the reference point location that affect reversely all the dataset;
- thermal effects on targets: the difference of temperature between summer and winter can cause seasonal cycle of dilatation and shrinkage of target, especially if is a metallic object.

These errors can be identified by sampling all high coherence (>0.9) radar targets from a selected stable area where LOS average velocities are in the range ±0.5 mm/year. The stable area can be choice

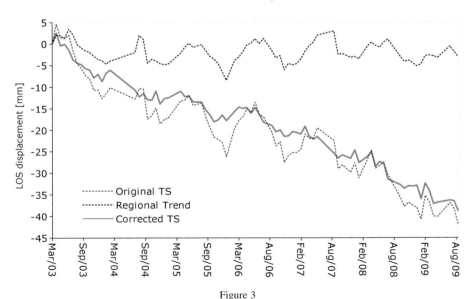

Figure 3
Removal of regional trend from TS data from the RADARSAT-1 SqueeSAR^{TM} ascending data available for Liguria Region, Italy. It is possible to appreciate the decrease on noise in de-trended time-series

Reprinted from the journal

for instance on the base of GPS data if available, or using ancillary geomorphological knowledge on the area. TS of these targets are averaged at each date of the monitoring period, and if the averaged TS reveals the presence of specific trends (e.g., seasonality), the whole dataset is probably affected by an artefact. It is possible to remove the artefact from the i-th TS of the DInSAR dataset of interest (i.e. TS_i), through simple subtraction of the averaged TS of the stable, coherent targets (i.e. TS_{Noise}). This allows to compute the corrected TS (i.e. TS_c):

$$TS_c = TS_i - TS_{Noise} \qquad (5)$$

Figure 3 shows an example of RADARSAT-1 TS ascending data processed with SqueeSAR^{TM} algorithm over Liguria Region. As clearly shown, the original TS data reveal strong seasonality. This trend is common to all the TS of this dataset, including very stable and high coherence PS and Distributed Scatterers (DS) of the dataset, so it is suggested that the observed seasonality relates to an artefact. The seasonal trend (black dot line) is obtained by averaging TS of stable (V_{los} in the range ± 0.5 mm/year) and high coherence (coher > 0.9) PS and DS of the SqueeSAR^{TM} dataset. Then the extracted seasonal

trend is subtracted from the original TS to remove the artefact, which helps to obtain less noisy TS (red line in Fig. 3).

3.2. Removing Single Date Anomalies

Another typology of error that can be observed during the analysis of DInSAR TS is related to anomalous displacement estimates recorded at a certain date of the monitoring period, and spatially diffused across the whole dataset of DInSAR targets. In order to detect such errors, stable (± 0.5 mm/year) and high coherence (>0.9) TS can be selected from the dataset, similarly to the above described approach for the removal of regional trends. If more than one-third of the selected TS shows high dispersion in the displacement value (e.g., $>\pm 5$ mm from TS regression line for X and C-Band; $>\pm 15$ mm for L-band) at the same date of acquisition, it is recommended to remove from the dataset the anomalous date to avoid misleading interpretations like unwrapping errors described in the next tools.

An example of this error was found in the RADARSAT-1 descending dataset "Dogliani" processed with SqueeSAR^{TM} for the Langhe Hills in NW Italy. In particular, on 07/01/2009 more than one-

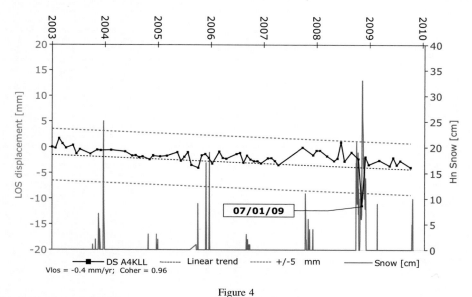

Figure 4

Example of stable TS (-0.4 mm/year) (Langhe Hill dataset) exceeding the ± 5 mm threshold at the date 7/01/2009, compared with snow height observed at ground (Hn) (Source Hydrological database ARPA Piemonte)

third of the selected TS exceed the ± 5 mm threshold. The anomalous values at that acquisition time might be related to snowfall occurred on the day of the SAR acquisition (see the sample PS in Fig. 4). Consequently it is suggested that this scene is removed and not included in the TS analysis.

3.3. Detect and Correct Possible Phase Unwrapping Errors

One of the limitations of DInSAR techniques is related to possible phase unwrapping errors caused by more than a quarter of the radar wavelength motions occurring between two successive

acquisitions or two close targets of the dataset (e.g., CROSETTO et al. 2010). We propose an empirical methodology to detect and mitigate potential phase unwrapping errors during the post-processing stage using a semi-quantitative approach on a simple TS plot.

Especially when working with landslides, it is possible to observe sudden motions occurring at specific dates of the monitoring interval (e.g., rapid movement along unstable slopes). Since these motions may be related to anomalies at specific acquisitions, it is necessary to remove other possible sources of errors from the dataset (as described in the previous section) before checking for unwrapping

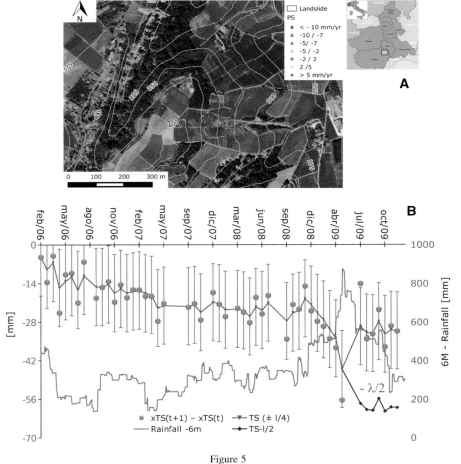

Figure 5

a Location of the case history; **b** The standard time-series with its *error bars* of ±λ/4 (14 mm) the *green dots* represent the differential displacement between two consecutive acquisition (xTS(t+1)−xTS(t)). If the the differential displacement is bigger than |λ/4| the *green dot* it is outside of TS *errors bars* (*red*) and it is possible to introduce a λ/2 correction (e.g. April–July 2009)

errors. If the absolute difference of the displacement (D) observed between two consecutive acquisitions (t_i and t_{i+1}) exceeds the phase ambiguity $\lambda/4$ (e.g. 14 mm for C-band data) it is possible to "jump" to the proposed replica (up or down) of TS that is placed at $\pm\lambda/2$.

$$\text{IF } D_{(t_{i+1})} - D_{(t_i)} > \lambda/4 \rightarrow D_{(t_{i+1})} - \lambda/2$$
(correction for decreasing TS)

$$\text{IF } D_{(t_{i+1})} - D_{(t_i)} < -\lambda/4 \rightarrow D_{(t_{i+1})} + \lambda/2$$
(correction for increasing TS)

Before to apply the jump it is important to take in account some limitations:

- in order to validate the jump to the replica it is necessary to have external sources of data, for instance, other monitoring measures, rainfall data, evidence of acceleration from field survey or literature.
- the tool can ease the correction of one unwrapping error.

Using a simple plot of TS it is possible to graphically see the conceptual model of this correction. In Fig. 5 an example of TS affected by this type of error is shown for a landslide in the Langhe hills (NW Italy). The PS and DS data were generated from RADARSAT-1 imagery processed with Squee-SAR^{TM} and are located close to the crown of the landslide reactivated in April 2009 (Fig. 5a). In Fig. 5b it is shown the time-series and the error bars of $\pm\lambda/4$ (±14 mm) and the real displacement between two consecutive acquisition (green dots). If the displacement is larger than $\lambda/4$ the green dots fall outside the errors bars and it can represent a possible case of phase unwrapping error. The TS studied from 2006 show weak movement until 2008 then from December 2008 it is possible to see a strong acceleration of the movement until April 2009 then followed by a positive Jump. Between April 2009—July 2009 the displacement it was larger than $+\lambda/4$. After a check with ground truth incandescence it was decided to apply a correction of $-\lambda/2$. From a qualitative point of view, with this correction the trend is more reliable with the possible landslides acceleration even if in terms of absolute movement

also this correction may underestimate movement because we do not know how many $\lambda/2$ jumps are occurred. The acceleration occurred from December 2008 to April 2009 well fit the cumulated rainfall and can explain the paroxysmal event occurred at the end of April 2009. This ground deformation behaviour better depicts the observation based on rainfall data and the expected motion of the landslide in response to the triggering factor. The landslide body has no PS because the movement was too rapid to guarantee sufficient phase correlation during the monitoring interval (on the order of several meters).

1. Averaging Time Series

When the distribution of annual velocities across the monitored area appears noisy and difficult to interpret, TS can be spatially averaged to better detect the general trends of deformations. This can be done, for instance, for the TS located in the same geomorphological unit, which fall within a single landslide or in a subsiding area which are thought to be characterised by similar ground motion velocities.

The averaging procedure helps to smooth the spatially variable velocities referring to single targets. Moreover, the comparison between the averaged TS and single PS or DS can be used to find processes possibly occurring at the local scale.

Figure 6 shows the case of the Mendatica landslide in the Ligurian Alps in Italy. It is possible to identify 33 PS and DS in the upper part of the landslides with velocity in the order of 8–12 mm/year. Clearly, the TS of single PS or DS present higher variance (5.8 mm/year) with respect to the averaged TS (2.5 mm/year), and the averaging procedure reveals the general trend of the TS that shows a strong acceleration of landslide motions in 2008.

4. TS classification, identification of trends and deviations

Following the reduction of noise and quality improvement (see Sect. 2), the analysis of DInSAR TS proceeds with the identification and classification of temporal trends in high to very high quality TS datasets. Attention should be paid, however, not to

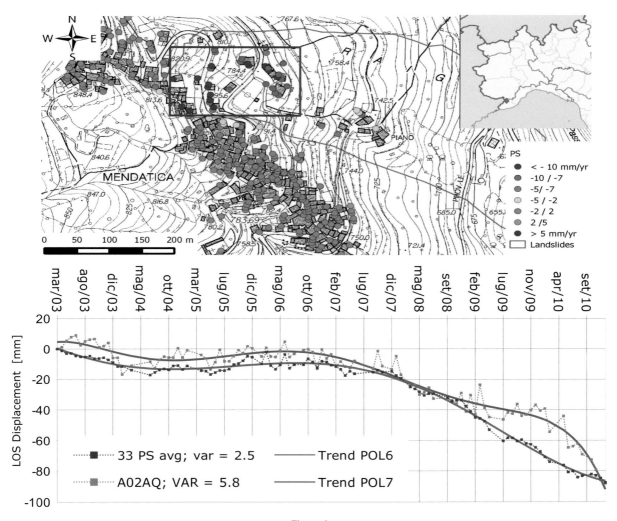

Figure 6
Mendatica landslides. Comparison between single TS and 33 averaged TS. It is possible to see the reduction of the noise and easier identification trends

misinterpret 'artificial' or residual trends by considering these as true ground movements. Identification of trend deviations in DInSAR TS is crucial to understand geological processes. The recent literature reports on some methodologies to overcome limits related to manual, visual analysis and classification of TS (e.g., BERTI *et al.* 2013; CIGNA *et al.* 2012). These methods can support radar-interpreters during their analysis of large DInSAR datasets, to identify critical areas of concern, non-linearity, acceleration/deceleration and, more generally, any deviations from a priori defined trends.

We implement here the following approaches (Table 10):

1. PS-Time: BERTI *et al.* (2013) developed 'PS-Time', an automatic classification tool for PS TS based on a conditional sequence of statistical tests. This tool allows identification of six trend types (i.e. uncorrelated, linear, bi-linear, quadratic, and discontinuous with constant and variable velocity), and additional parameters, such as the break date (date identifying the trend change), V_1, V_2 (the motion rate after and before the trend change), and the index BICW (showing the degree

Table 10

Comparative summary of the approaches of trend analysis of DInSAR time series tested in this research

Methodology	Objectives	Applications
PS-Time (BERTI *et al.* 2013)	Automated detection of trend typologies for each TS of the analyzed dataset, and additional parameters	Regional and local scale analysis of large DInSAR dataset, when the deformation behaviour and the date of the event of the studied process are unknown
Deviation Index, DI (CIGNA *et al.* 2012)	DI type 1 (DI1): Quantification of the deviation within TS after a certain date t_b, with respect to its prediction based on the 'historical' pattern	DI type 1 (DI1): Local and regional scale applications, when events of known date occurred and have changed the general trend of the TS
Mobile DI (TAPETE and CASAGLI, 2013)	DI type 2 (DI2): Quantification of the sudden trend change occurred at the date t_b, by evaluating the step recorded at t_b within the series, not necessarily associated with an overall trend change	DI type 2 (DI2): Local and regional scale applications, when events of known date occurred and have changed the TS locally and in the immediate of the event (e.g., sudden displacement), but not its general trend
	Retrieval of the curve of DI vs. t_b to identify peak values due to trend deviation or changes throughout the TS. In doing so, it complements the DI approach	Local and regional scale applications, when the deformation behaviour and the date of the event of the studied process are either known or unknown
Velocity time series	Evaluate the temporal variation of the velocity, by cutting TS into sub-intervals and re-computing step-wise velocities. May underline seasonal components	Local scale analysis, when long and not noisy TS are available. To reduce noise a spatial averaging of TS is recommended

of bi-linearity of TS). PS-Time can support both regional and local scale studies of large PS datasets, and be particularly efficient when the behaviour of the geological process under investigation is unknown. PS-Time is freely accessed at http://www.bigea.unibo.it/it/ricerca/pstime.

2. Deviation Index (DI) and mobile DI: CIGNA *et al.* (2012) developed the Deviation Index (DI), able to quantify trend deviations within DInSAR TS by reproducing the visual process of identification of trend changes. The first type of DI (i.e. DI1) quantifies the deviation recorded after a certain date t_b with respect to its prediction based on the linear regression of the TS records before t_b, whilst the second type (i.e. DI2) compares the TS behaviour before and after an event occurred at t_b, by measuring any displacements recorded at t_b and identifying its impact on the TS. The use of DI1 and DI2 can support both local and regional analyses of DInSAR TS, when an event is thought to have an impact on the ground motion trend, either permanently after a t_b, or only temporarily at t_b. Since the application of the DI requires t_b selection to cut the TS into two distinct subintervals to compare, building upon the original method by CIGNA *et al.* (2012), TAPETE and CASAGLI (2013) developed the "mobile DI" approach to identify objectively the correct break

date. This method consists in the computation of the DI across the entire monitoring period followed by analysis of DI variations as a function of changing t_b. The approach can be used when t_b cannot be defined a priori as no background information about the process to study is available. The main advantages of the mobile DI are both the confirmation of the suitability of a fixed t_b, and the identification of other possible dates acting as temporal breaks, which were not previously identified by visual inspection.

3. Velocity TS: With this tool it is possible to generate temporal series of velocity from DInSAR TS, by cutting the monitored period into regular sub-intervals and extracting partial linear regressions on the displacement data. Number and length of the sub-intervals can be determined by accounting for: (1) TS quality, which is influenced by temporal span and number of displacement records composing the sub-interval (generally, the shorter the length of the sub-intervals or lower the number of scenes, the higher the likelihood that the resulting series of velocities will appear noisy; as practical recommendation, TS averaging may be implemented beforehand, and sub-intervals should include at least 5–6 displacement records; see Sect. 2.3); and (2) the deformation trend of the studied phenomenon, that can be identified via

preliminary visual inspection of the TS (e.g., either long-term or season-based sub-intervals may be considered). We applied the above three methodologies to analyse two test areas in Italy: (1) Pontecurone, in Piemonte Region, NW Italy, affected by land subsidence, and (2) Crociglia landslide, NW Apennines. For both sites, two SAR data stacks processed by TRE S.r.l. with the SqueeSAR™ algorithm (FERRETTI *et al.* 2011) were employed:

- 89 RADARSAT-1 Standard Beam, ascending mode images (24/03/2003–22/05/2010);

- 84 RADARSAT-1 Standard Beam, descending mode images (28/04/2003–05/10/2009).

4.1. Subsidence Case History of Pontecurone

Pontecurone is a small rural village in the southwestern sector of the Po River Plain, located on the alluvial fan of Curone stream (Fig. 7), where thickness of alluvial sediments exceeds 150 m. Stratigraphic and geotechnical data for the region show a typical alluvial succession of gravel, sand (aquifer layer) and silt–clay (aquitard layers); in

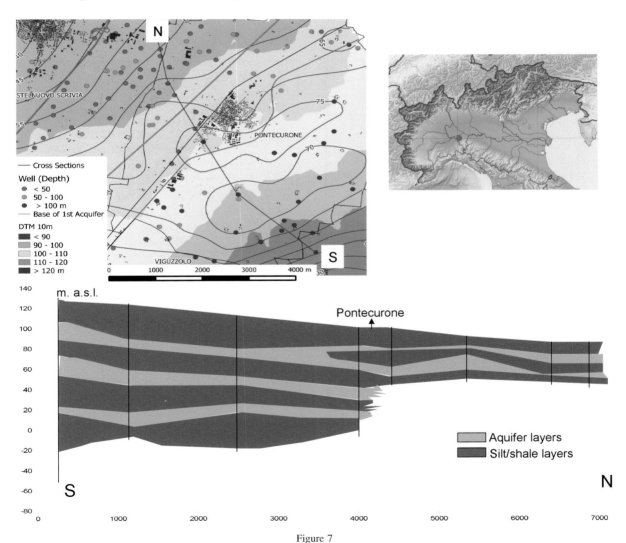

Figure 7
Simplified hydro-geological setting of Pontecurone. It is possible to see that the area at south of the village is characterized by the presence of deep borehole. Borehole data from Provincia Alessandria/Regione Piemonte Settore Acque

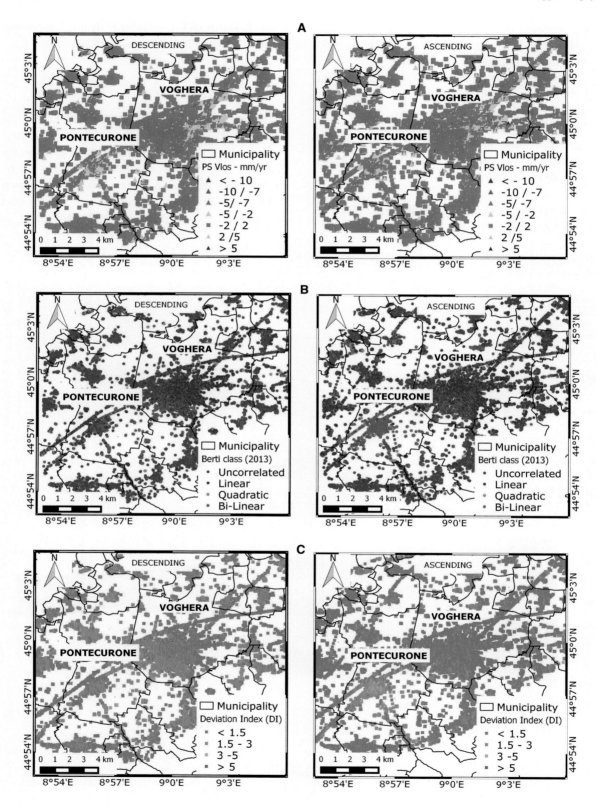

◄

Figure 8

RADARSAT-1 SqueeSAR™ time series over Voghera and Pontecurone. **a** LOS velocities for descending (*left*) and ascending (*right*) mode data processed with SqueeSAR™ by TRE S.r.l.; **b** Automated classification in descending (*left*) and ascending (*right*) mode, based on PS-Time classification tool by BERTI *et al.*, 2013; **c** Semi-automated classification descending (*left*) and ascending mode, based on the Deviation Index (DI) approach by Cigna et al. (2012)

Pontecurone clay sediments (at least 10 m thick) outcrop and locally confine the first aquifer (Fig. 7). In the upper part of Curone alluvial fan, several groundwater wells exploit confined aquifers, and are generally deeper than those in the northern sectors. Though there is little information available about water extraction, groundwater exploitation seems to be mainly related to agricultural irrigation.

Visual analysis of the annual LOS velocity (V_{LOS}) of both ascending and descending SqueeSAR™ RADARSAT-1 datasets, revealed general stability of the area, with only two confined zones in Pontecurone and Voghera showing motion rates down to −5 mm/year (Fig. 8a).

Automatic TS classification of over 28,000 points of the SqueeSAR™ ascending and 23,000 points of the descending TS datasets with PS-Time showed that in the area of Pontecurone most TS are classified as bi-linear, especially within the ascending dataset while, for the remainder of the dataset, the typology of TS are either linear or uncorrelated, and only a few are classified as quadratic (Fig. 8a). The BICW index showed for the area of Pontecurone values higher than 1.2, this indicating the occurrence of strongly bi-linearity, and identified the break date of the TS in the last part of 2008, i.e. 27/11/2008.

The computation of the DI1 confirms this evident change of trend in the TS of both ascending and descending data. By assuming November 2008 as the t_b for all the series, the resulting DI1 (Fig. 8c) show that the extension of the area affected by temporal deviations is by far wider than that depicted by the sole visual analysis of annual V_{LOS}. The DI1 results for the descending dataset (Fig. 8c left) highlight that large sectors of the scene recorded deviations as high as 1.5–3 times the respective predictions based on the temporal history of the area before t_b. More than half of the built-up area of Voghera, indeed, evidences

DI1 values of 1.5–3.0 and even higher than 3.0–5.0 for a small sector in the southwestern outskirts of the town. Pontecurone, Casei Gerola and Rivanazzano Terme show DI1 greater than 1.5. The DI1 computation for the ascending dataset (Fig. 8c right) also shows concentration of trend deviations over Pontecurone, with DI1 increasing by moving from the outskirts to the centre of the village.

It is worth noting that the ascending and descending mode TS are characterized by both different time lengths and different number of scenes composing the input satellite image stacks, as mentioned above. The use of a common t_b for both datasets creates TS subsamples with different lengths of the historical (i.e. pre-t_b) and updated (i.e. post-t_b) intervals in ascending and descending mode. In particular, while the length of the updated interval is 16-image long for the ascending data, this is only 10-image long for the descending one, hence the computation of DI1 for the latter is less reliable, due to the fewer number of scenes available after t_b.

To confirm the suitability of the fixed t_b and identify possible additional dates of trend deviation, the mobile DI was also calculated over the entire TS for both PS and DS of the ascending dataset over the village of Pontecurone where, as above discussed, a clear pattern was identified in the spatial distribution of the DI1 (Figs. 8c–9a). The example in Fig. 9 refers to the DS A4WEI in the northeastern quarter of the village where the highest values of DI1 at the selected t_b (2.8–5.3) were retrieved. The TS of the examined DS is characterized by an inversion of the trend (i.e. from moving away from the satellite to moving towards it) immediately after the t_b suggested by PS-Time and confirmed by the DI1 (Fig. 9c). The curve of the mobile DI1 gives an objective confirmation (Fig. 9d), with the main peak reaching 5.6 on 10/12/2008, and also adds evidence not retrieved by visual inspection of the TS. Indeed, after t_b another DI1 peak of ∼5.0 is identified on 11/11/2009 (Fig. 9d), i.e. in correspondence with a trend change occurred within the post-t_b sub-interval (Fig. 9c). This peak marks a further acceleration of the LOS displacements exactly 1 year after the occurrence of the main trend inversion (i.e. 10/12/2008). Moreover, an additional DI1 peak occurs at the end of 2007—early 2008, i.e. almost one year prior to the main trend

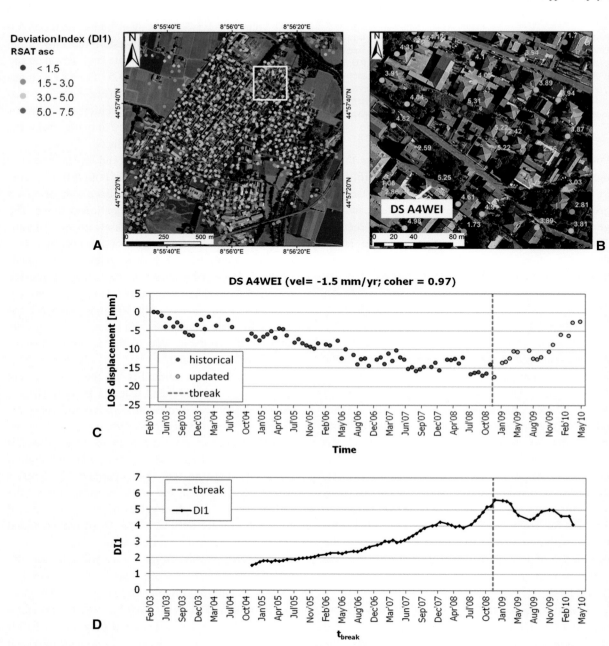

Figure 9

a Semi-automated classification of RADARSAT-1 SqueeSAR^TM ascending time series based on DI1 over the village of Pontecurone (cf. Fig. 8c overlapped onto Virtual Earth image (Bing Maps © 2013 Microsoft Corporation), with indication of the north–eastern quarter (see *white square*), where **b** the highest values of DI1 were retrieved (range: 2.8–5.3). **c** Time series and **d** graph of DI1 vs. t_b for the DS A4WEI. *Red dashed line* marks the a priori t_b in November 2008

inversion. Though less pronounced (DI1 of 4.2), this peak could indicate an early temporal break that sole visual inspection of the TS likely would have not identified.

The evidence discussed above with regard to DS A4WEI is also found within the TS of the neighbouring PS and DS, and therefore relates to instability affecting the entire local area in Pontecurone.

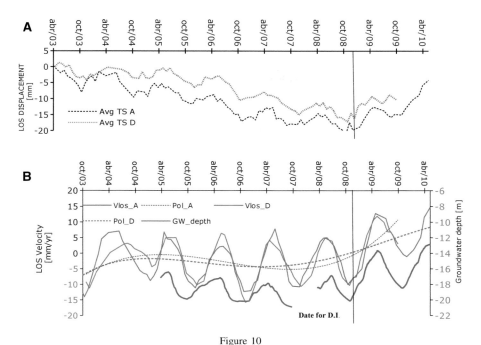

Figure 10

a Spatially averaged time series for ascending and descending geometry; **b** LOS velocity time series for Pontecurone, using 6-month long averaging window, and 1 year interpolation windows. Results are shown for both ascending and descending RADARSAT-1 datasets, and compared with ground water levels (source: Regione Piemonte–Settore Acque)

The computation of the TS of the LOS velocities was performed on the averaged TS of both the SqueeSAR^{TM} ascending and descending datasets, for the area where the DI1 spatial distribution showed the highest values. The velocity calculated for the two periods, preceding and following t_b, confirms observations based on the first two methodologies, and shows land subsidence at −3 to 05 mm/year rates occurring between 2003 and 2008, followed by 10 mm/year uplift (Fig. 10). By computing the velocity TS over 6-month long intervals, seasonal components can be recognised and, in particular, we identify negative LOS velocities until October/November, which are followed by periods of uplift in winter and early spring. Between 2003 and 2008, land subsidence seems stronger than uplift, whereas in the following period (between 2008 and 2010) the opposite is observed. The long-term trend across the full TS confirms the bi-linear trend detected by PS-Time automated classification, as well as the high DI1 values observed within this sector of the processed area.

The most likely cause of the observed seasonality is related to groundwater oscillations due to extraction from deep and confined aquifers for agricultural use (Fig. 10), consequent variation of pore water pressure and ground settlement and/or heave. Maximum water levels are reached in spring, while minimum levels in early autumn, after irrigation. Long-term oscillations are related to wet and dry periods, e.g., from 2003 to 2008 a drier period caused general decrease of water levels, whereas from 2009 to 2010 a wetter period caused general increase of piezometric levels.

4.2. Landslide, Case History of Crociglia

Crociglia landslide is a typical earth slide-earth flow that affects the weathered clay shale units widely outcropping in the NW Apennines. Borehole investigations suggested that the sliding surface is located at depths between 5 m and 15 m from the ground level, and recent field surveys evidenced presence of damage in residential buildings.

The analysis of RADARSAT-1 SqueeSAR^{TM} V_{LOS} in ascending and descending mode shows an active sector within the urban settlement (Fig. 11a). The landslide affects a west-facing, gentle slope

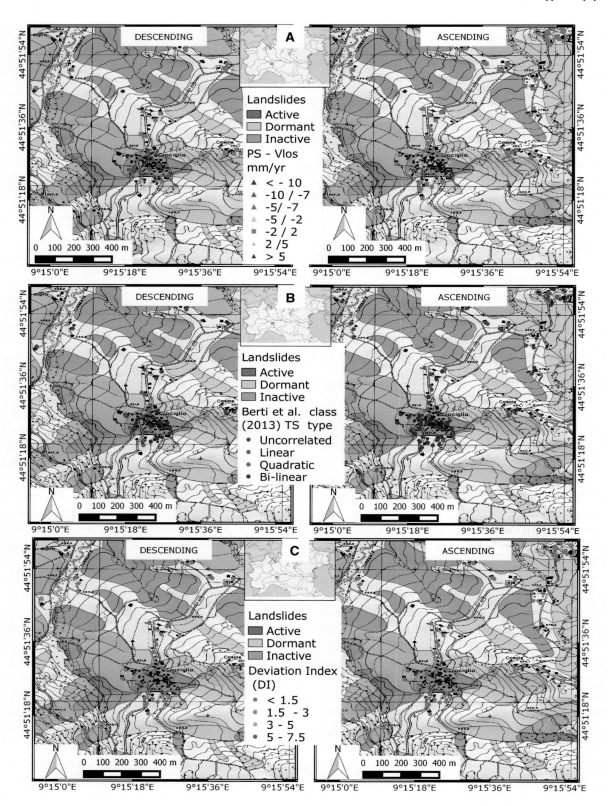

Figure 11
RADARSAT-1 SqueeSAR[TM] time series for Crociglia **a** LOS velocities based on (*left*) descending and (*right*) ascending mode data processed with SqueeSAR[TM] by TRE S.r.l.. **b** Automated classification of RADARSAT-1 SqueeSAR[TM] time series in (*left*) descending and (*right*) ascending mode, based on PS-Time classification tool by BERTI et al. (2013). **c** Semi-automated classification (*left*) descending and (*right*) ascending mode, based on the Deviation Index (DI) approach by Cigna et al. (2012)

(10°), and the main direction of the movement is horizontal, along the E-W direction, as confirmed by negative V_{LOS} in the descending dataset (i.e. -7 to -10 mm/year), and positive V_{LOS} in the ascending (i.e. $+4$ to $+6$ mm/year) dataset.

PS-Time automated classification of TS shows predominance of bi-linear, and linear behaviours only for a smaller portion of the TS, in both the ascending and the descending datasets (Fig. 11b). Bi-linear points extend to the south of the landslide boundaries, suggesting that instability involves a wider area and incorporates the southern sector of the village. The date recording the change in the TS trend is autumn 2008. As for Pontecurone, the BICW index shows values higher than 1.2, indicating strong bi-linearity.

The computation of DI1 was performed for 160 SqueeSAR[TM] TS in descending, and 230 TS in

ascending mode, by using the t_b of November 2008 for both the datasets. The DI1 maps (Fig. 11c) confirm spatially the extension of the deviations already evidenced by PS-Time, and DI1 values as high as 1.5–3.0 are observed, mostly within the mapped landslide boundaries, and part to the south, with DI1 peak of 3.4 in the descending dataset for a point just outside the landslide boundary.

The mobile DI approach shows that the TS of PS and DS within the landslide boundaries recorded similar deformation behaviour from the main scarp to the toe, thereby resulting in similar DI1 vs. t_b curves. Figure 12 shows the example of PS A75VJ from the descending dataset (which best depicts the occurred deformation due to the aspect of the slope), located in the lower part of the landslide. The TS is dominated by LOS motion away from the satellite (Fig. 12b), with -7.46 mm/year V_{LOS} and 0.97 coherence. Visual inspection of the TS highlights trend change only in the course of 2009, in correspondence with the last SAR acquisitions. Using a priori t_b on 30/11/2008 the DI takes on the value of 2.5 (Fig. 12a), whereas the mobile DI provides a clearer picture (Fig. 12c), by showing a DI1 increase during 2008 and the peak of 4.5 on 14/05/2009. The mobile DI

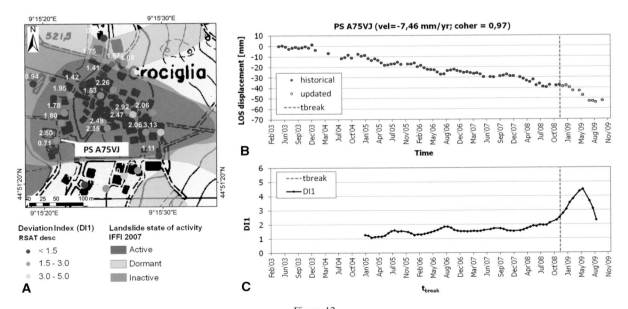

Figure 12
a Semi-automated classification of RADARSAT-1 SqueeSAR[TM] descending time series based on DI1 over Crociglia landslide (cf. Fig. 11a), superimposed on landslide inventory (IFFI 2007) and topographic map sheet Regione Lombardia B9C1 scale 1:10'000; **b** Time series and **c** graph of DI1 vs. t_b for the PS A75VJ. *Red dashed line* marks the a priori t_b in November 2008

calculation allows us to correct the interpretation and to detect objectively the trend change in 2009 that was already noted by visual inspection. The a priori t_b of 30/11/2008 therefore cannot be considered the ideal date to break the TS, though still critical, as the trend deviation was already in place. The features observed in the TS of PS A75VJ are also found for the neighbouring PS and DS, thus indicating similar deformation behaviour for all the points within the landslide boundaries.

The analysis of the TS velocity was based on averages of ~20 ascending and ~20 descending TS of points within the landslide boundaries and allowed identification of two main types of trends. The first depicts a long-term behaviour and confirms the evidences from the DI computation and PS-Time automated classification (Fig. 13a). The period between 2003 and 2008 shows V_{LOS} of −7 mm/year in the descending mode and +4 mm/year in ascending mode. The velocity TS based on 6 months-long intervals shows seasonal variations for both

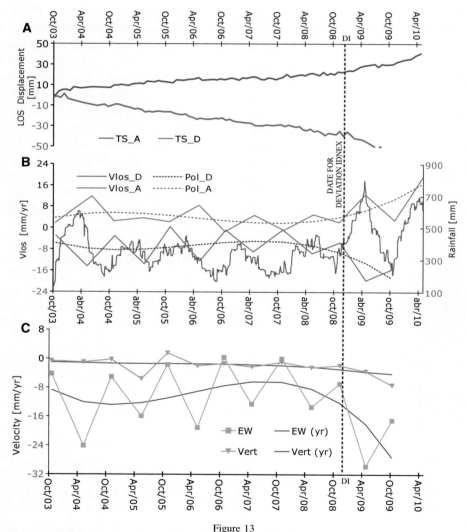

Figure 13

a Spatially averaged SqueeSAR[TM] time series for ascending and descending geometry of RADARSAT-1 datasets; **b** LOS velocity time series for Crociglia, using 6 months averaging window, and 1 year interpolation windows. Results are shown for both ascending and descending RADARSAT-1 datasets, and compared with rainfalls (source: Arpa Lombardia). **c** resolved E-W and vertical components of the velocity by combing ascending and descending RADARSAT-1 datasets acquisition geometries. Observed acceleration of the movements in winter-spring is followed by deceleration in summer/early autumn (Fig. 13b). We also computed the E–W and vertical component of the velocity (Fig. 13c). The projection of the 6-month V_{LOS} to the steepest slope direction for both geometries shows equivalent results. Average velocity during the winter is −20 mm/year, whereas velocity during the summer ranges between −7 and −4 mm/year. The agreement between ascending and descending datasets, also confirmed by analysis of ERS data, allows us to consider this result reliable also without comparison with other monitoring data. The physical explanation of the observed seasonal trend is most likely related to seasonal variations in saturation level of the soil, controlled by rainfall, evapotranspiration and snow melting. Long-term variations are thought to be related to rainfall, as the period between November 2008 and May 2010 was characterised by two rainy winter-spring periods. The main event occurred in April 2009, just after the t_b that was identified by both PS-Time and the DI approach, and the landslide motions mainly occurred in the areas nearby.

5. Conclusions

Deformation time series (TS) generated through advanced DInSAR techniques provide valuable information on the temporal evolution of geological processes as landslides and subsidence phenomena.

In this work we proposed and validated a methodology for the analysis of DInSAR TS, aimed at improving the understanding of the kinematical behaviour of geohazard processes and enabling also non-expert users to handle DInSAR deformation time series.

The proposed methodology consists of three main steps:

- Pre-processing (see Sect. 1): computation of the *SAR Dataset Quality Index* (SDQI) aimed at evaluating the quality of datasets prior to/in preparation to the trend analysis. Its assessment is based on five indexes associated to the characteristics of the available SAR datasets. SDQI was computed for several DInSAR datasets acquired by different SAR sensors and processed with advanced DInSAR techniques, i.e., PSInSAR™, SqueeSAR™, SBAS and SPN. The purposed SDQI can be applied to any other DInSAR techniques not tested in this work and it is an open instrument that can be discussed and improved by the community. The comparative tests showed that the optimal time-series should have the following features: a temporal span of 5–8 years, an almost complete series of images without long temporal gap (temporal baseline <180 days for L-band, <90 days for C-band; <45 days for X-band) and a good spatial baseline (<1000 m, <300 m; <200 m respectively for L, C and X band). The spatial resolution of sensor and the wavelength are also important, and SDQI approach also accounts for them.

- Post-processing (see Sect. 2): application of four approaches aimed at mitigating some of the errors and noise that can affect DInSAR TS and improving their quality. These approaches are based on empirical-stochastic observations and address the removal of regional trends and anomalous motion values, the correction of possible phase unwrapping errors and the noise reduction trough spatial averaging. However, such methods for TS quality improvement need strong control interaction and input by the DInSAR users and cannot be used with full automation. In particular, for the phase unwrapping error correction it is important to check the geomorphological reliability with ground truth data to avoid false positives. Application of these approaches showed good results for the case histories in our areas of interest.

- Detection and analysis of TS trends (see Sect. 3): three different methodologies for DInSAR TS trend analysis were compared with respect to a common dataset characterized by a very high quality (SDQI = 0.82) and when necessary, after quality improvement (as proposed in this work). These methodologies work on different aspects of the TS: (1) PS-Time automated classification is suitable when the deformation behaviour is unknown, and is able to suggest trend typology and other useful parameters (e.g., t_b, bi-linearity and seasonality indices); (2) Deviation Index or DI, improved with the mobile DI, is suitable to detect the geological processes when TS show trend changes in relation to specific events (e.g. sudden collapse, heavy rainfall, tectonic motion); and (3) the TS of velocity allows extraction of relevant temporal features (e.g., seasonality) at the local scale. The three methodologies were tested on two areas in Italy affected by urban subsidence and landslide processes, and showed agreement in picturing the motion behaviour of the investigated phenomena. In Pontecurone, the TS analysis allowed discovering of both seasonal and long-term ground oscillations, likely related to water level variations in the aquifers. In Crociglia, long-term and seasonal trend were detected, showing a correlation of landslide motions with rainfall records over 3–6 month long intervals.

The obtained results prove that the proposed methodology allows maximizing the information embedded in the DInSAR TS and encourage its implementation by end-users interested in improving the understanding of the deformation behaviour of geological processes.

Acknowledgments

The SAR data for Piemonte and Liguria regions in NW Italy were obtained through the project "Risk Nat" with Regione Liguria and ARPA Piemonte. The data over Oltrepò Pavese were derived from collaboration with Regione Lombardia. Both these data were processed by Tele-Rilevamento Europa - T.R.E. s.r.l. -. Datasets over Valle de Tena were derived from the SUDOE project in collaboration with IGME (*Instituto Geologico Y Minero de España*), Madrid, Spain, and processed by Altamira Information. SAR datasets over Umbria region, central Italy, were processed in the framework of the EC FP7 DORIS project (contract n. 242212). SAR datasets over Daunia Apennines, Southern Italy, were processed in the framework of contract with Italian Department of Civil Protection (contract n. 622).

REFERENCES

ARNAUD, A., ADAM, N., HANSSEN, R., INGLADA, J., DURO, J., CLOSA, J. and EINEDER, M. (2003), *ASAR ERS interferometric phase continuity.* Geoscience and Remote Sensing Symposium, 2003. IGARSS '03. Proceedings. 2003 IEEE International, 2, pp. 1133–1135, 21–25 July 2003.

BERARDINO, P., FORNARO, G., LANARI, R. and SANSOSTI, E. (2002), *A new algorithm for surface deformation monitoring based on small baseline differential interferograms.* Geoscience and Remote Sensing, IEEE Transactions, 40, no. 11, pp. 2375–2383.

BERTI, M., CORSINI, A., FRANCESCHINI, S., and IANNACONE, J. P. (2013), *Automated classification of Persistent Scatterers Interferometry time series*, Nat. Hazards Earth Syst. Sci., 13, 1945–1958, doi:10.5194/nhess-13-1945-2013.

BIANCHINI, S., CIGNA, F., RIGHINI, G., PROIETTI, C., and CASAGLI, N. (2012), *Landslide hotspot mapping by means of persistent scatterer interferometry.* Environmental Earth Sciences, 67, pp 1155–1172.

BLANCO-SANCHEZ, P., MALLORQUÍ, J.J., DUQUE, S. and MONELLS, D. (2008), *The Coherent Pixels Technique CPT): An Advanced DInSAR Technique for Nonlinear Deformation Monitoring,* Pure Appl. Geophys., 165, pp. 1167–1194.

BONANO, M., MANUNTA, M., MARSELLA, M., and LANARI, R. (2012). Long-term ERS/ENVISAT *deformation time-series generation at full spatial resolution* via *the extended SBAS technique.* International Journal of Remote Sensing, 33(15), 4756–4783.

BOVENGA, F., WASOWSKI, J., NITTI, D. O., NUTRICATO, R., and CHIARADIA, M. T. (2012). *Using COSMO/SkyMed X-band and ENVISAT C-band SAR interferometry for landslides analysis.* Remote Sensing of Environment, 119, 272–285.

CALÒ, F., ARDIZZONE, F., CASTALDO, R., LOLLINO, P., TIZZANI, P., GUZZETTI, F., LANARI, R., ANGELI M.-G., PONTONI, F. and MANUNTA, M. (2014). *Enhanced landslide investigations through advanced DInSAR techniques: The Ivancich case study,* Assisi, Italy. Remote Sensing of Environment, 142, 69–82.

CASU, F., MANZO, M., and LANARI, R. (2006). *A quantitative assessment of the SBAS algorithm performance for surface deformation retrieval from DInSAR data.* Remote Sensing of Environment, 102(3–4), 195–210. doi:10.1016/j.rse.2006.01.023.

CHEN, F., LIN, H., YEUNG, K., and CHENG, S. (2010), *Detection of slope instability in Hong Kong based on multi-baseline differential SAR interferometry using ALOS PALSAR data,* GIScience Remote Sens., 47, 208–220.

CIGNA, F., TAPETE, D., and CASAGLI, N., (2012): *Semi-automated extraction of Deviation Indexes (DI) from satellite Persistent Scatterers time series: tests on sedimentary volcanism and tectonically-induced motions,* Nonlin. Processes Geophys., 19, 643–655.

COLESANTI, C. and WASOWSKY, J. (2006) I*nvestigating landslides with space-borne Synthetic Aperture Radar (SAR) interferometry,* Eng Geol., 88, 173–199.

CROSETTO, M and CUEVAS, M. (2011). *Enabling Access to Geological Information in Support of GME.* EUROPEAN COMMISSION Seventh Framework Programme D3.6 PSI Validation Version 1 17th November 2011.

CROSETTO, M., MONSERRAT, O., IGLESIAS, R., and CRIPPA, B. (2010). *Persistent Scatterer Interferometry: Potential, limits and initial C- and X-band comparison.* Photogrammetric Engineering and Remote Sensing, 76 (9), 1061–1069.

DEVANTHÉRY, N., CROSETTO, M., MONSERRAT, O., CUEVAS-GONZÁLEZ, M., and CRIPPA, B. (2014). *An Approach to Persistent Scatterer Interferometry.* REMOTE SENSING, *6*(7), 6662–6679.

FARINA, P., COLOMBO, D., FUMAGALLI, A., MARKS, F. and MORETTI S. (2006) *Permanent Scatterers for landslide investigations: outcomes from the ESA-SLAM project* Engineering Geology, *88* (3–4), pp. 200–217.

FERRETTI, A., PRATI, C. and ROCCA, F., (2001) *Permanent Scatterers in SAR interferometry.* Geoscience and Remote Sensing, IEEE Transactions, *39*(1).pp 8–20.

FERRETTI, A., FUMAGALLI, A., NOVALI, F., PRATI, C., ROCCA, F. and RUCCI, A., (2011) *A New Algorithm for Processing Interferometric Data-Stacks: SqueeSAR.* IEEE T Geo-science and Remote Sensing, IEEE Transactions, *49* (9), pp.3460–3470.

FRANCESCHETTI, G., and LANARI, R. (1999) Synthetic aperture radar processing. CRC press.

GONZÁLEZ, P. J., and FERNÁNDEZ, J. (2011). *Error estimation in multitemporal InSAR deformation time series, with application to Lanzarote, Canary Islands.* Journal Of Geophysical Research: solid earth (1978–2012), 116(b10).

HANSSEN, R. (2001), *Radar Interferometry: Data Interpretation and ErrorAnalysis, 308* pp., Kluwer Acad., Dordrecht, Netherlands.

HERRERA, G., GUTIÉRREZ, F., GARCÍA-DAVALILLO, J. C., GUERRERO, J., NOTTI, D., GALVE, J. P., and COOKSLEY, G. (2013*). Multi-sensor advanced DInSAR monitoring of very slow landslides:* the Tena valley case study (central Spanish Pyrenees). Remote Sensing of Environment, 128, 31–43.

HOOPER, A. (2008). *A multi-temporal Insar method incorporating both persistent scatterer and small baseline approaches.* Geophysical Research Letters, *35*(16), L16302. doi:10.1029/ 2008GL034654.

MARINKOVIC, P. S., VAN LEIJEN, F., KETELAAR, G., and HANSSEN, R. F. (2005). R*ecursive data processing and data volume minimization for PS-InSAR.* In International geoscience and remote sensing symposium (Vol. *4*, p. 2697).

MEISINA, C., ZUCCA, F., NOTTI, D., COLOMBO, A., CUCCHI, A., SAVIO, G., GIANNICO, C. and BIANCHI M. (2008), *Geological interpretation of PSInSAR data at regional scale.* Sensors 8, 7469–7492.

METTERNICHT, G., HURNI, L., and GOGU, R. (2005), *Remote sensing of landslides: An analysis of the potential contribution to geo-*spatial systems for hazard assessment in mountainous environments. Remote Sensing of Environment 98, 284–303.

MILONE, G. and SCEPI, G. (2011) *A clustering approach for studying ground deformation trends in Campania Region through the use of PS-InSAR time series analysis,* J. Appl. Sci. Res., 11, 610–620.

RASPINI, F., LOUPASAKIS, C., ROZOS, D., ADAM, N. and MORETTI, S., (2014) - *Ground subsidence phenomena in the Delta municipality region (Northern Greece): Geotechnical modelling and validation with Persistent Scatterer Interferomet*ry. International Journal of Applied Earth Observation and Geoinformation, Vol. *28*, pp 78–89.

ROCCA, F. (2007). *Modeling interferogram stacks.* Geoscience and Remote Sensing, IEEE Transactions on, *45*(10), 3289–3299.

SOUSA, J. J., RUIZ, A. M., HANSSEN, R. F., BASTOS, L., GIL, A. J., GALINDO-ZALDÍVAR, J., and SANZ DE GALDEANO, C. (2010). *PS-InSAR processing methodologies in the detection of field surface deformation—Study of the Granada basin (Central Betic Cordilleras, southern Spain).* Journal of Geodynamics, *49*(3), 181–189.

SOWTER, A., BATESON, L., STRANGE, P., AMBROSE, K., and FIFIK SYAFIUDIN M. (2013) *DInSAR estimation of land motion using intermittent coherence with application to the South Derbyshire and Leicestershire coalfields.* Remote Sensing Letters *4*(10): 979–987.

TEATINI, P., TOSI, L., STROZZI, T., CARBOGNIN, L., CECCONI, G., ROSSELLI, R., and LIBARDO, S. (2012). *Resolving land subsidence within the Venice lagoon by persistent scatterer SAR interferometry.* Physics and Chemistry of the Earth, Parts A/B/C, 40, 72–79.

TAPETE, D., and CASAGLI, N. (2013), Testing Computational Methods to Identify Deformation Trends in RADARSAT Persistent Scatterers Time Series for Structural Assessment of Archaeological Heritage. In B. Murgante et al., (eds.), *Computational Science and Its Applications*—ICCSA 2013 (pp. 693–707): Springer Berlin Heidelberg.

TOMAS, R., HERRERA. G., LOPEZ-SANCHEZ, J.M., *et al.* (2010). *Study of the land subsidence in the Orihuela city (SE Spain) using PSI data: distribution, evolution and correlation with conditioning and triggering factors.* Engineering Geology, *115*, 105–121.

(Received March 16, 2014, revised March 9, 2015, accepted March 16, 2015, Published online March 27, 2015)

Reprinted from the journal

Pure Appl. Geophys. 172 (2015), 3107–3121
© 2014 Springer Basel
DOI 10.1007/s00024-014-0914-8

▌**Pure and Applied Geophysics**

Structure of Alluvial Valleys from 3-D Gravity Inversion: The Low Andarax Valley (Almería, Spain) Test Case

Antonio G. Camacho,[1] Enrique Carmona,[2,3] Antonio García-Jerez,[3] Francisco Sánchez-Martos,[4] Juan F. Prieto,[5] José Fernández,[1] and Francisco Luzón[2,3]

Abstract—This paper presents a gravimetric study (based on 382 gravimetric stations in an area about 32 km^2) of a nearly flat basin: the Low Andarax valley. This alluvial basin, close to its river mouth, is located in the extreme south of the province of Almería and coincides with one of the existing depressions in the Betic Cordillera. The paper presents new methodological work to adapt a published inversion approach (GROWTH method) to the case of an alluvial valley (sedimentary stratification, with density increase downward). The adjusted 3D density model reveals several features in the topography of the discontinuity layers between the calcareous basement (2,700 kg/m^3) and two sedimentary layers (2,400 and 2,250 kg/m^3). We interpret several low density alignments as corresponding to SE faults striking about N140–145°E. Some detected basement elevations (such as the one, previously known by boreholes, in Viator village) are apparently connected with the fault pattern. The outcomes of this work are: (1) new gravimetric data, (2) new methodological options, and (3) the resulting structural conclusions.

1. Introduction

The inverse gravimetric problem, namely the determination of a subsurface mass density distribution corresponding to an observed gravity anomaly, has an intrinsic non-uniqueness in its solution (e.g., Al-Chalabi 1971). Nevertheless, particular solutions can be obtained by including additional constraints on the model (geometry of the subsurface structure) and on the data parameters (statistical properties of the inexact data, e.g., Gaussian distribution of errors).

The shallow basin structures filled with light sedimentary material constitute a particularly interesting case of the gravity inversion, due to their geological interest. They cause a closed negative anomaly, which usually is modelled by considering an outcropping flat low-density body (homogeneous or stratified) limited by a concave shallow bottom (e.g., Leão et al. 1996). The usual inversion methods look for determining, in a non-linear approach, the bottom interface as defined by elementary cells. Rectangular prisms have been used widely to describe the model structure (e.g., Cordell and Henderson, 1968; Rama Rao et al., 1999).

The case of assuming a subsurface structure characterised by several sub-horizontal layers of prescribed density contrasts is more complex and involves a higher ambiguity. The main problem is to assign the features of the anomaly map as produced by irregularities on one or another interface. Traditional methodologies are mostly based on the calculation of the Fourier transform of the gravitational anomaly as the sum of the Fourier transform of powers of the perturbing interface topographies (e.g., Oldenburg, 1974; Chakraborty and Agarwal, 1992; Reamer and Ferguson, 1989).

Camacho et al. (2009, 2011a) describe a general method and a code ("GROWTH") to carry out a 3D gravity inversion in a not subjective approach, able to determine the geometry of the causative bodies. Camacho et al. (2011b) proposed a modification of the original methodology enabling one to deal with

[1] Instituto de Geociencias (CSIC-UCM), Facultad CC. Matemáticas, Plaza de Ciencias, 3, Ciudad Universitaria, 28040 Madrid, Spain. E-mail: antonio_camacho@mat.ucm.es

[2] Dpto de Química y Física, Universidad de Almería, Cañada de San Urbano s/n, 04120 Almería, Spain.

[3] Instituto Andaluz de Geofísica, Universidad de Granada, Campus Universitario de Cartuja, 18071 Granada, Spain.

[4] Dpto de Biología y Geología, Universidad de Almería, Cañada de San Urbano s/n, 04120 Almería, Spain.

[5] Dpto de Ingeniería Topográfica y Cartografía, ETSI Topografía, Geodesia y Cartografía, Universidad Politécnica de Madrid, Km 7.5 Autovía de Valencia, 28031 Madrid, Spain.

isolated bodies or stratified structures in a versatile form. The key idea is to modify the adjustment equations for isolated bodies (GROWTH method) by adding a weighting matrix enabling a shift of the adjusted anomalous masses closer to the discontinuity interfaces.

In the present paper we describe a new alternative possibility of obtaining stratified structures by means of the GROWTH methodology. The key idea is to change the prescribed density contrast at some stage during the growing process of modelling. It is combined with a suitable choice of the gravity constant offset. These new options allow modeling stratified structures as, for instance, alluvial valleys by means of 3D models. It will provide interesting results concerning the topography of the discontinuity surfaces, the presence of large faults, and the existence of the included particular anomalous bodies . These types of geological structures are composed fundamentally of sedimentary rocks and represent the material record in the form of rock layers or strata that once existed on the earth. Sedimentary rocks contain information about what earth surface environments were like in the past and can contain important natural resources. From the seismic engineering point of view, sedimentary materials can produce the so-called local seismic effects, generating the amplification of the seismic inputs and spectral resonances at the free surface of an alluvial valley (see, e.g., LUZÓN et al., 1995, 2004, 2009).

As a test example, the proposed gravity inversion approach is applied to a data set recently collected for studying the Low Basin of the River Andarax in the Betic Cordillera. This area is one of the most arid regions in Europe with very irregular precipitation. The intensive agricultural activity depends on the exploitation of the groundwater. We aim to determine a model of the subsurface mass distribution composed by some irregular strata.

2. Geological Setting

The tectonically active Betic Cordillera is a topographic manifestation of the collision between the African and Iberian plates. It consists of E–W and NE–SW trending mountain ranges separated by less elevated sedimentary basins. Betic basins were marine depocenters from late Miocene through Pliocene time, and emergent until upper Pliocene/lower Pleistocene time (SANZ DE GALDEANO and VERA, 1992).

The Low Basin of the River Andarax is located in the extreme south of the province of Almeria (Fig. 1) and coincides with one of the existing depressions in the Betic Cordillera. This valley, which is limited to the south by the Mediterranean Sea, is enclosed by Sierra Alhamilla (in the East), with its mainly metapelitic outcrops, and Sierra de Gádor (in the West), which constitutes a carbonate–dolomite massif with outcroppings of phyllites (Fig. 1). The depression is filled by post-orogenic detrital deposits of diverse lithology (marls, sandy silts, sands and conglomerates) with evaporite intercalations of gypsiferous nature. The mica-schists and quartzites are practically impervious, while the carbonate formation has high porosity and permeability values due to fissures and/ or karstification. The post-orogenic rocks exhibit great differences in permeability. The Miocene and Pliocene marly formations have very low permeability, whereas the Pliocene deltaic sediments and the Quaternary and Plio-quaternary formations are water-bearing. In accordance with this distribution, three hydrogeological units have been defined: Detrital Aquifer, Carbonate Aquifer, and Deep Aquifer (PULIDO-BOSCH et al., 1991; SÁNCHEZ-MARTOS, 1997).

The Low Andarax basin corresponds to an alluvial valley of about 250 km^2. It is located in one of the most arid regions in Europe, which is characterized by its low (200–350 mm/year) and irregular precipitation which falls mainly (70 %) in autumn and winter (MARTIN-ROSALES et al., 1996). This determines the pattern of groundwater exploitation supporting an intensive agricultural activity.

This basin is situated in an active seismic region, with the highest seismic hazard values in Spain, where shallow seismic series occur frequently (in fact, the valley is encircled by very near active faults systems). The intense tectonic activity of the zone (SANZ DE GALDEANO et al., 1985) favours an important geothermal activity, as demonstrated by the thermal springs of the Sierra Alhamilla (51.8 °C) and Alhama (40.8 °C). The tectonic activity of the area has

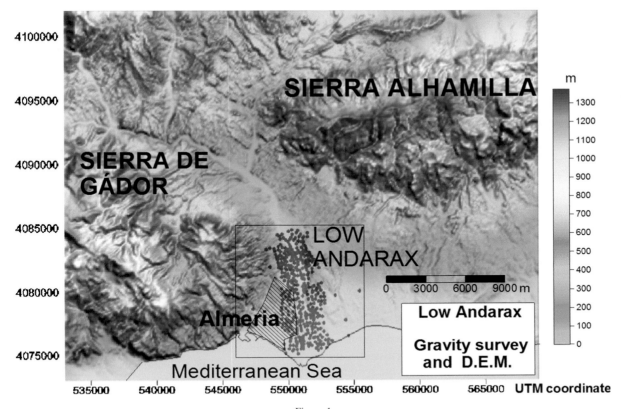

Figure 1

Sketch of the geographical location of the Low Andarax valley, enclosed by the Mediterranean Sea (south), Sierra Alhamilla (east), and Sierra de Gádor (west). The local Digital Elevation Model (DEM) and location of the gravity survey. The *striped area* indicates the city of Almería. The *small central rectangle* indicates the survey area for the next figures

affected the relationship between the different aquifer units. The main late Miocene to Quaternary tectonic structures in the southwestern side of the Alhamilla ridge, in the Almería–Níjar basin, are Pliocene–Quaternary high-angle normal faults striking NW–SE to NNW–SSE (MARTÍNEZ-MARTÍNEZ and AZAÑÓN, 1997). The fractures coincide with old faults from the Miocene period and have been reactivated during the Quaternary, with throws not higher than 10 m (VOERMANS and BAENA, 1983). These faults present many young echeloned scarps to the SW of Sierra Alhamilla (SANZ DE GALDEANO et al., 2010). The NW–SE striking fractures show a great influence on the topography, and are interpreted as deformations in the surface linked to a movement in depth conjugated to the Carboneras fault (PEDRERA et al., 2006).

Moreover, the seismic hazard is also reinforced because the landform is mainly composed of sedimentary materials, which produce the so-called local seismic effects, with the amplification of the seismic inputs and spectral resonances in the free surface of the valley.

Considering all those problems and characteristics, we aim to get new information about the subsurface 3D density structure of the Low Andarax valley (LAV) by using new geophysical data and new inversion methodology.

3. Methodology

We present in this section: (1) a brief description of the methodological principles of the previously published methodology for free inversion of isolated 3D bodies, and, (2), the modified version to account for the characteristics of basin structures with several discontinuity interfaces representing alluvial valleys environments.

3.1. General Approach

Large parts of the basic inversion methodology and associated mathematical concepts are described in CAMACHO et al. (2009, 2011a, b). We, therefore, summarize the key concepts here.

Suppose a data set is constituted of gravity values observed at n gravity benchmarks, irregularly distributed. Let (x_i, y_i, z_i), $i = 1, \ldots, n$, be the planar coordinates (UTM coordinates) and the altitudes of the gravity stations P_i and let Δg_i, be the respective gravity anomaly (Bouguer gravity anomaly). We must consider the gravity data as imprecise values whose uncertainties show Gaussian distribution, characterised by a covariance (n, n)-matrix, Q_D. Usually, we set $q_{ij} = 0$, for $i \neq j$, and $q_{ii} = e_i^2$, where e_i, $i = 1, \ldots, n$, are standard deviations of the gravity values.

The inversion process constructs a subsurface model defined by a 3-D aggregation of m parallelepiped cells, which are filled, in a "growth" process, by means of prescribed positive and negative density contrasts. The design equation to relate observables, i.e., the gravity anomaly Δg_i at n benchmarks (x_i, y_i, z_i), with modelling parameters and residuals v_i is :

$$\Delta g_i = \sum_{j \in J^+} A_{ij} \Delta \rho_j^+ + \sum_{j \in J^-} A_{ij} \Delta \rho_j^- + \delta g_{\text{reg}} + v_i, \quad (1)$$
$$i = 1, \ldots, n,$$

where A_{ij} is the vertical attraction for a unit density for the jth parallelepiped cell upon the ith observation point (e.g., PICK et al., 1973), $\Delta \rho_j^-$, $\Delta \rho_j^+$ are prescribed density contrasts (negative and positive fixed values) for the jth cell, J^+, J^- are sets of indexes corresponding to the cells filled with positive or negative density values, and δg_{reg} is a regional component composed of an offset regional value g_0 and a linear trend:

$$\delta g_{\text{reg}} = g_0 + g_x(x_i - x_M) + g_y(y_i - y_M), \quad (2)$$
$$i = 1, \ldots, n$$

x_M and y_M are average coordinates for the survey area, and g_x, g_y are unknown values for the horizontal gravity gradients.

Sets J^+ and J^- constitute the main unknown to be determined in the inversion approach. They design cells filled with positive and negative density contrast, then determine the geometry of the anomalous bodies in a non-linear relationship.

Following the general treatment of the least-squares inversion methods of TARANTOLA (1988), to solve the problem of non-uniqueness, we adopt a mixed minimization condition, based on model "fitness" (least square minimization of residuals) and model "smoothness" (l_2-minimization of total anomalous mass)

$$v^T Q_D^{-1} v + \lambda \quad m^T Q_M^{-1} m = \min, \quad (3)$$

where $m = (\Delta \rho_{1_1}, \ldots, \Delta \rho_m)^T$ (superscript T denotes transpose of a matrix) are density contrast values for the m cells of the model, $v = (v_1, \ldots, v_n)^T$ are residual values for the n data points, Q_D is an a priori covariance matrix for uncertainties of the gravity data, Q_M is an a priori covariance matrix for uncertainties of the model parameters, and λ is a factor for selected balance between fitness and smoothness of the model. For a problem without prior information about the model structure (CAMACHO et al., 2011b), we suggest taking a model covariance matrix Q_M given by a diagonal normalizing matrix of non-null elements that are the same as the diagonal elements of $A^T Q_D^{-1} A$. This covariance matrix allows getting inversion models located on suitable depths (see simulation tests in references), and it plays the role of the depth-weighting functions in the bibliography about gravity and magnetic data inversion (for instance LI and OLDENBURG 1998).

The problem of non-linearity of the system, with a large number of unknowns, is solved by a particular constructive process: the anomalous structures are formed by a nearly homogenous growth by cell addition, from previously adjusted "skeletal" structures, until the bodies attain a suitably developed size. The prismatic cells are systematically tested, step by step, with each prescribed density contrast, and then the best solution is adopted to grow anomalous bodies. The minimization fit conditions are applied for each growth step, and include a scale factor f, which relates the immature model to the global conditions concerning gravity fit and model size (mass and volume).

In practice, for an arbitrary $(k + 1)$th step, k prisms have been previously filled with the positive or negative fixed contrast values and the modelled gravity values will be g_i^c, $i = 1, \ldots, n$. Now, the process looks, throughout the m–k unchanged prisms, for one new prism to be modified. For that, for each jth unchanged prism, and for both the negative and positive prescribed density contrasts, the following equation system is considered:

$$g_i - (g_i^c + A_{ij}\Delta\rho_j)f - g_0 - g_x(x_i - x_M)$$
$$- g_y(y_i - y_M) = v_i, \quad i = 1, \ldots, n, \tag{4}$$

$$\boldsymbol{v}^T \boldsymbol{Q}_D^{-1} \boldsymbol{v} + \lambda f^2 \boldsymbol{m}^T \boldsymbol{Q}_M^{-1} \boldsymbol{m} = \text{min.}, \tag{5}$$

where $\Delta\rho_j$ are the prescribed values $\Delta\rho_j^+$ and $\Delta\rho_j^-$, and $f \geq 1$ is an unknown scale factor for fitting the modelled anomalies ($\Delta g_i^c + A_{ij}\Delta\rho_j$) to the observed anomalies (Δg). Then, the unknown parameters f, g_x and g_y are adjusted by solving the system (4) and (5), where the vector $\boldsymbol{m}$ of solutions now includes the values for the previously filled cells and the value $\Delta\rho_j$ that is being tested. Once the former linear equations have been solved, we can calculate the misfit value e_j^2 defined by

$$e_j^2 = \boldsymbol{v}^T \boldsymbol{Q}_D^{-1} \boldsymbol{v} + \lambda f^2 \boldsymbol{m}^T \boldsymbol{Q}_M^{-1} \boldsymbol{m} \tag{6}$$

as the parameter for the suitability of the jth prism and the adopted density contrast (positive $\Delta\rho_j^+$ or negative $\Delta\rho_j^-$). Then, the jth prism with a density contrast producing a minimum value of e_j^2 is selected to grow the anomalous body, adding its effect to the modelled Δg_i^c values (see CAMACHO et al., 2007, for details).

This process is repeated in a step-wise manner until a best fitting model is obtained. For each successive step, the scale value f decreases. The process stops when f approaches 1, resulting in the modelled 3-D structure for anomalous density and a final linear regional trend. CAMACHO et al. (2000, 2002) and GOTTSMANN et al. (2008) give some simulation examples showing the suitability of this 3D inversion approach while also pointing out some limitations.

For isolated anomalous bodies, we suggest including the parameter g_0 as an unknown parameter in the fit equations. The reference papers give details about this option. The adjusted value for g_0 will contribute to satisfy the minimization condition. For isolated bodies, we also suggest keeping the same anomalous density contrast ($\Delta\rho^+$ and $\Delta\rho^-$) across the

model growth. The resulting anomalous bodies will be homogeneous and comparable for each other.

In the present study we introduce two improvements in the methodology to allow for a suitable modeling of stratified erosional structures, as in the case of alluvial valleys. They are: (a) the adoption of a suitable gravity offset g_0, and (b) the adoption of stepped density contrasts.

3.2. Gravity Offset g_0

A value of g_0 resulting from a free adjustment according to the global minimization conditions (Eq. 4) will be not very different from a mean anomaly value. It will provide isolated anomalous bodies with good fitting. These isolated bodies will involve some depth inverse mass distribution: positive anomalous density upon a not anomalous medium, or negative anomalous density below a not anomalous medium.

For a realistic stratified structure (as the case of a basin) the density follows mostly a non-inverse distribution. The density increases with depth. Inverse distributions are possible, but not frequent. We could restrict our results to models with non-inverse density distribution only. It could be partially controlled with the gravity offset parameter g_0.

For a model cell j with (positive or negative) density contrast $\Delta\rho_j$ located just upon a cell k with (positive, negative or null) density contrast $\Delta\rho_k$, we define the inverse contribution C_j as $C_j = D_j S_j$, where $D_j = \Delta\rho_j - \Delta\rho_k$ if $\Delta\rho_j > \Delta\rho_k$ and $D_j = 0$ in other cases. S_j is the contact area between j and k cells. The sum of the inverse contributions

$$C = \sum_{j \in J^+, J^-} C_j p_j \Big/ \sum_j p_j,$$

$$\text{with } p_j = \frac{1}{n} \left(\sum_{i=1,\ldots,n} \text{dist}^2(j-\text{cell}, i-\text{benchmark}) \right)^{1/2} \tag{7}$$

gives an index of the mass inversion present in the model. For a value g_0 close to the mean anomaly, the model will be constituted mostly by isolated bodies. The index of mass inversion C will be high. If we try smaller g_0 values, the value C will decrease, and the model will offer larger positive masses in the bottom.

If, simultaneously, we limit the maximum model depth, the model becomes rather stratified. After some trials, and without another additional information, we can reach a suitable g_0 value that produces $C \approx 0$. The corresponding model will present a suitable mass/depth distribution.

3.3. Stepped Density Contrasts

Usually, to get homogeneous bodies for the anomalous structures, and without other previous information, we take the prescribed values $\Delta\rho_j^+$ and $\Delta\rho_j^-$ as constant values everywhere and at every time during the computation. Then, the model shows only one value for the anomalous density contrast everywhere. See simulation examples in CAMACHO et al. (2000, 2002).

The alternative approach we propose here is to construct models with a higher density contrast in their core (or their bottom, for stratified structures) and with a lighter density contrast for their periphery (or their top, for stratified structures). For that, we start the model growth (Fig. 2) with prescribed density contrasts $\Delta\rho_0^+$ and $\Delta\rho_0^-$ (for instance, 60 kg/m^3 in Figs. 2, 3). For the adjusted initial cell, the adjusted scale factor takes the value

f_0. The growth process continues by adding new filled cells to the anomalous model. The adjusted scale factor decreases rapidly for the initial steps (see Fig. 2). The density contrasts remains at their initial values $\Delta\rho_0^+$ and $\Delta\rho_0^-$. When the scale factor arrives at a value f_1, we introduce a new (smaller) density contrast $\Delta\rho_1^+$ (for instance, 30 kg/m^3 in Fig. 2) for the following cells in the model growth. The process continues with this new density contrast. The negative density contrast can change at this point (f_1) or in another independent moment.

The choice of the suitable value f_1 (or, better, f_1/f_0) for density change will be decided with regard to the resulting model, and the corresponding information from boreholes or geologic or seismic data. Figure 3 shows an example of model (1) with one density contrast (60 kg/m^3), and (2) with a change of density contrast (form 60 to 30 kg/m^3) at some step in the model growth (according Fig. 2). This concept can be extended to several successive density (decreasing) changes to produce a stratified model with several layers.

3.4. Synthetic Example

In previous papers (CAMACHO et al. 2000, 2002, and GOTTSMANN et al. 2008), we presented some simulation test examples corresponding to the general GROWTH methodology for gravity inversion. Now, this section shows a brief synthetic example to

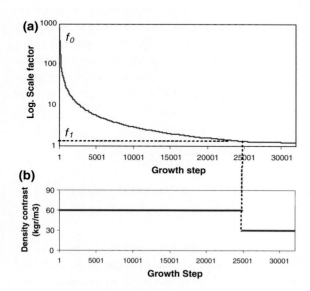

Figure 2
a Evolution of the scale factor during the model growth process, **b** stepped change of the density contrast according to the scale evolution to get a stratified inverse model

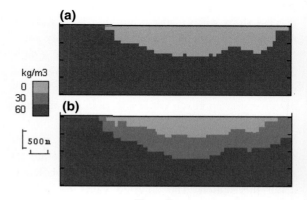

Figure 3
Effect of the stepped change of density contrast across the model growth according to Fig. 2. **a** Constant density contrast. **b** One step density contrast

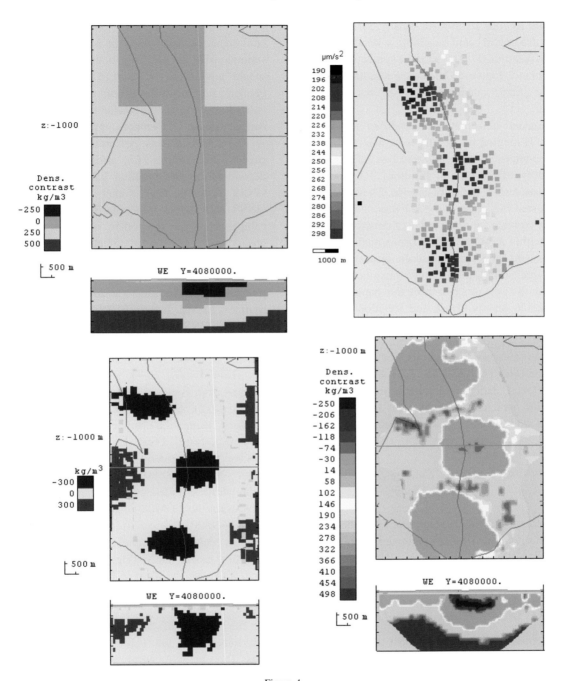

Figure 4

Modelling of a synthetic anomalous structure. **a** 3D synthetic model for a stratified basin structure. **b** Gravity anomaly corresponding to the synthetic structure for the application gravity points. **c** 3D model obtained by application of the general gravity inversion approach. **d** 3D model obtained by the modified inversion proposed in this paper

illustrate the effect of the new ways (adoption of a suitable gravity offset, and adoption of stepped density contrasts) corresponding to the study of an alluvial valley.

For higher homogeneity, we suppose the same area and the same distribution of gravity points as in the further application case (LAV, Sect. 4). Below this area we suppose a synthetic valley structure composed by four layers with density contrast -250, 0, 250, 500 kg/m^3 and reaching a depth of about $1{,}500$ m (see Fig. 4). The adopted synthetic value for g_0 is 0 μm/s^2. Figure 4b shows the simulated gravity anomaly field corresponding to the synthetic structure.

By means of direct application of the general GROWTH method for gravity inversion, with only a density contrast ±400 kg/m^3 and free adjustment of the gravity offset g_0, the resulting 3D model for anomalous density is that of Fig. 4c. The corresponding adjusted value for g_0 is 227 μm/s^2. This model could be suitable for other kinds of structures (isolate bodies within a rather homogeneous medium), but it is inadequate for a stratified basin context.

By application of the gravity inversion approach including the new procedures of this paper (adoption of a suitable gravity offset, and adoption of stepped density contrasts) and the same density contrast $(-250, 0, 250, 500$ kg/m$^3)$ then the resulting model (Fig. 4d) is clearly advantageous. It shows a stratified basin structure. The value for g_0 is now -37 μm/s^2. With a larger data set (greater coverage upon the anomalous structure) the fit with respect to the synthetic body would be even greater.

4. Application Test Case: The Low Andarax Valley (Almería, Spain)

As a test case, this new approach will be applied to get a structural 3D density model from the new gravity data observed in the LAV. In the following, we describe the data and the resulting model.

4.1. Data: Gravity, Positioning, and DEM

The gravimetric survey in the LAV was carried out during two field campaigns in July 2012 and May 2013. It consists of 382 gravimetric stations (see Fig. 1) covering the studied area, with a station spacing of about 200 m. A CG-5 Scintrex gravimeter was used for the field measurements. The total number of gravimetric observations was 437. The number of repeated stations was 11, and the number of observations at these base stations was 66.

Simultaneously, by using geodetic GPS TOPCON equipment, we determined the positions of the gravity stations with an accuracy better than ±5 cm (this would amount to about 0.15 μm/s^2). This GNSS geodetic support of the gravimetric observations was carried out using the Relative–Static method by carrier phase differences (HOFFMANN-WELLENHOF et al., 2008). It was performed collecting GNSS carrier observations for periods of 10–30 min, depending on the baseline length, and using two GNSS receivers. One of them remained collecting GNSS data in the same place during all the field campaign while the other one recorded GNSS observations on the gravimetric station simultaneously with the gravimetric observations.

For the GNSS data processing, an existing permanent GNSS station was added to the processing routines, ALME. It is an EUREF station operated by the IGN CORS network (PRIETO et al., 2000). GNSS data were processed using the International GNSS Service final combined solution's precise ephemerides (DOW et al., 2009) and calibration antennae patterns from National Geodetic Survey (MADER and BILICH, 2012). A L3 free fix ambiguity solution was obtained using the Hopfield tropospheric model (HOPFIELD, 1969). Final coordinates were computed relative to a geodetic reference system ETRS89, which uses the GRS80 ellipsoid (MORITZ, 1980). The orthometric altitudes were computed from ellipsoid ones using the EGM08-REDNAP geoid model from IGN (IGN, 2010).

This dataset is complemented with a digital elevation model, DEM (Fig. 1), for the surrounding area, with a step of 10 m × 10 m, and extended up to a radius of 20 km around the area (JUNTA DE ANDALUCÍA, 2005).

The process of data correction starts with the determination of the tidal correction for the gravity data. The obtained values range between -0.79 and $+1.10$ nμm/s^{2}l. Next, by using a global fit of the redundant observations, we obtained that the precision of the adjusted (relative) gravity values was ±0.38 μm/s^2. The adjusted instrumental drifts accounted for 8.2 ± 0.03 μm/s^2/day for July 2012 and 6.9 ± 0.09 μm/s^2/day for May 2013. Next, we carried out a determination of the gravimetric corrections due to terrain effects from the DEM and corresponding to the observation locations. The obtained values range between 264 and 1,469 µGal, for a reference terrain density value of 1,000 kg/m³ (this density value is only an initial value that will be changed during the inversion process).

Next, we determined the Bouguer gravity anomaly. First, computation of the normal gravity was referenced to GRS80 (MORITZ, 1980) and then, a free-air gradient -3.086 µm/s²/m. The Bouguer gravity correction was calculated using the average density of 2,300 kg/m³. This local value for average density was determined from the gravity data by looking for a minimum correlation between elevation and a gravity anomaly for the shortest wavelength components of topography. It is attained just by means of the further inversion process, according to an improvement

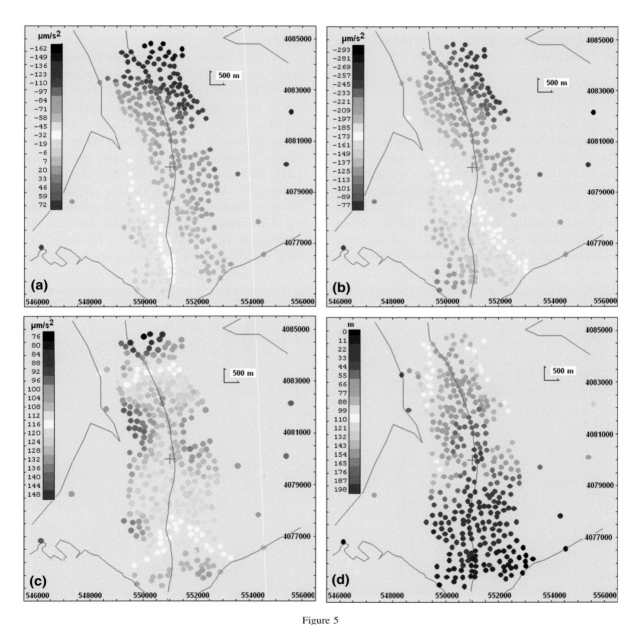

Figure 5

a Observed Bouguer anomaly (*colour* step 1,355 µGal). **b** Adjusted linear trend NE–SW (*colour* step 1,146 µGal). **c** Local anomaly (observed anomaly minus trend) (*colour* step 444 µGal). **d** Elevation of the gravity benchmarks (colour step 11.4 m). UTM coordinates in axes

developed by CAMACHO *et al.* (2007), which allows modifying the initial assumed terrain density.

By including the terrain correction for the gravity disturbances, we computed a Bouguer gravity anomaly (with terrain correction). The values of this Bouguer gravity anomaly (with topographic correction) have a dispersion of ± 34.18 µm/s^2, and a difference between extreme values of 225.93 µm/s^2. The corresponding anomaly map (Fig. 5a) shows the dominant presence of a clear ENE–WSW increasing regional trend, which does not allow distinguishing any local details.

4.2. Resulting Inverse Model

The first result of the inversion process is the simultaneous determination of a linear regional trend (Fig. 5b). In the LAV, this regional trend generates a gravity increase of 18,25 µm/s^2/km according to a ENE–WSW direction N51°E. CASAS and CARBÓ (1990) and GALINDO-ZALDÍVAR *et al.* (1997) showed sharp gravity gradients close to the coast in the area of Betics. TORNÉ and BANDA (1992) and GALINDO-ZALDÍVAR *et al.* (1997) explained these gradients as

due to a sharp local change of the crustal thickness. However, this trend can be related to the presence of dense limestones and dolomites of Sierra de Gádor in the western part of the valley.

After removing this regional trend, the resulting anomaly (local anomaly) (Fig. 5c) shows some local features, which should correspond to local anomalous density structures. The benchmark elevation values are shown in Fig. 5d. The inversion process uses both data sets to adjust a 3D structure for anomalous density.

The main decision for the gravity inversion is the choice of some density contrast for the model. For that, we follow the previous work by MARIN LECHADO (2005). He carried out gravimetric studies for two contiguous areas: Campo de Dalías (west) and Campo de Nijar (east). In his wide study he included seismic, magnetic, and borehole data, and geologic information. Based on this previous work (mainly in the part of Campo de Dalías), we select the following values to carry out the 3D gravity modelling of the basin: (1) limy basement: 2,700 kg/m^3; (2) intermediate sedimentary infill: 2,400 kg/m^3; (3) 2,250 kg/m^3; and (4) light deposits: 2,000 kg/m^3.

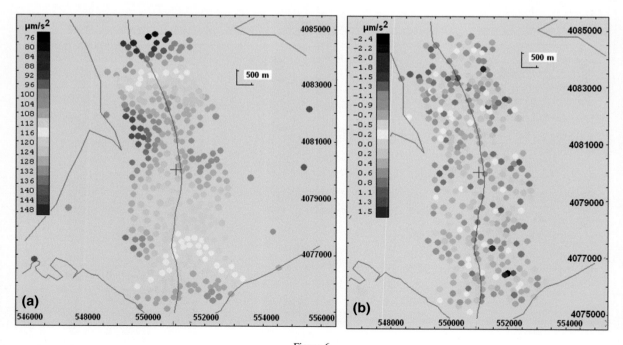

Figure 6

a Modelled anomaly after the gravity inversion (colour interval 4.29 µm/s^2). The fit to the observed anomalies (Fig. 4c) is very good. **b** Final residual values after the gravity inversion (colour interval 22 µGal). The standard deviation is about 60 µGal. UTM coordinates in the axes

So, we include the following anomalous density contrasts +450, +150, 0, −250 kg/m³ for the layered model. Moreover, MARIN LECHADO (2005) suggested depths for the basement of around 700 m. So we limit our model to a depth of 1.4 km.

Once those parameters are fixed, the inversion approach is nearly automatic. First, we select a 3D grid composed by 83,407 cells, with an average side of about 90 m. The resulting model is determined by a step-by-step aggregation of 1,000 cells filled with the prescribed density contrasts. It reproduces the observed anomaly well (Fig. 6a). The quality of the adjustment of the model can be assessed from the standard deviation of the final residuals, which is

0.6 µm/s² (Fig. 6b). These residuals, essentially uncorrelated, are produced by very local anomalies, slight imperfections in the topographic correction, or the errors in the altimetry or gravimetric observations.

Figure 7 shows the resulting 3D model for anomalous density by means of several cross-sections (horizontal sections, and WE and SN vertical profiles). This model shows a stratified structure composed by sub-horizontal layers. Figure 7 suggests the adjusted topography for the discontinuity surfaces S1 and S2 between the assumed media: (M1) +450 kg/m³ (basement 2,700 kg/m³), (M2) deep sedimentary infill +150 kg/m³ (2,400 kg/m³) and

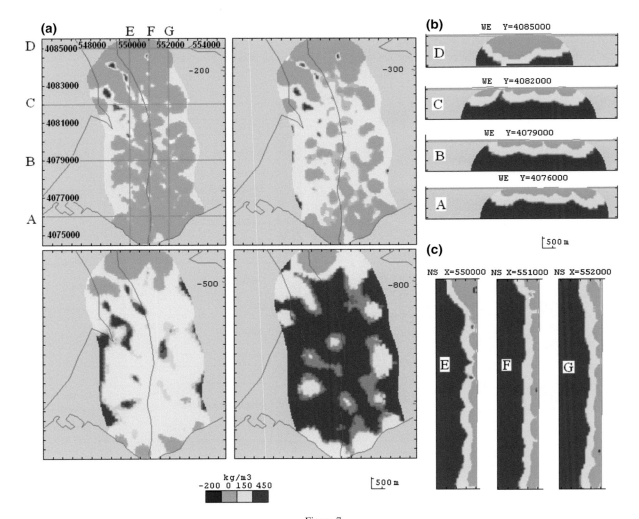

Figure 7
Inverse 3D model for anomalous density. Several cross-sections. **a** Horizontal sections (for 200, 300, 500, 800 m below sea level). **b** WE profiles. **c** NS profiles

157

(M3) shallow sediments 0 kg/m³ (2,250 kg/m³). Very shallow and light material, M4, −250 kg/m³ (2,000 kg/m³) appears only in few locations of the model due to the fact that distance between benchmarks (about 200 m) is greater than the layer thickness of about 100 m (MARIN LECHADO, 2005). We observe that the adjusted depths amount to about 700 m for S1 (basement) and to about 400 m for S2 (between M2 and M3, Neogene sediments) in the central portion of the low basin. A basement depth of ∼550 m was estimated in the South of the model in the test-site for the SCA method (GARCÍA-JEREZ, 2010), which shows high densities in that area beneath some interface lying between 400 and 600 m.

The morphology of the stratified structure mostly corresponds to a nearly flat low basin. However, depths of discontinuity surfaces oscillate showing a topography of lows and highs, suggesting some particular features, which could be correlated with structural peculiarities.

5. Discussion and Conclusions

The morphology of the LAV consists of a stratified structure composed by sub-horizontal layers, mostly corresponding to a nearly flat low basin. In the prospected area, the basement is found symmetrically to the river bed, being associated with the foothills of Sierra Alhamilla (eastern edge) and Sierra the Gádor (western edge). The near-surface sections (200–300 m) show a more heterogeneous density distribution that may correspond to greater variations in lithologies. Conversely, deeper densities become more uniform, meaning less diverse materials.

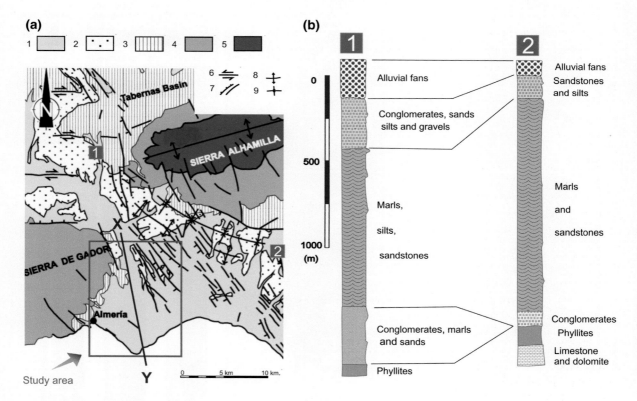

Figure 8

a Geological map of the Lower Andarax valley. *1* Quaternary detrital alluvial sedimentary deposits, *2* marls, sands and calcarenites (Pliocene), *3* marls, silts, and sandstones (middle Miocene), *4* Alpujárride complex (phyllites, limestones and dolomites), *5* Nevado–Filábride complex (metapelites), *6* strike slip fault, *7* normal fault, *8* antiform, *9* synform. The cross-section of this figure is marked (*X–Y*). Modified from PEDRERA *et al.* (2006). **b** Synthetic stratigraphic columns and their correlation in different sectors of the Almeria–Nijar Basin. Modified from MARIN LECHADO (2005)

We detect a sharp structure of low density in the most northern part of the gravity survey. It includes a deep hole in the basement filled by sediments. This is in agreement with the geological data by MARIN LECHADO (2005) (Fig. 8) showing the sedimentary basin fill (marls, silt sands and conglomerates) thickening to the north, and also with data from ground water wells (SÁNCHEZ-MARTOS, 1997). Nevertheless, we would need to extend the gravity survey to the North in future works to get a full coverage of this low and avoid the boundary effects which may distort the anomaly.

Some particular features have been inferred from the gravity data. For instance, we observe a particular crest (**V** in Fig. 9) of the layers, close to the village of Viator. This feature agrees with boreholes data and seismic noise surveys, which show a limestone-dolomitic mass at ∼200 m depth and a high-density anomaly. This place (near Viator village) is the only zone where boreholes shown in Fig. 9a reveal such a type of geological materials. This anomaly, which can be clearly seen in the 200–600 m depth sections, matches previous cross-sections proposed by SÁNCHEZ-MARTOS (1997) on the basis of borehole data. We have found a NW–SE trend for this anomaly and an approximate length of 2 km in horizontal sections down to 500 m depth. The model provides new information about similar structures located SE from anomaly V and east of the river channel. These areas have not been sufficiently explored with boreholes or they are very shallow. These anomalies are deeper than the Viator structure, being detected below a depth of 500 m.

Another interesting feature corresponds to certain structural alignments (lows) appreciated in the model for the basin. If we compare this map of alignments with a map of apparent faults in the area (Fig. 8a, MARIN LECHADO, 2005) some coincidence can be detected. It allows us to interpret these sharp model alignments as corresponding mostly to faults striking mainly NNW–SSE. Figure 10 shows they run on a horizontal profile obtained by inversion with the general GROWTH inversion method. Alignments following the NW–SE faults can be seen in these sections, according to the NNW–SSE normal faults shown by some authors (SANZ DE GALDEANO et al. 2010; PEDRERA et al., 2012). This N140–160E

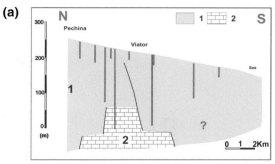

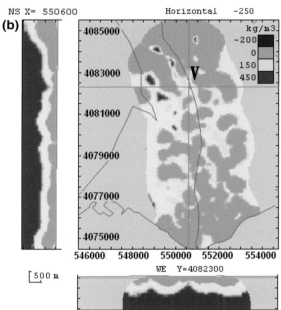

Figure 9
a Geological north–south section of the LAV and situation of the boreholes. *Legend 1* marls, sandy silts, sands and conglomerates. *2* Limestones and dolomites. **b** Horizontal (250 m depth) and vertical (NS and WE) cross-sections of the density model inverted from the gravimetric survey, showing a high density anomaly in Viator area. UTM coordinates

direction also coincides with old faults that occurred during the Miocene period (MARTÍNEZ-MARTÍNEZ and AZAÑÓN, 1997). The alignment N140°E close to the Sierra de Gádor foothills (NW of the prospected area) is the clearest one. It is also interesting to point out that the adjusted gravity regional trend follows the orthogonal course (51°N). This coincidence suggests some structural relation. Cross-sections of the density model also show probable SSW–NNE trends that have not been described in former works.

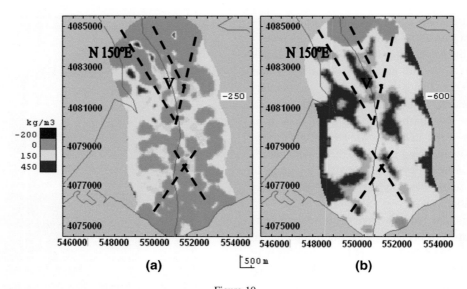

(a)   ⌈500 m   **(b)**

Figure 10
Some structural alignments (mainly N150°E) of lows in the inverse model. **a** Shallow section at 250 m depth below sea level. **b** Deep section at 600 m. *Letter V* indicates location of Viator village UTM coordinates

Acknowledgments

We thank S. Limonchi, A. Sánchez, and A. Jiménez for their help in the field campaigns. This work was supported by the Spanish research projects CGL2010-16250, and GEOSIR (AYA2010 17448), by the EU with FEDER, by the research team RNM-194, and the Water Resources and Environmental Geology Research Group (RNM-189) of Junta de Andalucía, Spain. A.G.-J. was supported by a Juan de la Cierva grant from the Spanish Government. This research is a contribution of the Moncloa Campus of International Excellence (UCM-UPM, CSIC).

REFERENCES

AL-CHALABI, M. (1971), *Some studies relating to non-uniqueness in gravity and magnetic inverse problem*, Geophysics, *36*, 835–855, doi:10.1190/1.1440219.

CAMACHO, A.G., MONTESINOS, F.G. & VIEIRA, R. (2000). *A 3-D gravity inversion by means of growing bodies*. Geophysics, *65*: 95–101.

CAMACHO, A.G., MONTESINOS, F.G. & VIEIRA, R. (2002). *A 3-D gravity inversion tool based on exploration of model possibilities*. Comput. Geosci, *28*, 191–204.

CAMACHO, A.G., NUNES, J.C., ORTIZ, E., FRANÇA, Z. & VIEIRA, R. (2007). *Gravimetric determination of an intrusive complex under the island of Faial (Azores). Some methodological improvements*. Geophys. J. Int. *171*, 478–494.

CAMACHO, A.G., FERNÁNDEZ, J., GONZÁLEZ, P.J., RUNDLE, J.B., PRIETO, J.G., ARJONA, A. (2009). *Structural results for La Palma Island using 3-D gravity inversion*. Journal of Geophysical Research, *114*, B05411, doi:10.1029/2008JB005628.

CAMACHO, A.G., GOTTSMANN, J. & FERNÁNDEZ, J. (2011a). *The 3-D gravity inversion package GROWTH2.0 and its application to Tenerife Island, Spain*. Comput. Geosci, *37* (2011) 621–633.

CAMACHO, A.G., FERNÁNDEZ, J. & GOTTSMANN, J. (2011b). *A new gravity inversion method for multiple sub-horizontal discontinuity interfaces and shallow basins*. J. Geophys. Res., *116*, B02413, doi:10.1029/2010JB008023.

CASAS, A & CARBÓ, A. (1990). *Deep structure of the Betic Cordillera derived from the interpretation of a complete Bouguer anomaly map*. J. Geodynamics, *12* (2–4), 137–147.

CHAKRABORTY, K., and B. N. P. AGARWAL (1992), *Mapping of crustal discontinuities by wavelength filtering on the gravity field*, Geophys. Prospect., *40*, 801–822, doi:10.1111/j.1365-2478.1992.tb00553.x.

CORDELL, L. AND HENDERSON, R. G. (1968). *Iterative three-dimensional solution of gravity anomaly data using a digital computer*, Geophysics, *33*, 596–601, doi:10.1190/1.1439955.

DOW, J., NEILAN, R. E. and RIZOS, C. (2009). "*The International GNSS Service in a changing landscape of Global Navigation Satellite Systems*". Journal of Geodesy, *83*(3–4), pp. 191–198.

GARCÍA-JEREZ, A. (2010) Desarrollo y evaluación de métodos avanzados de exploración sísmica pasiva. Aplicación a estructuras geológicas locales del sur de España. PhD. Thesis, Universidad de Almería, Spain.

GALINDO-ZALDÍVAR, J., JABALOY, A., GONZÁLEZ-LODEIRO, F., ALDAYA, F., (1997). *Crustal structure of the central Betic Cordillera (SE Spain)*. Tectonics *16*, 18–37.

GOTTSMANN, J., CAMACHO, A.G., MARTI, J., WOOLLER, L., FERNÁNDEZ, J., GARCIA, A. AND RYMER, H. (2008). *Shallow structure beneath the Central Volcanic Complex of Tenerife from new* gravity data: *Implications for its evolution and recent reactivation.* Phys. Earth Planet. Int., *168*, 212–230.

HOFMANN-WELLENHOF, B., LICHTENEGGER, H. AND WASLE, E. GNSS—Global Navigation Satellite Systems. GPS, GLONASS, Galileo & more. (SpringerWienNewYork, Wien, Austria, 2008).

HOPFIELD, H.S. (1969). *Two-quadratic tropospheric refractivity profile for correction satellite data.* Journal of Geophysical Research, *74*(18), 4487–4499.

IGN—INSTITUTO GEOGRÁFICO NACIONAL. El Nuevo modelo de geoide para España EGM08-REDNAP (Centro de Observaciones Geodésicas, Instituto Geográfico Nacional, Madrid, 2010).

JUNTA DE ANDALUCÍA (2005). Modelo Digital del Terreno de Andalucía. Relieve y Orografía. ISBN: 84-96329-34-8. Sevilla.

LEÃO, J. W. D., MENEZES, P. T. L., BELTRAO, J. F. AND SILVA, J. B. C. (1996). *Gravity inversion of basement relief constrained by the knowledge of depth at isolated points.* Geophysics, *61*, 1702–1714, doi:10.1190/1.1444088.

LI, Y and OLDENBURG, D.W: (1998) *3-D inversion of gravity data.* GEOPHYSICS, VOL. *63*, 109–119.

LUZÓN, F., AOI S., FÄH D. and SÁNCHEZ-SESMA F.J. (1995). *Simulation of the seismic response of a 2D sedimentary basin: A comparison between the Indirect Boundary Element Method and a Hybrid Technique.* Bull. Seism. Soc. Am. Vol. *85*, pp. 1501–1506.

LUZÓN, F., L. RAMÍREZ, F. J. SÁNCHEZ-SESMA and A. POSADAS (2004). *Simulation of the seismic response of sedimentary basins with vertical constant-gradient of velocity.* Pure and Applied Geophysics, Vol. *161*, 1533–1547.

LUZÓN, F., F.J. SÁNCHEZ-SESMA, J.A. PÉREZ-RUIZ, L. RAMÍREZ, and A. PECH (2009). *In-plane seismic response of inhomogeneous alluvial valleys with vertical gradients of velocities and constant Poisson ratio.* Soil Dynamics and Earthquake Engineering, doi:10.1016/j.soildyn.2008.11.007.

MADER, G. and BILICH, A.L. (2012). *Absolute Antenna Calibration at the US National Geodetic Survey.* AGU Fall Meeting. San Francisco, 3–7 December.

MARIN LECHADO, C. (2005). Estructura y evolución tectónica reciente del Campo de Dalías y de Níjar en el contexto del límite meridional de las Cordilleras Béticas orientales. PhD Thesis. Universidad de Granada.

MARTÍNEZ-MARTÍNEZ, J.M., AZAÑÓN, J.M. (1997). *Mode of extensional tectonics in the southeastern Betics (SE Spain). Implications for the tectonic evolution of the peri-Alborán orogenic system.* Tectonics 16, 205–225. doi:10.1029/97TC00157.

MARTIN-ROSALES, W., PULIDO-BOSCH, A., VALLEJOS, A. & LÓPEZ-CHICANO, M (1996) *Extreme rainfall in Campo de Dalías and Southern edge of Sierra de Gádor (Almería).* Geogaceta, *20* (6), 1251–1254.

MORITZ, H. (1980). *Geodetic Reference System 1980,* Bulletin Géodésique, *54*(3), pp. 251–265.

OLDENBURG, D. W. (1974), *The inversion and interpretation of gravity anomalies,* Geophysics, *39*, 526–536, doi:10.1190/1.1440444.

PEDRERA , A., MARIN-LECHADO, C., GALINDO-ZALDIVAR, J., RODRI-GUEZ-FERNANDEZ, L.R., RUIZ-CONSTAN, A. (2006): Fault and fold interaction during the development of the Neogene-Quaternary Almeria–Nijar basin (SE Betic Cordilleras). In: C. MORATTI, A. CHALUAN (eds.), Tectonics of the Western Mediterranean and North Africa. Geological Society, London, Special Publications, 217–230. doi:10.1144/GSL.SP.2006.262.01.13.

PEDRERA, A., GALINDO-ZALDÍVAR, J., MARÍN-LECHADO, C., GARCÍA-TORTOSA, F.J., RUANO, P., LÓPEZ GARRIDO, A.C., AZAÑÓN, J.M., PELÁEZ, J.A. y GIACONIA, F. (2012), *Recent and active faults and folds in the central-eastern Internal Zones of the Betic Cordillera,* Journal of Iberian Geology 38, 191–208.

PICK, M., J. PICHA, and V. VYSKÔCIL (1973), Theory of the Earth's Gravity Field, 538 pp., Elsevier, Amsterdam.

PRIETO, J., SÁNCHEZ-SOBRINO, J.A. and QUIRÓS, R. (2000). Spanish National GPS Reference Station Network (ERGPS). Boletín Real Instituto y Observatorio de la Armada, 3/2000.

PULIDO-BOSCH, A., SÁNCHEZ MARTOS, F., MARTÍNEZ VIDAL, J.L., NAVARRETE, F. (1991). *Characterization of the overexploitation in the middle and lower Andarax (Almería, Spain).* XXIII IAH Congress Proc., Vol. *I*, 563–569.

RAMA RAO , P., K. V. SWAMY, and I. V. RADHAKRISHNA MURTHY (1999), *Inversion of gravity anomalies of three-dimensional density interfaces,* Comput. Geosci., *25*, 887–896, doi:10.1016/S0098-3004(99)00051-5.

REAMER, S. K., and J. F. FERGUSON (1989), *Regularized two-dimensional Fourier gravity inversion method with application to the Silent Canyon caldera, Nevada,* Geophysics, *54*, 486–496, doi:10.1190/1.1442675.

SÁNCHEZ-MARTOS, F. (1997). Estudio hidrogeoquímico del Bajo Andarax (Almería). PhD Thesis. University of Granada, Spain, 290 pp.

SANZ DE GALDEANO, C., J. RODRÍGUEZ FERNÁNDEZ, and A. C. LÓPEZ GARRIDO. (1985). *A strike-slip fault corridor within the Alpujarra Mountains (Betic Cordilleras, Spain).* Geologische Rundschau, *74*, 641–675.

SANZ DE GALDEANO, C. AND VERA, J.A. (1992). *Stratigraphic record and palaeogeographical context of the Neo-gene basins in the Betic Cordillera, Spain.* Basin Research, *4*: 21–36. doi:10.1111/j.1365-2117.1992.tb00040.x.

SANZ DE GALDEANO, C., SHANOV, S., GALINDO-ZALDIVAR, J., RADULOV, A., NIKOLOV, G. (2010): *A new tectonic discontinuity in the Betic Cordillera deduced from active tectonics and seismicity in the Tabernas Basin.* Journal of Geodynamics 50, 57–66. doi:10.1016/j.jog.2010.02.005.

TARANTOLA, A. (1988). *The inverse problem theory: Methods for data fitting and model parameter estimation.* Elsevier, Amsterdam, 613 pp.

TORNÉ, M. AND BANDA, E. (1992). *Crustal thinning from the Betic Cordillera to the Alboran Sea.* Geo-Mar. Lett., *12*, 76–81.

VOERMANS, F. AND BAENA, J. (1983). Memoria y hoja geológica de Almería (1043) 1:50.000. IGME. Madrid. 53 p.

(Received February 28, 2014, revised July 7, 2014, accepted July 22, 2014, Published online August 17, 2014)

Reprinted from the journal

Pure Appl. Geophys. 172 (2015), 3123–3137
© 2014 Springer Basel
DOI 10.1007/s00024-014-0985-6

| Pure and Applied Geophysics

Characterization of Cavities Using the GPR, LIDAR and GNSS Techniques

MIGUEL ANGEL CONEJO-MARTÍN,[1] TOMÁS RAMÓN HERRERO-TEJEDOR,[1] JAVIER LAPAZARAN,[2] ENRIQUE PEREZ-MARTIN,[1] JAIME OTERO,[2] JUAN F. PRIETO,[3] and JESÚS VELASCO[3]

Abstract—The study of the many types of natural and manmade cavities in different parts of the world is important to the fields of geology, geophysics, engineering, architectures, agriculture, heritages and landscape. Ground-penetrating radar (GPR) is a noninvasive geodetection and geolocation technique suitable for accurately determining buried structures. This technique requires knowing the propagation velocity of electromagnetic waves (EM velocity) in the medium. We propose a method for calibrating the EM velocity using the integration of laser imaging detection and ranging (LIDAR) and GPR techniques using the Global Navigation Satellite System (GNSS) as support for geolocation. Once the EM velocity is known and the GPR profiles have been properly processed and migrated, they will also show the hidden cavities and the old hidden structures from the cellar. In this article, we present a complete study of the joint use of the GPR, LIDAR and GNSS techniques in the characterization of cavities. We apply this methodology to study underground cavities in a group of wine cellars located in Atauta (Soria, Spain). The results serve to identify construction elements that form the cavity and group of cavities or cellars. The described methodology could be applied to other shallow underground structures with surface connection, where LIDAR and GPR profiles could be joined, as, for example, in archaeological cavities, sewerage systems, drainpipes, etc.

Key words: Cavities, underground cellars, GNSS, LIDAR, GPR, geodetection.

1. Introduction

The study of the different types of natural and manmade cavities is important to the fields of geology, geophysics, engineering, architecture, agriculture, heritages and landscapes (LÓPEZ-GETA 2002; DEPARIS et al. 2008; JOL 2009; PETTINELLI et al. 2011). Geodetection is a noninvasive technique that is suitable for the accurate location of buried structures underground. Traditionally, this kind of underground cavity is characterized by topographic work, normally using laser imaging detection and ranging (LIDAR) techniques. As these techniques do not detect hidden structures, ground-penetrating radar (GPR) has also been used in cavity characterization by different authors. We add the integration of LIDAR with GPR data, joined by Global Navigation Satellite System (GNSS) positioning, to achieve the characterization of structures not detected by LIDAR.

GPR systems work by propagating a radio wave through the ground. The propagation characteristics of the ground medium determine the propagation velocity of the electromagnetic wave (EM velocity) and its attenuation as the wave propagates. Except for a few magnetic minerals, propagation in most ground media depends mainly on their electrical properties (e.g., LORENZO 1996; REPPERT and MORGAN 2000; LAPAZARAN 2004). Laser scanning is also a suitable measuring technique for monitoring the deformations of certain structures over time to ensure structural safety and guarantee the control of the structure (KEUMSUK et al. 2007; BURNS et al. 2012).

The most commonly used techniques in geomorphology are GPR, seismic refraction and direct current (DC) electrical resistivity. These techniques are useful in answering unanswered questions in geomorphological research regarding the thickness of

---

[1] Departamento de Ingeniería Cartográfica, Geodesia y Fotogrametría, Expresión Gráfica, EUIT Agrícola, Universidad Politécnica de Madrid, Ciudad Universitaria S/N, 28040 Madrid, Spain. E-mail: miguelangel.conejo@upm.es; tomas.herrero.tejedor@upm.es; enrique.perez@upm.es
[2] Departamento de Matemática Aplicada, ETSI de Telecomunicación, Universidad Politécnica de Madrid, Avenida Complutense 30, 28040 Madrid, Spain. E-mail: javier.lapazaran@upm.es; jaime.otero@upm.es
[3] Departamento de Ingeniería Topográfica y Cartografía, ETSI Topografía, Geodesia y Cartografía, Universidad Politécnica de Madrid, Carretera de Valencia, km 7, 28031 Madrid, Spain. E-mail: juanf.prieto@upm.es; jesus.velasco@upm.es

sediments and internal structures. SCHROTT and SASS (2008) report that the use of a single geophysical technique or a single interpretation tool is not recommended for many geomorphological surface and subsoil conditions as it may lead to substantial errors in interpretation. Due to modifications in the physical properties of the subsoil material (e.g., sediments, water content), in many cases only a combination of more than one geophysical method can give sufficient vision to avoid misinterpretation. Similarly, the use of geophysical methods has made it possible to identify the size, depth, shape and direction of difficult-to-access zones such as areas concealed behind walls or where there have been landslides within the cavities (JOL 2009).

The joint use of GPR techniques, LIDAR and GNSS has been little used. Recently, a combination of GPR techniques and electrical resistivity tomography (ERT) has been applied to reveal archaeological structures in a mausoleum (NUZZO et al. 2009) and to detect a bronze foundry complex in the Acropolis (LEOPOLD et al. 2011). SCHEIB et al. (2008) applied a combination of methodologies to characterize the geomorphology of a small mountain basin in Scotland. A terrestrial LIDAR was used to create a digital terrain model (DTM) on the surface and GNSS as a complement to other techniques. DEPARIS et al. (2008) used combined LIDAR, GPR and ERT data to assess the stability of cliffs. More recently, KEUSCHNIG et al. (2010) and HARTMEYER and KEUSCHNIG (2012) used a multidisciplinary approach based on GPR, ERT and LIDAR to investigate the stability responses of rock faces to climate change in high mountain areas. These techniques are also widely used in subsoil research for other purposes such as archaeology (CONYERS and GOODMAN 1997; PETTINELLI et al. 2011), underground water contamination (LÓPEZ-GETA 2002), the general location of structures and anomalies in the subsoil, and localization and mapping of underground urban services (YOUNG and LORD 2002; FRANCESE and MORELLI 2006; FRANCESE et al. 2009). RODRÍGUEZ-GONZÁLVEZ et al. (2014) developed a spatial information system in underground cavities that is able to detect and document the various elements with an array of cartographic products using LIDAR, GPR and ERT techniques. However, they do not calculate the

precision of the detections, and the integration of the LIDAR and GPR registration is not calibrated. In this study, the authors estimate the EM velocity based on prior knowledge of a depth anomaly, information that is not always available.

In this study, we apply GPR surface prospecting and underground LIDAR scanning together with GNSS positioning to enable the determination and location of internal cavity structures, one of the aims of this research. To achieve this, we describe a novel methodology for estimating EM velocity based on its direct calibration by 'tuning' it using LIDAR-DTM data in an iterative process. We also explain an application test case, including data collection with the three sensors (GPR, LIDAR and GNSS), their processing and results, following a description of the methods used. The results are analyzed and discussed.

2. Methods

The GPR technique uses high-frequency electromagnetic waves to obtain structural information of the subsoil. The electromagnetic pulse is emitted from the transmitter antenna and travels through the subsoil at a speed determined by the dielectric properties of the subsoil materials. The pulse is reflected because of the lack of homogeneity (e.g., at a layer limit) and is received by a second antenna on the receiver, measuring the travel time. This EM velocity is conditioned by the dielectric permittivity of the soils (VASUDEO et al. 2009).

The dielectric permittivity of the medium (F m^{-1}) can be expressed as:

$$\varepsilon = \varepsilon_0\, \varepsilon_r, \tag{1}$$

where ε_0 is the dielectric permittivity in space ($\varepsilon_0 = 8.854187817 \times 10^{-12}$ F m^{-1}) and ε_r is the relative permittivity of the medium (dimensionless).

The attenuation (dB) of the wave propagated to a depth z and returned can be characterized as:

$$N_A = \frac{17.372\, z\, \omega}{c} \sqrt{\frac{\varepsilon_r}{2}\left[\sqrt{1 + \left(\frac{\sigma}{\omega\varepsilon}\right)^2} - 1\right]}$$
$$= 17.372\, \frac{z}{\delta}, \tag{2}$$

where $c = 2.99792458 \times 10^8$ m s^{-1} is the EM velocity in vacuum; σ is the conductivity of the medium (S m^{-1}) and ω the angular frequency of the wave ($\omega = 2\pi f$, where f is the wave's frequency). The parameter δ is the penetration depth, also known as the skin depth, and represents the depth at which the amplitude of the wave is reduced to $1/e$ of its initial value.

The EM velocity in a dielectric is given by:

$$v = \frac{c}{\sqrt{\frac{\varepsilon_r}{2}\left[\sqrt{1 + \left(\frac{\sigma}{\omega\varepsilon}\right)^2} + 1\right]}}. \quad (3)$$

When σ is small as compared with $\omega\varepsilon$, the EM velocity can be expressed as:

$$v = \frac{c}{\sqrt{\varepsilon_r}}. \quad (4)$$

The attenuation is clearly dependent on frequency: the higher the frequency, the lower the penetration of the wave. However, the EM velocity is not dependent on frequency. In low conductivity media, the permittivity is the parameter controlling both the attenuation and the EM velocity: the higher the permittivity, the lower both the velocity and the penetration of the wave. Similarly, when the medium has high conductivity, both the velocity and penetration are reduced. The ground tends to have low conductivity, but it can be greater in a wet terrain, particularly if minerals are present (Table 1).

A GPR recording is a time sequence of received amplitudes and requires the knowledge of the EM velocity to locate these detections at a depth. The bibliography includes several well-known methods for measuring EM velocity (e.g., DOBRIN and SAVIT 1988; BALANIS 1989; LORENZO 1996; YILMAZ 2001; LAPAZARAN 2004; JOL 2009). There is no easy way of implementing these methods with reliable results; there are persistent problems with the shape, steepness or positioning of the reflector, and the hypotheses on soil homogeneity are not always achieved.

Some methods aim to obtain the effective parameters of the medium in the laboratory using ground samples. Other methods try to determine prior environmental stratigraphy and its relation to reflection times (FERNÁNDEZ PASTOR 2007). For instance,

Table 1

Propagation parameters of different media (modified from BENSON et al. 1983; DAVIS and ANNAN 1989; PÉREZ 2001; DANIELS et al. 2004)

Material	Effective permittivity, ε_{efr}	Conductivity, σ (mSm^{-1})	EM velocity, v (m µs^{-1})
Air	1	0	300
Distilled water	80–88	0.01	33
Fresh water	80–88	0.1–10	33
Salt water (and marine)	80–88	4,000	10
Polar snow	1.4–3	–	190–250
Pure ice	3.17–3.19	0.02–0.003	168–169
Permafrost	1–10	0.01–15	80–300
Concrete	4–10	10^{-9}–10^{-2}	95–150
Limestone dry-wet	**4–16**	10^{-5}–25	**75–150**
Schist dry-wet	**5–15**	1–100	**77–134**
Granite dry-wet	**4–15**	10^{-9}–1	**110–130**
Sand dry-saturated	**3–30**	10^{-7}–1	**55–174**
Silt dry-saturated	**5–30**	1–100	**63–100**
Clay dry-saturated	**4–50**	0.25–>1,000	**60–170**

The typical soils that might be found above Spanish cellars are highlighted in bold

LÓPEZ-PIÑERO et al. (1998) used local time-domain reflectometry (TDR) to estimate the depth of penetration of the radar signal. They found that the data significantly differ when using computed values of ε_r instead of the commonly accepted values calculated in laboratory research with radar.

EM velocity is highly dependent on the soil type, as shown in Table 1. This dependence is also very closely correlated with soil porosity—particularly when it is not consolidated—its wetness and the concentration of dissolved mineral salts. Even when the macroscopic composition of the soil medium is known, a measurement of the EM velocity is required.

In this article, we propose a method to estimate the EM velocity in the ground by matching two profiles corresponding to the same place although obtained using two different techniques: GPR from the surface and LIDAR from inside and outside the cavities, integrating both systems with GNSS techniques. Once the EM velocity is known, the GPR profile will be properly processed and migrated, and detections will also show the hidden cavities and structures by comparing them with the visible internal structures detected by LIDAR.

We compute the EM velocity by superimposing two longitudinal profiles along the same site with different measurement techniques. One is recorded by means of GPR and the other using LIDAR, both integrated with GNSS techniques. We estimate the EM velocity in the ground required to locate the GPR detection of hidden cavities and old hidden structures from the cellar by fitting them in our test case. This is a novel method for tuning the EM velocity through the conjunction of these technologies.

The emergence of new terrestrial laser scanner devices in the field of measurement has increased the possibility of obtaining more accurate and complete 3D models of the objects. Data acquisition with laser scanning devices is also very quick. However, particular care should be taken during the analysis, processing and modeling phases of the laser scanner data, which tend to have a high presence of noise that needs to be removed before interpretation. If the object is complex, multiple LIDAR records must be made and georeferenced, identifying at least three homologous points in the overlapping area of two consecutive LIDAR records. The subsequent data-processing phase must calculate the three rotations and the three translations in order to refer each pair of LIDAR records to a single reference system. The scheme of the methodology is shown in Fig. 1 and is explained throughout this article.

The main problem when processing LIDAR records is to refer all the point clouds of the different scenes obtained to a single coordinate system. The first phase of processing the LIDAR data involves identifying the relationship between the different coordinate systems in each scene materialized by a point cloud. In order to refer each pair of LIDAR records to a single reference system, the three rotations and three translations can be calculated using the following equation (BORNAZ and RINAUDO 2004):

$$\begin{pmatrix} x \\ y \\ z \end{pmatrix} = \begin{pmatrix} x_0 \\ y_0 \\ z_0 \end{pmatrix} + R_{\omega\phi\kappa} \begin{pmatrix} D\sin V\cos H \\ D\sin V\sin H \\ D\cos V \end{pmatrix}, \quad (5)$$

where (x, y, z) represent the coordinates of a point cloud referred to the single reference system, and $(x_0,$

$y_0, z_0)$ are the three translations to refer the reference point of the scanner to the single reference system. These translations are actually the reference point coordinates of the scanner referred to the single reference system. $R_{\omega\phi\kappa}$ is the rotation matrix between the measuring system of the scanner and the single reference system. D is the distance from the scanner reference point to the point measured on the terrain. V and H are respectively the zenithal and horizontal angles to the terrain point recorded by the scanner.

The rotation matrix and the translations for each couple of scanner scenes in Eq. (5) were calculated using the scanned control points (spheres) common to each pair of scenes. Thus, in each scanned scene during the field phase, measurement was guaranteed of at least three control points (spheres) between every two overlapping scenes in their common scanned area. These spheres naturally remained motionless during the LIDAR registration phase.

Once each set of transformations and rotations, joining each pair of scenes, had been computed, all the recorded points in the different scenes became relative to a single coordinate system, forming a single point cloud. We subsequently performed a final coordinate transformation to refer all the registered points to the ETRS89 reference system using the records of the spheres. The positions of these spheres were determined during the field phase using GNSS techniques in the ETRS89 reference system.

3. Application Test Case: Atauta (Soria, Spain)

As a test case, our methodology has been applied to a specific type of cavity used for underground wine cellars. The underground cellars found in different parts of Spain are part of a scattered agricultural landscape, which today is sometimes in disrepair and often at risk of disappearing. The observation and detection of both the outside and underground parts of the cavity are essential for compiling an inventory of the rural heritage (FUENTES-PARDO and GUERRERO 2006). We study the characterization of cavities by GPR, LIDAR and GNSS of a group of wine cellars located in Atauta (Soria, Spain), in the Duero River corridor, shown in Fig. 2. This is a unique architectural complex built on a smooth hillock, as shown in Fig. 3.

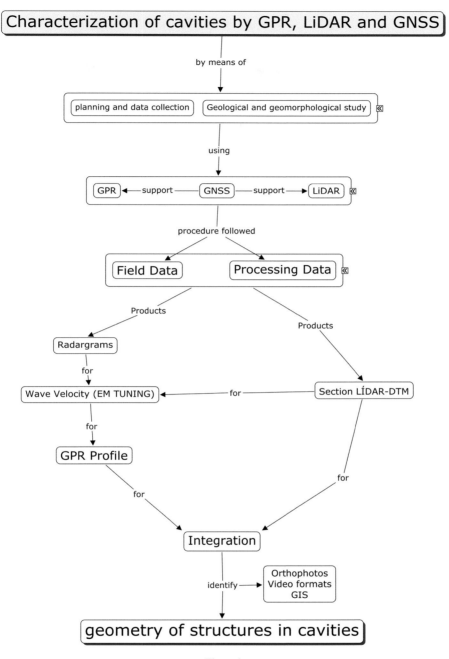

Figure 1
Diagram of the procedures used to characterize cavities by the GPR, LIDAR and GNSS techniques

The cavities in the study are usually formed by a cave or cellar dug into the natural soil below ground level where wine is produced and stored (OCANA and GUERRERO 2006). Each cavity has a single entrance that allows access through a vault or tunnel to its interior spaces. The entrance is north-facing in order to facilitate ventilation, and there are normally one or more chimneys, known by the Spanish term "*zarceras*."

In this type of underground cavity, external vibrations are not transmitted to the interior. The humidity is very high with minor variations. The

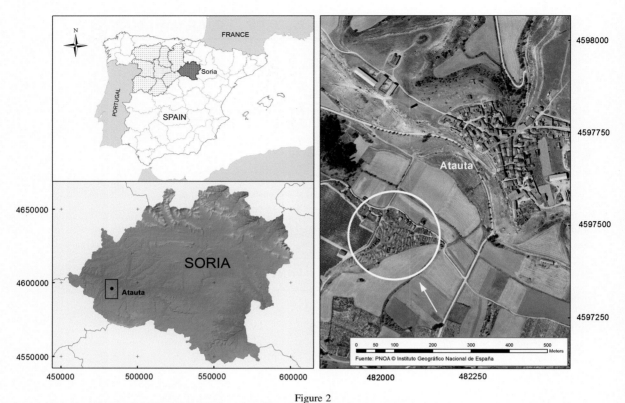

Figure 2
Location of the study area in the province of Soria in central Spain. The *arrow* on the orthoimage in the *right inset* indicates the different cavities in the wine cellars associated with the town of Atauta (orthoimage from the National Geographic Institute)

Figure 3
Aerial view of the underground cavities in the study showing their similar orientations and a single entry for the entire set of cavities

summer heat absorbed by the surroundings is released during the winter, and the cold accumulated during the winter is released during the summer. The temperature inside is almost constant throughout the year, a feature in common with other types of cavities. Certain conditions of humidity and temperature can be obtained in the cavity by varying the depth and type of soil. Thus, the data on the volume of earth

supported by the wine cellar and the air contained inside are directly related to its hygrothermal behavior and its influence on the atmosphere in the cavity (SILVIA and IGNACIO 2005).

The study area, Atauta, is located 950 m above sea level. It is set in an open valley, oriented east-west with prevailing westerly winds. The temperature range from day to night is around 20 °C from fall to spring. The study area belongs to the large morpho-structural domain of the Duero Basin.

NOZAL and HERRERO (2005) set the lithological description as alluvial sediments over limestone. The sediments are mainly silt, clay, and silty and clayey sand in the study area. They also show similar materials with a presence of carbonate and terrigenous sediment representing alluvial sedimentary environments. Several studies confirm this geological variety, and there are records of between 20 and 30 different types of soils. Soils are characterized by very sandy or calcareous rock bottom agglomeration from 70 to 80 cm deep. The predominant soils in the areas where the cavities are located contain a high percentage of fine particles with low or no plasticity. No significantly coarse particles were found in the soils (CAÑAS *et al.* 2012).

4. Field Data Collection and Data Processing

4.1. Field Data Collection

The radar data were collected using a Malå Ramac GPR system (Malå Geoscience, Malå, Sweden) with unshielded antennae of 200-MHz and 100-MHz RTA (Rough Terrain Antenna). The antenna configuration was collinear following the profile direction (parallel end-fire) in all cases. The 200-MHz prospecting was done by two operators. One operator carried the antennae [both the radar and the real time kinematic (RTK) GPS antennae], while the other worked the computer. However, the 100-MHz prospecting only required one operator because of the flexible snake-like design of the 100-MHz RTA antennae, which allows the antenna to be maneuvered easily. The 200-MHz profiles were measured with an antenna separation of 0.6 m, sampling frequency of >2,000 MHz and recording of one trace each 0.1 s. The 100-MHz profiles were obtained with an antenna separation of 2.2 m (fixed by the antenna system), sampling frequency >1,000 MHz and recording one trace each 0.1 s. The horizontal speed of the GPR over the surface of the terrain was slower than 0.5 ms^{-1}. This slow horizontal speed was chosen in order to ensure good GPR horizontal resolution and good accuracy in the trace positioning by GNSS.

The prospecting was done in February 2013 after a series of rainy days. We measured along two profiles with GPR as shown in Fig. 4: one along the longitudinal axis of a cavity and the other crossing several almost parallel cellars. Both profiles were measured using two radar frequencies: 100 and 200 MHz.

Simultaneously, a Faro focus 3D instrument was used to obtained laser scanner scenes of the interior and exterior of the cavity. This equipment has ±2 mm precision when measuring distances and an angle resolution of 2.0×10^{-4} radians. Its range is

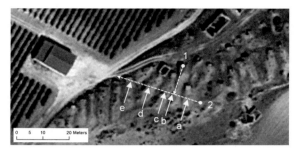

Figure 4
Localization of longitudinal profile 1 over cellar *b* and profile 2, transversal to cellars *a*, *b*, *c*, *d* and *e*

from 0.6 to 120 m. In order to the georeference, at least three homologous points in the overlapping area of two consecutive LIDAR records (a set of white spheres) were used. These spheres were correctly positioned on each LIDAR record, and their positions were then measured with GNSS techniques (see Fig. 5). The spheres were made of plastic and had a 7.25-cm radius.

LIDAR station records were planned on the ground. Special care was taken when placing the spheres (control points of overlap between scenes; see Fig. 6). We also made sure all scans overlapped and covered the full study area. The result of each scan is a complete record of a point cloud of the scanned object. Each point has (*x, y, z*) coordinates on an instrumental reference system. Several scans were required to completely cover the cavities in this study. In this case we used a horizontal resolution of 8.0×10^{-4} rad and a vertical resolution of 16.0×10^{-4}, which means about 26 million points per scan. The total amount of time required for all scans was about 45 min.

The GNSS geodetic control was applied to support the observations of terrestrial LIDAR and GPR and determine the position of the points observed with these two systems both horizontally and vertically. These positions must be obtained on the European Terrestrial Reference System 1989 (ETRS89) geodetic reference system, with sufficient accuracy.

The geodetic control was performed in February 2013. We used three GX1230GG Leica GNSS dual-frequency geodetic receivers equipped with a Leica AX1202GG geodetic antenna. A 1-h observation

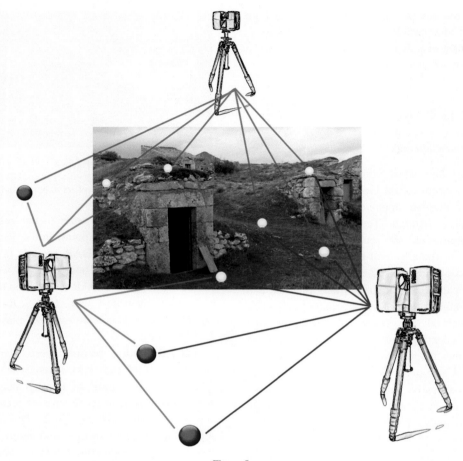

Figure 5
Field methodology for LIDAR scanning. Each scene included the scanning of several spheres. These spheres are also recorded on the connecting scene. This methodology assures the spheres as control points in order to join the two scenes. Later, the positions of these control points were recorded by GNSS

session was performed using the static method for the relative phase difference (HOFMANN-WELLENHOF *et al.* 2008) with one of the receivers. A second receiver was used simultaneously to observe various control points in order to georeference the terrestrial LIDAR observations. Global Navigation Satellite System GNSS-RTK (RIZOS 2002) technology was selected for this purpose. To perform this task, spherical white targets were placed at suitable points to be registered with the LIDAR system at a later stage. Essential and significant parts of the cellars' masonry construction were recorded using the same system.

Also using GNSS-RTK technology, the third receiver was simultaneously coupled to the GPR system to send the position obtained by RTK through a NMEA protocol to the GPS processor. The position

of the GPS antenna with respect to the antennae of the GPR equipment was properly calibrated in order to avoid systematic offsets in the position of the antennae.

4.2. Data Processing

The radar profiles were processed using Reflexw software (SANDMEIER 2012). Previous to the signal processing, the profiles were prepared ("pre-processed", as referred in Fig. 7) by introducing the correct coordinates in the trace headers. These coordinates were obtained after the GNSS postprocessing. The matching of each trace with its corrected coordinates was done though the time code correspondence. Once the GPR profiles had been prepared

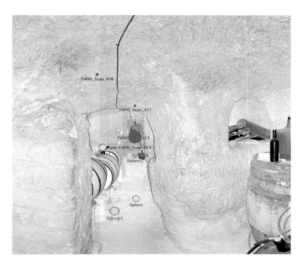

Figure 6
Part of a LIDAR scene recorded in the cavity, showing the position of the spheres in the scene. These spheres act as control points to join the different scenes in the subsequent phase of LIDAR data processing

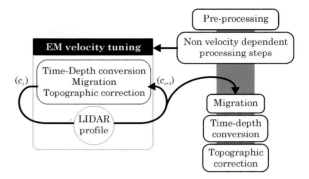

Figure 7
Flowchart of the radargram processing steps detailing the tuning of the EM velocity through a comparison between the processed GPR profile and the LIDAR profile

for processing, we applied some non-velocity-dependent processing steps. This includes as first step the statics correction of each trace in order to align the arrival of the direct wave along the profile. This processing step was necessary when using the 200-MHz antennae, as it had some vertical movement during the profiling. A second static correction step was then applied to state the correct zero time. The next processing steps were: DC and low frequency removal (dewow filtering); generating a profile of equally space traces by removals and interpolations; 2D filtering to remove the direct wave and the rest of the statics; amplitude scaling in the form of a time-dependent exponential gain; band-pass filtering to reduce both the high- and low-frequency noise. As explained below, the value of the EM velocity required for the next processing steps was obtained by an iterative tuning procedure that adjusts the radar and LIDAR profiles. Once the EM velocity has been fixed, the velocity-dependent processing steps must be applied in order to finish the processing of the GPR profile. Consequently, we applied migration (Stolt f-k), conversion from times to depths and the introduction of the surface topography at zero depth.

As stated, we developed a novel methodology in order to obtain the EM velocity as a result of the combination of the LIDAR processed profile and the

processing of the GPR profile in an iterative tuning procedure. It starts after the non-velocity-dependent processing steps (DC removal, filtering, statics adjustment and time-dependent gain). An expected value of the EM velocity for the ground type should be selected as the starting value (c_i in Fig. 7). Then we convert time to depth and compare depths of reflectors registered by both GPR and LIDAR. In these comparisons, we could determine whether the used velocity is slower or faster than the real one. As depth is proportional to EM velocity, using the relation between the obtained depth and the LIDAR-measured depth, we obtain the relation between the used velocity and a new and better estimate (c_{i+1}) of the real EM velocity. However, as the migration effects depend on the velocity used, the comparison of both profiles is more accurate if the LIDAR profile is compared with the migrated and topographically corrected profile using this velocity. The process must be iterated and tuned until no change is needed. We obtained it testing three velocities.

Referring to the LIDAR data processing, various filters were applied to correct the noisy points and eliminate any non-natural objects that were scanned such as the control points (spheres). These filtering operations were designed to eliminate scattered and excessively distant points and points with noise (SOTOODEH 2006). The software used for these operations was SCENE 5.1 from FARO Technologies, Inc. The point cloud was subsequently processed with Bentley Pointools V8i from Bentley Systems, Inc. A registration function algorithm was chosen as a criterion for classifying the point cloud. This algorithm performs a classification based on the position of a point at the time of registration (SITHOLE and VOSSELMAN 2004). The final outcome of this phase is a complete 3D Digital Terrain Model (DTM) for the inner and outer cavity, referred to ETRS89.

This DTM was used to obtain LIDAR profiles for comparison with the GPR records. Within this DTM, a thin section was defined consisting of two vertical parallel planes spaced a few centimeters apart. The terrain profile was generated by selecting the points recorded between the two vertical planes (see Fig. 8).

To compile the orthophotos, we used the corresponding orthographic photograph generator algorithm implemented on FARO SCENE 5.1 software. This algorithm uses all the scanned content (both points and objects) of a crop box to create an orthophoto (see Fig. 9). As the data sets recorded with LIDAR systems are three dimensional, more than one plane of image rectification can be used to generate the corresponding orthophoto. To create each orthophoto, items and objects were projected on a specific plane of projection. This projection plane is the backplane of the table and depends on the current viewpoint and viewing direction in the active 3D view (PESCI et al. 2007). Viewpoints have been chosen perpendicular to the main axes of the cavity structures.

Finally, we made virtual reality recreations to assist in representing and identifying details inside and outside the cavities. The final products of these recreations consist of static images, video formats

generated with rotational movements that show different parts of the cavities integrated on the spatial information system.

Data recorded with GNSS receivers were subsequently processed with Leica Geo Office software. Data from the Continuously Operating Reference Station (CORS) GNSS regional network, along with their coordinates in the ETRS89 system (BLANCO 2010), were added to this process. IGS final solution precise ephemeris (DOW et al. 2009) as well as absolute antenna calibrations from the National Geodetic Survey (NGS) (BILICH and MADER 2010) were added to this process. The longest GNSS vectors were about 14 km. Ambiguities were fixed using the ionosphere free L3 combination with the Hopfield tropospheric model (HOPFIELD 1969). The position of the static GNSS receiver was assigned a standard deviation of 0.6 cm horizontally and 1.1 cm vertically.

These static GNSS vectors used to join the geodetic control to the ETRS89 system were included along with the GNSS-RTK vectors to form a geodetic network, which was then processed using GeoLab software. Information on the variance-covariance matrices from the static method process was incorporated into this network adjustment along with those computed for the GNSS-RTK points. The estimated standard errors of the instrumentation involved were also incorporated into the network adjustment. Geoid undulations were included to compute the orthometric heights, all from the local earth gravitational model EGM08-REDNAP (IGN 2010), which ensures

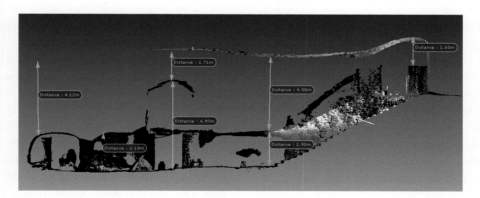

Figure 8
Longitudinal profile of the cavity obtained with LIDAR. This profile is obtained from the DTM generated with the different LIDAR records made

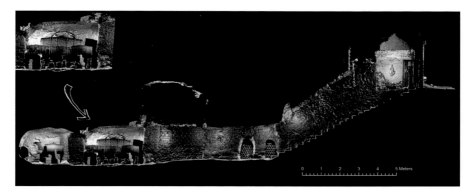

Figure 9
Orthophotograph of the cavity studied in this article based on the DTM

an average accuracy of 3.8 cm relative to the mean sea level datum in Alicante.

Each of the solutions was independently weighted with an appropriate variance factor for each set of its data, with successive iterations following the procedure of ALTAMIMI *et al.* (2011). The two independent solutions were the complete set of static GNSS vectors and the set of RTK vectors computed in the field. A final solution was then computed forming a unique network constrained to ETRS89 CORS regional network stations using the weighted combination of the two independent solutions. Control points used to support observations of terrestrial LIDAR and radar GPR were no further than 40 m from the fixed receiver, and the average standard deviation obtained during this final network adjustment phase of RTK was 0.9 cm horizontally and 1.7 cm in altitude.

5. Analysis and Discussion

A GPR profile was initially obtained following the longitudinal axes of a cavity. A LIDAR scene was then recorded of the surface where the GPR profile was made and of the internal part of the cavity. Then we created a common profile using both techniques joined using GNSS techniques, as discussed above. Special care was taken when connecting the indoor and outdoor LIDAR scenes in order to obtain a unique product for the study zone. Both the GPR and LIDAR measurement techniques were accurately positioned in order to ensure a valid EM velocity

value. This was the main reason for selecting a differential carrier-phase GNSS method to join the GPR and LIDAR records. The EM velocity was obtained as a result of both measurements during the processing of the radargram (Fig. 7). This was the final product of an iterative velocity tuning loop that joins the GPR detections with the LIDAR profiles of both the surface and the interior. This joint was obtained after the migration and topographic-correction processing steps of the radargram.

The penetration of the GPR wave in the soil was not as deep as we expected a priori. There is a wide variability in the wave-propagation parameters through the ground, depending on the soil mixture and air and water contents, as shown in Table 1. Although we used two antennae in order to have different penetration capabilities (larger in 100 MHz) and resolution (better in 200 MHz), in both cases penetration fell short of our expectations and the cavity floors were scarcely detected in the radargrams. We attribute this to the high conductivity of the soil, as it was an artificial coverage of sandy and calcareous rock particles with a significant percentage of fine particles of silt, clay, and silty or clayey sand, and its humidity was also very high because of the previous days' rain.

As the profiles were compared combining outside GPR and inside LIDAR measurements, and because LIDAR also gives the real positions of the detections, EM velocity can be tuned in order to match LIDAR positions with GPR detections using the iterative EM velocity tuning method described above. We used the ceiling of the stairway as the main contact for the comparison of GPR and LIDAR images, although the lateral structures of the main dome also contributed to this tuning. The coupling of the GPR and LIDAR results enables us to estimate a value of 130 m μs^{-1} for the EM velocity in the ground medium, which is appropriate for this mixture of wet silt, sand, clay and limestone. This velocity value has been determined for local soil conditions at the time of registration of the GPR data and was therefore applied to the other GPR record made under the same soil conditions.

This technique is based on the same principles as transillumination, but it has fewer needs and requires less time. The common midpoint (CMP) technique requires large wave penetration, as travel times grow with antennae offset, and accurate in situ positioning of the GPR transmitter and receiver. Following this combined GPR-LIDAR technique, the measurement of EM velocity can be obtained even using GPR with joint antennae (e.g., snake-like or shielded antennae). The involved measurement is noninvasive since only LIDAR works inside of the cavity, and no physical installation is needed (e.g., to support an antenna in contact with the ceiling of the cavity dome, at ca. 5 m over the cavity floor, as needed for transillumination). No in situ accurate positioning is required in this technique. However, the positioning of GPR traces and reflectors is measured with the accuracy of GNSS and LIDAR, respectively, once their data have been processed.

Figure 10 shows the 100-MHz GPR profile obtained from itinerary 1 (Fig. 4), both with and without interpretation. The former is shown in the upper panel, superimposing the transparent GPR profile on the LIDAR profile. The interior profile of the wine cellar detected by LIDAR is represented by dotted lines (hidden cavities are not detected). The GPR detection is shown by dashed lines and the road level by a horizontal dash-dotted line. As we could verify in situ, the dome is an artificial piece covering the cavity, from which it can be seen in the figure that rests on the road level, at the center of the profile. A resonance effect can be seen in the stairway zone produced by multiple reflections between the stair treads and the ceiling. Some resonance effect is also visible within the dome cavity. Some spaces and structures not visible from the inner cavity, and consequently nondetectable by the LIDAR, are detected at the GPR profile. Examples of this are a discontinuity over the structure that supports the roof stairs or the chimney pipe (the almost vertical thick line at the left of the dome in Fig. 10) and a wide cavity or hidden structure around it. However, an uncertainty of ca. 0.5 m must be taken into account because of the limited resolution capability of the 100-MHz GPR. Consequently, differences less than 0.5 m between GPR and LIDAR detections must be assumed as within the GPR uncertainty.

Using the EM velocity obtained from the longitudinal profile shown in Fig. 10 (itinerary 1 of Fig. 4), the domes of nearby cavities (less than 30 m away) were detected using a 200-MHz GPR profile (itinerary 2 in Fig. 4). These cavity domes are shallower than 2 m, as shown in Fig. 11.

6. Summary and Conclusions

We present a methodology to determine the EM velocity of the ground because of a combination of LIDAR DTM with GPR profiles. This constitutes a novel methodology to estimate the EM velocity of the ground in GPR profiles. This technique is easier and faster as compared with other classical techniques such as common mid-point or transillumination. It also has fewer needs for the GPR, as it can be implemented using a GPR with joint antennae (e.g., snake-like or shielded antennae) using standard common offset cables, and propagation paths are shorter than in CMP. The accurate positioning required for GPR and LIDAR profiles is not required in situ, but a posteriori, after their processing. This methodology is suitable for use in other shallow underground spaces with access from the surface, such as other natural cavities, archaeological cavities, sewerage systems, drainpipes or other multipurpose constructed underground spaces.

We conducted a survey using GPR assisted with LIDAR and GNSS techniques over some cavities used for underground wine cellars in Atauta (Soria, Spain) with basically 1 h of fieldwork. This survey allowed the correct detection of the inner structures of cavities in high resolution and with great accuracy. GNSS techniques provided an accuracy of centimeters to the positioning of both the GPR profiles and

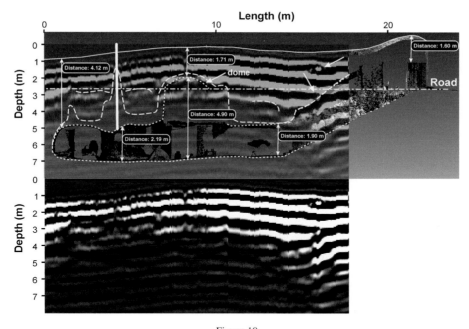

Figure 10
The *lower panel* shows a GPR profile (longitudinal profile 1 over cellar *b* in Fig. 4); the *upper panel* shows the same transparent GPR profile superimposed on the LIDAR profile, matching both surfaces and detections, because of an iterative EM velocity tuning (explained in Fig. 7). Several distances are measured. LIDAR measurement is marked by *dotted lines*; *dashed lines* represent some GPR detections of cavities and structures above the cellar. The *horizontal dash-dotted line* represents the road level

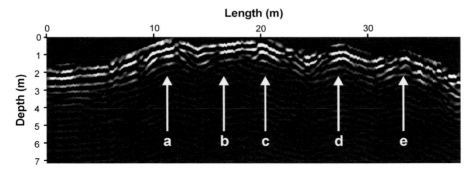

Figure 11
Radargram showing the different domes over the cavities, distributed as shown in Fig. 4

LIDAR DTM. Two different GPR antennae—200 and 100 MHz—were used, detecting the presence of structures such as the entrance beam, chimney and other nearby entrances.

Our methodology provided good results of the EM velocity estimate, even though some cavity floors were almost undetected in the radargrams because of the high conductivity of the wet soil. The ground over the cavity was an artificial porous covering of sandy and calcareous rock particles with a high percentage of fine particles, and it was wet because of a recent rain.

This project may contribute to the declaration of underground cellars as elements of cultural interest by the Comisión de Patrimonio Cultural de Castilla y León—Junta de Castilla y León (Heritage Department of the Regional Government of Castile-Leon).

Acknowledgments

The research in this work is part of the research and development project financed by the Spanish Ministry of Education and Science (2008), "The underground construction wine cellar. Construction eco-systems for quality wines" (BIA2004-03266). The authors would like to thank Topcon Positioning Spain for both their help with hardware and support for this work. This study could not have been done without the cooperation of the municipality of Atauta, Spain, and its mayor. The authors' work has also been sponsored by the International Campus of Excellence CEI-Moncloa. We also thank the editor, Francisco Luzón, two anonymous reviewers for their many suggestions to improve the manuscript, and Pru Brooke-Turner for revising the English version of this article.

REFERENCES

ALTAMIMI, Z., COLLILIEUX, X., and MÉTIVIER, L. (2011), *ITRF2008: an improved solution of the international terrestrial reference frame*, Journal of Geodesy, *85*(8), 457–473.

BALANIS, C. A., *Advanced engineering electromagnetics*. Chapter 10: Spherical Transmission Lines and Cavities (John Wiley & Sons., New York 1989).

BENSON, R. C., GLACCUM, R. A., NOEL, M. R., *Geophysical Techniques for Sensing Buried Wastes and Waste Migration* (Environment Monitoring Systems Laboratory, Office of Research and Development, U.S. Environmental Protection Agency, 1983).

BILICH, A. and MADER, G., GNSS Absolute Antenna Calibration at the National Geodetic Survey, In *23rd International Technical Meeting of the Satellite Division of the Institute of Navigation* (ION, Portland, OR, September 21–24, 2010) pp. 1369–1377.

BLANCO, M., *Cálculo de coordenadas de las estaciones de la red GNSS de Castilla y León* [Computing coordinates of the GNSS stations network of Castilla y León] (ITACYL, Junta de Castilla-Leon, Valladolid, Spain 2010) (in Spanish).

BORNAZ, L., and RINAUDO, F., Terrestrial Laser Scanning Data Processing, In *20th ISPRS Congress*, (ISPRS, Istanbul, Turkey, July 12–23, 2004) pp. 514–520.

BURNS, B., CLARK, W.W. and McMICHAEL, I., Modeling GPR Data from LIDAR Soil Surface Profile, In SPIE Defense, Security, and Sensing (International Society for Optics and Photonics, 2012) pp. 835712–835712-9.

CAÑAS, I., CID-FALCETO, J., and MAZARRÓN, F. R. (2012), *Bodegas subterráneas excavadas en tierra: Características de los suelos en la Ribera del Duero (España)*. [Underground cellars excavated on earth: Soil characteristics in the Ribera del Duero] Informes de la Construcción, *64*(527), 287–296 (in Spanish).

CONYERS, L., and GOODMAN, D., *Ground-penetrating radar: An introduction for archaeologists*, (US/Mountain Book. AltaMira Press, Walnut Creek, CA 1997).

DANIELS, D. J., and INSTITUTION OF ELECTRICAL, E., *Ground penetrating radar* (Institution of Electrical Engineers, London 2004).

DAVIS, J. L., and ANNAN, A. P. (1989), *Ground-penetrating radar for high-resolution mapping of soil and rock stratigraphy*, Geophysical Prospecting, *37*(5), 531–551.

DEPARIS, J., FRICOUT, B., JONGMANS, D., VILLEMIN, T., EFFENDIANTZ, L., and MATHY, A. (2008), *Combined use of geophysical methods and remote techniques for characterizing the fracture network of a potentially unstable cliff site (the 'Roche du Midi', Vercors massif, France)*, Journal of Geophysics and Engineering, *5*(2), 147–157.

DOBRIN, M. B., and SAVIT, C. H., *Introduction to Geophysical Prospecting* (McGraw-Hill, Singapore, 1988).

DOW, J., NEILAN, R. E., and RIZOS, C. (2009). *The International GNSS Service in a changing landscape of Global Navigation Satellite Systems*, Journal of Geodesy, *83*(3–4), 191–198.

FERRÁNDEZ PASTOR, F. J., *Deriva frecuencial de la transmisión electromagnética por efecto del medio*, [Drift frequency of the electromagnetic transmission due to medium effect] (PhD Thesis, Universidad de Alicante, 2007), (in Spanish).

FRANCESE, R. and MORELLI, G., New perspectives in buried utility detection and mapping with a Multi-Scan GPR System. In *19th EEGS Symposium on the Application of Geophysics to Engineering and Environmental problems* (EEGS, Palermo, Italy, April 2006).

FRANCESE, R. G., FINZI, E., and MORELLI, G. (2009), *3-D high-resolution multi-channel radar investigation of a Roman village in Northern Italy*, Journal of Applied Geophysics, *67*(1), 44–51.

FUENTES-PARDO, J. M., and GUERRERO, I. C. (2006), *Subterranean wine cellars of Central-Spain (Ribera de Duero): An underground built heritage to preserve*, Tunnelling and Underground Space Technology, *21*(5), 475–484.

HARTMEYER, I., KEUSCHNIG, M., and SCHROTT, L. (2012), *A scale-oriented approach for the long-term monitoring of ground thermal conditions in permafrost-affected rock faces, Kitzsteinhorn, Hohe Tauern Range, Austria*. Austrian Journal of Earth Sciences., *105*(2), 128–139.

HOFMANN-WELLENHOF, B., LICHTENEGGER, H., and WASLE, E., *GNSS - Global Navigation Satellite Systems: GPS, GLONASS, Galileo, and more* (Springer, Vienna, Austria 2008).

HOPFIELD, H. S. (1969), *Two-quartic tropospheric refractivity profile for correcting satellite data*, Journal of Geophysical Research *74*(18), 4487–4499.

IGN, *El Nuevo modelo de geoide para España EGM08-REDNAP* [The new geoid model EGM08-REDNAP to Spain] (Instituto Geográfico Nacional, Madrid, Spain 2010) (in Spanish).

JOL, H. M., *Ground penetrating radar: theory and applications* (Elsevier Science, Amsterdam, Netherlands 2009).

KEUSCHNIG, M., OTTO, J. and SCHROTT, L. (2010), *Application of GPR on rough terrain surfaces for monitoring issues using a simple ropeway system*. Geophysical Research Abstracts, *12*, EGU2010-3573.

KEUMSUK, L., TOMASSO, M., AMBROSE, W. A., *et al.* (2007), *Integration of GPR with stratigraphic and LIDAR data to investigate behind-the-outcrop 3D geometry of a tidal channel reservoir analog, upper Ferron Sandstone, Utah*. Leading Edge, *26*, 8, 994–998.

LAPAZARAN, J. J., *Técnicas de procesado de datos de georradar y su aplicación al estudio del régimen termodinámico de los glaciares fríos y politérmicos* [GPR data processing techniques and its application to the study of the thermodynamic regime of cold and polythermal glaciers] (PhD Thesis, Universidad Politécnica de Madrid, Madrid 2004) (in Spanish).

LEOPOLD, M., GANAWAY, E., VOLKEL, J., HAAS, F., BECHT, M., HECKMANN, T. *et al.* (2011), *Geophysical Prospection of a Bronze Foundry on the Southern Slope of the Acropolis at Athens, Greece.* Archaeological Prospection, *18*(1), 27–41.

LORENZO, E., *Prospección geofísica de alta resolución mediante Geo-Radar. Aplicación a obras civiles* [High resolution geophysical prospection by Geo-Radar. Civil works applications] (CEDEX, Madrid, Spain 1996) (in Spanish).

LÓPEZ-GETA, J.A., Los acuíferos de la provincia de Jaén. In *I Jornadas sobre el Presente y Futuro de las Aguas Subterráneas en la Provincia de Jaén* [The aquifers in the province of Jaén] (Hidrogeología y aguas subterráneas, 7. IGME, Madrid, Spain, 2002, ISBN. 84-7840-472-4, pp. 15-27) (in Spanish).

LÓPEZ-PIÑEIRO, A., GARCÍA-NAVARRO, A., and COLLINS, M. E. (1998), *Estimación de la profundidad de penetración del Radar (GPR) a partir de medidas reflectométricas TDR "in situ"* [Estimating Radar (GPR) penetration depth from TDR reflectometric measurements "in situ"], *Edafología*, 5, 11 (in Spanish).

NOZAL, F. and HERRERO, A. (2005), *El Mioceno del borde meridional del Corredor Aranda de Duero-Burgo de Osma (SE Cuenca del Duero) [The Miocene of the southern edge of Aranda de Duero Runner-Burgo de Osma (SE Duero Basin)]*, *Rev. Soc. Geol. de España*, 18 (1–2), 21–37 (in Spanish).

NUZZO, L., LEUCCI, G., and NEGRI, S. (2009), *GPR, ERT and Magnetic Investigations Inside the Martyrium of St. Philip, Hierapolis, Turkey.* Archaeological Prospection, 16(3), 177–192.

OCANA, S. M., and GUERRERO, I. C. (2006), *Comparison of analytical and on site temperature results on Spanish traditional wine cellars.* Applied Thermal Engineering, 26(7), 700–708.

PÉREZ, M. V., *Radar de subsuelo. Evaluación para aplicaciones en arqueología y en patrimonio histórico-artístico* [Underground radar. Evaluation for applications in archeology and historical-artistic heritage] (PhD Thesis, Universitat Politècnica de Catalunya, Barcelona, Spain 2001) (in Spanish).

PETTINELLI, E., BARONE, P. M., MATTEI, E., and LAURO, S. E. (2011), *Radio wave techniques for non-destructive archaeological investigations.* Contemporary Physics, *52*(2), 121–130.

PESCI, A., FABRIS, M., CONFORTI, D., LODDO, F., BALDI, P., and ANZIDEI, M. (2007), *Integration of ground-based laser scanner and aerial digital photogrammetry for topographic modelling of Vesuvio volcano.* Journal of Volcanology and Geothermal Research, *162*(3–4) 123–138.

REPPERT, P. M., MORGAN, F. D., and TOKSÖZ M. N. (2000), *Dielectric constant determination using ground-penetrating radar reflection coefficients*, Journal of Applied Geophysics, *43*, 189–197.

RIZOS, C. (2002), *Network RTK Research and Implementation – A Geodetic Perspective*, Journal of Global Positioning System, *1*, 144–150.

RODRÍGUEZ-GONZÁLVEZ, P., MUÑOZ-NIETO, A., GOZALO-SANZ, I., MANCERA-TABOADA, J., GONZÁLEZ-AGUILERA, D., CARRASCO-MORILLO, P. (2014), *Geomatics and Geophysics Synergies to Evaluate Underground Wine Cellars*, International Journal of Architectural Heritage: Conservation, Analysis, and Restoration, *8*(4), 537–555.

SANDMEIER, K.J., *Reflexw 2D processing software (version 7.0)* (Sandmeier Scientific Software, Karlsruhe, Germany 2012). Retrieved from http://www.sandmeier-geo.de/reflexw.html.

SCHEIB, A., ARKLEY, S., AUTON, C., BOON, D., EVEREST, J., KURAS, O., *et al.* (2008), *Multidisciplinary characterisation and modelling of a small upland catchment in Scotland*, Quaestiones Geographicae, *27*A(2), 45–62.

SCHROTT, L., and SASS, O. (2008), *Application of field geophysics in geomorphogy: Advances and limitations exemplified by case studies*, Geomorphology, 93(1–2), 55–73.

SILVIA, M. O., and IGNACIO, C. G. (2005), *Comparison of hygrothermal conditions in underground wine cellars from a Spanish area*, Building and Environment, *40*(10), 1384–1394.

SITHOLE, G., VOSSELMAN, G. (2004), *Experimental comparison of filter algorithms for bare-earth extraction from airborne laser scanning point clouds*, ISPRS Journal of Photogrammetry and Remote Sensing, *59*(1–2), 85–101.

SOTOODEH, S. (2006), *Outlier detection in laser scanner point clouds*, International Archives of Photogrammetry, Remote Sensing and Spatial Information Sciences, *36*(5), 297–302.

VASUDEO, A. D., KATPATAL, Y. B., and INGLE, R. N. (2009), *Uses of dielectric constant reflection coefficients for determination of ground-penetrating radar*, World Applied Sciences Journal, *6*(10), 1321–1325.

YILMAZ, Ö., *Seismic data analysis: processing, inversion, and interpretation of seismic data* (Society of Exploration Geophysicists, Tulsa OK 2001).

YOUNG, R. and LORD, N. (2002), *A hybrid laser-tracking/GPS location method allowing GPR acquisition in rugged terrain*, The Leading Edge, *21*(5), 486–490.

(Received March 10, 2014, revised October 30, 2014, accepted November 7, 2014, Published online November 25, 2014)

Reprinted from the journal

Pure Appl. Geophys. 172 (2015), 3139–3162
© 2015 Springer Basel
DOI 10.1007/s00024-015-1067-0

Pure and Applied Geophysics

Estimation of Seismic and Aseismic Deformation in Mexicali Valley, Baja California, Mexico, in the 2006–2009 Period, Using Precise Leveling, DInSAR, Geotechnical Instruments Data, and Modeling

Olga Sarychikhina,[1] Ewa Glowacka,[1] Braulio Robles,[2] F. Alejandro Nava,[1] and Miguel Guzmán[3]

Abstract—Ground deformation and seismicity in Mexicali Valley, Baja California, Mexico, the southern part of the Mexicali-Imperial valley, are influenced by active tectonics and human activity. In this study, data from two successive leveling surveys in 2006 and 2009/2010 are used to estimate the total deformation occurred in Mexicali Valley during 2006–2009. The leveling data span more than 3.5 years and include deformation from several natural and anthropogenic sources that acted at different temporal and spatial scales during the analyzed period. Because of its large magnitude, the aseismic anthropogenic deformation caused by fluid extraction in the Cerro Prieto geothermal field obscures the deformation caused by other mechanisms and sources. The method of differential interferograms stacking was used to estimate the aseismic (interseismic tectonic and anthropogenic) components of the observed displacement, using SAR images, taken in 2007 during a period when no significant seismicity occurred in the study area. After removing the estimated aseismic signal from the leveling data, residual vertical displacement remained, and to identify possible sources and mechanisms of this displacement, a detailed analysis of records from tiltmeters and creepmeters was performed. The results of this analysis suggest that the residual displacement is mainly caused by moderate-sized seismicity in the area of study. Modeling of the vertical ground deformation caused by the coseismic slip on source fault (primary mechanism) of the two most important earthquakes, May 24, 2006 (Mw = 5.4) and December 30, 2009 (Mw = 5.8), was performed. The modeling results, together with the analysis of geotechnical instruments data, suggests that this moderate-sized seismicity influences the deformation in the study area by coseismic slip on the source fault, triggered slip on secondary faults, and soft sediments deformation.

Electronic supplementary material The online version of this article (doi:10.1007/s00024-015-1067-0) contains supplementary material, which is available to authorized users.

[1] División de Ciencias de la Tierra, CICESE, Carretera Ensenada-Tijuana #3918, Zona Playitas, C. P. 22860 Ensenada, Baja California, Mexico. E-mail: osarytch@cicese.mx; osarytch@yahoo.com

[2] Instituto Mexicano de Tecnología del Agua (IMTA), Paseo Cuauhnáhuac 8532, Col. Progreso, C. P. 62550 Jiutepec, Morelos, Mexico.

[3] Facultad de Ingeniería, Universidad Autónoma de Baja California, Boulevard Benito Juárez s/n, C. P. 21280 Mexicali, Baja California, Mexico.

Key words: Ground displacement, Leveling surveys, Differential interferograms stacking, Tiltmeters, Creepmeters, Moderate-sized seismicity.

1. Introduction

Ground deformation is the surface expression of various physical processes which include, among others, earthquakes, volcanic eruptions, landslides, and subsidence. Ground deformation can be due to natural or anthropogenic factors, or to a combination of both. The type, spatial extent, and temporal pattern of the ground deformation largely depend on the type and magnitude of the driving processes. Monitoring of ground deformation is required to provide a better understanding of its causes, triggering factors, and mechanisms; this understanding is essential for making informed assessments and mitigation of natural and anthropogenic hazards.

Mexicali Valley, located in northeastern Baja California, Mexico (Fig. 1), is a region where significant and rapid ground deformation has become increasingly evident in many urban and rural areas. The ground deformation in this region is caused not only by a variety of natural processes, such as earthquakes, continuous tectonic deformation, and volcanic activity, but also by human activity, mainly that of geothermal fluid extraction in the Cerro Prieto Geothermal Field (CPGF).

Ground deformation in Mexicali Valley has been monitored by repeated precise leveling ground surveys (e.g., García 1978; Lira and Arellano 1997; Glowacka et al. 1999, 2006, 2012) and GPS (Lira 1999a), and by a network of geotechnical instruments

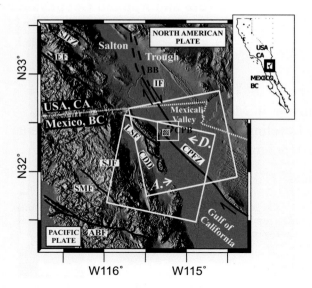

Figure 1

Regional map of northern Baja California, Mexico, and southern California, USA. The Shuttle Radar Topography Mission Digital Elevation Model is used as background. *Large white rectangles* indicate the spatial coverage of the Envisat ASAR images; *D* indicates descending track, *A* indicates ascending track, and *white arrows* indicate the radar view direction for each track. The *smaller white open rectangle* represents the study area. The smallest, *white filled rectangle* indicates the Cerro Prieto Geothermal Field site. The principal tectonic faults and structures are also indicated: *ABF* Agua Blanca Fault, *BB* Brawley Basin, *CDD* Cañada David Detachment, *CPB* Cerro Prieto Basin, *CPFZ* Cerro Prieto Fault Zone, *EF* Elsinore Fault, *IF* Imperial Fault, *LSF* Laguna Salada Fault, *SJF* Sierra Juarez Fault, *SJFZ* San Jacinto Fault Zone, *SMF* San Miguel Fault. Modified from SUÁREZ-VIDAL et al. (2008)

that includes tiltmeters and creepmeters (GLOWACKA et al. 2002, 2010a, b). These ground-based techniques provide accurate and precise information, but only at a number of measuring points in a deforming surface. Repeated ground surveys are time-consuming, expensive, and restricted by available resources. For these reasons, for analyzing the active deformation in a given area, it is convenient to integrate geodetic and geotechnical measurements with data from Differential Synthetic Aperture Radar Interferometry (DInSAR). DInSAR is a space-based geodetic technique widely applied in studies of earth-surface deformation (e.g., MASSONNET and FEIGL 1998; Bürgmann et al. 2000; ZHOU et al. 2009). This technique has the benefits of high spatial (tens of meters) and temporal (months) resolution, large spatial coverage (thousands of km^2), and centimeter-scale

accuracy (GABRIEL et al. 1989; BÜRGMANN et al. 2000; HANSSEN 2001). However, the capacity of DInSAR to resolve time-dependent displacement is limited by the availability of the images, and by the temporal and geometric separation of the images.

In this study, data from two successive leveling surveys in 2006 and 2009/2010 are used to estimate the total deformation occurred during the 2006–2009 period, and data from advanced DInSAR stacking, geotechnical instruments, and seismic catalogues are used to interpret the deformation pattern observed from the leveling surveys. Modeling of ground deformation caused by coseismic slip on the source faults of the two most significant earthquakes which occurred during the analyzed period was used to evaluate the contribution of coseismic slip to the total observed displacement field.

2. Description of the Study Area

Mexicali Valley (the southern part of the Mexicali-Imperial valley) is located within an extremely active tectonic region, in the boundary between the Pacific and North American plates (Fig. 1). This region features a wide zone of transform faults that belong to the San Andreas fault system, accommodating an interplate relative motion rate of ~45 mm/yr (BENNETT et al. 1996).

Along Mexicali Valley, the major Imperial-Cerro Prieto fault system extends in a northwest–southeast direction. The sedimentary extensional zone that connects these faults is known as the Cerro Prieto pull-apart basin (LOMNITZ et al. 1970; ELDERS et al. 1984; LIPPMANN et al. 1984). Several normal faults, oblique to the major faults, have been formed within the basin (Fig. 2) as a consequence of the tensional stress regime imposed by the Cerro Prieto-Imperial fault system (ELDERS et al. 1984; LIPPMANN et al. 1984; GONZÁLEZ 1999; SUÁREZ-VIDAL et al. 2008). This pull-apart basin has a sediment cover that thickens from around 2.5 km at the western margin to about 5.2 km at the eastern margin (LIRA 2005).

Because of its location on the border between the North American and Pacific plates, Mexicali Valley is a seismically active area. Historically, the most relevant large-magnitude ($M \geq 6$) earthquakes in the

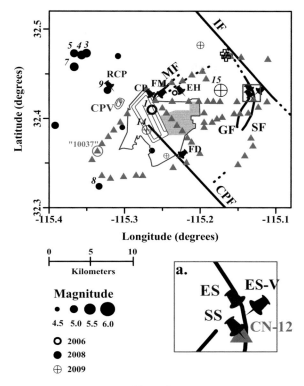

Figure 2

Map of the study area with *circles* showing the epicenters of earthquakes with $M \geq 4.5$ occurred in Mexicali Valley during 2006–2009, from RESNOM and RANM (see Table 1); the numbers correspond to the event numbers in Table 1. *Thick solid black lines* are known surface traces of tectonic faults. *CPF* Cerro Prieto fault, *IF* Imperial fault, *SF* Saltillo fault, *MF* Morelia fault, *GF* Guerrero Fault. *Dotted thick black lines* are proposed surface fault traces based on mapped fissure zones from GONZÁLEZ et al. (1998), LIRA (2006), SUÁREZ-VIDAL et al. (2007, 2008), and GLOWACKA et al. (2006, 2010c). *Black crosses mark* the location of sites where liquefaction was observed after the December 2009 earthquake sequence. The *gray thick line* frames the limits of the Cerro Prieto Geothermal Field. *Gray triangles* are leveling benchmarks used in this study. A *gray circle* encloses the location of the reference "10037" benchmark. *Black pins* tilted to the right indicate the location of the vertical creepmeters, while *gray pins* tilted to the *left* indicate the location of the tiltmeters. The borders of the evaporation pond and the Cerro Prieto volcano (CPV) are also shown (*thin gray lines*). The *gray filled* polygon encloses the extraction wells area. The injection wells are localized within the evaporation pond. The *gray rectangle* indicates the area shown below in the *rectangle labeled* (**a**), which shows a close-up of the area where the geotechnical instruments are located on the Saltillo fault. ES-V is the vertical creepmeter; ES and SS are biaxial surface tiltmeters; CN-12 is the leveling benchmark

region occur along the main Imperial and Cerro Prieto faults, outside the zone of extensional deformation (FREZ and GONZÁLEZ 1991), while most

of the shallow-depth, lower-magnitude ($M \leq 5.5$) events are associated with minor normal faults in the zone of extensional deformation.

Thick sedimentary sequences, extensional tectonics, high heat flow, active faults, and underground water flow from the Colorado river created the conditions for the formation of the geothermal reservoirs which constitute the CPGF. The CPGF began generating power in March 1973; it is the oldest and largest Mexican geothermal field in operation and the second largest in the world. At present, the CPGF is capable of generating 720 MW of electric power from geothermal fluids extracted from reservoirs up to 3,000 m deep (CFE 2010).

Deep fluid extraction in the CPGF has influenced deformation in the area by accelerating the subsidence and by accelerating and/or inducing slip on the tectonic faults in Mexicali Valley (GLOWACKA et al. 1999). A comparison of leveling data (GLOWACKA and NAVA 1996; Gloacka et al. 1999) and DInSAR (CARNEC and FABRIOL 1999; HANSSEN 2001; SARYCHIKHINA et al. 2011) with the CPGF extraction history, suggests that the current vertical displacement, attaining up to ∼18 cm/year, is mainly related to the fluid extraction. However, the field observations and geotechnical instruments data reveal that the geometry of the subsiding area is controlled by tectonic faults (GLOWACKA et al. 1999, 2010a, c; SUÁREZ-VIDAL et al. 2008). Modeling of the tectonic subsidence (SARYCHIKHINA 2003; GLOWACKA et al. 2005) and anthropogenic subsidence (GLOWACKA et al. 2005; SARYCHIKHINA et al. 2011) confirmed that the shape of the subsidence basin is controlled by tectonic faults, while most of the subsidence rate is anthropogenic.

SARYCHIKHINA et al. (2011) compared the subsidence rate obtained by DInSAR for the December 2004–December 2005 period with the 1994–1997 and 1997–2006 leveling results (GLOWACKA et al. 1999, 2006) in order to evaluate the changes in the spatial pattern and rate of land subsidence. This comparison revealed that the magnitude of maximum subsidence in the CPGF extraction zone did not change during this later period; however, its locus migrated to the northeast. The subsidence in the area between the eastern limits of the CPGF and the Saltillo fault (Fig. 2) increased by a factor of ∼1.5 between 1997 and 2005. The authors found that the changes in the ground deformation pattern and rate of the subsidence are correlated in time with the development of the CPGF.

It is well accepted that earthquakes can be induced or triggered by fluid injection and fluid and material extraction (e.g., McGarr et al. 2002). Earthquakes and subsidence related to oil, gas, and geothermal fluids extraction have been reported (e.g., Grasso et al. 1992; Segall 1989, 1992; Mossop and Segall 1997). In the Geysers geothermal field region, seismicity became more frequent since commercial exploitation of the field began in 1960 and increased with increasing field development (e.g., Eberhart-Phillips and Oppenheimer 1984; Majer et al. 2007). A study by Brodsky and LaJoie (2013) in the Salton Sea geothermal field shows that, after correcting for the aftershock activity, the net fluid volume (extracted-injected) correlates with seismicity.

As mentioned before, Mexicali Valley is characterized by a high seismicity level related to its location at the boundary between two tectonic plates. Analysis of the seismicity close to the CPGF suggests the possibility that seismicity in Mexicali Valley is stimulated by fluid extraction and injection in this geothermal field; this possibility was first proposed by Majer and McEvilly (1982) after a production increase between 1978 and 1981 apparently caused an increase in microseismicity. Fabriol and Munguía (1997) observed the case of a single earthquake which correlated in time and space with a sharp increase in fluid injection at a well located less than 1 km from the epicenter, and Fabriol and Glowacka (1997), based on the earthquake catalogue of USGS-Caltech for the 1988–1996 period and the recordings of the local array since 1994, proposed that a mid-range effect could exist between the increase of injected volume of fluid in winter and the seismicity occurring during the following summer. Glowacka and Nava (1996) suggested that both local seismicity increases and the Imperial Valley (ML = 6.6, 15 October 1979), Victoria (ML = 6.1, 9 June 1980), and Cerro Prieto (ML = 5.4, 7 February 1987) earthquakes could have been triggered by sustained, large increases in the extraction rate at the CPGF in 1979 and 1986, respectively. These authors analyzed pressure changes, seismic diffusivity, and delay times between extraction increases and earthquake

occurrences and found, not conclusive, but supporting statistical and spatio-temporal observations, that these earthquakes could have been triggered by fluid extraction. Urban and Lermo (2012) analyzed epicenters from Fabriol and Munguía (1997) and Rebollar et al. (2003) and found that the seismicity in the CPGF is clustered mainly in the exploitation area and at depths within 1 km of the injection and production intervals, suggesting that 45 % of events in the analyzed period are induced by the injection or production operations.

Using a model of rectangular tensional cracks, to represent aquifers in the sedimentary rock environment of Mexicali Valley, based on the hydrological model of the field from Lippmann et al. (1991), Glowacka et al. (2005) evaluated the Coulomb stress changes caused by fluid extraction and found that under fluid extraction conditions, Coulomb stresses increase by more than 0.1 bar/year for all faulting types, a value that, according to Harris (1998), is enough to trigger an earthquake.

Two processes are taking place in our area of the study. One is the process of stress accumulation and relaxation due to tectonic deformation on the plate boundary. In the case of Mexicali Valley, this process is responsible for strike slip earthquakes on the main faults and normal earthquakes on the faults that limit the Cerro Prieto pull-apart basin (e.g., Suárez-Vidal et al. 2008), for coseismic related deformation, and possibly for interseismic deformation. Using GPS data from Bennett et al. (1996) and supposing an elastic half-space, Glowacka et al. (2005) estimated that the maximum subsidence related to the aseismic (interseismic) tectonic dispersion in the Cerro Prieto Pull-apart basin can be of the order of 4–6 mm/yr. Every tectonic earthquake which occurs on a normal fault will cause subsidence due to the displacement on the source fault. Additionally, any earthquake in the region can cause subsidence by shaking the soft, poorly compacted, young sediments of the Colorado Delta and triggering slip on the secondary nearby faults.

The second process is related to the anthropogenic activity in the CPGF. This activity can induce or trigger earthquakes and cause subsidence. Subsidence can be a continuous process, or occur in the form of slip events, as observed on the Saltillo fault that limits the subsidence basin to the east and southeast, as proposed by GLOWACKA *et al.* (2010c). From the point of view of deformation surveys (e.g., leveling survey, DInSAR), anthropogenic subsidence is not distinguishable from tectonic interseismic subsidence, but since interseismic tectonic subsidence is estimated to be less than 6 mm/year, while the observed subsidence is of the order of dozens of centimeters per year, we can suppose that almost all (96 %, according to GLOWACKA *et al.* 2005) of the interseismic subsidence is of anthropogenic origin.

According to SEGALL (1989, 1992), stress changes caused by fluid extraction will modify the Coulomb stress in such a way as to favor the occurrence of normal and strike slip earthquakes on the flanks of reservoir. Since normal and strike slip earthquakes are typical for pull-apart basins, extraction will favor these earthquakes and the coseismic subsidence related to them. However, fluid injection, which diminishes the effective stress on faults, will favor any kind of earthquakes and cause subsidence related to the mechanism of these earthquakes. Additionally, any earthquake, whether induced, triggered, or naturally occurring, will cause subsidence by shaking the soft, poorly compacted, sediments of the Colorado Delta.

Finally, although part of the seismicity observed in the analyzed area during the 2006–2009 period can be of anthropogenic origin, there is no way to identify it. Similarly, part of the interseismic subsidence is of tectonic origin, but as mentioned before, this part is a very small portion.

In the following analysis we will refer to the deformation associated with earthquakes coseismic deformation and deformation in soft sediments as well as slip on secondary faults triggered by earthquakes, as seismic deformation, without specification of it origin and/or causes; while deformation not specifically related to seismicity, i.e., interseismic tectonic deformation, and anthropogenic subsidence and fault creep, will be referred to as aseismic deformation.

3. Seismicity During the Analyzed Period

The seismicity data in the study area for 2006–2009 was obtained from the seismic catalogues of the Northwestern Seismic Network of Mexico (RESNOM), the Northwestern Accelerograph Network of Mexico (RANM), the Southern California Seismic Network (SCSN), and the Global Centroid-Moment-Tensor Project (CMT), yielding information about the location and, if possible, the focal mechanism of earthquakes with magnitude greater than 4.5. The minimum threshold for the earthquake search was chosen based on the lowest limit of the coseismic detection capabilities of the DInSAR technique (EARLE and COGBILL 2002; MELLORS *et al.* 2004; LOHMAN and SIMONS 2005; DAWSON and TREGONING 2007; DAWSON *et al.* 2008). There are no studies about the sensitivity of DInSAR for detection and characterization of small and moderate sized events in the studied area. For this reason, we used studies about theoretical estimates of the DInSAR detection threshold (EARLE and COGBILL 2002; DAWSON and TREGONING 2007) which state that earthquakes smaller than about 4.5, depending on the depth and focal mechanism, are unlikely to produce enough surface deformation to be consistently detectable using DInSAR. We also based our thresholds on studies about the actual moderate sized earthquakes observed by DInSAR in Southern California (MELLORS *et al.* 2004), Zagros Mountains (LOHMAN and SIMONS 2005), and Western Australia (DAWSON *et al.* 2008).

In general, the rate of seismicity in the studied area during this period has remained high, including numerous swarms and mainshock–aftershock sequences; a total 16 earthquakes of moderate magnitude ($4.5 \leq M < 6$), including nine earthquakes with $M \geq 5.0$ (Table 1), occurred during this period. Note that there were no earthquakes with $M \geq 4.5$ in this area during 2007. The epicenters are shown in Fig. 2.

There were three important seismic sequences, theoretically capable of producing significant surface deformation, during the analyzed period in the study area: the May 2006 foreshock–mainshock–aftershock sequence, the February 2008 seismic swarm, and the December 2009 foreshock–mainshock–aftershock sequence.

Between May 22 and 28, 2006, a series of earthquakes occurred in Mexicali Valley. The stronger event of the sequence, Mw = 5.4, occurred on May 24 at 04:20 h (UTC); it had normal mechanism according to CMT and was related to the Morelia

Table 1

Hypocenter parameters for the earthquakes with M ≥ 4.5 occurred in Mexicali Valley in 2006–2009 from RESNOM, RANM (italic font) and SCSN (Mw values)

No.	Date and time of origin		Latitude (°)	Longitude (°)	Depth (km)	Magnitude (Mw)	FM
	mm/dd/yyyy	hh:mm:ss					
1	*05/24/2006*	*04:20:26*	*32.410*	*−115.264*	*4*	*5.4*	
2	*05/28/2006*	*11:55:25*	*32.429*	*−115.234*	*4*	*4.6*	
3	02/09/2008	07:12:06	32.473	−115.351	9	5.1	
4	02/11/2008	18:29:32	32.471	−115.357	7	5.1	
5	02/12/2008	04:32:38	32.473	−115.367	9	5.0	
6	02/12/2008	09:27:21	32.470	−115.309	4	4.7[a]	
7	02/19/2008	22:41:29	32.458	−115.367	8	5.1	
8	02/20/2008	01:28:52	32.324	−115.334	7	4.8	
9	02/22/2008	19:31:17	32.432	−115.323	9	4.9	
10	02/22/2008	19:33:53	32.390	−115.303	6	4.6[a]	
11	09/05/2008	21:54:32	32.364	−115.264	6	4.7[a]	
12	11/20/2008	19:23:01	32.392	−115.392	9	4.9[a]	
13	04/12/2009	03:19:25	32.358	−115.245	4	4.6[a]	
14	09/19/2009	22:55:19	32.387	−115.271	2	5.2	
15	12/30/2009	18:48:58	32.432	−115.173	9	5.8	
16	12/30/2009	18:53:24	32.482	−115.200	8	4.9[a]	

The available focal mechanisms (FM) from the Global CMT catalogue are shown

[a] Local magnitude (from RESNOM)

fault, which is a SE dipping normal fault that trends obliquely to the major faults in the area (Fig. 2). This event was felt at regional distances and produced some cracking on roads and structural damage to constructions in the epicentral area. The mainshock produced a surface rupture over more than 5 km, with maximum vertical displacement up to 30 cm and horizontal motion less than 4 cm (SUÁREZ-VIDAL et al. 2007); it also caused ground accelerations up to 0.5 g (4.9 m/s^2) (MUNGUÍA et al. 2009), and groundwater level changes of up to 6 m in the observation wells (SARYCHIKHINA et al. 2009).

In February 2008 an earthquake swarm including four earthquakes Mw ≥ 5.0 occurred north of the Cerro Prieto geothermal field. The focal mechanism of the six strongest earthquakes (4.8 < Mw ≤ 5.1) in the swarm is available from the CMT catalogue. These six earthquakes had similar right-lateral, strike-slip mechanisms with a small normal component (about 30 %), suggesting tectonic movement consistent with activity on the Cerro Prieto fault and with the trans-tensional tectonics of the area.

The December 2009 foreshock–mainshock–aftershock sequence had an Mw = 5.8 mainshock on December 30 that was widely felt in northeastern Baja California, northwestern Sonora, southern California, and southwestern Arizona. The strike-slip mainshock caused some ground failure and liquefaction in the vicinity of the epicenter (Fig. 2). Both the northwest-striking focal plane of the mainshock and the north–northwest alignment of the aftershocks were consistent with faulting along the Imperial fault.

4. Total Vertical Ground Deformation in 2006–2009 from Leveling Surveys

Leveling surveys measure elevation differences between benchmarks. By repeating surveys, highly accurate (mm scale) measurements of elevation changes (vertical displacements) over time can be obtained.

Leveling measurements in Mexicali Valley began in the 1960's, as part of the Cerro Prieto geothermal field preparations (VELASCO 1963). These measurements have been repeatedly carried out up to the present, mainly by the Mexican Federal Electricity Commission (Comisión Federal de Electricidad, CFE), for monitoring land elevation in the CPGF and surrounding area, with varying frequency, precision, coverage, and density (GARCÍA 1978; GRANNELL et al. 1979; DE LA PEÑA 1981; WYMAN 1983; LIRA and ARELLANO 1997; GLOWACKA

et al. 1999; LIRA 1999b), and as surveys for tectonics or earthquake studies (DARBY et al. 1981; DE LA PEÑA 1981; DARBY et al. 1984; LIRA 1996, 1999c; GONZÁLEZ et al. 1998).

In this study, we use the measurements from two successive first-order, second-class leveling surveys (5 mm/km$^{1/2}$ accuracy). The data were collected during campaigns in February–May 2006, before the May 2006 earthquakes sequence (GLOWACKA et al. 2006), and January–February 2010 (GLOWACKA et al. 2012). Only benchmarks measured by both surveys were taken into account. After discarding the data from benchmarks with evidently erroneous readings, this dataset comprises 58 benchmarks. The "10037" benchmark is the reference point for both leveling surveys and is located in the SW part of the study area, near the foothills of the Cucapah Mountains, and is assumed to be stable up to the time of the April 4, 2010 El Mayor-Cucapah earthquake. The location of the benchmarks used in this study is shown in Fig. 2.

The leveling values of elevation changes that occurred between these leveling surveys were interpolated between benchmarks using ordinary kriging with a linear variogram and no drift (CRESSIE 1990; Surfer, version 9, Golden Software, Golden, CO, USA). A regular grid with 0.0009° (100 m) x-spacing and 0.011°(100 m) y-spacing was generated. This grid was used to create the contour map of total ground deformation[1] occurred between the leveling surveys (Fig. 3). Because the extrapolated values outside the network are unreliable, only ground deformation observed inside the leveling network is analyzed and discussed in this study.

Two regions of high subsidence occur within the larger E-W oriented elliptical zone shown in Fig. 3. The maximum subsidence occurred between the leveling surveys is ∼0.60 m and is located in the area between the eastern limits of the CPGF and the Saltillo fault. A local maximum with ∼0.55 m subsidence is located in the CPGF extraction wells zone, near the N-NE limit of CPGF. A similar subsidence pattern was previously observed by GLOWACKA et al. (1999, 2005) and SARYCHIKHINA et al. (2011). From

---

[1] All contour maps: Fig. 3, Fig. 5c, Figs. 6a, b, Fig. 9a–c, were obtained using grids generated by the procedure described above.

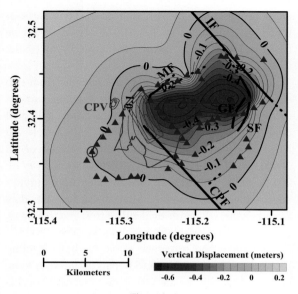

Figure 3

Contour map of total vertical displacement measured by the differential leveling surveys in the 2006–2009 period. Contours are drawn every 0.05 meters. *Negative* values indicate ground subsidence. *Brown triangles* are leveling benchmarks used in this study. A *brown circle* encloses the location of the reference "10037" benchmark. The borders of the evaporation pond, the CPGF limits, and the CPV (*gray lines*) are superposed on the map for reference. Faults notation as in Fig. 2

analyses of the subsidence occurred during 1994–1997 and 1997–2005, when the seismicity was low, GLOWACKA *et al.* (1999, 2005) and SARYCHIKHINA *et al.* (2011) interpreted the western subsidence local maximum as directly related to the extraction, and the eastern maximum as an effect of fluid migration from the recharge zone.

The leveling data analyzed here span more than 3.5 years when seismicity was relatively high, so that they may contain deformation from a superimposition of natural and anthropogenic sources that act at different times and scales during the analyzed period. In what follows we tried to estimate the contribution of different sources to the deformation observed by leveling.

5. *Estimation of Ground Deformation Using the DInSAR Technique*

Differential synthetic aperture radar interferometry is a well-established, space-based geodetic technique for mapping ground deformation caused by

natural (e.g., MASSONNET *et al.* 1993, 1995; AMELUNG *et al.* 2000; FIALKO and SIMONS 2001; STRAMONDO *et al.* 2005; GONZÁLEZ-ORTEGA *et al.* 2014) and anthropogenic (e.g., MASSONNET *et al.* 1997; CARNEC and FABRIOL 1999; CARNEC and DELACOURT 2000; FIALKO and SIMONS 2000; RAUCOULES *et al.* 2003; GONZÁLEZ *et al.* 2012) sources. The standard DInSAR configuration (conventional DInSAR) exploits the phase difference between two complex SAR images acquired over the same area at two different times from slightly different positions (MASSONNET and RABAUTE 1993; GENS and VAN GENDEREN 1996). After removing the effects of the earth's curvature and topography, the DInSAR image (also called differential interferogram) provides a measurement, along the radar line-of-sight (LOS), of possible ground deformation occurred between the acquisitions. The application of conventional DInSAR is limited by temporal and geometrical decorrelation phenomena (ZEBKER and VILLASENOR 1992) and by phase distortions introduced by atmospheric effects.

One way to overcome the two shortcomings of conventional DInSAR, low coherence over long temporal separations and the atmospheric influence, is by stacking differential interferograms (stacking DInSAR). Stacking of differential interferograms aims to combine the information from several differential interferograms in order to extract common information. The most basic procedure is to compute integer linear combinations of interferograms (HANSSEN 2001) or a temporal averaging of multiple interferograms, increasing the signal (ground deformation) to noise ratio. The main assumption is that the deformation phase is highly correlated and the error terms (atmosphere, signal noise, orbit error slope, and nonlinear ground displacements) are uncorrelated between independent pairs. If a reasonable coherence level can only be obtained over short time periods, then several short temporal baseline differential interferograms of successive periods can be summed to obtain the ground deformation over the whole period covered by the data. When several differential interferograms are summed and divided by the total (cumulative) time interval of all interferograms in years, an average annual deformation rate can be obtained (MELLORS and BOISVERT 2003).

In this study, 55 Single Look Complex (SLC) Envisat (Environmental satellite) ASAR (Advanced Synthetic Aperture Radar) images of the study area, taken during February 2006–March 2010 and obtained from the European Space Agency (ESA) as part of ESA CAT-1 project (ID-C1P3508), were used (Fig. 4). The Envisat ASAR, which began operation in 2002, is a C-band (5.66 cm) sensor with average (scene center) incidence angle of 23° and 35 days revisit period. Because of the steep incidence angle of Envisat radar, the interferograms are mostly sensitive to the vertical component of the displacement vector.

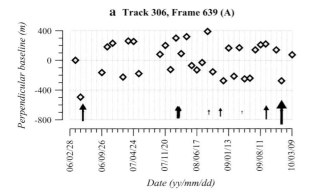

a Track 306, Frame 639 (A)

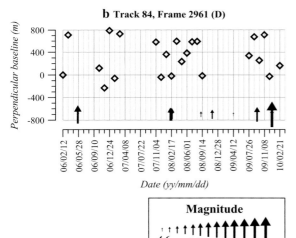

b Track 84, Frame 2961 (D)

Magnitude

4.6   5.0   5.4   5.8

Figure 4
Acquisition data versus perpendicular baseline for Envisat ASAR images (*diamonds*) acquired over the study area during the analyzed period from ascending (**a**) and descending (**b**) tracks. *X* axis minor ticks mark 35-day intervals, which is the revisiting period of the Envisat satellite. The *black arrows* indicate the occurrence time of moderate sized seismic events (4.5 ≤ M < 6.0), listed in Table 1

Among the images, 32 were from ascending track 306, frame 639 (Fig. 4a), and 23 from descending track 84, frame 2961 (Fig. 4b). Data from ascending and descending passes have different imaging geometry (sight direction, see Fig. 1), providing two linearly independent LOS measurements. The spatial coverage of these images is presented in Fig. 1.

The Gamma ISP and DIFF/GEO Software Packages (WEGMÜLLER and WERNER 1997) were used to calculate differential interferograms from the SLC data. All possible combinations with temporal baseline up to 350 days and perpendicular baselines less than 1,000 meters were computed for each track. Envisat/DORIS precise verified orbital data (VOR) (ZANDBERGEN et al. 2003) and the 3-arcsec Shuttle Radar Topography Mission (SRTM) digital elevation model were used during processing to apply orbital and topographic phase corrections.

All interferograms were multi-looked four times in range and 20 times in azimuth, resulting in a final pixel size of approximately 100 m by 100 m. The linear phase trend due to orbit error was removed from the differential interferograms via baseline residual correction. Adaptive filtering (GOLDSTEIN and WERNER 1998) was applied to each interferogram to reduce the noise. Interferograms were unwrapped using the minimum cost flow (MCF) algorithm (CONSTANTINI 1998; CONSTANTINI et al. 1999). Areas with coherence less than 0.5 in the filtered differential interferograms were masked out before unwrapping. Each differential interferogram was visually examined to identify problems caused by decorrelation and atmospheric effects.

As reported in the previous DInSAR studies of Mexicali Valley ground deformation (CARNEC and FABRIOL 1999; HANSSEN 2001; SARYCHIKHINA et al. 2011), the vegetation cover of a great part of the studied area makes application of the conventional DInSAR technique very difficult. The temporal decorrelation in the areas covered by agricultural fields around the CPGF increases considerably for interferometric pairs that span more than 3 months and makes the DInSAR phase of long time period interferograms difficult to measure and unwrap. Unfortunately, it was impossible to form the dataset of coherent short temporal baseline (35–140 days) successive periods differential interferograms with

187

Table 2

Parameters of the interferometric pairs used in this study for estimation of the aseismic component of vertical ground displacement through differential interferograms stacking

No.	Master (yyyy/mm/dd)	Slave (yyyy/mm/dd)	$B_\perp$ (m)	Btemp (days)
Ascending pass, track 306, frame 639				
1	2006/12/05	2007/02/13	−455	70
2	2006/12/05	2007/03/20	32	105
3	2007/02/13	2007/03/20	487	35
4	2007/02/13	2007/04/24	−5	70
5	2007/02/13	2007/05/29	51	105
6	2007/03/20	2007/04/24	−5	35
7	2007/03/20	2007/05/29	−437	70
8	2007/04/24	2007/05/29	56	35
9	2007/05/29	2007/10/16	258	140
10	2007/10/16	2007/11/20	119	35
11	2007/10/16	2007/12/25	−205	70
12	2007/11/20	2007/12/25	−325	35
Descending pass, track 84, frame 2961				
1	2006/12/24	2007/01/28	−840	35
2	2006/12/24	2007/03/04	−54	70
3	2007/01/28	2007/03/04	785	35
4	2007/11/04	2007/12/09	−610	35
5	2007/11/04	2008/01/13	−220	70

reasonable coherence level, covering the whole studied period, for both ascending and descending datasets because of some gaps in data acquisition within the studied period (Fig. 4), geometrical decorrelation of some short time period differential interferograms, and presence of strong decorrelation in the interferograms spanning the May 2006 earthquake sequence, caused by the strong deformation gradient in the near-field of the coseismic fault (SARYCHIKHINA *et al.* 2009). Thus, it was impossible to obtain the total ground deformation occurred in the studied period using conventional DInSAR technique and stacking of successive differential interferograms method.

In what follows we will try to use the DInSAR technique to estimate the contribution of different mechanisms (aseismic and seismic) to the total observed ground deformation.

To estimate the aseismic deformation occurring within the analyzed period, we used averaging (stacking) of several short time period differential interferograms acquired between December 2006 and December 2007. Based on the absence of moderate to large earthquakes in 2007, it seems reasonable to assume that this stacking contains information about the aseismic displacement rate. GLOWACKA and NAVA

(1996) and GLOWACKA *et al.* (1999) found that the subsidence rate changed after sustained extraction increases. Based on this phenomenon and on the essentially constant geothermal fluid extraction rate during 2006–2009 (CFE 2010), we can assume that the aseismic displacement rate was constant (or without considerable variation) during the whole studied period.

Twelve ascending and five descending differential interferograms, processed from images acquired between December 2006 and December 2007, were selected for their quality, in terms of high coherence and low noise level, for further stacking (Table 2). Additional figures showing the differential interferograms selected for stacking are available in the electronic edition of this article (Figures E1 and E2).

For each track, the corresponding set of unwrapped differential displacement was averaged together (stacked) to obtain a final estimate of the linear yearly displacement rate. Before stacking, the interferograms were referenced to a common region (32 × 32 pixels) and were shifted accordingly to set the reference phase to zero. The common reference region includes the location of reference benchmark "10037" from the 2006 and 2009/2010 leveling surveys. For the descending track, pixels that were decorrelated on any interferogram were not included in the stack, and for the ascending track, pixels that were decorrelated on four or more interferograms (on ≥25 % of used interferograms) were not included in the stack. This was done in order to preserve more data-bearing pixels; the

processing difference between tracks is due to the different number of interferograms available for each stack. The displacement rate (cm/year) maps from ascending and descending datasets are shown in Figs. 5a, b, respectively.

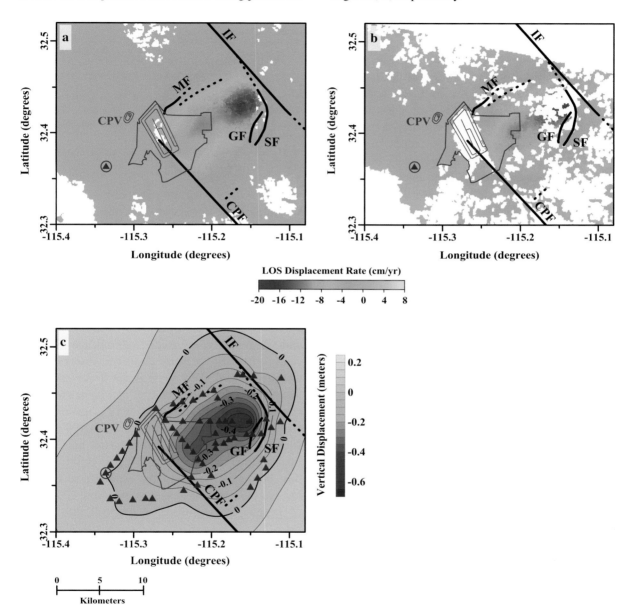

Figure 5

Top Maps of LOS displacement rate (cm/year) for 2007, obtained using the stacking technique on Envisat ASAR images from ascending (**a**) and descending (**b**) tracks. *Negative values* indicate increase in the LOS range. **c** Contour map of aseismic vertical displacement that occurred between the leveling surveys estimated from stacking DInSAR data. Contours are drawn every 0.05 m. *Negative* values indicate ground subsidence. *Brown triangles* are leveling benchmarks used in this study. A *brown circle* encloses the location of the reference "10037" benchmark. The borders of the evaporation pond, the CPGF limits, and the CPV (*gray lines*) are superposed on the maps for reference. Faults notation as in Fig. 2

The similarity between ascending and descending displacement rate maps, along with the known vertical displacement from ground-based measurements, including leveling surveys, suggest that the observed LOS displacements may be interpreted as reflecting mostly vertical surface displacement. Assuming vertical displacement only, the LOS displacement rate from both tracks was projected onto the vertical direction using the incident angle of the Envisat ASAR. The root mean square error (RMSE) of the vertically projected LOS displacement rates between the two passes is approximately 1.3 cm/year. The vertical displacement rate from ascending and descending tracks was averaged in order to reduce the error caused by the presence of some horizontal component of the displacement. Finally, the value of the averaged vertical displacement rate was estimated at the location of every benchmark. If a pixel value was indeterminate, the value from the nearest pixel (distance less than 500 m) was used instead. The resulting values were multiplied by the time (years) occurred between leveling surveys at each benchmark to obtain the aseismic vertical displacement. The contour map of estimated aseismic vertical displacement that occurred between the leveling surveys is shown in Fig. 5c.

According to GLOWACKA et al. (2005), as mentioned in chapter 2, the maximum possible aseismic tectonic vertical displacement in the Cerro Prieto pull-apart center is less than 0.6 cm/year, which is even less than the 1.3 cm/year RMSE calculated above, so that the possible contribution of continuous tectonic interseismic displacement plays a minor role.

A comparison of the volumes of the total observed subsidence and of its aseismic component was done using two corresponding regular grid files and the grid volume tool of the Surfer software. The Surfer grid volume tool determines the volume between two surfaces, given as grid files, or as a grid file and a surface with constant z value. Surfer uses three different methods: Trapezoidal Rule, Simpson's Rule, and Simpson's 3/8 Rule (Surfer User's guide, chapter 19, 443–459) to determine volumes. The difference in the volume calculations by the three different methods measures the accuracy of these calculations. In this study, the surface with constant level of zero subsidence was taken as upper surface

Table 3

Subsidence volume calculated for the total observed vertical displacement and for its aseismic component using the grid volume tool of Surfer software

Subsidence volume for			
Total observed vertical displacement		Aseismic component	
(dlong·dlat·m)*	(m³)	(dlong·dlat·m)*	(m³)
5.828E−03**	5.822E+06	4.023E−03**	4.019E+06

* The units of calculated by Surfer subsidence volume depend on the units of used grids. In this case, the three dimensions are degrees longitude (X value) by degrees latitude (Y value) by meter of vertical displacement (Z value)

** The average value of subsidence volume calculated by three different methods is presented

and the grid of total observed vertical deformation and of aseismic component was taken as lower surface for volume calculation for both cases. The results of the calculation are presented in Table 3. Only the volumes of the subsiding parts are compared. Taking into account the estimated values of subsidence volume (Table 3) and assuming negligible tectonic interseismic displacement contribution to the total observed deformation, we estimated that the anthropogenic aseismic subsidence represents ~70 % of the total subsidence observed in the analyzed period. This value is the lower amount of the possible real percentage of anthropogenic subsidence. It is less than the 95 % estimated by GLOWACKA et al. (2005) for the 1994–1997 period, and less than the 82–90 % estimated by CAMACHO IBARRA (2006) for the 1977–2001 period. We attribute this discrepancy between the estimates to the high level of moderate-sized seismicity during 2006-2009.

Despite the usefulness of DInSAR data to estimate the aseismic, mainly anthropogenic, component of displacement, we could not use this technique for coseismic displacement estimation. As was reported by SARYCHIKHINA et al. (2009), the lack of short time spanning interferometric pairs covering the occurrence of the May 2006 earthquake sequence, and the strong deformation gradient in the near-field of coseismic faulting, caused strong decorrelation in the processed differential interferograms. The complicated deformation field produced by the February 2008 earthquake swarm, with predominantly horizontal displacement (inferred from the earth-quakes focal mechanism; see Table 1) make the vertical displacement component difficult to estimate without coseismic deformation modeling (SARYCHIKHINA *et al.* 2012). The predominantly horizontal component of displacement for the December 2009 earthquake sequence (Table 1) and the strong decorrelation of the differential interferograms spanning this event did not allow estimation of the vertical coseismic displacement component from DInSAR data (SARYCHIKHINA *et al.* 2012).

6. Estimation and Analysis of Seismic Displacement

Figure 6a shows the residual vertical displacement obtained by removing the estimated aseismic, mainly anthropogenic, component from the total displacement measured by the leveling surveys. The residual displacement shown in Fig. 6 is shaped somewhat like a bone, with a maximum (labeled $1'$) in the western part (labeled 1) south from Morelia fault and another maximum (labeled $2'$) in the eastern part (labeled 2) over the Saltillo fault. Both maxima are ~ 0.3 m. As can be seen, the residual vertical displacement is not negligible and may be caused by moderate-sized earthquakes occurred in the study area during the analyzed period. As mentioned in chapter 2, an earthquake can produce vertical ground displacement through several mechanisms: coseismic displacement on primary, source fault, triggered slip on secondary nearby faults and triggered soft sediments deformation.

In what follows we will try to identify the sources and mechanisms of residual displacement using the displacement history from geotechnical instruments data and from modeling of coseismic displacement on the source fault.

6.1. Geotechnical Instruments Data

Since 1996, geotechnical instruments installed by CICESE have operated in Mexicali Valley for continuous recording of deformation phenomena. During 2006–2009, the REDECVAM (Red de Deformaciones de la Corteza en el Valle de Mexicali)

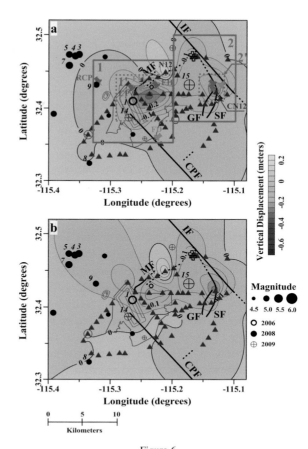

Figure 6

a Contour map of residual vertical displacement obtained after removal of the aseismic component (Fig. 5c) from the observed vertical displacement. **b** Contour map of residual vertical displacement obtained after removal from the observed vertical displacement of the estimated aseismic component (Fig. 5c) and the displacement caused by coseismic slip on the source fault of the May 24, 2006 (Mw = 5.4) and December 30, 2009 (Mw = 5.8) earthquakes (Fig. 9c). Contours are drawn every 0.05 meters. *Negative* values indicate ground subsidence. Numbers 1, $1'$ and 2, $2'$ indicate areas with high residual vertical displacement described in text. *Brown triangles* are leveling benchmarks used in this study. A *brown circle* encloses the location of the reference "10037" benchmark. Numbered benchmarks are those that present local or individual high residual vertical displacement. *Orange pins* indicate the location of the vertical creepmeters. *Magenta pins* indicate the location of the tiltmeters. The borders of the evaporation pond, the CPGF limits, and the Cerro Prieto volcano (*gray lines*) are superposed on the map for reference. The epicenters of the earthquakes with M ≥ 4.5 occurred in Mexicali Valley during 2006–2009, from RESNOM and RANM (see Table 1), are shown as *circles*; each number corresponds to the event number in Table 1. Thick *solid black lines* are known surface traces of tectonic faults. *Black crosses* mark the location of sites where liquefaction was observed after the December 2009 earthquake sequence

Table 4

Operational history of the geotechnical instruments used in this study

Instrument name	Instrument type	Registration periods	Period used
ES-V	Vertical creepmeter	02.1996–07.2009	02.1996–07.2009
FM	Vertical creepmeter	04.2004–08.2007	01.2006-08.2007
FD	Vertical creepmeter	12.2007–present	12.2007–12.2009
CP	Tiltmeter in vault	08.1998–08.2007	01.2006-08.2007
RCP	Tiltmeter in well	03.2003–present	01.2006–12.2009
EH	Tiltmeter in well	03.2003–12.2012	01.2006–12.2009
ES	Tiltmeter in vault	07.1998–07.2003	07.1998–12.2009
		02.2008–present	
SS	Tiltmeter in vault	02.2008–present	04.2008–12.2009

network included three creepmeters and seven tilt-meters; all instruments have sampling intervals in the 1–20 min range (NAVA and GLOWACKA 1999; GLO-WACKA *et al.* 2002, 2010b, c). Creepmeters (Geokon model 4420) were installed on a vertical plane, perpendicular to the faults, in order to record vertical displacement. Biaxial tiltmeters (Applied Geome-chanics, models 711, 712, and 722) were installed in shallow vaults or wells, close to the faults. Table 4 lists the eight instruments, and lists the periods when the instruments were recording, and Fig. 2 shows the instrument locations. We discarded data from two instruments that presented discontinuities or were very noisy. We also discarded the noisiest component of all tiltmeter data. The eliminated component was more sensitive to tilt changes caused by water irrigation and/or nearby construction work, than to coseismic or subsidence changes.

Two instruments have the longest recording period: the ES-V creepmeter and the ES tiltmeter, which were installed in 1996 and 1998, respectively, on the Saltillo fault (Fig. 2a) as a part of the REDECVAM network. The creepmeter operating vertically at the Saltillo fault recorded continuous vertical creep and slip-predictable episodic slip events, with 5.3 –7.3 cm/year vertical displacement rate (GLOWACKA *et al.* 2010c), west side down (Fig. 7a). Since the Saltillo fault constitutes the southeastern border of the Cerro Prieto pull-apart basin, this displacement was interpreted as a bound-ary effect of the anthropogenic subsidence caused by fluid extraction in the CPGF (GLOWACKA *et al.* 1999, 2010c). The ES-V creepmeter stopped operating a few months before the December 2009 earthquake

sequence, and one tiltmeter, ES, that stopped working in 2003, was replaced in 2008, with some small changes in the benchmark surroundings.

The displacement recorded by ES-V is shown in Fig. 7a, b, and 7c, and the ES "*Y*" north–south component is shown in Fig. 7b and c. The records show a change in the vertical displacement rates across the fault (along the 3 m span of the creepmeter) from 5.3 cm/year (for 1996–2003) to 7.3 cm/year (2003–2009) (Fig. 7a). The change in the vertical displacement rate observed around 2003 was interpret-ed as a result of production changes in the CPGF around year 2000, delayed by the process of poroelastic diffusion (GLOWACKA *et al.* 2010c). The deformation rate change is reflected also as different tilt rates recorded by the ES tiltmeter, shown in Figs. 7b and c.

Deformation is not continuous, but occurs in steps (episodic slip events), recorded by the creepmeters and tiltmeters, separated by months of quiescence; large events account for about 50 % of the vertical displacement. Aseismic, episodic slip events have amplitudes of 1–3 cm and durations of 1–3 days (NAVA and GLOWACKA 1999; GLOWACKA *et al.* 2002). A somewhat similar behaviour, but in the lateral direction, has been observed on the strike slip faults in California (e.g., BILHAM and BEHR 1992; WEI *et al.* 2009, 2013).

Figure 8a, b, c, and d shows displacement and seismicity reported during 2006, 2007, 2008, and 2009, respectively. The coseismic displacement, appearing as a discontinuity in the displacement graph, can be seen for some earthquakes in some of the instrumental records. In particular, in Fig. 8a, discontinuities in the CP-X, RCP-X, FM-V, and EH-

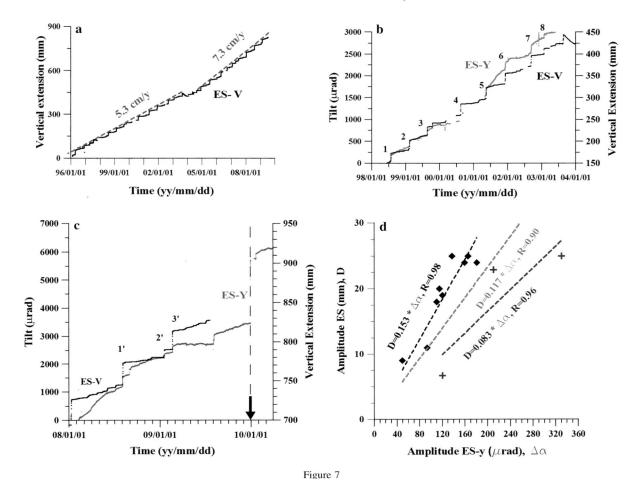

Figure 7

a Geotechnical instruments records spanning the 2006–2009 period. ES-V is the vertical creepmeter. ES-Y is the record of the north–south component of the biaxial surface tiltmeter ES. **b** Geotechnical instruments records during the 1998–2003 period. *Numbers* indicate the significant slip events recorded by both instruments. **c** Geotechnical instruments records during the 2008–2009 period. *Numbers* indicate the significant slip events recorded by both instruments; the *black arrow* indicates the occurrence time of the December 2009 mainshock event. **d** Amplitudes, D, of slip events recorded by the ES-V creepmeter vs. tilt change, Δα, recorded by the ES-Y tiltmeter for 1998–2003 in *black diamonds* and *black line*, for 2008–2009 with *dark grey crosses* and *dark grey line*, the *light grey line* corresponds to all events recorded during the 1998–2009 period. R is the coefficient of determination

X records,[2] and a small tilt change in the RCP-X record can be observed for the May 24, 2006 (Mw = 5.4), Morelia fault earthquake. In 2007 (Fig. 8b) all instruments show continuous data, which coincide with a lack of significant seismicity, while in Fig. 8c, the RCP-X record shows a permanent displacement corresponding to the February 2008 earthquake swarm (with four earthquakes with

magnitude larger than 5) and the November 20, 2008 (M = 4.9) earthquake.

During 2009 EH-X and FD-V present discontinuities related to the September 19 (Mw = 5.2), earthquake, while ES-Y, SS-X, present discontinuities at the time of the December 30 (Mw = 5.8) earthquake. RCP-X has small tilt changes during both these 2009 earthquakes, and FD shows a slip event triggered by the December 30, 2009 (Mw = 5.8) event.

From analyzing the discontinuities in record continuity for instruments in the different areas (1, 1′, 2, or 2′) where instruments are located, we can make the following observations:

---

[2] Hereafter, "X" in the instruments name means that the east–west tilt component is analyzed, "Y" means the north–south tilt component is analyzed, and "V" indicates that the instrument is a creepmeter that records the vertical component of ground displacement.

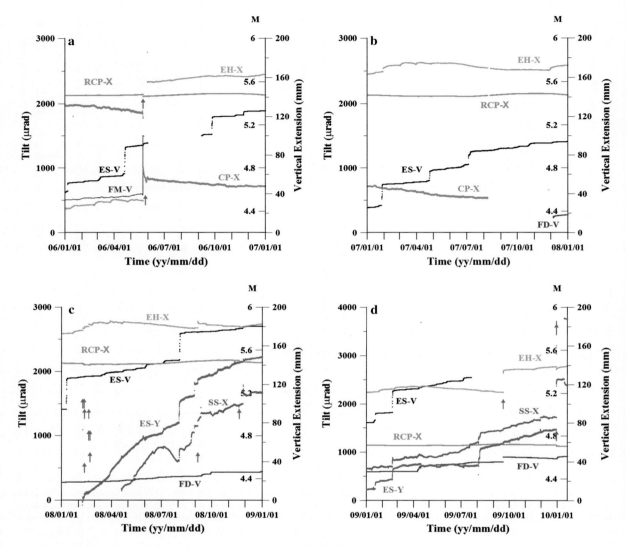

Figure 8

Vertical extension recorded on the ES-V, FD-V, and FM-V creepmeters, and tilt recorded on the CP, RCP, EH, and SS, *X* component, and ES, *Y* component tiltmeters, during 2006 (**a**) 2007 (**b**), 2008 (**c**) y 2009 (**d**) Tilt and extension values are relative. The creepmeter FM-V was separated from its benchmark during the 2006 earthquake and the value of coseismic displacement of 20–25 cm was estimated by SARYCHIKHINA *et al.* (2009). *Red arrows* indicate earthquakes listed in Table 1, with magnitude scales on the *right-hand axis*. The coseismic tilt change recorded by EH-X during the 2006 earthquake was out of range; and ES-Y stopped working 2 h and 29 aftershocks, after the December 30, 2009 earthquake

1. The May 24, 2006 (Mw = 5.4) earthquake affected all of area 1, including the 1′ subarea, where CP, RCP, EH, and FM are located, and caused record discontinuities at the time of earthquake occurrence. Since ES-V did not present record discontinuities, we can conclude that this event did not significantly affect the 2′ area.

2. The northern part of area 1 was affected by the February 2008 earthquake swarm, when the RCP record was slightly disrupted at the time of earthquakes occurrence, while ES (area 2′), FD (area 1, south) and EH (1′) records did not present discontinuities.

3. Area 1, western part, was slightly affected by the November 20, 2008 (M = 4.9) earthquake, at the RCP location, while EH, and FD (1′ and southern 1), and ES-V, ES, and SS in area 2′ were not affected.

4. The September 19, 2009 (Mw = 5.2) earthquake affected all of area 1, where FD, EH and RCP are located, and caused record discontinuities at the time of earthquake occurrence, while SS and ES, located in area 2′, did not present record discontinuities.

5. Areas 2 and 1 were affected by the December 30, 2009 (Mw = 5.8) earthquake, when ES and SS were most affected, while EH, RCP, and FD were only slightly moved.

Finally, based on geotechnical instruments records, we can summarize that the residual displacement in area 1 is mainly due to the May 24, 2006 (Mw = 5.4) earthquake, while the rest of the displacement was caused by the February 2008 earthquake swarm (northern part), and the November 20, 2008 (M = 4.9), September 19, 2009 (Mw = 5.2), and December 30, 2009 (Mw = 5.8) earthquakes.

The local residual ground uplift at the N12 benchmark (Fig. 6) is ∼0.18 m; however, no large residuals are found at the benchmarks closer to N12, which are located ∼1 km away from it. For this reason we assume that the uplift at N12 is due to benchmark instability.

Areas 2 and 2′, as can be seen from the geotechnical instruments data, were influenced only by the December 30, 2009 (Mw = 5.8) earthquake. In what follows we will try to find out whether the value of the observed 2′ anomaly can be explained using the values of the coseismic discontinuities observed in the instrumental recordings.

Residual ground subsidence of 0.34 m was measured at the CN12 benchmark located close to the Saltillo fault, on its western downthrown block. The total vertical deformation at this location for the period between the leveling surveys is −0.48 m, while that estimated from the DInSAR data aseismic displacement at this location is about −0.17 m. However, this last estimation may contain serious errors due to coherence loss in this area (Fig. 5b). Another way to evaluate aseismic displacement close to CN12 is to analyze the vertical displacement recorded by the ES-V creepmeter on the Saltillo fault. Taking into account that the vertical subsidence in ES-V is equal to 7.3 cm/year (Fig. 7a), which,

according to GLOWACKA et al. (2010c) is anthropogenic, the total aseismic subsidence close to the CN12, during the span of the leveling study (3.5 years), should be ∼0.25 m. This means that ∼0.23 m of the observed subsidence is due to other causes.

In what follows we will try to find out whether the 0.23 m of CN12 subsidence could be related to the December 30, 2009 (Mw = 5.8) earthquake.

As mentioned before, the ES-V creepmeter stopped operating a few months before the December 30, 2009 (Mw = 5.8) earthquake, but the ES tiltmeter recorded a 2,250 μrad of coseismic tilt change in the "Y" (north–south) component. The displacement and episodic slip events recorded by the ES-V and ES "Y" (north–south) component for the 1998–2003 period are shown in Fig. 7b, and those for 2008–2009 period in Fig. 7c, d show the relationship between the amplitudes of slip events recorded by the creepmeter versus those recorded by the tiltmeter at ES, for periods 1998–2003, 2008–2009, and for the entire dataset. We chose the "Y" (north–south) tilt component of the ES tiltmeter because it is the one that most resembles the creepmeter records. A least-squares fit to the slip, D, vs. the associated tilt change, Δα, yields the relation

$$D = C\Delta\alpha \qquad (1)$$

where C equals 0.153 for 1998–2003, 0.083, for 2008–2009, and 0.117 for the entire dataset.

For coseismic tilt change Δα = 2250 μrad, the corresponding coseismic slip (west side down) on the Saltillo fault, employing the appropriate C values, would be D = 344 mm for 1998–2003, 187 mm for 2008–2009, and 263 mm for the entire dataset. Since there were only three slip events during the 2008–2009 period, the statistics for this period are unreliable, thus we consider that the 0.26 m value obtained using the C value estimated for the entire dataset should be more appropriate. This value is similar to the 0.23 m of residual subsidence observed at the CN12 benchmark (located 525 m to the south from the ES-V), so that we can conclude that the residual deformation on zone 2′ is caused by slip on the Saltillo fault triggered by the December 30, 2009 (Mw = 5.8) earthquake.

6.2. Modeling of Coseismic Vertical Displacement Caused by Slip on the Source Fault

The analysis of geotechnical instruments data shows that the residual vertical displacement may be related to several moderate-sized earthquake occurrences.

We have modeled the coseismic displacement caused by the coseismic slip on the source fault (primary mechanism) of the two largest earthquakes: May 24, 2006 (Mw = 5.4) and December 30, 2009 (Mw = 5.8). The modeling was done considering a finite rectangular fault with uniform slip embedded in an isotropic elastic medium and the FORTRAN77 program EDCMP software (WANG et al. 2003). A shear modulus (rigidity) of 10 GPa and a Poisson's ratio of 0.25 were assumed for the elastic half-space (see SARYCHIKHINA et al. 2009 for details about estimation of the values of these elastic parameters).

For the May 24, 2006 earthquake we used the source parameters estimated by SARYCHIKHINA et al. (2009) using forward modeling of surface deformation data and static volume strain change (inferred from coseismic changes in groundwater level). The preferred fault model estimated by SARYCHIKHINA et al. (2009) has strike = 48°, rake = 89°, and dip = 45°, length = 5.2 km, width = 6.7 km, and 34 cm of uniform slip (Table 5).

The parameters of the source fault of the December 30, 2009 earthquake were based on the empirical relationships of WELLS and COPPERSMITH (1994), and the observed focal mechanism (Table 5); the fault was centered at the hypocentral location.

Using the models of source fault parameters, the coseismic vertical displacement at each leveling benchmark location was calculated for both earthquakes (Fig. 9a, b). Figure 9c presents the contour map of the summed displacement caused by these two earthquakes.

As can be seen in Fig. 9a, and as would be expected, the May 24, 2006 earthquake produced significant vertical ground deformation in the study area, due to its normal mechanism and shallow hypocentral depth. The strike-slip December 30, 2009 earthquake produced vertical ground deformation of the order of mm (Fig. 9b) comparable with the accuracy of the leveling survey. Thus, the coseismic slip on the source fault of the December 30, 2009 earthquake could not be responsible for the residual deformation in Area 2.

6.3. Interpretation of Residual Seismic Displacement

The residual vertical displacement obtained after removal from the observed vertical displacement of the estimated aseismic component (Fig. 5) and of the displacement caused by coseismic slip on the source fault of the May 24, 2006 (Mw = 5.4) and December 30, 2009 (Mw = 5.8) earthquakes (Fig. 9c) is shown in Fig. 6b. Compared with that in Fig. 6a, the residual vertical displacement in the western part of the study area (labeled 1) was partially reduced by extraction of the coseismic displacement caused by the May 24, 2006 earthquake. The residual vertical displacement with 0.15 m maximum still persists in this area. The analysis of the geotechnical instruments records suggests that deformation in this part of study area 1 is mainly due to the May 24, 2006 (Mw = 5.4) earthquake. However, it also suggests that some displacement may have been caused by the February 2008 earthquake swarm (northern part),

Table 5

Source fault parameters for the May 24, 2006 and December 30, 2009 earthquakes. NP1 and NP2 refer to nodal planes 1 and 2 from the USGS Body Wave Moment Tensor Solutions. The north–northwest alignment of the aftershocks implies that nodal plane 2 (highlighted in gray) is the fault plane and is consistent with faulting along the Imperial fault

Source	Strike (°)	Rake (°)	Dip (°)	Length (km)	Width (km)	Slip (cm)
May 24, 2006 Mw = 5.4 earthquake						
SARYCHIKHINA et al. (2009)	48	−89	45	5.2	6.7	34
December 30, 2009 Mw = 5.8 earthquake						
USGS Body Wave (NP1)	63	−10	90			
Moment Tensor Solutions (NP2)	153	−180	80			
WELLS and COPPERSMITH (1994) for Mw = 5.8 strike-slip mechanism				9.7	5.9	33

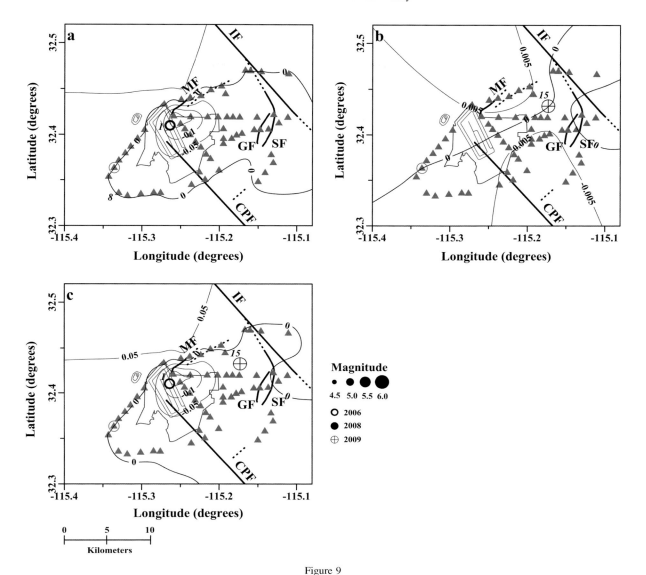

Figure 9

Contour map of vertical ground displacement caused by coseismic slip on the source fault of **a** May 24, 2006 (Mw = 5.4) earthquake and **b** December 30, 2009 (Mw = 5.8) earthquake. **c** Total displacement caused by these two earthquakes. Contours are drawn every 0.05 meters in (**a**) and (**c**), and every 0.005 meters in (**a**). *Negative* values indicate ground subsidence. The epicenters of the May 24, 2006 and December 30, 2009 earthquakes are shown as *circles*; the number corresponds to the event number in Table 1. *Gray triangles* are leveling benchmarks used in this study. A *gray circle* encloses the location of the reference "10037" benchmark. The borders of the evaporation pond, CPGF limits and the CPV (*gray lines*) are superposed on the maps for reference. Faults notation as in Fig. 2

and by the November 20, 2008 (M = 4.9), September 19, 2009 (Mw = 5.2) and December 30, 2009 (Mw = 5.8) earthquakes. From the hypocentral location of the mentioned earthquakes, and from their focal mechanisms and magnitudes we can suggest that the primary coseismic mechanism (coseismic slip on the source fault) of these earthquakes is not responsible for the residual

displacement observed in this zone. However, these earthquakes probably may have influenced the vertical deformation field in this area by a secondary coseismic mechanism: earthquake triggering of soft sediments deformation and slip on nearby located faults. Some of the residual vertical displacement in this part of the study area can be due to unmodeled complexity of the source fault of the May 24, 2006

197

earthquake or to benchmark instability, as suggested by SARYCHIKHINA *et al.* (2009).

There are no important differences between Fig. 6a and b in the eastern part of the study area (labeled 2 and 2′), which indicates that there was no significant contribution of coseismic slip on the source fault of the December 30, 2009 (Mw = 5.8) earthquake to the total observed vertical displacement field. However, the coseismic signal observed on the geotechnical instruments data suggests that the residual deformation signal in Area 2′ is related to the December 30, 2009 earthquake, and was caused by triggered slip on the Saltillo fault. The observed liquefaction north of the December 30, 2009 earthquake epicenter suggests that the residual vertical displacement in Area 2 could have been caused by earthquake-triggered soft sediments deformation.

7. Discussion

In this study, an integrated analysis of satellite observations, ground-based geodetic and geotechnical measurements, and seismicity data was performed in order to identify (estimate) the contribution of different sources and mechanisms to the observed vertical deformation field. The area of study is characterized by complex ground deformation, reflecting a combination of anthropogenic and natural causes. The overlapping of different sources of deformation complicates the interpretation of the ground deformation phenomenon.

The study underlines the importance of better spatial–temporal data coverage of monitored phenomena to improve the knowledge of its causal sources and mechanisms. The terrestrial observations in a repeat survey mode (e.g., precise leveling survey) are typically too sparse in both temporal and spatial resolution. Both spatial and temporal resolution of ground deformation monitoring could be considerably improved by the use of DInSAR and/or Advanced DInSAR techniques due to their high spatial resolution and coverage, with short repeat observation cycle (to attain good temporal resolution). However, the capacity of DInSAR to resolve time-dependent deformation is limited by availability of the images, by the temporal and geometric

separation of the images, and by the magnitude (gradient) of the ground deformation. This study highlights the importance of using the geotechnical instrumentation in the ground deformation monitoring scheme that, in this study, allows to establish a correlation between the observed ground deformation and seismicity.

In order to characterize ground deformation in the study area better, a monitoring scheme has to be designed. In general, and as this study shows, a complete multidisciplinary approach is the appropriate monitoring scheme. To achieve a better identification and quantification of the contribution of each individual deformation factor to the total ground deformation field in Mexicali Valley what is needed is a dense coverage networks of seismic and geotechnical instruments (and/or permanent GPS stations), supported by detailed fault survey, and frequent SAR images acquisition for DInSAR (and/or A-DInSAR) processing, controlled and rectified by sporadic leveling surveys. In addition, the cross-comparison of different datasets will provide validation of the obtained measurements.

Seismicity in Mexicali Valley occurs both tectonically and as a consequence of injection and extraction of the fluids in the CPGF. In this study, we estimate the seismic deformation without specification of its origin and/or causes. Open access to the pressure and volume of injected and extracted fluids data for the CPGF, coupled with seismic data from monitoring system with adequate sensitivity and high location accuracy, will significantly improve our understanding of influence of geothermal fluids production to the subsidence and seismicity of the area of study. The influence of the anthropogenic stress field should not be ignored in future seismic hazard assessments of the Valle de Mexicali.

The electricity production at the CPGF is of vital importance for the development of Baja California, and the economic pressures will surely result in future increased production and consequently, increased influence of this anthropogenic activity on the ground deformation and seismicity of Mexicali Valley. Thus, the seismic/deformation/production monitoring and analysis will be of great importance for geological, environmental, seismic hazard assessment, and mitigation actions.

8. Conclusions

This paper presented results of ground deformation monitoring in Mexicali Valley from 2006 to 2009. The leveling data span more than 3.5 years and include both aseismic and seismic deformation. The DInSAR stacking technique was used to estimate the aseismic component of observed displacement using SAR images acquired in 2007 when there was no significant seismicity in the study area. The combination of precise leveling data and DInSAR data allowed us to obtain the map of vertical displacement caused by the sources probably not related to anthropogenic activity. The detailed analysis of geotechnical instrument records allowed us to tentatively identify the sources of this displacement as being mainly caused by moderate-sized seismicity occurred in the area of study. Based on geotechnical instruments records analysis and results of modeling we can suggest that the seismicity influences the deformation in the study area through several mechanisms, including coseismic slip on source fault, triggered soft sediments deformation and slip on secondary faults. Still, at least 70 % of the total subsidence observed during 2006–2009 is anthropogenic.

Acknowledgments

Images from the European Space Agency's Envisat satellite were used to generate the interferometric data. The data were obtained as part of the ESA Cat-1 Project (ID—C1P3508).This research was sponsored in part by CONACYT project number 105907-F, CONAGUA agreement GRPBC-CICESE-01, 2005 and 2009, and CICESE internal funds. The authors gratefully acknowledge the contributions of Carlos Iván Hernández Gutiérrez (UAS) who participated in the preliminary part of this study thanks to a "Programa Delfín" summer scholarship. E.G is grateful to Dr O. Lazaro-Mancilla for his aid during the fieldwork after the December 30, 2009 earthquake.The authors thank the two anonymous reviewers for their constructive comments and valuable suggestions that allowed us to improve and clarify the manuscript.

REFERENCES

AMELUNG, F., JONSSON, S., ZEBKER, H., and SEGALL, P. (2000), *Widespread uplift and 'trapdoor' faulting on Galapagos volcanoes observed with radar interferometry*, Nature (407), 993–996.

BENNETT, R. A., RODI, W., and REILINGER, R. E. (1996), *Global Positioning System constraints on fault slip rates in southern California and northern Baja, Mexico*. J. Geophys. Res. *101* (B10), 21943–21960.

BILHAM, R., and BEHR, J. (1992), *A 2-layer model for aseismic slip on the Superstition Hills fault, California*, Bull. Seismol. Soc. Am. *82*, 1223–1235.

BRODKSY, E., and LAJOIE, L.J. (2013), *Anthropogenic Seismicity Rates and Operational Parameters at the Salton Sea Geothermal Field*, Science *341*(6145), 543–546, doi:10.1126/science. 1239213.

BÜRGMANN, R., ROSEN, P. A., and FIELDING, E. J. (2000), *Synthetic aperture radar interferometry to measure Earth's surface topography and its deformation*, Annu. Rev. Earth Planet. Sci. *28* (1), 169–209.

CAMACHO IBARRA, E. (2006), Análisis de la deformación vertical del terreno en la región de confluencia del sistema de fallas Cerro Prieto-Imperial en el periodo 1962–2001, M.Sc. Thesis, Department of Geophysics, CICESE, Ensenada, B.C., Mexico, 135 pp.

CARNEC, C., and FABRIOL, H. (1999), *Monitoring and modeling land subsidence at the Cerro Prieto geothermal field, Baja California, Mexico, using SAR interferometry*, Geophys. Res. Lett. *26*(9), 1211–1214.

CARNEC, C., and DELACOURT, C. (2000), *Three years of mining subsidence monitored by SAR interferometry, near Gardanne, France*, J. Appl. Geophys. *43*(1), 43–54.

CFE (2010) *Cerro Prieto Geothermal Field*. CFE, Residencia General de Cerro Prieto, 46 pp.

COSTANTINI, M. (1998), *A novel phase unwrapping method based on network programming*, IEEE Trans. Geosci. Remote Sens. *36*(3), 813–821.

COSTANTINI, M., FARINA, A., and ZIRILLI, F. (1999), *A fast phase unwrapping algorithm for SAR interferometry*, IEEE Trans. Geosci. Remote Sens. *37*(1), 452–460.

CRESSIE, N. (1990), *The origins of Krigin*, Math. Geol. *22*(3), 239–252.

CHRISTIANSEN, L. B., HURWITZ, S., and INGEBRITSEN, S.E. (2007), *Annual modulation of seismicity along the San Andreas Fault near Parkfield, CA*, Geophys. Res. Lett. *34*, L04306, doi:10. 1029/2006GL028634.

DARBY, D., GONZÁLEZ, J., and LESAGE, P. (1984), *Geodetic studies in Baja California, Mexico, and the evaluation of short-range data from 1974 to 1982*, J. Geophys. Res. *89*(B4), 2478–2490.

DARBY, D., NYLAND, E., SUÁREZ-VIDAL, F., CHAVEZ, D., and GONZÁLEZ, J. (1981), *Strain and displacement measurements for the June 9, 1980 Victoria, Mexico, Earthquake*, Geophys. Res. Lett. *8*(6), 549–551.

DAWSON, J., and TREGONING, P. (2007), *Uncertainty analysis of earthquake source parameters determined from InSAR, A simulation study*, J. Geophys. Res. *112*, B09406, doi:10.1029/2007JB005209.

DAWSON, J., CUMMINS, P., TREGONING, P., and LEONARD, M. (2008), *Shallow intraplate earthquakes in Western Australia observed by Interferometric Synthetic Aperture Radar*, J. Geophys.Res. *113*, B11408, doi:10.1029/2008JB005807.

DE LA PEÑA, L. A. (1981), *Results from the first order leveling surveys carried out in the Mexicali Valley and at the Cerro Prieto field, Baja California*. In: Proceedings of the Third Symposium on the Cerro Prieto Geothermal Field, Baja California, Mexico, Lawrence Berkeley Laboratory, Berkeley, California, pp.281–291.

EARLE, P. S., and COGBILL, A. H. (2002), Potential of InSAR for routine earthquake analysis. ftp://hazards.cr.usgs.gov/Earle/Papers/InSAR_routine.pdf.

EBERHART-PHILLIPS, D., and D. H. OPPENHEIMER (1984), *Induced seismicity in The Geysers GeothermalArea, California*. J.Geophys. Res., *89*, 1191–1207

ELDERS, W. A., BIRD, D. K., WILLIAMS, A. E., and SCHIFFMAN, P. (1984), *Hydrothermal-flow regime and magmatic heat source of the Cerro Prieto geothermal system, Baja California, Mexico*, Geothermics, *13*, 27–47.

FABRIOL, H., and GLOWACKA, E. (1997), *Seismicity and Fluid Reinjection at Cerro Prieto Geothermal Field: Preliminary Results*. In: Proceedings of 22nd Workshop on Geothermal Reservoir Engineering, Stanford University, Stanford, CA, 11–17.

FABRIOL, H., and MUNGUÍA, L. (1997), *Seismic Activity at the Cerro Prieto Geothermal Area (Mexico) from August 1994 to December 1995, and Relationship with Tectonics and Fluid Exploitation*, Geophys. Res. Lett. *24*(14), 1807–1810.

FIALKO, Y., SIMONS, M. (2000), *Deformation and seismicity in the Coso geothermal area, Inyo Country, California: Observation and modeling using satellite radar interferometry*, J. Geophys. Res. *195*(B9), 21781–21793.

FIALKO, Y., and SIMONS, M. (2001), *Evidence for on-going inflation of the Socorro magma body, New Mexico, from interferometric synthetic aperture radar imaging*, Geophys. Res. Lett. *28*, 3549–3552.

FREZ, J., and GONZÁLEZ, J. J. (1991), Crustal structure and seismotectonics of northern Baja California. In: Dauphin J. P. and Simoneit B. R. T. (eds.) The Gulf and Peninsular Province of the Californias. American Association of Petroleum Geologists, 261–283.

GABRIEL, A. G., GOLDSTEIN, R. M., and ZEBKER, H. A. (1989), *Mapping small elevation changes over large areas: Differential radar interferometry*, J. Geophys. Res. *94*, 9183–9191.

GARCÍA, J. R. (1978), Estudios de nivelación de primer orden en Cerro Prieto. In: Proceedings of the First Symposium on the Cerro Prieto Geothermal Field, Baja California, México, pp.148–150.

GENS, R., and VAN GENDEREN, J. L. (1996), *SAR interferometry, issues, techniques, applications*, Int. J.Remote Sens. *17*(10), 1803–1835.

GLOWACKA, E., and NAVA, F. A. (1996), *Major earthquake in Mexicali Valley, Mexico, and Fluid Extraction at Cerro Prieto Geothermal Field*, Bull. Seismol. Soc. Am. *86*, 93–105.

GLOWACKA, E., GONZÁLEZ, J., and FABRIOL, H. (1999), *Recent vertical deformation in Mexicali Valley and its relationship with tectonics, seismicity, and the exploitation of the Cerro Prieto geothermal field, Mexico*, Pure Appl. Geophys. *156*(4), 591–614.

GLOWACKA, E., NAVA, F. A., DE COSSIO, G. D., WONG, V., and FARFAN, F. (2002), *Fault slip, seismicity, and deformation in Mexicali Valley, Baja California, Mexico, after the M 7.1 1999 Hector Mine earthquake*, Bull. Seismol. Soc. Am., *92*(4), 1290–1299.

GLOWACKA, E., SARYCHIKHINA, O., and NAVA, F. A. (2005), *Subsidence and stress change in the Cerro Prieto Geothermal Field, B.C., Mexico*, Pure Appl. Geophys. *162*, 2095–2110.

GLOWACKA, E., SARYCHIKHINA, O., SUÁREZ, F., MENDOZA, R., and NAVA, F.A. (2006), Estudio geológico para definir la zona de hundimiento con el fin de relocalización del canal Nuevo Delta en el Valle de Mexicali, Informe Técnico, CICESE, México, 505 pp.

GLOWACKA, E., SARYCHIKHINA, O., SUÁREZ, F., NAVA, F. A, FARFAN, F., DE COSSIO BATANI, G. D., and GARCIA ARTHUR, M. A. (2010a), Anthropogenic subsidence in the Mexicali Valley, B.C., Mexico, caused by the fluid extraction in the Cerro Prieto geothermal Field and the role of faults. In: Proceedings of the World Geothermal Congress (WGC) 2010, Bali, Indonesia, CD.

GLOWACKA, E., SARYCHIKHINA, O., NAVA, F. A., SUAREZ, F., RAMÍREZ, J., GUZMAN, M., ROBLES, B., FARFAN, F., and DE COSSIO BATANI, G. D. (2010b), *Continuous monitoring techniques of fault displacement caused by geothermal fluid extraction in the Cerro Prieto Geothermal Field (Baja California, Mexico)*. In: Correón-Freyre D., Cerca M. and Galowey D. (eds.) Land Subsidence, Associated Hazards and the Role of Natural Resources. IAHS Publ., *339*, pp.326–332.

GLOWACKA, E., SARYCHIKHINA, O., SUÁREZ, F., NAVA F. A., and MELLORS, R. (2010c), *Anthropogenic subsidence in Mexicali Valley, Baja California, Mexico, and slip on the Saltillo fault*. Environ. Earth Sci. *59*(7), 1515–1524.

GLOWACKA, E., SARYCHIKHINA, O., ROBLES, B., SUÁREZ, F., RAMÍREZ, J., and NAVA, F. A. (2012), Estudio geológico para definir la línea de hundimiento cero y monitorear la subsidencia de los módulos 10, 11 y 12 en el Valle de Mexicali, en el distrito de riego 014, Rio Colorado, B.C. Reporte Técnico Final, Convenio con CONAGUA, CICESE, Mexico, 560 pp.

GLOWACKA, E., SARYCHIKHINA, O., MÁRQUEZ RAMÍREZ, V.H., NAVA PICHARDO, F.A., GARCIA ARTHUR, M.A., FARFAN SANCHEZ, F.J., and OROZCO LEON, L.R. (2013), *Seismicity, deformation and fluid extraction in the Cerro Prieto Geothermal field during 1973–2009*. In: 2013 Annual Meeting of the Seismological Society of America (SSA), Salt Lake City, Utah, Seismol. Res. Lett. *84*(2), 384.

GOLDSTEIN, R. M., and WERNER, C. L. (1998), *Radar interferogram filtering for geophysical applications*, Geophys. Res. Lett. *25*(21), 4035–4038.

GONZÁLEZ-ORTEGA, A., FIALKO, Y., SANDWELL, D., NAVA PICHARDO, F.A., FLETCHER, J., GONZÁLEZ-GARCÍA, J., LIPOVSKY, B., FLOYD, M., and FUNNING, G. (2014), *El Mayor-Cucapah (Mw 7.2) earthquake: Early near-field postseismic deformation from InSAR and GPS observations*, J. Geophys. Res., B *119*(2), 1482–1497.

GONZÁLEZ, J., GLOWACKA, E., SUÁREZ, F., QUIÑONES, J.G., GUZMÁN, M., CASTRO, J.M., RIVERA, F., and FÉLIX, M.G. (1998), Movimiento reciente de la Falla Imperial, Mexicali, B. C, Ciencia para todos Divulgare, Universidad Autónoma de Baja California, *6*(22), 4–15.

GONZÁLEZ, M. (1999), *Actualización del modelo del basamento en el campo geotérmico de Cerro Prieto, BC, México*, Geotermia *15*(1), 19–23.

GONZÁLEZ, P.J., TIAMPO, K.F., PALANO, M., CANNAVÓ, F., and FERNÁNDEZ, J. (2012), *The 2011 Lorca earthquake slip* distribution controlled by groundwater crustal unloading, Nat. Geosci. *5*, 821–825.

GRANNELL, R. B., TARNMAN, D. W., CLOVER, R. C., LEGGEWIE, R. M., ARONSTAM, P. S., KROLL, R. C., and EPPINK, J. (1979), Precision gravity studies at Cerro Prieto. In: Proceedings of the Second Symposium on the Cerro Prieto Geothermal Field, Baja California, Mexico, Lawrence Berkeley Laboratory, Berkeley, California, pp. 329–331.

HANSSEN, R.F. (2001) *Radar Interferometry: Data Interpretation and Error Analysis*. Kluwer Academic Publishers, Dordrecht, The Netherlands, 328 pp.

HARRIS, R. A. (1998), *Introduction to special section: Stress triggers, stress shadows, and implications for seismic hazard*, J. Geophys. Res. *103*(B10), 24347–24358.

LIPPMANN, M. J., GOLDSTEIN, N. E., HALFMAN, S. E., WITHERSPOON, P. A. (1984), *Exploration and development of the Cerro Prieto geothermal field*, J. Petrol. Tech. *36*(9), 1579–1591.

LIPPMANN, M.J., TRUESDELL, A.H., MAÑÓN, A.M. and HALFMAN, S.E. (1991), *A review of the hydrogeologic-geochemical model for Cerro Prieto*, Geothermics, *20*, 39–52.

LIRA, H. (1996), Resultados del monitoreo de desplazamiento de la falla Cerro Prieto en 1996. Informe Técnico RE 21/96, Comisión Federal de Electricidad, Residencia de Estudios, México.

LIRA, H. (1999a), Resultados de la nivelación de precisión realizada con GPS, en 1998, en el campo geotérmico de Cerro Prieto. Informe Técnico RE 16/98, Comisión Federal de Electricidad, Residencia de Estudios, México, 30 pp.

LIRA, H. (1999b), *Monitoreo de la Subsidencia en el Campo Geotérmico de Cerro Prieto, B.C., México*, Geotermia, Revista Mexicana de Geoenergia *15*(1), 31–38.

LIRA, H. (1999c) Resultados del monitoreo de desplazamiento de la falla Cerro Prieto en 1998. Informe Técnico RE05/99. Comisión Federal de Electricidad, Residencia de Estudios. México, 6 pp.

LIRA, H. (2005) *Actualización del modelo geológico conceptual del yacimiento Geotérmico de Cerro Prieto*, Geotermia, Revista Mexicana de Geoenergia *18*(1), 37–46.

LIRA, H. (2006), Características del sismo del 23 de Mayo de 2006, Informe RE-023/2006, Comisión Federal de Electricidad, Residencia de Estudios, México.

LIRA, H., and ARELLANO, J. F. (1997), Resultados de la nivelación de precisión realizada en 1997, en el campo geotérmico Cerro Prieto. Informe Técnico RE 07/97, Comisión Federal de Electricidad, Residencia de Estudios, México, 25 pp.

LOHMAN, R. B., and SIMONS M. (2005), *Locations of selected small earthquakes in the Zagros Mountains*, Geochem. Geophys. Geosyst. 6, Q03001, doi:10.1029/2004GC000849.

LOMNITZ, C., MOOSER, F., ALLEN, C. R., BRUNE, J. N., and THATCHER, W. (1970), *Seismicity and tectonics of the northern Gulf of California, Mexico: preliminary results*, Geofísica Internacional *10*, 37–48.

MAJER, E. L., and McEVILLY, T.V. (1982), *Seismological Studies at the Cerro Prieto Geothermal Field, 1978–1982*, In: Proceedings of the Fourth Symp. on the Cerro Prieto Geothermal Field, Baja California, Mexico, Comisión Federal de Electricidad, 145–151.

MAJER, E. L., BARIA, R., STARK, M., OATES, S., BOMMER, J., SMITH, B., and ASANUMA, H. (2007), *Induced Seismicity Associated with Enhanced Geothermal Systems*, Geothermics, *36*, 185–222.

MASSONNET, D., and RABAUTE, T. (1993), *Radar interferometry, limits and potential*, IEEE Trans. Geosci. Remote Sens. *31*(2), 455–464.

MASSONNET, D., ROSSI, M., CARMONA, C., ADRAGNA, F., PELTZER, G., FEIGL, K., and RABAUTE, T. (1993), *The displacement field of the Landers earthquake mapped by radar interferometry*, Nature *364*, 138–142.

MASSONNET, D., BRIOLE, P., and ARNAUD, A. (1995), *Deflation of Mount Etna monitored by spaceborne radar interferometry*, Nature, *375*, 567–570.

MASSONNET, D., HOLZER, T.L., VADON, H., 1997, *Land subsidence caused by the East Mesa geothermal field, California, observed using SAR interferometry*: Geophys. Res. Lett., *24* (8), 901–904.

MASSONNET, D., and FEIGL, K. L. (1998), *Radar interferometry and its application to changes in the Earth's surface*, Rev. Geophys. *36*(4), 441–500.

McGARR, A., SIMPSON, D., and SEEBER, L. (2002), *Case histories of induced and triggered seismicity*, In: International Handbook of Earthquake and Engineering Seismology 2002, 81A, 647–661.

MELLORS, R. J., MAGISTRALE, H., EARLE, P., and COGBILL, A. (2004), *Comparison of four moderate-size earthquakes in southern California using seismology and InSAR*, Bull. Seismol. Soc. Am. *94*, 2004–2014.

MOSSOP, A., and SEGALL, P. (1997), *Subsidence at The Geysers geothermal field, N. California from a comparison of GPS and leveling surveys*, Geophys. Res. Lett. *24* (14), 1839–1842.

MUNGUÍA, L., GLOWACKA, E., SUÁREZ-VIDAL, F., LIRA-HERRERA, H., and SARYCHIKHINA, O. (2009), *Near-Fault Strong Ground Motions Recorded during the Morelia Normal-Fault Earthquakes of May 2006 in Mexicali Valley, BC, Mexico*, Bull. Seismol. Soc. Am. *99* (3), 1538–1551.

NAVA, F. A., and GLOWACKA, E. (1999), *Fault slip triggering, healing, and viscoelastic after working in sediments in the Mexicali-Imperial Valley*, Pure Appl. Geophys. *156*, 615–629.

PENNINGTON, W. D., and DAVIS, S.D. (1986), *The evaluation of seismic barriers and asperities caused by the depressuring of fault planes in oil and gas fields of South Texas*, Bull. Seism. Soc. Am. *76*, 923–948.

RAUCOULES, D., MAISONS, C., CARNEC, C., LE MOUELIC, S., KING, C., and HOSFORD, S. (2003), *Monitoring of slow ground deformation by ERS radar interferometry on the Vayvert Salt Mine (France). Comparison with ground based measurements*, Remote Sens. Environ. *88* (4), 468–478.

REBOLLAR, C. J., REYES, L. M.., QUINTANAR, L., and ARELLANO, J. F. (2003), *Stress Heterogeneity in the Cerro Prieto Geothermal Field, Baja California, Mexico*, Bull. Seismol. Soc. Am. *93*, 783–794.

ROSEN, P. A., HENSLEY, S., JOUGHIN, I. R., Li, F. K., MADSEN, S. N., RODRIGUEZ, E., and GOLDSTEIN, R. M. (2000), *Synthetic aperture radar interferometry*. In: Proceedings of the IEEE, 88, pp.333–382.

SARYCHIKHINA, O. (2003) Modelación de subsidencia en el campo geotérmico Cerro Prieto. MSc. Thesis, CICESE, México, 101 pp.

SARYCHIKHINA, O., GLOWACKA, E., MELLORS, R., VÁZQUEZ, R., MUNGUÍA, L., and GUZMÁN M. (2009), *Surface Displacement and Groundwater Level Changes Associated with the 24 May 2006 Mw 5.4 Morelia Fault Earthquake, Mexicali Valley, Baja California, Mexico*, Bull. Seismol. Soc. Am. *99*(4), 2180–2189.

SARYCHIKHINA, O., GLOWACKA, E., MELLORS, R., and SUÁREZ-VIDAL, F. (2011), *Land subsidence in the Cerro Prieto Geothermal Field, Baja California, Mexico, from 1994 to 2005. An integrated analysis of DInSAR, leveling and geological data*, J. Volcanol. Geoth. Res. *204*, 76–90.

SARYCHIKHINA, O., GLOWACKA, E., SUAREZ, F., and HINOJOSA, A. (2012), Analysis of ground deformation related to moderate-size

Reprinted from the journal

earthquakes using InSAR; application for the seismicity in the Mexicali Valley, Baja California, Mexico. Book of abstracts 33rd General Assembly of the European Seismological Commission (GA ESC 2012), 19-24 August 2012, Moscow, 126–127.

SEGALL, P. (1989), *Earthquakes triggered by fluid extraction*, Geology *17*, 942–946.

SEGALL, P. (1992), *Induced stresses due to fluid extraction from axisymmetric reservoirs,* Pure Appl. Geophys *139*, 535–560.

STRAMONDO, S, MORO, M, TOLOMEI, C, CINTI, F.R., and DOUMAZ F. (2005), *InSAR surface displacement field and fault modelling for the 2003 Bam earthquake (southeastern Iran)*, J. Geodyn. *40*, 347–353.

SUÁREZ-VIDAL, F., MUNGUÍA-OROZCO, L., GONZÁLEZ-ESCOBAR, M., GONZÁLEZ-GARCÍA, J., and GLOWACKA, E. (2007), *Surface rupture of the Morelia fault near the Cerro Prieto Geothermal Field, Mexicali, Baja California, Mexico, during the Mw 5.4 earthquake of 24 May 2006*, Seismol. Res. Lett. *78*(3), 394–399.

SUÁREZ-VIDAL, F., MENDOZA-BORUNDA, R., NAFARRETE-ZAMARRIPA, L., RÁMIREZ, J., and GLOWACKA, E. (2008), *Shape and dimensions of the Cerro Prieto pull-apart basin, Mexicali, Baja California, México, based on the regional seismic record and surface structures*, Int. Geol. Rev. *50*(7), 636–649.

URBAN, E., and LERMO, J.F. (2012), Relationship Of Local Seismic Activity, Injection Wells And Active Faults In The Geothermal Fields Of Mexico, In: Proceedings of Thirty-Seventh Workshop on Geothermal Reservoir Engineering, Stanford University, Stanford, CA, SGP-TR-194

VELASCO, J. (1963), Levantamiento Gravimétrico, Zona Geotérmica de Mexicali, Baja California, Consejo de Recursos Naturales no Renovables, 55 pp.

WEGMÜLLER U., and WERNER C. L. (1997) *Gamma SAR processor and interferometry software*. In: Proceedings of the 3rd ERS Symposium, Eur. Space Agency Spec. Publ., ESA SP-414, pp. 1686–1692.

WEI, M., SANDWELL, D., and FIALKO, Y. (2009), *A silent Mw 4.7 slip event of October 2006 on the Superstition Hills fault, southern California*, J. Geophys. Res., *114*, B07402. doi:10.1029/2008JB006135.

WEI, M., KANEKO, J., LIU Y., and McGUIRE, J. J. (2013), *Episodic fault creep events in California controlled by shallow frictional heterogeneity*, Nature Geoscience *6*, 566–570. doi:10.1038/ngeo1835.

WELLS, D. L., and COPPERSMITH, K. J. (1994), *New Empirical Relationship among Magnitude, Rupture Width, Rupture Area, and Surface Displacement*, Bull. Seism. Soc. Am., *84*(4), 974–1002.

WYMAN, R.M. (1983), Potential Modeling of Gravity and Leveling Data over Cerro Prieto Geothermal Field. M.Sc. Thesis, California State University, Long Beach, CA, USA, 79 pp.

ZANDBERGEN, R., OTTEN, M., RIGHETTI, P.L., KUIJPER, D., and DOW, J.M. (2003), *Routine operational and high-precision orbit determination of Envisat*, Adv. Space Res. *31*, 1953–1958.

ZEBKER, H.A., and VILLASENOR, J. (1992), *Decorrelation in Interferometric Radar Echoes*, IEEE Trans. on Geoscience and Remote Sensing *30*(5), 950–959.

ZHOU, X., CHANG, N.-B., and LI, S. (2009), *Applications of SAR Interferometry in Earth and Environmental Science Research*, Sensors *9*, 1876–1912. doi:10.3390/s90301876.

(Received February 26, 2014, revised February 10, 2015, accepted March 4, 2015, Published online May 8, 2015)

Pure Appl. Geophys. 172 (2015), 3163–3177
© 2014 Springer Basel
DOI 10.1007/s00024-014-0883-y

| Pure and Applied Geophysics

Source Parameters of Earthquakes Recorded Near the Itoiz Dam (Northern Spain)

A. Jiménez,[1] J. M. García-García,[2,3] M. D. Romacho,[2,3] A. García-Jerez,[2,3] and F. Luzón[2,3]

Abstract—We calculate the source parameters and attenuation from earthquakes recorded near the Itoiz dam, from 2004 to 2009, with magnitudes ranging between 1.2 and 5.2. We use a Genetic Algorithm in order to fit the three-component P-wave spectra with the spectral level, corner frequency, and attenuation factor as searching parameters. The obtained moments range from 1.72×10^{11} to 2.65×10^{15} Nm, the radii span from 0.09 to 1.00 km, and the stress drops vary from 0.006 to 29.462 MPa. The maximum value for the Q attenuation factor is 794, and the minimum value is 53. We find a good agreement between empirical and theoretical relationships between moment and magnitude. There seems to be a breakdown of self-similarity, but it could be due to the method used. We group the data by means of a Self-Organizing Map and the clusters found are related by their magnitude, and not by other considerations.

Key words: Source parameters, Reservoir induced seismicity, Genetic Algorithm, Self-Organizing maps.

1. Introduction

The Itoiz dam is located in the Jaca-Pamplona basin, in the framework of the South Pyrenean Zone (Fig. 1). On September 18, 2004, a 4.6 mbLg earthquake was widely felt in the western Pyrenees in the region around Pamplona. Preliminary locations reported an epicenter <20 km ESE of Pamplona and close to the Itoiz reservoir, which started impounding in January 2004. This earthquake produced a considerable social alarm about the seismic hazard in the region induced by the impoundment of the reservoir. Our goal is to better understand the mechanisms of the seismicity around the Itoiz dam, and give some insight about its origin. In order to do so, we compare the seismic parameters in that area with other areas where no induced seismicity is involved.

The region has been extensively studied since the occurrence of the 18 September, 2004 event. For example, Ruiz et al. (2006) analyzed the seismic series corresponding to the 18 September 2004 earthquake; Luzón et al. (2009, 2010) and Durá-Gómez and Talwani (2010) studied the effect of the pore pressure due to the impoundment of the reservoir. Santoyo et al. (2010) calculated the stress produced by the reservoir near Itoiz. They showed that the aftershock sequence was indeed produced by the stress transfer caused by the 18 September 2004 earthquake. Rivas-Medina et al. (2011) calculated the seismic hazard near the Itoiz dam. In previous works (Jiménez et al., 2009; Jiménez and Luzón, 2011, 2012), we studied the main seismic clusters near the Itoiz dam. We did not find any difference between the 18 September 2004 cluster and the others, except for a higher fractal dimension of the epicentre distribution.

After the 18 September 2004 event, new stations have been deployed in order to monitor the seismic activity in the area. Among them, the University of Almería installed five broadband stations near the reservoir, in order to study the seismicity around the Itoiz dam from 2008 to 2010. With those data, and the ones obtained from the Instituto Geográfico Nacional (IGN), we analyze the source parameters from 45 events since 2004.

It is usually established that reservoir induced seismicity presents lower stress drops than the tectonic earthquakes (Abercrombie and Leary, 1993). Their compilation study included hydraulic fracturing in Fenton Hill, NM (Fehler and Phillips, 1991) and coring-induced earthquakes from Manitoba, Canada

---

[1] School of Environmental Sciences, University of Ulster, Cromore Rd, Coleraine BT52 1SA, Northern Ireland, UK. E-mail: a.jimenez@ulster.ac.uk

[2] Department of Chemistry and Physics, University of Almería, Carretera de Sacramento s/n, Cañada de San Urbano, 04120 Almería, Spain.

[3] Instituto Andaluz de Geofísica, Granada, Spain.

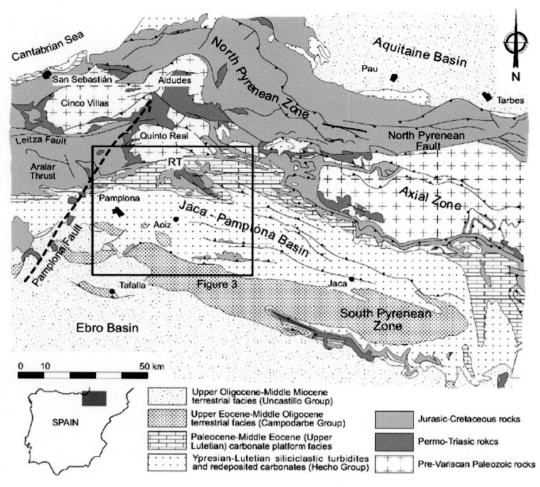

Figure 1
Geological setting (RUIZ *et al.*, 2006)

(GIBOWICZ *et al.*, 1990). In both studies the stress drops were on average one order of magnitude lower ($\Delta\sigma \leq 10$ MPa) than those of naturally occurring earthquakes. This is still observed when the compilation is corrected to use a consistent method for all analyses using earthquake spectra. More recently MANDAL *et al.* (1998) analyzed a set of M 1.4–4.7 induced earthquakes at the Koyna-Warna reservoirs in India and found shallow earthquakes ($1 \leq z \leq 4$ km) with very low stress drops, $\Delta\sigma \leq 2$ MPa. However, some works do not find any difference between natural and induced events (TOMIC *et al.*, 2009).

Another interesting question is the breakdown of self-similarity in small earthquakes. For large earthquakes, it is established that the stress drop is constant with the moment, but previous investigations found that the corner frequency of small events increases very slowly or becomes constant with decreasing seismic moment below a certain value, implying a breakdown in the self-similarity of the rupture processes for small earthquakes (Jin *et al.*, 2000). Analyzing earthquakes of magnitude in the range 2.5–7, ATKINSON (1993, 2004) noted that the stress drop increases with magnitude (from about 0.3 to 10 MPa) until $M = 4$, and then above that magnitude it appears to have a relatively constant value in the range 10–20 MPa. Atkinson concluded that the low stress drops for small events really reflect low high-frequency spectral amplitudes and not a bandwidth limitation. Other studies, using borehole data, revealed that the apparent corner frequencies of microearthquakes are significantly higher than those estimated from surface recordings (HAUKSSON et al., 1987; ABERCROMBIE, 1995). The apparent high-frequency limit to corner frequency observed using seismograms recorded at the surface may be due mainly to high attenuation in the weathered, near to the surface material, as concluded by FRANKEL and WENNERBERG (1989) and ABERCROMBIE (1995, 1997), BINDI et al. (2001), among others. If there were a breakdown in similarity, it would mean a minimum size for the source dimension, of around 100 m. In the same way we have a maximum size given by the elastic properties of the lithosphere (ARCHULETA et al., 1982), so that it would not be a self-similar rupture process (MAI and BEROZA, 2000; SCHOLZ, 1990).

In the present work, we obtain the source parameters (moment, radius, and stress drop) from 45 earthquakes recorded near the Itoiz dam (Northern Spain), by using Genetic Algorithms (GA), following JIMÉNEZ et al. (2005). We use three-component P-wave records. Then we analyze the results, and we compare them with other events occurring in a region (the Granada basin, in Southern Spain) where we are sure that no induced seismicity is involved.

2. Data Selection and Analysis

The digital data used in this work were recorded by seven short period (1 Hz) seismic stations and two accelerometers (from the Instituto Geográfico Nacional, IGN, Spain), five broadband instruments and four accelerometers (from our research group). We use the three components from all instruments. The dataset comprises 45 low to moderate local earthquakes recorded by 18 three-component seismic stations from 2004 to 2009 in the Itoiz zone (Navarra, Spain). In Fig. 2 we show the map with all the events analyzed and the locations of the stations. All stations are located in hard rock. As STEIDL et al. (1996) pointed out, site effects can be present. In the method section we will explain how those effects are treated. In Fig. 3 we show the water level at the Itoiz dam, the number of earthquakes and the magnitudes of the seismicity recorded in the region.

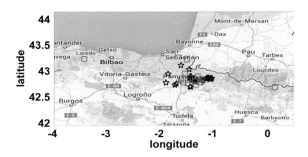

Figure 2
Location of the analyzed events (*stars*) and the stations (*red squares*)

The data set was selected on the basis of a high signal to noise ratio, accurate epicentral location, and sufficient separation between P and S phases. In the location program, a number is assigned to indicate the quality of the time series. The higher the number, the clearer the starting points of both P and S waves, and the better the epicentral location too. We only chose the higher numbers, that is, the events with a clear impulsive P wave, with low error in its starting point, and clear variation in amplitude and phase of the S wave. We need sufficient separation between the phases in order to have a P wave not contaminated with the S wave. The recording affected by saturation effects and other problems were discarded. The body wave magnitude (mbLg) of the selected events, ranged between 1.2 and 5.2.

The location accuracy relied upon a very good reading of the first P-wave arrival and the relative large number of stations used. We used a window length that includes the maximum amplitude of the P-wave and avoids overlapping with other types of waves. The range of the window length is from 0.27 s for small events (magnitude 1.3) to 3.56 s for the bigger events. Most of the events have depths of <10 km (between 1 and 11 km), and so they are considered shallow events. The hypocentral distances are between 0.6 and 174 km. At this distance range, the seismic records are mainly influenced by the near to the surface structure of the crust because most source-station ray paths are confined to the crust (FERNÁNDEZ et al., 2010). In FERNÁNDEZ et al. (2010) they speak about a range of distances between 0 and 100 km. For larger distances, the effect of the near mantle can affect the values for the Q factors.

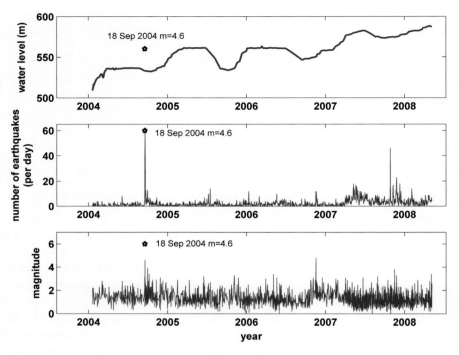

Figure 3
Water level and seismicity near the Itoiz dam over time

However, we are confident that this is the range of most of our distances between source and station. The events are recorded very close to the stations in most of the cases. We only have two stations that are not very close (see Fig. 2). It should be noted that the impulsive P-wave arrivals are characteristic to all selected events.

All the spectra studied from P-waves were calculated using a Fast Fourier Transform. First, the signal was base-line corrected subtracting the average of all points of the record. The time series were windowed from the start of the phase by using a both ends 10 % cosine taper. The resultant seismogram was padded with zeros to an integer power-of-2 prior to the FFT. Signal windows of varying lengths were tested in order to select a length that would avoid contamination from other phases and maintain the resolution and stability of the spectra (GARCÍA et al., 1996). In GARCÍA et al. (1996), they describe the method, following PROCHÁZ-KOVÁ (1970). We would think that taking a small window length we would have no contamination between phases. However,

that gives a poor resolution in the spectrum. So we make a trade off between window length and contamination. We will choose the minimum length with the maximum energetic content and with less contamination from other phases. Thus, we try different window lengths and compare the spectrum. Whenever we find that the spectra have the same shape (flat part and decay after the corner frequency), we choose that window length.

The sampling rate is of 200 samples per second for the accelerometer recording and a rate of 100 samples per second for velocity recordings. The spectra were subsequently corrected for instrumental response. Prevention of noise windows of equal length than the signal window were used to compute the signal noise relation (SNR) and only records with a SNR higher than 3 were retained. Spectral amplitudes were divided by the appropriate power of the frequency to give a displacement spectrum. A total of 360 P-wave spectra are used in this study. Figure 1 shows the distribution of the epicentres of the considered earthquakes.

3. Method

Following ABERCROMBIE (1995), and after correcting for the geometrical spreading and instrumental response, the problem is solved for the spectral level Ω_0 and the corner frequency f_c, while simultaneously correcting for path-averaged attenuation Q, by fitting P-wave and S-wave spectra with the ω^{-2} source model proposed by BOATWRIGHT (1978). The spectral shape is described by:

$$\Omega(f) = \frac{\Omega_0 e^{-\pi f t/Q}}{\left[1 + \left(\frac{f}{f_c}\right)^4\right]^{1/2}}, \tag{1}$$

where Ω_0 is the spectral level, f_c is the corner frequency, Q is the path-averaged attenuation factor, and t is the travel time. With the values of Ω_0 and f_c, the seismic source parameters can be obtained from the well-known relationships established for a circular fault (BRUNE, 1970, 1971; HANKS and WYSS, 1972). Additional information is the knowledge of the attenuation factor Q. It represents the loss of the energy of the waves along the path between the source and the station, and it can give useful information about the structure of the earth (DEL PEZZO et al., 1991).

We use a Q independent of the frequency. FEHLER and PHILLIPS (1991) demonstrated that variation of Q from 600 to 3,000 changed the corner frequency by <15 % from the result obtained by assuming a constant average Q. A more complex attenuation function does not result in a better fit to the data, and produces multiple equivalent solutions in the sense that they have the same fitting accuracy (JIMÉNEZ et al., 2005). Among these other functions, we tried to include the kappa factor (GARCÍA GARCÍA et al., 2004) accounting for the site effect. No improvement was found in using it, and a constant Q was the better choice. So, the Q values account for the attenuation along the path, and the site effect at middle to high frequencies can not be distinguished from the noise. In the aforementioned article (JIMÉNEZ et al., 2005), where we used the same method as the one we use here, we calculated the seismic parameters for a set of earthquakes that were the same as the work in (GARCÍA et al., 1996) They performed the Snoke's method (SNOKE, 1987). Both methods gave similar results. Thus, we are confident that the method is stable.

We use Genetic Algorithms (GA) in order to fit the three parameters that describe the spectrum. GA are methods of global optimization, which have proven effective when the models are described by a few parameters, the problem is nonlocal (the global optimum is needed, but there are many local optima) and nonlinear, and there is no a priori knowledge of the behavior of the function. In geophysics, and particularly in seismology, many problems often have such features. The basic principles of GA were established by HOLLAND (1975), and are well described, for example, in GOLDBERG (1989), DAVIS (1991), MICHALEWICZ (1992) or REEVES (1993).

The GA used in this paper has been implemented in JIMÉNEZ et al. (2005). The definitive search strategy chosen was selection by rank wheel, crossover based on fitness, and replacement by rank wheel. This strategy is moderately elitist, because the best individuals are easily selected, but there is low selection pressure. The simple GA was improved with the reinitialization of the population when the convergence stays blocked (JIMÉNEZ et al., 2005).

The algorithm was tested with synthetic data and applied to the spectra of 13 earthquakes in Granada (Southern Spain). The obtained parameters are the corner frequency, the spectral level, and the attenuation along the path. HOUGH et al. (1989) showed that the trade-off between Q and the corner frequency is strong when a wide range of values for both parameters are used to fit the data. So then it is important to use a good optimization tool to calculate those parameters. That's the reason for using a powerful tool such as a GA. In JIMÉNEZ et al. (2005) we compared the implemented GA with the commonly used Nelder–Mead algorithm (TOMIC et al., 2009), and found that our technique is more appropriate for this task. For further details, see JIMÉNEZ et al. (2005).

For obtaining the source parameters, the best fit was chosen for introducing its spectral parameters to the relationships given by BRUNE (1970) and HANKS and WYSS (1972):

$$M_0 = \frac{4\pi\rho v^3 \Omega_0 d}{KR}, \tag{2}$$

$$r = \frac{kv}{f_c}, \tag{3}$$

$$\varDelta\sigma = \frac{7M_0}{16r^3} \qquad (4)$$

being ρ the density of the medium (2.7 g/cm^3), v the velocity [5.5 km/s for the P-wave, following Ruiz et al. (2006)], d being the hypocentral distance, accounting for the geometrical spreading; K being the wave amplification at the free surface; and R being the radiation pattern coefficient for the P-waves and S- waves. Since the focal mechanism could not be determined (the spectra correspond to small earthquakes), the rms averages $R(P) = 0.52$ and $R(S) = 0.63$ were used (AKI and RICHARDS, 1980). The k is a factor that depends on the model. If we use Brune's model, this factor is 0.372, but if we use Madariaga's model (MADARIAGA, 1976), it is 0.21. So, the radii are 1.76 times larger for the BRUNE (1970) model, and the stress drop is 5.6 times larger for the MADARIAGA (1976) model. The SATO and HIRASAWA (1973) model lies in between. All three models are in common usage, and the model used must be considered when comparing results from different studies (TOMIC et al., 2009).

In order to remove the effect of the free surface in a three-component record, we follow the method described in SHIEH (1995). The motion of the particle we have in the record of the window where we suppose the P wave is the composition of the incident P-wave, the reflected wave, and the SV-wave produced by the P incident. So, by means of the equations in AKI and RICHARDS (1980), we have the horizontal and vertical components of the motion. By dividing the horizontal into the vertical component of the records, and comparing this quantity to the theoretical one, we can calculate the angle of incidence and then remove the free surface effect (SHIEH, 1995). In that way, we can calculate the spectrum of the incident P-wave. So, we first calculate the spectrum with the free surface effect removed and then fit the spectrum with the GA to obtain the source parameters. This is a very important step, since we only want the contribution of the P wave, and not that of the SV-wave, that contaminates the spectrum. We performed the same transformation for S-waves, but the separation of the SV and SH waves was very problematic, so that they were highly contaminated by other phases. Thus, we only used the P-wave for our calculations.

Each real spectrum was introduced in the GA implemented with the following considerations for the fitting parameters. First, the spectral level limits for the search in that parameter were obtained by visualizing the spectrum. Although the source spectrum computed from numerical data might present large uncertainties, we gave a wide range for the searching limits for the spectral level (1 or 2 orders of magnitude, depending on the spectrum). Second, the corner frequency were searched between 0.1 and 25 Hz, which is a reasonable value for the magnitudes involved. Finally, a maximum of 800 was introduced for the Q values after consulting several studies made in the zone (PUJADES et al., 1990; GONZÁLEZ, 2001). The resolution for each parameter (spectral level, corner frequency, and Q factor) was of 5, 6, and 7 bits, respectively. That is, 2 to the power of 5, 6 and 7. The population (number of models in each generation) and number of generations were 20 and 100, respectively. The algorithm was executed three times, and the best fit was selected for calculating the source parameters. We found that with three times only we obtained consistent results, so we did not increase the number of runs and, therefore, the computation time.

For each event, the average values for seismic moment, source radius, and stress drop were computed. Calculations were made using three component P-wave records. The average values were estimated as (ARCHULETA et al., 1982):

$$\bar{x} = \text{antilog}10\left(\frac{1}{N}\sum_{i=1}^{N}\log10x_i\right), \qquad (5)$$

where N is the number of stations used. The reason is that the errors associated with Ω_0 and r are log-normally distributed. The standard deviation of the logarithm, SD($\log\bar{x}$) was estimated by calculating the variance of the individual logarithms about the mean logarithm:

$$\text{SD}(\log10\bar{x}) = \left(\frac{1}{N-1}\sum_{i}^{N}[\log10x_i - \log10\bar{x}]^2\right)^{1/2},$$
$$(6)$$

and multiplicative error factors, Ex, were calculated as:

$$\text{Ex} = \text{anti log }10(\text{SD}(\log 10x)). \qquad (7)$$

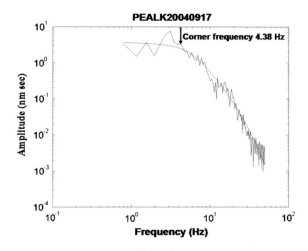

Figure 4
Example of a fitted spectrum. In *red* the fitted spectrum, and in *blue* the real one. The corner frequency is plotted too

4. Results and Discussion

In Fig. 4 we show an example of a fitted spectrum. All the spectra had good visual correspondence between the fitted line and the experimental data. In Table 1 we write the results for the source parameters of all the events, with their corresponding multiplicative errors, when possible (when more than two stations recorded the earthquake). The higher errors correspond to the stress drops, because they are a derived quantity, calculated from the moment and radius. The averaged error for the moment is 2.5, and for the radius is 1.83, which are very good. Note that the smallest error in logarithmic scale is 1, which means that all the results were the same. The corner frequencies span from 1.66 to 23.44 Hz. So, the sampling rate of 100 and 200 samples/s are appropriate for the studied data set. Note that we use Brune's model, the P-wave velocity is set at 5.5 km/s, and the minimum source radius is 0.09 km.

The obtained moments range from 1.72×10^{11} to 2.65×10^{15} Nm, with an average value of 8.51×10^{13} Nm; the radii span from 0.09 to 1.00 km, with an average value of 0.32 km. Finally, the stress drops vary from 0.006 to 29.462 MPa with an average value of 1.076 MPa.

The values we find are very similar to the ones obtained by TUSA and GRESTA (2008) in southern Sicilia, with tectonic events, and magnitudes ranging from 0.7 to 4.6. The stress drops found by PREJEAN and ELLSWORTH (2001) in the Long Valley Caldera, Eastern California, range from 0.01 to 30 MPa, with magnitudes spanning from 0.5 to 5, a result which is very similar to the one obtained in the present study.

As we mentioned before, MANDAL et al. (1998) analyzed a set of M 1.4–4.7 induced earthquakes at the Koyna-Warna reservoirs in India and found shallow earthquakes ($1 \leq z \leq 4$ km) with very low stress drops, $\Delta\sigma \leq 2$ MPa. The depths of the events in our data set vary from 0 to 11.1 km, so they are shallow earthquakes too. In Table 1 we can observe that 96 % of the events have stress drops lower than 2 MPa, if we use the BRUNE (1970) model, and 80 % are lower than 2 MPa if we use the MADARIAGA (1976) model. In TOMIC et al. (2009) they find higher stress drops, because they use the MADARIAGA (1976) model. For events with magnitude ranging from 0.3 to 3.1, HUA et al. (2013) found very low values for the static stress drops (from 0.01 to 0.26 MPa). It has to be noted that the highest stress drop is found for the 5.2 event, and it is much higher than the stress drops found for the other earthquakes (29.462 MPa for the 5.2 earthquake, and 96 % of events are below 2 MPa). As can be seen in Table 1, the multiplicative error factors for the 5.2 events are low (2.19, 1.48, 1.54, for the moment, radius, and stress drop), so we are confident that the high value for its stress drop is accurate.

Now we will analyze the results obtained for the attenuation factors. It is important to remember that we obtain the Q factors for each path, in order to compare our results with other attenuation studies done in the area of interest. In Fig. 5 we show the Q factors as a function of the distance. Note that the Q factor is calculated for each path, so no error can be associated. As stated before, each fitted spectrum is the result of three calculations of the parameters with the GA, so that the obtained parameters are very stable. The maximum value is 794, and the minimum value is 53 with an average value of 444. This means that the limits imposed in the search (between 10 and 800) were reasonable, and in agreement with previous studies in the region (PUJADES et al., 1990; GONZÁLEZ, 2001). We find that, in general, the attenuation factor increases up to an epicentral

Table 1

Source parameters of the analyzed events

nsis	Date	Magnitude (mbLg)	Moment (Nm)	Radius (km)	Stress drop (MPa)	EM	Er	Es	Number of stations
1	20040916191706	3.2	8.65E+13	0.31	1.217	5.06	2.38	32.77	3
2	20040917025856	3	7.43E+12	0.21	0.346	2.47	3.08	11.82	2
3	20040918125218	5.2	2.65E+15	0.34	29.462	2.19	1.48	1.54	3
4	20040918125507	3.2	1.98E+14	0.24	6.532	–	–	–	1
5	20040918152547	2.5	1.54E+13	0.73	0.018	–	–	–	1
6	20040918195828	3.1	2.60E+13	1.00	0.011	–	–	–	1
7	20040919054023	2.8	6.73E+13	0.27	1.440	–	–	–	1
8	20040930130907	3.9	1.90E+14	0.38	1.506	1.73	2.81	15.09	3
9	20041007061630	3.4	4.55E+13	0.46	0.207	3.73	2.01	12.97	3
10	20041020164719	2.7	3.31E+13	0.34	0.355	1.76	2.27	17.48	3
11	20041023174246	2.9	1.34E+13	0.31	0.189	6.15	1.80	35.93	3
12	20041109112619	2.9	9.75E+12	0.27	0.218	2.14	2.88	11.13	2
13	20060111031417	1.5	1.72E+11	0.19	0.012	–	–	–	1
14	20060111033502	1.2	1.80E+11	0.14	0.030	–	–	–	1
15	20060111142103	1.5	4.64E+11	0.09	0.305	–	–	–	1
16	20060111143211	1.4	2.39E+11	0.12	0.066	–	–	–	1
17	20060111184704	1.6	2.04E+11	0.16	0.021	–	–	–	1
18	20070411014540	2.9	1.51E+13	0.36	0.141	4.42	1.69	2.18	3
19	20070514125550	2.3	3.74E+12	0.17	0.342	–	–	–	1
20	20070608123322	1.9	7.94E+11	0.10	0.321	–	–	–	1
21	20070731214029	2.7	3.45E+12	0.47	0.015	1.87	1.54	6.87	2
22	20070829075318	3.1	1.39E+13	0.34	0.151	1.91	1.91	5.25	4
23	20071015205657	2.8	5.31E+12	0.20	0.306	2.69	2.10	8.56	5
24	20071122112044	2.3	3.51E+12	0.29	0.065	1.23	1.99	6.49	3
25	20071122185450	2.1	1.56E+12	0.44	0.008	2.17	1.78	3.71	3
26	20080426135505	2.8	7.40E+12	0.33	0.093	2.32	1.57	3.89	4
27	20080517041011	2.3	2.21E++12	0.32	0.030	3.12	1.51	3.54	4
28	20080526160213	2.6	2.37E+13	0.34	0.258	5.67	1.38	2.17	2
29	20080604210825	2.9	1.78E+13	0.32	0.234	1.45	1.35	1.70	2
30	20080622074755	2.4	3.29E+12	0.32	0.042	–	–	–	1
31	20080625010713	2.1	7.12E+11	0.27	0.016	3.27	1.89	4.41	4
32	20080627195904	2	4.36E+11	0.32	0.006	–	–	–	1
33	20080709082814	2	1.99E+12	0.21	0.093	1.82	1.25	3.58	2
34	20080725212446	2.7	1.41E+13	0.46	0.065	4.26	2.88	8.18	4
35	20080915041001	2.4	1.86E+12	0.15	0.234	1.13	1.41	2.48	2
36	20080916070024	2.8	4.91E+12	0.32	0.065	1.33	1.89	7.78	4
37	20080920161022	2.4	2.98E+12	0.22	0.124	1.46	1.49	4.63	3
38	20080928135600	2.4	1.47E+12	0.31	0.022	1.90	1.12	2.29	3
39	20081010133108	2.9	8.43E+12	0.19	0.516	1.94	2.27	11.67	5
40	20081126063000	2.2	2.95E+12	0.20	0.162	1.58	1.51	2.35	3
41	20081211210757	2.8	4.62E+13	0.28	0.950	2.12	1.31	1.77	4
42	20090121200701	2.9	1.18E+14	0.42	0.700	2.06	1.61	3.18	8
43	20090127212208	2.5	5.46E+13	0.73	0.063	–	–	–	1
44	20090701150326	2.6	1.02E+14	0.33	1.200	1.43	1.05	1.25	2
45	20091024200524	2.6	2.41E+13	0.33	0.282	1.08	1.43	3.12	2

EM, Er, and Es are the multiplicative errors in moment, radius and stress drop, respectively

distance of around 50 km, indicating that the scattering of the waves reaches the Moho, which is very deep in this area. It is around 46–48 km (PEDREIRA et al., 2003).

PUJADES et al. (1990), and more recently GONZÁLEZ (2001), calculated the Q-coda values for this region. The latter obtained $Q_0 = 53$ and $v = 1.12$ for the ELIZ station (from the IGN), the closest to our area

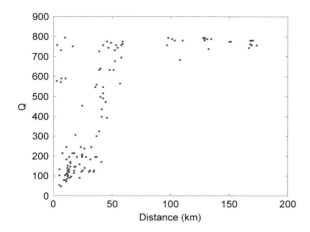

Figure 5

Attenuation factors vs. distance between event and station. Each *point* represents only one path

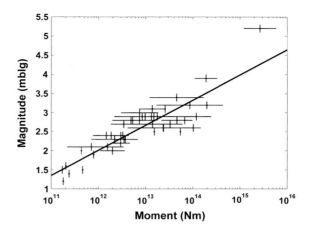

Figure 6

Relationship between magnitude and moment, with their corresponding errors. We also plot the line which best fits the data. The *dots with only vertical error bars* were obtained by analyzing just one spectrum

of study, assuming a relationship between Q and the frequency given by:

$$Q = Q_0 \left(\frac{f}{f_0}\right)^v. \tag{8}$$

If we use the averaged value of 444 (the average value we obtained) for all the frequencies, and assuming the same v, we would obtain $Q_0 = 26$, also very low. Note that our distances are lower (they use data up to 300 km), and so the value for the attenuation should be lower too (ZELT *et al.*, 1999). If we suppose an exponent $v = 0.85$, close to the value of the EGRA station, also used for studying the attenuation in the Pyrenees by GONZÁLEZ (2001), the value for Q_0 would be 53. So that would be the range of parameters for the attenuation we found, in good agreement with the studies in the region. In any case, as GONZÁLEZ (2001) says, the relationship in Eq. (8) does not hold for a long range of frequencies as we used in our calculations. Also, note that we use a constant value for all the frequencies so that the comparisons have to be taken carefully.

Now we are analyzing the relationship between the different magnitudes calculated. In first place, in Fig. 6 we plot the moment versus the magnitude of the events. The best fit for the line is $M_L = (0.66 \pm 0.05) \log M_0 - (5.9 \pm 0.7)$, with correlation coefficient of 0.88. This relationship is exactly the one expected theoretically ($M_L = 2/3 \log M_0 - 6$). This is if the moment is expressed in Nm, as

we have done (DEICHMANN, 2006). The significance of the obtained relationship is of 5.89σ. This means that our calculation for the seismic moment is theoretically consistent with the magnitude given by the IGN.

In order to test the hypothesis of constant stress drop, in Fig. 7 we show the relationship between source radius and seismic moment. The lines represent the contours with constant stress drops. As we can see, the stress drops vary, and have a high dispersion in their values. It is a similar result to the one obtained by TUSA and GRESTA (2008). We also did not find a relationship between the corner frequency and moment. There is a trend of increasing moment with decreasing corner frequency, but the correlation coefficient is very low. Note that the errors in the source radii are low (around 1 or 2 in Table 1), which means that the corner frequencies have been correctly calculated. The large errors in the stress drops are a consequence of being a parameter calculated from both moment and radius, with their associated errors.

In Fig. 8 we see that the stress drop increases with increasing moment. The best fit gives $\log \Delta\sigma = (0.81 \pm 0.08) \log M_0 - (5.4 \pm 1.1)$ with a correlation coefficient of 0.77, and a significance of 6.4σ. We did not use the data with no error in the fit, because they were recorded only in one or two stations. The trend with respect to the magnitude is also positive as JOST *et al.* (1998) found in injection-induced earthquakes of magnitude ranging between -2 and 0. ONCESCU

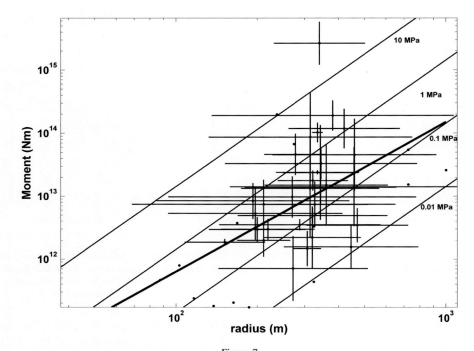

Figure 7
Radius vs. moment with their *errors bars*. The *lines* are contours of equal stress drops in MPa. The best fit is also presented, with a *thicker line*. The *dots with no error bars* were obtained by analyzing just one spectrum

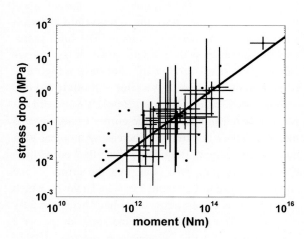

Figure 8
Moment vs. stress drop, with the corresponding *error bars*. The line is the best fit, without including the data with only one or two records. The *dots with no error bars* were obtained by analyzing just one spectrum

et al. (1994), HAAR *et al.* (1984) and HUA *et al.* (2012, 2013) found that stress drop increased with moment too.

In our data, a trend of increasing stress drop with increasing seismic moment is evident, suggesting a

self-similarity violation. The breakdown in similarity that we observe is very similar to the one ATKINSON (1993, 2004) found in northeastern America for events having Mw less than about 4, and the one found by TUSA and GRESTA (2008) with earthquakes in southern Sicily. However, CASTRO *et al.* (1995), for earthquakes with magnitude between 2.8 and 4.8, found that the stress drop decreases with increasing moment. DROUET *et al.* (2005) found that constant stress drop is valid for the Pyrenees in the range of magnitude 2.7–5.4, which is the opposite we found. DROUET *et al.* (2008) found a slight tendency to higher stress drops for higher moments in the Western Pyrenees, but they interpreted it as an effect of the finite-frequency bandwidth that they analyzed. MORI and FRANKEL (1990) and SMITH and PRIESTLEY (1993) said that there is no apparent correlation between stress drop with moment with depth. With events of magnitude between 3.3 and 4.9, we did not find any relationship between stress drop and depth. This is in agreement with the results in MOYA *et al.* (2000).

IDE *et al.* (2003) thought that the use of constant Q and/or propagation effects could introduce artificial size dependence in apparent stress measurements. They used data from boreholes, and could calculate Empirical Green Functions to correct by path and site effects. We do not have borehole data, and we can not calculate appropriate Empirical Green Functions, because most of the events have a low magnitude. We, thus cannot use their method in order to compare the results with our assumption of constant Q.

Now we will compare our results near the Itoiz dam with others obtained in the Granada basin. We know that there is not reservoir induced seismicity in the Granada basin. The source parameters were analyzed in García et al. (1996), with earthquakes of magnitudes ranging from 1 to 3.5. We will classify the data into clusters, for both regions, the Itoiz dam and the Granada basin, by means of Self-Organizing Maps (SOM), and then we will compare the two results.

The SOM algorithm is a convenient, unsupervised learning method, which is widespread in various scientific fields [see references in Kohonen (2001)]. They provide a way of representing multidimensional data in much lower dimensional spaces. This is usually in one or two dimensions. This process, of reducing the dimensionality of vectors, is essentially a data compression technique known as *vector quantisation*. In addition, the Kohonen technique creates a network that stores information in such a way that any topological relationships within the training set are maintained. One of the most interesting aspects of SOM is that they learn to classify data *without supervision*. The way SOM go about reducing dimensions is by producing a map of usually one or two dimensions, which plot the similarities of the data by grouping similar data items together. So SOM accomplish two things, they reduce dimensions and display similarities.

SOM have already been applied, in different contexts, for active seismic data sets (Essenreiter et al., 2001; Klose, 2006; De Matos et al., 2007; Higgins et al., 2011) and in seismology (Maurer et al., 1992; Musil and Plešinger, 1996; Tarvainen, 1999; Plešinger et al., 2000; Allamehzadeh and Mokhtari, 2003; Ozerdem et al., 2006; Esposito et al., 2008; Carniel et al., 2009; Köhler et al., 2009, 2010). We will use them as a classification tool where the variables are the moment and the stress drop. We

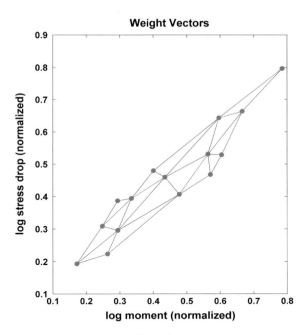

Weight Vectors

Figure 9
Results of the SOM algorithm to the data in the Granada basin. In the x axis we represent the rescaled logarithm (in base 10) of the moment from 0 to 1, and in the y axis the rescaled logarithm of the stress drop

use the MATLAB functions for obtaining our clusters. The variables are the logarithm of the moment and the stress drop, divided by the sum of the logarithms in each case. In Figs. 9 and 10 we can see the obtained clusters. Each node represents a cluster found by the algorithm. The lines represent the nearest nodes in the clustering algorithm. We observe that the clusters are ordered by magnitude, when the moment increases (and not by other factors such as location or time) with increasing stress drop. It is in agreement with the previous results for the relationship between moment and stress drop. So there seems to be a clustering of the events depending on the magnitude only, for both regions. This dependence on the magnitude only for the Itoiz region was also found in Jiménez and Luzón (2011), where we analyzed the seismicity near the Itoiz dam by means of weighted complex networks. This comparison is interesting because we are analyzing two different kind of seismic sets, with supposedly different seismicity origins. SOM are classifiers that are not influenced by a priori knowledge of the data. So, the fact that both regions present the same behaviour

Weight Vectors

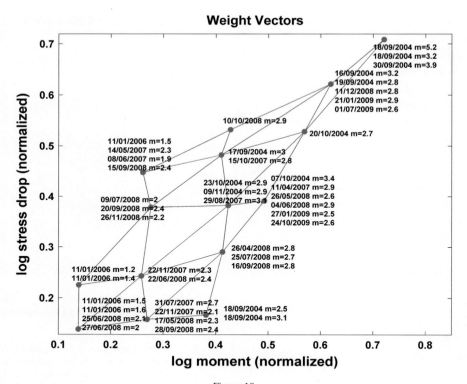

Figure 10
Results of the SOM algorithm to the data in the Itoiz dam. In the *x* axis we represent the rescaled logarithm (in base 10) of the moment from 0 to 1, and in the *y* axis the rescaled logarithm of the stress drop

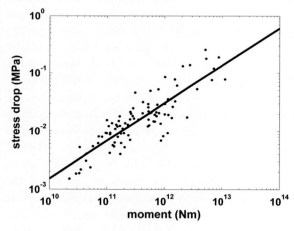

Figure 11
Moment vs. stress drop, for the data in García *et al.* (1996). The *line* is the best fit

the data. This is something that could indicate an effect of the redistribution of stresses with time near the Itoiz dam, due to the impoundment of the reservoir.

In Fig. 11 we show the data from García *et al.* (1996). There we see an increase of the stress drop with increasing magnitude, as we found in the present work.

5. Conclusions

We have analyzed the seismicity near the Itoiz dam by calculating the source parameters and attenuation from earthquakes recorded from 2004 to 2009 with magnitudes between 1.2 and 5.2. We assume a BOATWRIGHT (1978) model for the source, and a spectral falloff with Q not dependent on the frequency for the attenuation. The parameters to be found by the GA are the spectral level, the corner frequency, and the attenuation factor (JIMÉNEZ *et al.*,

with respect to the relationship between stress drops and moments is very important. It is also very interesting to see that time does not influence the relationship between moment and the stress drop in

2005). With those data, we calculate the source parameters following BRUNE (1970, 1971). The obtained moments range from 1.72×10^{11} to 2.65×10^{15} Nm, the radii span from 0.09 to 1.00 km, and the stress drops vary from 0.006 to 29.462 MPa. The maximum value for the Q attenuation factor is 794, and the minimum value is 53. The values we find are very similar to the ones obtained by TUSA and GRESTA (2008) in southern Sicily, with tectonic events, and the stress drops are not particularly low, as some authors think it is supposed to be for seismicity induced by the injection of fluid. The classification by means of SOM gives us clusters that depend on the magnitude, and not other considerations, in agreement with the results in JIMÉNEZ and LUZÓN (2011). It is remarkable that there is the coincidence between the theoretical and empirical relationship for the obtained moments and the magnitude. We found an increase of the stress drop with increasing moment to have high significance, which is a breakdown of the self-similarity in the process. As IDE *et al.* (2003) pointed out, the use of constant Q and/or propagation effects can introduce artificial size dependence in apparent stress measurements.

Acknowledgments

We wish to thank the Confederación Hidrográfica del Ebro (CHE) for allowing us to use their facilities and their seismic data. We also thank the Instituto Geográfico Nacional (IGN), Spain. This work was partially supported by the Secretaría General para el Territorio y la Biodiversidad part of the Ministerio de Medio Ambiente, Rural y Marino, Spain, under Grant 115/SGTB/2007/8.1, by EU with FEDER and by the research team RNM-194 of the Junta de Andalucía, Spain. The work of AJ is supported by a Juan de la Cierva grant, from the Spanish Government.

REFERENCES

ABERCROMBIE, R. E. (1995), *Earthquake source scaling relationships from 1 to 5 ML using seismograms recorded at 2.5 km. depth*, J. Geophys. Res. *100*, 24015–24036.

ABERCROMBIE, R. E. (1997), *Near surface attenuation and site effects from comparison of surface and deep borehole recordings*, Bull. Seism. Soc. Am. *87*, 731–744.

ABERCROMBIE, R.E. and Leary, P. (1993), *Source parameters of small earthquakes recorded at 2.5 km depth, Cajon Pass, Southern California: implications for earthquake scaling*, Geophys. Res. Lett., *20*, 1511–1514.

AKI, K. and RICHARDS, P. G. (1980), *Quantitative seismology*, W. H. Freeman, San Francisco.

ALLAMEHZADEH, M. and MOKHTARI, M. (2003), *Prediction of Aftershocks Distribution Using Self- Organizing Feature Maps (SOFM) and Its Application on the Birjand-Ghaen and Izmit Earthquakes*. JSEE: Fall 2003, *5*(3), 1–15.

ARCHULETA, R., CRANSWICK, E., MUELLER, C. and SPUDICH, P. (1982), *Source parameters of the 1980 Mammoth Lakes, California, earthquake sequence*, J. Geophys. Res. *87*, 4595–4607.

ATKINSON, M. G. (1993), *Earthquake source spectra in eastern north America*, Bull. Seism. Soc. Am., *83*, 1778–1798.

ATKINSON, M. G. (2004), *Empirical attenuation of ground-motion spectral amplitudes in southeastern Canada and the northeastern United States*, Bull. Seism. Soc. Am. *94*, 1079–1095.

BINDI, D., SPALLAROSSA, D., AUGLIERA, P. and CATTANEO, M. (2001), *Source parameters from the aftershocks of the 1997 Umbria - Marche (Italy) seismic sequence*. Bull. Seism. Soc. Am., *91*, 448–455.

BOATWRIGHT, J. (1978), *Detailed spectral analysis of two small New York state earthquakes*, Bull. Seism. Soc. Am. *68*, 1117–1131.

BRUNE, J. N. (1970), *Tectonic stress and the spectra of seismic shear waves from earthquakes*, J. Geophys. Res. *75*, 4997–5009.

BRUNE, J. N. (1971), *Correction to Tectonic stress and the spectra of seismic shear waves from earthquakes by J. N. Brune*, J. Geophys. Res., *76*, 5002.

CARNIEL, R. BARBUI, L. and MALISAN, P. (2009), *Improvement of HVSR technique by self-organizing map (SOM) analysis*, Soil Dynamics and Earthquake Engineering, *29*(6), 1097–1101, doi:10.1016/j.soildyn.2008.11.008.

CASTRO, R.R., MUNGUÍA, L. and BRUNE, J.N. (1995), *Source spectra and site response from P and S waves of local earthquakes in the Oaxaca, México subduction zone*. Bull. Seism. Soc. Am, *85*, 923–936.

DAVIS, L. (*ed.*) (1991). Handbook of Genetic Algorithms, *Van Nostrand Reinhold, New York.*

DE MATOS, M., OSORIO, P. and JOHANN, P. (2007), *Unsupervised seismic facies analysis using wavelet transform and self-organizing maps*, Geophysics, *72*, 9–21. doi:10.1190/1.2392789.

DEL PEZZO, E., DE MARTINO, S., DE MIGUEL, F., IBÁÑEZ, J. M. and SORGENTE, S. (1991), *Characteristics of the seismic attenuation in two tectonically active zones of Southern Europe*, Pure Appl. Geophys., *135*, 91–106.

DEICHMANN, N. (2006), *Local Magnitude, a Moment Revisited*, Bull. Seism. Soc. Am., *96*(4), 1267–1277, doi:10.1785/0120050115.

DROUET, S., SOURIAU, A., and COTTON, F. (2005), *Attenuation, seismic moment, and site effects for weak-motion events: application to the Pyrenees*, Bull. Seism. Soc. Am., *95*(5), 1731–1748. doi:10.1785/0120040105.

DROUET, S., CHEVROT, S., COTTON, F., and SOURIAU, A. (2008), *Simultaneous inversion of source spectra, attenuation parameters, and site responses: application to the data of the French Accelerometric Network*, Bull. Seism. Soc. Am. *98*, 198–219.

DURÁ-GÓMEZ, I. and TALWANI, P., (2010), *Reservoir-induced seismicity associated with the Itoiz Reservoir, Spain: a case study*, Geophys. J. Int., *181*, 343–356, 2010, doi:10.1111/j.1365-246X.2009.04462.x.

ESPOSITO, A., GIUDICEPIETRO, F., D'AURIA, L., SCARPETTA, S., MAR-
TINI, M. COLTELLI, M. and MARINARO, M. (2008), *Unsupervised
neural analysis of very-long-period events at Stromboli volcano
using the self-organizing maps*, Bull. Seism. Soc. Am., 98(5),
2449–2459, doi:10.1785/0120070110.

ESSENREITER, R., KARRENBACH M. and TREITEL, S. (2001), *Identification and classification of multiple reflections with self-
organizing maps*, Geophys.Prospect., 49(3), 341–352, doi:10.
1046/j.1365-2478.2001.00261.x.

FEHLER, M., and PHILLIPS, W.S. (1991), *Simultaneous inversion for
Q and source parameters of microearthquakes accompanying
hydraulic fracturing in granitic rock*, Bull. seism. Soc. Am, 81,
553–575.

FERNÁNDEZ, I., CASTRO, R. and HUERTA, C. (2010), *The spectral
decay parameter kappa in Northeastern Sonora, Mexico*, Bull.
Seism. Soc. Am., 100(1), 196–206, doi:10.1785/0120090049.

FRANKEL, A., and WENNERBERG, L. (1989), *Microearthquake spectra
from the Anza, alifornia, seismic network: site response and
source scaling*, Bull. Seism. Soc. Am. 79, 581–609.

GARCÍA GARCÍA, J. M., ROMACHO M. D. and JIMÉNEZ, A. (2004),
*Determination of near-surface attenuation, with κ parameter, to
obtain the seismic moment, stress drop, source dimension and
seismic energy for microearthquakes in the Granada Basin
(Southern Spain)*, Phys. Earth Planet. Int. 141, 9–26.

GARCÍA, J. M., VIDAL, F., ROMACHO, M. D., MARTÍN-MARFIL, J. M.,
POSADAS, A. and LUZÓN, F. (1996), *Seismic source parameters for
microearthquakes of the Granada basin (southern Spain)*, Tectonophysics, 261, 51–66.

GIBOWICZ, S. J., HARJES, H. P., and SHÄFER, M. (1990), *Source
parameters of seismic events at Heinrich Robert mine, Ruhr
basin, Federal Republic of Germany: evidence for non double-
couple events*, Bull. Seism. Soc. Am. 80, 1157–1182.

GOLDBERG, D. E. (1989). Genetic Algorithms, *in Search, Optimization and Machine Learning, Addison-Wesley, Reading, MA*.

GONZÁLEZ, J. (2001), Estructura anelástica de coda-Q en la Penín-
sula Ibérica, PhD Thesis, Departament d'Enginyeria del Terreny,
Cartogràfica i Geofísica Universitat Politècnica de Catalunya.

HAAR, L.C., FLETCHER, J.B., and MUELLER, C.S. (1984), *The 1982
Enola, Arkansas, swarm and scaling of ground motion in the
eastern United States*. Bull. Seism. Soc. Am., 74, 2463–2482.

HANKS, Th. C., and WYSS, M. (1972), *The use of body-wave spectra
in the determination of seismic-source parameters*, Bull. Seism.
Soc. Am. 62, 561–589.

HAUKSSON, E., TENG, T. and HENYEY, T. L. (1987), *Results from a
1500 m deep, three-level downhole seismometer array: site
response, low Q values, and fmax*, Bull. Seism. Soc. Am., 77,
1883–1904.

HIGGINS, M., WARD, C. and DE ANGELIS, S. (2011), Determining an
Optimal Seismic Network Configuration Using Self-Organizing
Maps, in *Advances in Artificial Intelligence Lecture Notes in
Computer Science*, Vol. 6657/2011, 170–173, doi:10.1007/978-
3-642-21043-3_20.

HOLLAND, J. (1975). Adaptation, *in Natural and Artificial Systems,
University of Michigan Press, Ann Arbor*.

HOUGH, S. E., JACOB, K., BUSBY, R., and FRIBERG, P. (1989), *Ground
motion from a magnitude 3.5 earthquake near Massena, New
York: evidence for poor resolution of corner frequency for small
events*, Seism. Res. Lett., 60, 95–1000.

HUA, W., CHEN, Z., ZHENG, S. (2012), *Source Parameters and
Scaling Relations for Reservoir Induced Seismicity in the*

Longtan Reservoir Area. Pure and Applied Geophysics, 170(5),
767–783.

HUA, W., ZHENG, S., YAN, C., CHEN, Z. (2013), *Attenuation, Site
Effects, and Source Parameters in the Three Gorges Reservoir
Area, China*. Bulletin of the Seismological Society of America,
103(1), 371–382.

IDE, S., BEROZA, G. C., PREJEAN, S.G., and ELLSWORTH, W.L. (2003),
*Apparent break in earthquake scaling due to path and site effects
on deep borehole recordings*, J. Geophys. Res., 108, B52271,
doi:10.1029/2001JB001617.

JIMÉNEZ, A., GARCÍA, J. M., and ROMACHO, M. D., (2005), *Simultaneous Inversion of Source Parameters and Attenuation Factor
Using Genetic Algorithms*. Bull. Seism. Soc. Am. 94,
1401–1411, doi:10.1785/0120040116.

JIMÉNEZ, A., TIAMPO, K. F., POSADAS, A. M., LUZÓN, F. and DONNER,
R. (2009), *Analysis of complex networks associated to seismic
clusters near the Itoiz reservoir dam*. The European physical
journal. Special topics, 181–195, doi:10.1140/epjst/e2009-
01099-1.

JIMÉNEZ, A., and LUZÓN, F. (2011), *Weighted complex networks
applied to seismicity: the Itoiz dam (Northern Spain)*. Nonlinear
Processes in Geophysics, 18, 477–487, doi:10.5194/npg-18-477-
2011.

JIMÉNEZ, A. and LUZÓN, F. (2012), Diffusion Entropy Analysis and
Hurst exponent near the Itoiz dam, *in Handbook on the Classi-
fication and Application of Fractals, (Nova Science Publishers,
Inc 2012)*, ISBN: 978-1-61324-198-1, pp. 115–131.

JIN, A., MOYA, C. A. and ANDO, M. (2000), *Simultaneous determination of site responses and source parameters of small
earthquakes along the Atotsugawa fault zone, central Japan*,
Bull. Seism. Soc. Am., 90, 1430–1445.

JOST, M. L., BÜSSELBERG, T., JOST, O., and HARJES, H.P. (1998),
*Source parameters of injection-induced microearthquakes at
9 km depth at the KTB deep drilling site, Germany*, Bull. Seism.
Soc. Am., 88(3), 815–832.

KLOSE, C. (2006), *Self-organizing maps for geoscientific data
analysis: geological interpretation of multidimensional geophysical data*, Comput. Geosci., 10(3), 265–277, doi:10.1007/
s10596-006-9022-x.

KOHONEN, T. (2001), *Self-Organizing Maps*, Springer Series in
Information Sciences, Vol. 30, Third Extended Edition, 501 pp,
Springer Berlin, Heidelberg, New York, 1995, 1997, 2001.

KÖHLER, A., OHRNBERGER, M., and SCHERBAUM, F. (2009), *Unsupervised feature selection and general pattern discovery using
Self-Organizing Maps for gaining insights into the nature of
seismic wavefields*, Comput. Geosci., 35(9), 1757–1767, doi:10.
1016/j.cageo.2009.02.004.

KÖHLER, A., OHRNBERGER, M. and SCHERBAUM, F. (2010), *Unsupervised Pattern Recognition in Continuous Seismic Wavefield
Records using Self-Organizing Maps*, Geophys. J. Int., 182,
1619–1630. doi:10.1111/j.1365-246X.2010.04709.x.

LUZÓN, F., GARCÍA-JEREZ, A., SANTOYO, M. A., and SÁNCHEZ-SESMA,
F. J. (2009), A hybrid technique to compute the pore pressure
changes due to time varying loads: application to the impounding
of the Itoiz reservoir, northern Spain, in Poromechanics-iv, edited
by H. Ling, A. Smyth and R.Betti, Destech Publications, Inc.,
Lancaster, Pennsylvania, ISBN: 978-1-60595-006-8, 1109–1114.

LUZÓN, F., GARCÍA-JEREZ, A., SANTOYO, M. A., and SÁNCHEZ-SESMA,
F. J. (2010), *Numerical modelling of pore pressure variations
due to time varying loads using a hybrid technique: the case of*

the Itoiz reservoir (Northern Spain), Geophys. J. Int., *180*, 327–338, doi:10.1111/j.1365-246X.2009.04408.x.

MADARIAGA, R. (1976), *Dynamics of an Expanding Circular Fault*, Bull. Seism Soc. Am., *66*, 639–666.

MAI, P.M. and BEROZA, G.C. (2000), *Source scaling properties from finite-fault rupture models.* Bull. Seism. Soc. Am., *90*, 604–615.

MANDAL, P., RASTOGI, B.K. and SARMA, C.S.P. (1998), *Source parameters of Koyna earthquakes, India*, Bull. seism. Soc. Am., *88*, 833–842.

MAURER, W., DOWLA, F. and JARPE, S. (1992), Seismic event interpretation using self-organizing neural networks, in *Proceedings of the International Society for Optical Engineering (SPIE)*, Vol. 1709, 950–958, doi:10.1117/12.139971.

MICHALEWICZ, Z. (1992). Genetic Algorithms + Data Structures = Evolution Programs, *Springer-Verlag, Berlin Heidelberg*.

MORI, J., and FRANKEL, A. (1990), *Source parameters for small events associated with the 1986 north Palm Springs, California, earthquakes determined using empirical Green's functions*, Bull. Seism. Soc. Am. *80*, 278–295.

MOYA, A., AGUIRRE J. and IRIKURA, K. (2000), *Inversion of source parameters and site effects from strong ground motion records using genetic algorithms.* Bull. Seism. Soc. Am., *90*, 977–992.

MUSIL, M. and PLEŠINGER, A. (1996), *Discrimination between local microearthquakes and quarry blasts by multi-layer perceptrons and Kohonen maps*, Bull. Seism. Soc. Am., *86*(4), 1077–1090.

ONCESCU, M.C., CAMELBEECK, T. and MARTIN, H. (1994), *Source parameters for the Roermond aftershocks of April 13–May 2, 1992 and site spectra for P and S waves at the Belgian seismic network*, Geophys. J. Inter. *116*, 673–682.

OZERDEM, M. S., USTUNDAG, B. and DEMIRER, R. M. (2006), *Self-organized maps based neural networks for detection of possible earthquake precursory electric field patterns*, Advances in Engineering Software *37*(4), 207–217, doi:10.1016/j.advengsoft.2005.07.004.

PEDREIRA D., PULGAR, J. A. GALLART, J. and DÍAZ, J. (2003), *Seismic evidence of Alpine crustal thickening and wedging from the western Pyrenees to the Cantabrian Mountains (north Iberia)*, J. Geophys. Res., *108*(B4), 2204, doi:10.1029/2001JB001667.

PLEŠINGER, A., RŮŽEK, B. and BOUŠKOVÁ, A. (2000), *Statistical interpretation of WEBNET seismograms by artificial neural nets*, Studia Geophysica et Geodaetica, *44*(2), 251–271, doi:10.1023/A:1022119011057.

PREJEAN, S. G., and ELLSWORTH, W. L. (2001), *Observations of earthquake source parameters at 2 km depth in the Long Valley Caldera, Eastern California*, Bull. Seism. Soc. Am. *91*, 165–177.

PROCHÁZ-KOVÁ, D. (1970), *Properties of earthquake spectra as a function of length of record*, Travaux de L'Inst. Geophys. de L'Académie Tchécosl. Sc., *352*, 167–179.

PUJADES, L. G., CANAS, J. A. EGOZCUE, J. J. PUIGVI, M. A. GALLART, J. LANA, X. POUS, J. and CASAS, A. (1990), *Coda Q distribution in the Iberian Peninsula*, Geophys. J. Int. *100*, 285–301.

REEVES, C. (1993). Modern Heuristic Techniques for Combinatorial Problems, *Blackwell Scientific Publications*.

RIVAS-MEDINA, A., SANTOYO, M. A. LUZÓN, F. BENITO, B. GASPAR-ESCRIBANO, J. M. and GARCÍA-JEREZ, A. (2011), *Seismic Hazard and ground motion characterization at the Itoiz dam (Northern Spain)*, Pure Appl. Geophys., doi:10.1007/s00024-011-0405-0.

RUIZ, M., GASPA, O. GALLART, J. DÍAZ, J. PULGAR, J. A. GARCÍA-SANSEGUNDO, J. LÓPEZ-FERNÁNDEZ, and C. GONZÁLEZ-CORTINA, J. M. (2006), *Aftershocks series monitoring of the September 18, 2004 3 M = 4.6 earthquake at the western Pyrenees: A case of reservoir triggered seismicity?*, Tectonophysics, *424*, 223–243, doi:10.1016/j.tecto.2006.03.037.

SANTOYO, M. A., GARCÍA-JEREZ, A. and LUZÓN, F. (2010), *A sub-surface stress analysis and its possible relation with seismicity near the Itoiz Reservoir, Navarra, Northern Spain*, Tectonophysics, *482*, 205–215, doi:10.1016/j.tecto.2009.06.022.

SATO, T., and HIRASAWA, T. (1973), *Body wave spectra from propagating shear* cracks, J. Phys. Earth, *21*, 415–431.

SCHOLZ, C.H. (1990), *The mechanics of earthquakes and faulting.* Cambridge University Press, Cambridge.

SHIEH, C. F (1995), Study on the free surface coupling effect of seismic waves. *TAO*, 6, 197–207.

SMITH, K.D., and PRIESTLEY, K.F. (1993), *Aftershocks stress release along active fault planes of the 1984 Round Valley, California, earthquake sequence applying time – domain stress drop method*, Bull. Seism. Soc. Am., *83*, 144–159.

SNOKE, J. A. (1987), *Stable determination of (Brune) stress drops*, Bull. Seism. Soc. Am. 77, 530–538.

STEIDL, J. H., TUMARKIN, A. G. and ARCHULETA, R. J. (1996), *What is a reference site?*, Bull. Seism. Soc. Am., *86*, 1733–1748.

TARVAINEN, M. (1999), *Recognizing explosion sites with a self-organizing network for unsupervised learning*, Phys. Earth planet. Int., *113*(1–4), 143–154.

TOMIC, J., ABERCROMBIE R. E. and DO NASCIMENTO, A. F. (2009), *Source parameters and rupture velocity of small M ≤ 2.1 reservoir induced earthquakes* Geophys. J. Int., *179*, 1013–1023 doi:10.1111/j.1365-246X.2009.04233.x.

TUSA, G. and GRESTA, S. (2008), *Frequency-Dependent Attenuation of P Waves and Estimation of Earthquake Source Parameters in Southeastern Sicily, Italy*, Bull. Seism. Soc. Am., *98*(6), 2772–2794, doi:10.1785/0120080105.

ZELT, B. C., DOTZEV, N.T. ELLIS, R.M. and ROGERS, G.C. (1999), *Coda Q in Southwestern Columbia, Canada*, Bull. Seism. Soc. Am., *89*, 1083–1093.

(Received February 13, 2014, revised June 5, 2014, accepted June 16, 2014, Published online June 28, 2014)

Reprinted from the journal

Pure Appl. Geophys. 172 (2015), 3179–3188
© 2015 Springer Basel
DOI 10.1007/s00024-014-1018-1

Identification of T-Waves in the Alboran Sea

Enrique Carmona,[1,2] Javier Almendros,[1,3] Gerardo Alguacil,[1,3] Juan Ignacio Soto,[4,5]
Francisco Luzón,[1,2] and Jesús M. Ibáñez[1,3]

Abstract—Analyses of seismograms from ~1,100 north-Moroccan earthquakes recorded at stations of the Red Sísmica de Andalucía (Southern Spain) reveal the systematic presence of late phases embedded in the earthquake codas. These phases have distinctive frequency contents, similar to the P and S spectra and quite different to the frequency contents of the earthquake codas. They are best detected at near-shore stations. Their amplitudes decay significantly with distance to the shoreline. The delays with respect to the P-wave onsets of the preceding earthquakes are consistently around 85 s. Late phases are only detected for earthquakes located in a small region of about 100 × 60 km centered at 35.4°N, 4.0°W near the northern coast of Morocco. Several hypotheses could, in principle, explain the presence of these late phases in the seismograms, for example, the occurrence of low-energy aftershocks, efficient wave reflections, or Rayleigh waves generated along the source-station paths. However, we conclude that the most-likely origin of these phases corresponds to the incidence of T-waves (generated by conversion from elastic to acoustic energy in the north-Moroccan coast) in the southern coast of the Iberian Peninsula. T-waves are thought to be generated by energy trapping in low-velocity channels along long oceanic paths; in this case, we demonstrate that they can be produced in much shorter paths as well. Although T-waves have been already documented in other areas of the Mediterranean Sea, this is the first time that they have been identified in the Alboran Sea.

1. Introduction

T-waves are generated when acoustic energy is trapped within low-velocity channels in the sea that act as wave-guides and permit the propagation of acoustic waves to long distances without significant energy loss. The position and extent of these low-velocity channels depend on the physical properties of the sea water. For example, for most oceans the position is at depths of 700–1,300 m, but for the Mediterranean Sea it is only at a depth of 100 m (Porter 1979; Salon et al. 2003). The phenomenon of T-wave generation has been used for underwater communications or to detect submarine volcanic eruptions that otherwise would pass unnoticed. In the case of earthquakes, the propagation of seismic energy across marine paths often results in the generation of T-waves (Okal and Talandier 1997; Okal 2008; Ito et al. 2012). The exact mechanism is poorly known, but implies energy conversions between elastic and acoustic waves at the shores and multiple reverberations with the sea floor.

Since the first operations of the Red Sísmica de Andalucía (RSA), a regional network located in Southern Spain, it had been recognized that many earthquakes from Northern Morocco were closely followed by a small-amplitude event. These secondary events were most clearly seen at the southern stations of the network, but their exact nature and origin have not yet been established. Technological improvements of the RSA and the large number of north-Moroccan earthquake waveforms currently available in the RSA data base encouraged us to pursue an investigation about the origin of these waves.

---

[1] Instituto Andaluz de Geofísica, Universidad de Granada, Campus de Cartuja, C/Profesor Clavera 12, 18071 Granada, Spain. E-mail: ecarmona@ugr.es

[2] Departamento de Química y Física, Universidad de Almería, Cañada de San Urbano s/n, 04120 Almería, Spain.

[3] Departamento de Física Teórica y del Cosmos, Universidad de Granada, Campus de Fuentenueva, Edificio Mecenas, 18071 Granada, Spain.

[4] Facultad de Ciencias, Departamento de Geodinámica, Universidad de Granada, Avda. Fuentenueva s/n, 18071 Granada, Spain.

[5] Instituto Andaluz de Ciencias de la Tierra (CSIC-Universidad de Granada), Avda. de las Palmeras 4, 18100 Armilla, Granada, Spain.

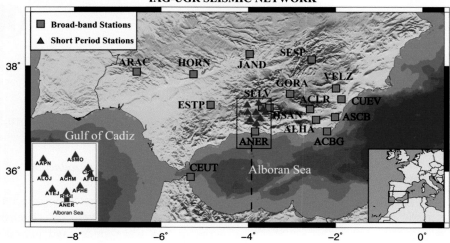

Figure 1

Map of the Alboran Sea region, showing the position and configuration of the Red Sísmica de Andalucía (RSA) in Southern Spain. *Triangles* represent vertical component, short period stations. *Squares* mark the position of three-component broadband stations. *Labels* are shown for stations mentioned in the text. The *dashed line* marks the profile shown in Fig. 8

2. Instruments

The RSA (Fig. 1) is a regional seismic network developed and maintained by the Instituto Andaluz de Geofísica (IAG) at the University of Granada, Spain (MORALES *et al.* 2007). The RSA is composed of a regional broad-band network and a short-period network (Fig. 1). The broad-band network currently consists of fifteen seismic stations with triaxial seismometers (STS-2) characterized by a flat response up to 120 s. They are sampled at 50 samples per second (sps) and digitized with a 24-bit Earth Data PS2400 converter. A GPS receiver is used for time synchronization. The short-period network is composed of eight short-period stations (ALGUACIL *et al.* 1990). Stations are situated at distances of up to 60 km from the central recording site located at the IAG in Granada. All instruments are vertical-component, short-period seismometers with a natural frequency of 1 Hz, except for station CRT, which is a triaxial seismometer with an extended response up to 10 s. Data are transmitted via radio telemetry. Until 1988 the only recording devices were thermal paper drums; after 1988, digital recording was introduced using a 14-bit A/D converter at 100 sps, which was upgraded to 16 bit on 2012.

The main objectives of the RSA are (1) the monitoring of the microseismicity in the Granada depression (Southern Spain) and surrounding areas, and (2) to record on-scale small and moderate earthquakes produced in the Iberia-Magreb plate contact.

3. Observations and Data Analysis

Visual inspection of RSA records reveals that many of the earthquakes generated in northern Morocco are closely followed by a small-amplitude event or wave packet embedded in the earthquake coda (Fig. 2). These phases have high-frequency spectral contents, similar to the spectra of the P- or S-waves but quite different to the spectrum of the rest of the coda. To investigate the characteristics of these secondary waves, we selected a data set of 1,110 earthquakes comprising all large earthquakes produced in northern Morocco and the Alboran Sea, and recorded by the RSA between 1984 and 2013. The selection criteria were the following: (1) epicenter location placed between latitudes of 34.5 and 36.5°N and between longitudes of 2 and 6°W; and (2) magnitude higher than 3.0.

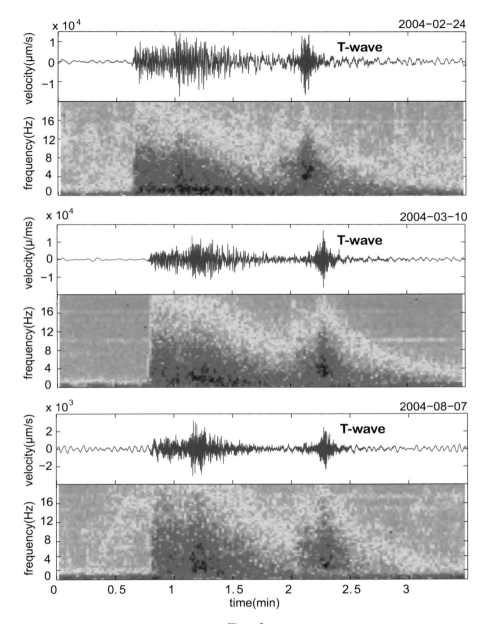

Figure 2

Examples of three seismograms recorded at station ANER showing a clear late phase at the end of the earthquake coda. Spectrograms showing the different spectral content of these late phases compared to the late earthquake coda

Late wave packets are clearly detected in only ∼ 30 % of the studied events. Figure 3 shows the epicentral location of the earthquakes with and without these late phases. These locations have been estimated using the RSA. For earthquakes prior to approximately 2000, they might be biased due to the small aperture of the network compared to the distance to the source (CALVERT *et al.* 1997). However, these errors are not crucial for our study since we are not interested in their precise locations but in their waveforms at the RSA stations.

In the studied earthquake dataset, we found the presence of late phases in the seismograms independent of the earthquake magnitude. However, the

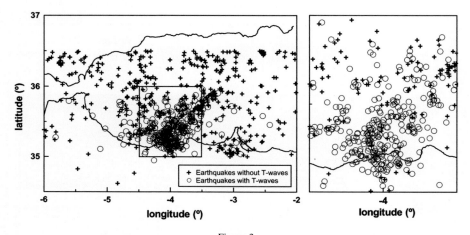

Figure 3
Epicentral map of the 1,110 earthquakes selected for this study. *Circles* represent earthquakes where late phases are clearly identified at the
ATEJ and ANER stations, while *crosses* show the position of earthquakes lacking clear late phases

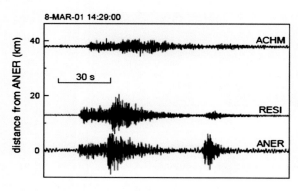

Figure 4
Examples of the vertical-component waveforms for a single
earthquake at stations ANER, RESI, and ACHM, located at
different distances from the coast. Note the rapid attenuation of the
wave packet that arrives after the earthquake coda

epicentral location seems to determine if this late
phase occurred or not. They appear only in events
located within a small region of about 100×60 km
and centered at 35.4°N and 4.0°W, near the southern
coast of the Alboran Sea. This was the site of the
1994 and 2004 Al-Hoceima seismic series (CALVERT
et al. 1997; EL ALAMI *et al.* 1998).

Late events following the earthquake codas have
been identified at most of the stations. The largest
amplitudes and earliest arrival times of the late events
are registered in the near-shore stations (e.g., ANER,
ATEJ, RESI, and APHE; Fig. 1). It is generally
observed that these phases show a rapid attenuation
with distance from the coast (Fig. 4).

We studied the attenuation of the late phases
using the spectral ratio technique. We selected two
stations (ATEJ and ALOJ) aligned in a NS direction. We found that the Q value can be represented
by the empirical law $Q = 25 \times f^{1.0}$. IBÁÑEZ *et al.*
(1990) and DE MIGUEL *et al.* (1992) investigated the
seismic wave attenuation in this same area, using
surface waves and coda waves. They found that the
quality factor can be represented by $Q_{Lg} = (105 \pm
25) f^{(0.8 \pm 0.14)}$. This indicates that the quality factors
obtained here are definitely smaller for the late
phases than for other seismic phases.

This difference in seismic attenuation explains
why late phases are recorded clearly at near-shore
stations, and gradually disappear at stations farther
from the coast. However, there is evidence suggesting
that distance to the coast is not the only factor controlling the late phase amplitudes. For example,
station ACBG is close to the shoreline and located at
a similar distance from the Al-Hoceima region as is
the ANER station (Fig. 1). However, late phases are
clearly observed in ANER seismograms but they are
difficult to identify in ACBG seismograms. This
observation indicates that the detection of late phases
in the seismograms is somehow affected by the
source-station azimuth.

The arrival times of the late phases are difficult to
determine precisely, since they are usually recorded
with low signal-to-noise ratios due to the presence of
the coda. We used a technique based on the

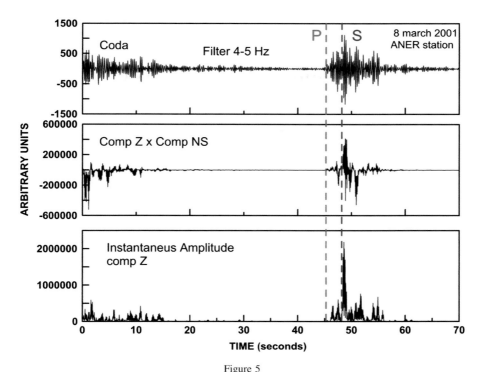

Figure 5
Sketch of the instantaneous amplitude technique used to identify the first arrival of late phases (*red dashed line*) at station ANER. The *blue dashed line* shows the detection of the TS-wave (see text for explanations)

instantaneous amplitude of the filtered seismograms (BUTTKUS 2000) which enhances high-frequency arrivals. We determined the first arrival times of the late phases at every station where they were observed (Fig. 5). The results show that delays between the P-wave onsets and the late phases are approximately constant. For example, they range between 75 and 95 s at the ATEJ station, with an average of 85 s (Fig. 6). If we consider the ATEJ station as reference, the average delay between P and TP phases is 1 s longer for the APHE station and 5 s longer for the ALOJ station.

Using the first arrival times of the late phases to the RSA stations, we were able to apply simple location procedures to estimate the positions of the source areas. We attempted to locate the late phases whose arrivals had been clearly determined at a minimum of four RSA stations. The results of this procedure are shown in Fig. 7. Even if the location errors might be relatively large, due to the uncertainty of the phase picking procedure, we can establish that all sources are located in a region near the northern shore of the Alboran Sea. The set of T-wave arrivals

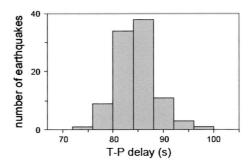

Figure 6
Histogram of the delay times between the P-wave of the preceding earthquake and the late phases following the earthquake coda at station ATEJ

that have been located in this study come from a zone with an estimated area of about 15 km × 20 km (in longitude and latitude, respectively) to the south of the ANER station. Late phases sometimes show a definite inner structure, with a waveform that resembles that of an earthquake. This fact is clearest at the ANER station that is nearest the source. Particle motion analyses at this station show complex patterns that are difficult to interpret due to the low

223

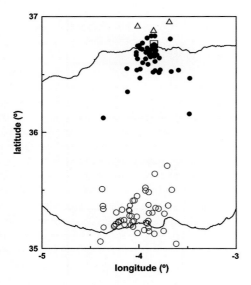

Figure 7

Apparent centers of the acoustic-to-seismic wave conversions at the near station shore of the Alboran Sea (*black dots*) for 53 selected north-Moroccan earthquakes (*open dots*). Triangles and *square* show the positions of RSA stations (see Fig. 1)

signal-to-noise ratios. In any case, at least two phases can be distinguished (Fig. 5), which might correspond to arrivals of P- and S-waves. However, the delays between these two phases are approximately constant and do not seem to increase with the epicentral distance. Alternate explanations have to be proposed to explain the presence of these inner phases in the late wave packets.

4. Discussion

4.1. Origin of the Late Phases

The observations described previously could be explained under different scenarios. Since the time delay between the late phases and the P-wave onsets is commonly large (around one and half minutes), it can be argued that the late phases are in fact low-energy aftershocks. However, as these secondary events occur systematically at a fixed time after most of the earthquakes in the Al-Hoceima region, their interpretation as aftershocks is unlikely. A much broader distribution of delays with no clear peaks should be expected under this scenario. In consequence, we will assume hereafter that the late wave

packets are in some way a consequence of the preceding earthquakes instead of independent events.

Another interpretation of the origin of the late phases could be related to reflection of the body waves from a preceding earthquake in a sharp velocity contrast. Part of the seismic energy generated at the source would travel directly to the RSA stations, while the other might propagate to this hypothetical velocity contrast and be reflected, arriving in the RSA stations in a time determined by the velocity of the medium and the distance to the reflector. This situation could explain the observed differences in the frequency content between the late phase and the coda. However, the large delays between the arrivals of the body waves and the late phases impose an unrealistically distant position of this hypothetical reflector from the Alboran Sea area. For example, assuming a homogeneous medium with a wave velocity of 6 km/s, the reflector should be located at ∼250 km. The waves would experience attenuation through 500 km, and in consequence, they would likely disappear before their arrival to the RSA seismic station.

Another explanation for the late waves could be the generation of Rayleigh waves along the path between the source and the stations. This interpretation has been previously suggested in the region (CHOURAK *et al.* 2001). To fit the large delays of about 85 s between the assumed Rayleigh waves and the P-wave onsets, the former must be generated in the near-source (Al-Hoceima) region and propagate as Rayleigh waves through the Alboran Sea Basin. For the average distance of 180 km between the source region and the ATEJ station, for example, and assuming an average velocity of 6 km/s for the direct P-wave, we determined that these Rayleigh waves should propagate at ∼1.5 km/s. Since their frequency contents are centered in the range 3–5 Hz, the wavelengths should be as small as a few 100 m. The structure of the Alboran Sea is complex, with major sedimentary (∼12 km) accumulations in the West (SOTO *et al.* 2008, 2010). It is difficult to envisage a significant propagation of Rayleigh waves through these sedimentary layers, with important volumes of under-compacted (and fluid-rich) sediments, without disappearing due to attenuation. In fact, efficient attenuation of surface waves in the Alboran Sea area

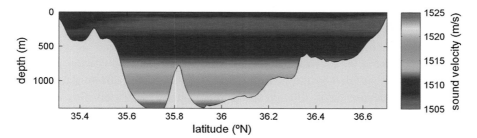

Figure 8

SN profile across the Alboran Sea (along the 4°W meridian) showing acoustic velocities in the water column and a vertically-exaggerated section of the seafloor topography. *Diagram* constructed using the Ocean Data View database (from the U.S. NODC World Ocean Atlas 2009) and the software ODV (version 4.0)

has been already documented (Calvert *et al.* 2000), which renders unrealistic the origin of the late waves as Rayleigh waves. Another argument that makes this interpretation even more unlikely comes from their directivity. Late phases are detected only at particular stations, for earthquakes from a certain region and for certain source-station geometries. The efficient generation of surface waves necessary to produce Rayleigh waves with enough energy to reach the RSA stations is at odds with the lack of late wave packets at the near-coast south-eastern stations and the rest of the inland stations. The rapid attenuation of the late phase with distance across the RSA, evidenced in Fig. 4, cannot be accomplished by Rayleigh waves that have already travelled 180 km without any significant energy loss.

4.2. Late Phases as T-waves in the Alboran Sea

The fact that the velocity that best fits the delay with the P-wave onset is 1.5 km/s yields a clue towards the solution. This is precisely the acoustic velocity of sound in water. Consequently, a reasonable explanation of the origin of the late phases is that they could be earthquake-generated T-waves. T-waves are produced when acoustic energy is trapped within low-velocity channels in the sea, because they act as wave guides and permit the propagation of energy to long distances without significant losses (i.e., Tolstoy and Ewing 1950). The position and extent of these low-velocity channels, also called SOFAR channels, depend on physical properties of the sea water. For example, most oceans have a SOFAR channel located at depths of 700–1,300 m, while for the central Mediterranean

Sea it is only at 100 m depth (Porter 1979; Salon *et al.* 2003). This is also confirmed by our study of the sonic velocities along a NS section in the Alboran Sea. Using the Ocean Data View database and software (version 4.0, http://odv.awi.de/) we can identify a shallow SOFAR channel in the water column of the Alboran Sea at a depth of 100–150 m (Fig. 8).

T-waves can be generated in oceans not only by propagation of seismic energy but also by other processes like submarine landslides or underwater volcanic eruptions (Norris and Johnson 1969; Talandier and Okal 1987). The exact mechanism of T-wave generation is still under debate, but it implies energy conversions between elastic and acoustic waves at the shores and/or multiple reverberations with the sea floor (Talandier and Okal 1998; De Groot-Hedlin and Orcutt 1999; Okal 2008). T-wave records with great content at high frequencies and large amplitudes have been described in oceanic trajectories (De Groot-Hedlin and Orcutt 2001; Huang *et al.* 2013). T-waves with spectral content at low frequencies have also been reported for deep earthquakes by other authors (e.g., Okal 2001, 2008). In our case, the earthquakes in the Alboran Sea are mostly superficial and the T-waves generated have of large high-frequency content and amplitudes exceeding even the amplitude of the original earthquake. T-waves are most likely generated when seismic waves propagating from the Al-Hoceima source-region to the Alboran Sea are converted into acoustic waves. These waves propagate within a SOFAR channel in the Alboran Sea with negligible energy loss. When this energy reaches the near-shore station, in the Spanish coast, it is converted back to seismic energy that is finally recorded at the RSA stations.

The converted waves propagate from the north shore of the Alboran Sea to the RSA stations following a very shallow path, which explains the high attenuation illustrated in Fig. 4. This attenuation is higher than the attenuation suffered by the P- and S-waves that travel along deeper paths through a more consolidated medium. As a consequence, inland stations like ESTP, SELV, ASMO, or AAPN rarely show clear late phases.

The sources of the late phases depicted in Fig. 7 represent the acoustic-to-seismic conversion areas. They correspond to the regions where the acoustic energy (T-waves) generated in the southern shoreline of the Alboran Sea hits the coast of the Iberian Peninsula and is transformed into elastic energy. These areas may not be truly point sources, since the conversions probably occur along segments of the coast. But if they are not large compared to the network aperture, these locations can give us an estimate of where the acoustic waves are converted back to seismic waves.

A possible interpretation of the inner phases observed in the T-wave packets (Fig. 5) is that P- and S-waves are generated when the acoustic energy traveling as T-waves along the low-velocity channel in the Alboran Sea is converted to elastic energy at the near-shore station. However, models of T-wave propagation demonstrate that, for gentle slopes, the converted waves are most likely composed of mixed surface waves (STEVENS et al. 1999) instead of well-defined body waves. Moreover, if P- and S-waves were generated, the S–P delays of the converted waves should increase with an increasing conversion area-station distance, but this trend is not clear in our data. An alternate explanation is that two kinds of T-waves are generated at the near-source shore, one due to the arrival of P-waves and the other related to an S–P conversion of S-waves (TALANDIER and OKAL 1998; OKAL 2008). These TP and TS acoustic waves both propagate along the SOFAR channel in the Alboran Sea, reach the near-shores station, and are converted to seismic energy, producing two different phases in the seismograms. The delay between these two phases would then be produced during the propagation of P- and S-waves between the source and the near-source coast. T-waves are very sensitive to the source-station geometry, since only quasi-horizontal motions would penetrate the SOFAR channel (TALANDIER and OKAL 1998).

The surface water in the Strait of Gibraltar consists of eastward-flowing Atlantic water (Global Ocean Associates 2014). At depth, there is dense, high-salinity westward flowing water. The water between these two layers is a mixing zone. Depending on the winds, the non-tidal water at the surface flows at velocities between two and four knots. The non-tidal currents below the surface flow to the west with speeds decreasing with depth (ROSS et al. 2000). The presence of these strong currents near the Strait of Gibraltar would disturb the formation of a stable SOFAR channel, which prevents the observation of T-waves at stations on the north-western shores of the Alboran Sea (GARRET et al. 1990; VARGAS-YÁÑEZ et al. 2002). However, we have detected twenty T-wave records generated by earthquakes located in the coasts of Almería and Granada (SW Spain) and registered at the CEUT station. The average delay of the T-wave arrival from the P phase is of about 120 s. Finally, the Alboran Island edifice would also partially disturb T-wave propagation to stations from the Al-Hoceima region to SE Spain.

The Al-Hoceima area is probably the most seismically active area in the Alboran sea, increasing the likelihood of occurrence and detection of T-waves. Nevertheless, according to the map in Fig. 3 it can be seen that many earthquakes occur in other areas such as the Nador area, where no event produced detectable T-waves. Probably, the combined effect of complex land-sea interface geometry, sea-floor topography, and changing currents could explain the limited size of the north-Moroccan region where earthquakes produce T-waves (Fig. 3).

5. Conclusion

We have found evidences of the generation of T-waves and their propagation along shallow paths in the Alboran Sea. Although most studies on this subject refer to T-waves detected after long oceanic paths on the order of hundreds or even thousands of km (OKAL and TALANDIER 1997; PISERCHIA et al. 1998; TALANDIER and OKAL 1998; OKAL 2008; HUANG et al.

2013), T-wave detections on relatively short paths have been also documented (LIN 2001).

We demonstrate that the late phases following some north-Moroccan earthquakes are seismic waves converted from T-waves. This is the first time that T-waves are documented for such short paths in the Mediterranean Sea, although other studies ascertained the presence of T-waves in other areas of the Mediterranean. For example, SHAPIRA (1981) found T-waves recorded at several stations in Israel caused by offshore explosions. They also recorded T-waves caused by the Zemmouri (Argelia) earthquake in 2003 (ALASSET *et al.* 2006). More recently, D'ANNA *et al.* (2008) and MANGANO *et al.* (2011) reported T-wave arrivals produced the October 10, 2007 earthquake in Grece (ML = 4.5) and registered by Hydrophones and OBS in the Ionian Sea. The fixed delay between the preceding earthquakes and the late phases made us exclude their origin as aftershocks. These late phases have been previously interpreted as Rayleigh waves (CHOURAK *et al.* 2001), although we conclude that this interpretation is not adequate. Other explanations such as the reflection in some strong velocity contrast are excluded as well.

Some authors propose that P- and S-waves hitting the near-source shore may both produce T-waves that propagate to the near-shore station where they are converted back into elastic waves. In this interpretation, what we identify as P- and S-waves within the T-phase are not real P and S phases but artifacts due to the delay induced by the generation of T-waves from P- and S-waves near the source.

Acknowledgments

We thanks Rosa Martín, Antonio García-Jerez, Flor de Lis Mancilla, and Benito Martín for their assistance producing this paper. This work was partially supported by Project CGL2010-16250 of the Spanish Ministry of Science and Technology, and by the Junta de Andalucía groups RNM-194, RNM-104, and RNM-376.

REFERENCES

ALASSET, P., HÉBERT, H., MAOUCHE, S., CALBINI, V., MEGHRAOUI, M. (2006). *The tsunami induced by the 2003 Zemmouri earthquake (Mw = 6.9, Algeria): modelling and results*. Geophys. J. Int., 166, 213–226. doi:10.1111/j.1365-246X.2006.02912.x.

ALGUACIL, G., GUIRAO, J. M., GÓMEZ, F., VIDAL, F., DE MIGUEL, F. (1990). Red Sísmica de Andalucía (RSA): a digital PC-based seismic network, in Seismic networks and rapid digital data transmission and exchange, Notebooks of the European Center for Geodynamics and Seismology 1, pp. 19–27, Luxemburg.

BUTTKUS, B. (2000). Spectral Analysis and Filter Theory in Applied Geophysics, Chapter 2.7, p 38. Springer, Berlin.

CALVERT, A., GÓMEZ, F., SEBER, D. BARAZANGI, M., JABOUR, N., IBENBRAHIM, A., DEMNATI, A. (1997). *An integrated geophysical investigation of recent seismicity in the Al Hoceima region of North Morocco*. Bull. Seism. Soc. Am., 87, 637–651.

CALVERT, A., SANDVOL, E., SEBER, D., BARAZANGI, M., VIDAL, F., ALGUACIL, G., JABOUR, N. (2000). *Propagation of regional seismic phases (Lg and Sn) and Pn velocity structure along the Africa-Iberia plate boundary zone: tectonic implications*. Geophys. J. Int., 142, 384–408.

CHOURAK, M., BADAL, J., CORCHETE, V., SERÓN, F. J. (2001). *A survey of the shallow structure beneath the Alboran Sea using Rg-waves and 3D imaging*. Tectonophysics, 335, 255–273.

D'ANNA, G., MANGANO, G., D'ALESSANDRO., A., D'ANNA, R., PASSAFIUME, G., SPECIALE, S. (2008). First long time OBS campaign in the Ionian Sea. *Rapporti Tecnici IGNV*. no 72. (http://www.earth-prints.org/handle/2122/4049).

DE GROOT-HEDLIN, C. D., ORCUTT, J. A. (1999). *Synthesis of earthquake-generated T-waves*, Geophys. Res. Lett., 26, 1227–1230.

DE GROOT-HEDLIN, C., ORCUTT, J. A. (2001). *Excitation of T-phases by seafloor scattering*, J. Acoust. Soc. Am., 109, 1944–1954.

DE MIGUEL, F., IBÁÑEZ, J.M., ALGUACIL, G., CANAS, J.A., VIDAL, F., MORALES, J., PEÑA, POSADAS, A.M., LUZÓN, F. (1992). *1–18 Hz Lg attenuation in the Granada Basin (southern Spain)*. Geophys. J. Int. 111, 270–280.

EL ALAMI, S. O., TADILI, B. A., CHERKAOUI, T. E., MEDINA, F., RAMDANI, M., AÏT BRAHIM, L., HARNAFI, M. (1998). *The Al Hoceima earthquake of May 26, 1994 and its aftershocks: A seismotectonic study*, Annali di Geofisica, 41, 519–537.

GARRETT, C., THOMPSON, K., BLANCHARD, W. (1990): *Are changes in the sea level drop through the Strait of Gibraltar due to hydraulic state flips?* Nature, 348, 292.

Global Ocean Associates. An Atlas of Internal Solitary-like Waves and their Properties (2014). 2nd. Edition. Prepared under contract with the Office of Naval Research. http://www.internalwaveatlas.com (Accessed in March 2014).

HUANG, B., SHIH, M., LAI, Y., CHEN, K., HUANG, W, LIU, Ch. (2013). *Observations of Earthquake-Generated T-Waves in the South China Sea: Possible Applications for Regional Seismic Monitoring*. Terr. Atmos. Ocean. Sci., 24(1) 1. doi:10.3319/TAO.2012.10.09.01.2013.

IBÁÑEZ, J.M., DEL PEZZO, E., DE MIGUEL, F., HERRAIZ, M., ALGUACIL, G., MORALES. (1990) *Depth-dependent seismic attenuation in the Granada zone (Southern Spain)*. Bull. Seism. Soc. Am., October 80(5), 1232–1244.

ITO, A., SUGIOKA, H., SUETSUGU, D., SHIOBARA, H., KANAZAWA, T., FUKAO, Y. (2012). *Detection of small earthquakes along the Pacific-Antarctic Ridge from T-waves recorded by abyssal ocean-bottom observatories*. Mar. Geophys. Res. doi:10.1007/s11001-012-9158-0.

LIN, C. H. (2001). *T-waves excited by S-waves and oscillated within the ocean above the southeastern Taiwan forearc*. Geophys. Res. Lett., 28, 3297–3300.

MANGANO, G., D'ALESSANDRO, A., D'ANNA, G. (2011). Long term underwater monitoring of seismic areas: design of an Ocean Botton Seismometer with Hydrophone and its perfomance evaluation. OCEANS, 2011 IEEE-Spain. doi:10.1109/Oceans-Spain.2011.6003609.

MORALES, J., ALGUACIL, G., MARTÍN, J.B., MARTOS, A. (2007). *The Instituto Andaluz de Geofísica-Universidad de Granada seismic network in Southern Spain*. Orfeus Newsletter. *7*(2).

NORRIS, A., JOHNSON, R. H. (1969). *Submarine volcanic eruptions recently located in the Pacific by SOFAR hydrophones*, J. Geophys. Res., *74*, 650–664.

OKAL, E. A., TALANDIER, J. (1997). *T waves from the great 1994 Bolivian deep earthquake in relation to channeling of S wave energy up the slab*, J. Geophys. Res., *102*, 27421–27437.

OKAL, E.A. (2001). *"Deteched" deep earthquakes: are they really?.* Phys. Earth Planet Int., *127*, 109–143.

OKAL, E.A. (2008). *The generation of T waves by earthquakes.* Advances in Geophysics, *49*, chapter 1.

PISERCHIA, P. F., VIRIEUX, J., RODRIGUES, D., GAFFET, S., TALANDIER, J. (1998). *Hybrid numerical modelling of T-wave propagation: application to the Midplate experiment*, Geophys. J. Int., *133*, 789–800.

PORTER, R. P. (1979). Dispersion of axial SOFAR propagation in the western Mediterranean, J. Acoust. Soc. Am., *53*, 181–191.

ROSS, T., GARRETT, C, and LE TRAON, P.-Y. (2000). *Western Mediterranean sea-level rise: changing exchange flow through the Strait of Gibraltar.* Geophys Res Lett, *27*, 2949–2952. doi:10.1029/2000GL011653.

SALON, S., CRISE, A.,PICCO, P., DE MARINIS,E., GASPARINI, O. (2003). *Sound speed in the Mediterranean Sea: an analysis from a climatological data set.* Annales Geophysicae *21*, 833–846.

SHAPIRA, A. (1981). *T phases from underwater explosions off the coast of Israel.* Bull. Seism. Soc. Am., *71*(4), 1049–1059.

SOTO, J.I., FERNÁNDEZ-IBÁÑEZ, F., FERNÀNDEZ, M. and GARCÍA-CASCO, A. (2008*). Thermal structure of the crust in the Gibraltar Arc: influence on active tectonics in the Western Mediterranean.* Geochemistry, Geophysics, Geosystems, *9.* doi:10.1029/2008GC002061.

SOTO, J.I., FERNÁNDEZ-IBÁÑEZ, F., TALUKDER, A.R., and MARTÍNEZ-GARCÍA, P. (2010). Miocene shale tectonics in the northern Alboran Sea (Western Mediterranean). In: Shale Tectonics (Ed., L. Wood). AAPG Memoir *93*, 119–144.

STEVENS, J. L., BAKER, G. E., COOK, R. W., D'SPAIN, G. L., BERGER, L. P., DAY, S. M. (1999). *Empirical and numerical modeling of T-phase propagation from ocean to land*, Eos Trans. Am. Geophys. U., *80*, F658 (abstract).

TALANDIER, J., OKAL, E. A. (1987). *Seismic detection of underwater volcanism: the example of French Polynesia*, Pure Appl. Geophys., *125*, 919–950.

TALANDIER, J., OKAL, E. A. (1998). *On the mechanism of conversion of seismic waves to and from T waves in the vicinity of island shores*, Bull. Seism. Soc. Am., *88*, 621–632.

TOLSTOY, I., EWING, W. M. (1950). *The T phase of shallow-focus earthquakes*, Bull. Seism. Soc. Am., *40*, 25–52.

VARGAS-YÁÑEZ, M., PLAZA, F., GARCÍA-LAFUENTE, J., SARHAN, T., VARGAS, J.M., and VÉLEZ-BELCHI, P. (2002): *About the seasonal variability of the Alboran Sea circulation.* Journal of Marine Systems, *35*, 229–248.

(Received March 15, 2014, revised December 11, 2014, accepted December 15, 2014, Published online January 17, 2015)

Pure Appl. Geophys. 172 (2015), 3189–3228
© 2014 Springer Basel
DOI 10.1007/s00024-014-0916-6

An Overview of Geodetic Volcano Research in the Canary Islands

José Fernández,[1] Pablo J. González,[2] Antonio G. Camacho,[1] Juan F. Prieto,[3] and Guadalupe Brú[1]

Abstract—The Canary Islands are mostly characterized by diffuse and scattered volcanism affecting a large area, with only one active stratovolcano, the Teide–Pico Viejo complex (Tenerife). More than 2 million people live and work in the 7,447 km² of the archipelago, resulting in an average population density three times greater than the rest of Spain. This fact, together with the growth of exposure during the past 40 years, increases volcanic risk with respect previous eruptions, as witnessed during the recent 2011–2012 El Hierro submarine eruption. Therefore, in addition to purely scientific reasons there are economic and population-security reasons for developing and maintaining an efficient volcano monitoring system. In this scenario geodetic monitoring represents an important part of the monitoring system. We describe volcano geodetic monitoring research carried out in the Canary Islands and the results obtained. We consider for each epoch the two main existing constraints: the level of volcanic activity in the archipelago, and the limitations of the techniques available at the time. Theoretical and observational aspects are considered, as well as the implications for operational volcano surveillance. Current challenges of and future perspectives in geodetic volcano monitoring in the Canaries are also presented.

Key words: Canary islands, volcanic activity, volcano geodesy, deformation and gravity change, space and terrestrial techniques, deformation modeling and inversion techniques.

1. Introduction

Large volcanic eruptions are usually associated with polygenetic volcanoes where recurrent events occurin the same volcanic edifice. The area to be monitored in these cases can be relatively well-defined. However, a challenging scenario occurs when diffuse and scattered volcanism affects a large area. Volcanism in the Canary Islands is mostly characterized by the second scenario (Fig. 1) where there is only an active stratovolcano (Teide–Pico Viejo complex, Tenerife), but with historical activity predominantly occurring in more frequent monogenetic cones along elongated rifts (Fernández et al., 1999, 2003).

As evidenced by the impact of previous eruptions (e.g., 1971 Teneguía eruption in La Palma; 2011–2012 El Hierro submarine eruption), one must consider the population of more than 2 million people living in the archipelago (population density in its 7,447 km²is three times that of Spain), the level of economic activities and the recent growth in exposure. In consequence, apart from the purely scientific reasons, there are economic and security issues for the population necessitate development and maintenance of an efficient volcano monitoring system. Traditional geochemical and geophysical observations of seismic, hydrological or fumarolic activity have proved to be very useful in volcano monitoring, and are usually the core of operative volcano monitoring systems. Seismic observation involves distributing a series of seismic sensors around the volcano, or area to be monitored, and automatically sending log data to a control center that stores and processes it (Quaas et al. 1996). Seismic monitoring is useful for locating volcanic and/or tectonic sources and monitoring and evaluating their temporal evolution. Perhaps the closest example of this surveillance has been the locating of the El Hierro submarine volcanic eruption that began in October 2011 (http://www.ign.es/ign/resources/volcanology/HIERRO.html). Geodetic monitoring is an integral part of this monitoring system, complementing different seismic, geophysical and geochemical techniques (see e.g.,

---
[1] Institute of Geosciences (IGEO) (CSIC, UCM), Plaza de Ciencias 3, 28040 Madrid, Spain. E-mail: jft@mat.ucm.es
[2] Institute of Geophysics and Tectonics, School of Earth and Environment, University of Leeds, Leeds LS2 9JT, UK.
[3] Department of Engineering Surveying and Cartography, School of Engineering Surveying, Geodesy and Cartography, Technical University of Madrid, Madrid, Spain.

SIGURDSSON *et al.* 2000; DZURISIN 2007; PÉREZ and HERNÁNDEZ 2008).

This review describes the evolution of geodetic research with emphasis on volcano monitoring in the Canary Islands. We review the main methodological and observational results achieved since the twentieth century, as well as implications of these results on operational volcano surveillance in every epoch. We also present some new unpublished deformation results for the Tenerife and Fuerteventura Islands.

2. Geological and Volcanic Framework

The Canary Islands are located in the northwestern part of the Nubian (African) tectonic plate, relatively far from major plate boundaries and close to the thickened Western African craton continental lithosphere. The Canary Islands are a group of seven major islands forming a rough East–West trending archipelago (Fig. 1). The latter forms part of a large group of magmatic plateaus, together with the Selvagen Islands, several seamount complexes and the Madeira group (CGMW 2010). This magmatic

province sits on a Jurassic-age (150–180 Ma) oceanic lithosphere, implying a relatively cold, strong and thick lithosphere. The archipelago is located on the transitional zone, a passive margin, between oceanic and continental crust (e.g., BANDA *et al.* 1981; SURIÑACH 1986). These conditions, in conjunction with a slow moving plate, have important implications on partial melting of the mantle underneath the Canary Islands and contribute to a geodynamic scenario that has been extremely complex to uncover.

Since inception of the hot spots theory (MORGAN 1971) the Canary Islands have been attributed to this origin. However, some unusual features in their geochemistry, the influence of tectonics and lithosphere structure (absence of gravity and bathymetry anomalies) raised questions on the full validity of this hypothesis, calling for hybrid geodynamic scenarios (ANGUITA and HERNAN 2000 and references therein).

The Canary Islands show an age progression in dated rocks, from Fuerteventura–Lanzarote in the east to El Hierro Island in the west (Fig. 1). In general, older islands have clearer evidence of longer erosional periods and flank collapse processes. Indeed, the easternmost and oldest islands of

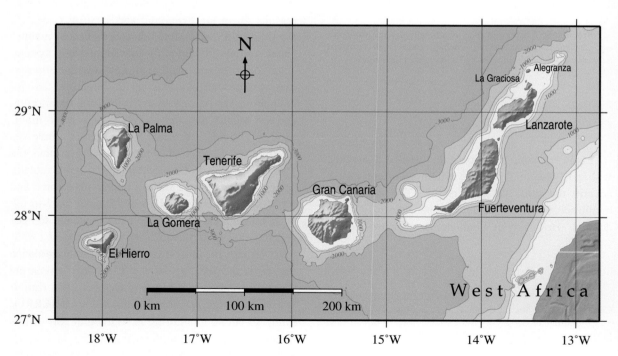

Figure 1
Location map of the Canary Islands and its position relative to the West African coast

Fuerteventura and Lanzarote are the lowest with elevations below 1,000 m. Their sub-aerial volcanism started around 20–15 Ma with little evidence of scattered-diffuse Quaternary volcanism. Gran Canaria, Tenerife and La Gomera could be classified as being of an intermediate-age with Miocene ages of 14.5, 11.9, 9.4 Ma, respectively (Hoernle and Carracedo 2009). This subgroup is the largest by volume, with Tenerife rising ∼7,000 m from the surrounding seafloor. A unique characteristic is the voluminous erupted phonolite deposits, a highly differentiated and relatively high-silica content rock in Gran Canaria and Tenerife. In Gran Canaria and La Gomera more deeply eroded, outcrop-intrusive complexes and dike swarms (e.g., Vallehermoso and Tejeda, respectively) (Ancochea et al. 2003). Finally, the youngest group is La Palma and El Hierro which started forming 4 Ma and only emerged above sea level in the last 2 million years. Both Islands are in a highly rapid growing stage with frequent basaltic eruptions along structural rifts.

In contrasting to the hot spot theory, recent Holocene volcanism is spread over the archipelago, except for La Gomera and Fuerteventura. Lanzarote, Tenerife, La Palma and El Hierro have erupted consistently during the last 500 years. A total of 19 historical subaerial eruptions have been recorded in the Canary Islands (Siebert et al. 2011). The most voluminous eruption took place from 1730 to 1736 in Lanzarote (second oldest), the last on-shore eruption was at Teneguía Volcano (La Palma) in 1971 (Romero 2000) and the most recent was submarine on the island of El Hierro in 2011–2012 (Carracedo et al. 2012; López et al. 2012; González et al. 2013).

Regarding the kinematics of the archipelago, some moderate earthquakes have been registered in the Canary Islands region, which could be attributed to diffuse tectonic activity, e.g., the M5.2 1989 Tenerife–Gran Canaria earthquake (Mezcua et al. 1992) and the M6.2 1959 Atlantic ocean earthquake (Wysession et al. 1995). However, there is no evidence of significant crustal strain accumulation around the Canary Islands. Moreover, considering the age of the volcanism, and the location of the volcanic seamounts and islands, the observed age progression could be explained by rotation of the Nubian plate with respect to a fixed point beneath the lithosphere mantle (supporting a mantle plume). The location and age of volcanism can be described with an Euler rotation pole at approximately 56.8 N, 45.8 W, with an angular rate of ∼0.20°/Ma using ages from 0 to 35 Ma. Conversely, using ages for the interval 35–64 Ma the African plate was rotating with respect to the fixed point with a different Euler pole located at 35.8 N, 45.8 W (Geldmacher et al. 2005).

Another important observation regarding the kinematics is the vertical (uplift-subsidence) long-term motions. As part of growth of an intraplate oceanic island the lithosphere under the mantle source should undergo relative uplift followed by subsidence after passing though the mantle source, eventually leading to formation of a seamount (guyots). None of the Canary Islands show significant subsidence processes, although guyots can be observed in bathymetric maps around the archipelago (Fig. 1). In contrast, most of the islands show a significant uplift history with outcropping seamount volcanoes in La Palma and Fuerteventura (Hoernle and Carracedo 2009), but whether it reflects endogenous growth or regional uplift has not been resolved yet. On shorter time scales, in Lanzarote, Fuerteventura, and Tenerife there are also Quaternary-dated raised beaches at different elevations (Zazo et al. 2002; Kröchert et al. 2008).

3. Volcano Geodetic Studies in the Canaries During the Twentieth Century

Until recent decades Canarian volcanism has only been studied using geological, geophysical and geochemical techniques with little attention being paid to geodetic measurements, especially for volcano monitoring aspects. Since the 1980s several projects included the application, study, development and/or validation of different geodetic techniques and observation methods for volcanic activity monitoring. The research has covered the observational and theoretical aspects of detecting and interpreting deformation and gravity changes, as well as methods to determine crustal structure from gravity observations, a basic tool for interpretation of observed gravity anomalies.

The temporal evolution and spatial coverage of research in geodetic volcano monitoring had two basic constraints: (a) the level of volcanic activity in the archipelago and (b) limitations of the techniques available at the time. The level of volcanic activity was greatly reduced after the Teneguía eruption on La Palma in 1971 (ARAÑA and FÚSTER 1974; HERNÁNDEZ-PACHECO and VALS 1982). Generally, low levels of detected volcanic activity result in fewer efforts being made to investigate the background levels of geophysical, geochemical and geodetic parameters. Accordingly, in the last decades of the twentieth century scientific and observational works focused on the islands of greatest potential risk (Tenerife, La Palma) or offering the best facilities for developing and maintaining research instruments (Lanzarote). Furthermore, the characteristics of classical geodetic observations and instruments (leveling, triangulation, trilateration, gravimetry, etc.) available at that time (late 1970s to 2000) (FERNÁNDEZ *et al.* 1999) made it time-consuming and expensive to cover the whole surface of the islands.

On account of all those constraints, investigations using classical geodetic techniques were conducted with very limited spatial coverage, trying to reduce the observational costs to a reasonable amount consistent with the level of volcanic activity and the real risk. This is the case of the Las Cañadas Caldera in Tenerife, Cumbre Vieja in La Palma and the Geodynamic Station in Lanzarote.

3.1. Earliest Studies

Scientific work done before the 1980s that can be considered related, at least in part, to study and understanding of volcanic activity in the Canary Islands concentrated on the design, construction and observation of a geodetic network for the realization of a geodetic reference system; also included was a gravity survey with structural objectives in and around several islands.

3.1.1 Geodetic Reference Frame

The first geodetic project with scientific quality in the Canary Islands dated back to the 1920s and sought to materialize a geodetic reference frame. In 1923 (GIL MONTANER 1929a), the *Instituto Geografico y Estadístico* (IGE) designed a geodetic network covering the whole Archipelago (observed between 1925 and 1928) using classical observation techniques of angles and distances (TORROJA 1926). The project included a large network with first-order specifications covering the seven main islands and other filling networks inside the islands with second and third order specifications (CATURLA 1996). The spatial configuration of this network is shown in Fig. 2.

GIL MONTANER (1929b) reported the geodetic link of the Canarian archipelago with the African continent at Cabo Juby (Morocco). All observations were completed in 1930 (CATURLA 1996) and the triangulation chain was finally adjusted by means of a least squares fit in 1936 (IGC 1938). Astronomical coordinates from the trigonometric point at Pico de las Nieves in Gran Canaria were used to start the computations. This is the reason why the geodetic reference frame embodying this network is known as datum "Pico de las Nieves 1930." The Greenwich meridian was selected as the origin of latitudes, the first time this meridian was used in Spain, along with the Hayford ellipsoid (HAYFORD 1910). In 1968 a joint computation of the main network of 1930 was performed together with the islands' second and third order networks (IGC 1938) to obtain a single framework for the entire archipelago. This new calculation, which did not include any new observations, constitutes the geodetic frame named "Pico de las Nieves 1968" (PN68).

In the period 1976–1978 the *Instituto Geográfico Nacional* (IGN) conducted doppler satellite observations for the TRANSIT (Navy Navigation Satellite System, NNSS) system (CATURLA 1978). The geodetic points of La Laguna in Tenerife, Maspalomas in Gran Canaria and Toston in Fuerteventura were joined by very long baselines with Villafranca and San Fernando stations on the Iberian Peninsula. These geodetic observations were the first to tie the islands with the mainland. Coordinate calculations were performed in the WGS72 geodetic reference system achieving a submeter accuracy.

In 1982 the IGN began to define a new geodetic network reusing some ancient monuments and densifying the previous networks using classical observations by triangulation and distance

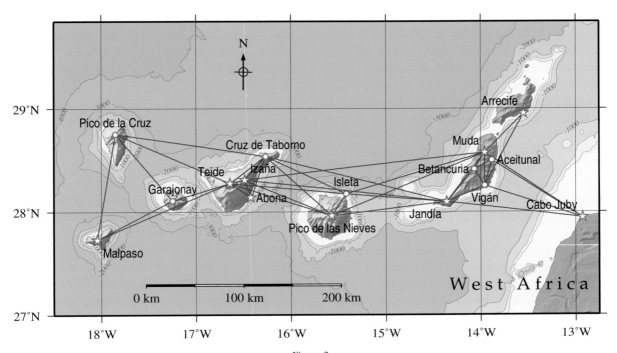

Figure 2
Schematic graph of the first geodetic network link of the Canary Islands, made in the 1930s. *Stars* denote stations where astronomical observations were made. *Circles* represent the trigonometric first order stations. *Lines* represent geodetic observations between different stations

measurement with laser geodimeter (CATURLA 1996). The Tenerife, La Palma, La Gomera and El Hierro island networks were thus completed, forming the geodetic frame called "Pico de las Nieves 1984" (PN84). During the 1990s observation of the networks in Gran Canaria, Lanzarote and Fuerteventura was completed. The networks of these three islands were no longer computed in the new geodetic frame PN84 because a new GNSS geodetic network was projected for the Canary Islands (BARBADILLO-FERNÁNDEZ and QUIRÓS-DONATE 1996).

The GPS network consisted of a regional network linking the seven main islands and the corresponding inner geodetic networks in each of them. The regional network consisted of 12 stations spread throughout the islands starting with the IGS CORS station in Maspalomas (MASP), and is computed (BARBADILLO-FERNÁNDEZ and QUIRÓS-DONATE 1996; CATURLA 1996) under ITRF93, epoch 1994.9. Immediately thereafter a GNSS network was observed and computed for each island, making a total of 296 stations that form the geodetic frame "REGCAN95". Using these

GNSS stations a recalculation of the former networks of angular observations and distances previously observed was performed (CATURLA 1996), thus completing the new geodetic network for the Canary Islands. Orthometric heights were computed in this phase for each geodetic point by the trigonometric leveling method. The IGN also conducted precise leveling profiles throughout the islands, referring to mean sea level in each of the islands. CATURLA and PRIETO (1996) computed transformation parameters between the four different geodetic frames that have existed in the Canary Islands since 1930.

3.1.2 Structural Gravimetry

In 1965, 1967 and 1968 the Department of Geophysics at Imperial College, University of London conducted the first marine and terrestrial seismic and gravimetric observation experiments in the Canary Islands. In 1965 ground gravity data were measured on the islands of Lanzarote, Gran Canaria, Tenerife and El Hierro thereby facilitating mapping of the Bouguer Anomaly and allowing implementation of structural models for the islands of Tenerife and Lanzarote. In 1967 and 1968 the first off-shore campaigns of marine gravimetry and seismic refraction were carried out along 2,650 km of profiles. Although some data were taken in the eastern part of the archipelago, efforts focused more on the western part; gravity maps were subsequently published for the areas of Gran Canaria, Tenerife, La Gomera, La Palma and El Hierro. The main objective of this work was to discern whether they were really oceanic Islands or rather were part of the African continent.

MacFarlane and Ridley (1968) were the first to interpret on-shore gravity anomalies in Tenerife and, based on a small number of stations, they identified a strong positive gravimetric anomaly south of Teide. This maximum was interpreted as being due to a very dense, conical-shaped intrusive body (whose density increases with depth from 2,800 to 3,100 kg/m^3) stretching from the Mohorovičić discontinuity (the Moho) to about 4 km below the surface. They also detected in the gravity map the pattern of three major fracture systems at 120° to each other, of which two were already known from geological field studies.

For Lanzarote Island the first interpretation of terrestrial gravity anomalies was described by Mac-Farlane and Ridley (1969). They also consider preliminary seismic data (Dash and Bosshard 1969) and data from a marine gravity profile near Lanzarote. In their work the transition position of the islands between the continental crust and the oceanic crust is concluded. The presence of a large high-density intrusive body beneath the center-south of the island and the existence of the dominant structural direction ENE are also inferred.

Using the Imperial College seismic refraction data and gravimetry Bosshard and MacFarlane (1970) presented a comprehensive study of mantle depth (giving values of 12 km west of La Palma and El Hierro, 13.9 km south of La Gomera and Tenerife, 15 km north of Tenerife and Gran Canaria and 21–22 km under the continental shield), cortical thickness indicative of isostatic compensation and structural models for Tenerife and Gran Canaria. They conclude that the crust in El Hierro, La Palma, La Gomera and Tenerife is essentially oceanic, while Gran Canaria is located in the transition zone

between oceanic crust and continental crust. They also conclude that the islands are not part of the African continent, but are independent volcanic edifices emerged in NE–SW fracture zones.

3.2. 1980s and 1990s: The Rise of Modern Volcano Geodesy in the Canary Islands

3.2.1 Structural Gravimetry

In 1987 the United States Geological Survey (USGS) conducted a marine gravity campaign in the Canary Islands as part of a series of studies of the East Atlantic margin. This work was conducted with good precision and the studied area is limited to a few kilometers in the sea around each island (Folger et al. 1990). After that, various geophysical studies including marine gravimetry have been carried out in and around the archipelago. For example, in 1993 the University of Oxford conducted a marine seismic survey that included gravimetric observations along some profiles in the archipelago (Watts et al. 1993).

Watts (1994) compiled data from 44 sea profiles and also included land data from western Africa obtained by the University of Leeds in an area including the Canary Islands. He studied the lithospheric flexure in the archipelago and by comparing seismic and free-air anomaly gravity data and elastic models estimated values for the elastic thickness of the lithosphere and various results concerning lithospheric flexure. He concluded that the observed thinning of the oceanic lithosphere in the region would be produced by thermal disturbances that come from a mantle plume. Despite the lack of a long wavelength topographic bulge, he reaches the conclusion that the plume may be quite narrow.

Ranero et al. (1995) obtained a free air anomaly map for various profiles and in combination with seismic data also obtained a thinning of the crust in the area, concluding that the most likely origin of the Canary Islands is a hot spot.

Subsequently, Watts et al. (1997) again worked on the same compilation of free-air gravity anomalies, also using seismic reflection data to compare the observed data with gravimetric ones calculated using a 2-D model of Tenerife Island. They present a two-dimensional interpretive model of the flanks of

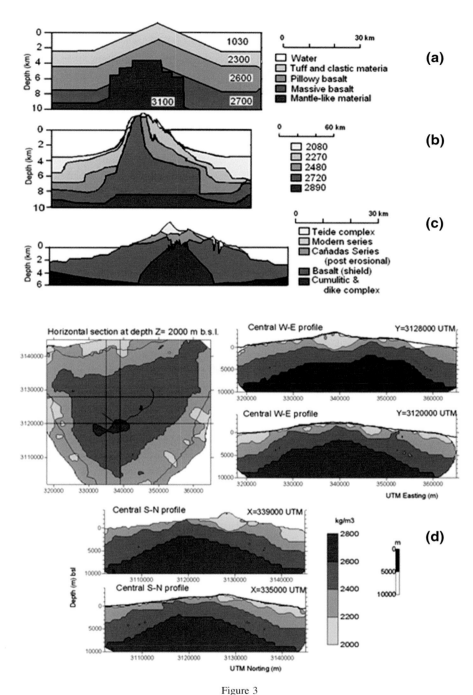

Figure 3
Some published models obtained from gravimetric data (and seismic and geological information) presenting the Tenerife structure as composed of subhorizontal layers. **a** Modified from main characteristics of BOSSHARD and MACFARLANE (1970); **b** modified main characteristics of WATTS *et al.* (1997); **c** modified from main characteristics of FERNÁNDEZ *et al.* (2009); and, **d** from CAMACHO *et al.* (2011a) according to a 3D non-linear inverse approach

Tenerife and adjacent crust. Their NE–SW profile shows a volcanic 2,720 kg/m^3 density core over which are layers of density from 2,480 kg/m^3 at the bottom to 2,270 kg/m^3 in the top layer. They also identified around the island a flexion moat filled with sediments (with a density of 2,080 kg/m^3).

CANALES and DAÑOBEITIA (1998), working on a new compilation of bathymetric and gravity data from marine profiles in the area, conclude the existence of a subsurface load correlated with the surface volcanic loads. Surface expression of the subsurface load is an oceanic swell detected in the anomalous shallow basement and in the residual gravity/geoid anomaly. Yet JUNG and RABINOWITZ (1986), FILMER and McNUTT (1989) and WATTS (1994) suggest the geophysics of the archipelago and the intervening seas are characterized by the absence of a bathymetric swell or a geoid high.

With respect to the terrestrial gravimetric data, since the 1970s different gravimetric surveys have been carried out to better define local terrestrial anomalies or highlight certain structural details in Lanzarote (SEVILLA and PARRA 1975), Tenerife (VIEIRA et al. 1986; CAMACHO et al. 1991; ABLAY and KEAREY 2000; ARAÑA et al. 2000), Fuerteventura (MONTESINOS, 1999), Gran Canaria (ANGUITA et al. 1991; CAMACHO et al. 2000). Advances in this line of research has been marked by three factors: (1) the emergence of GPS positioning equipment; (2) the availability of digital mapping; and (3) the increasing computing power allowing gravity inversion approaches to be more ambitious.

Indeed, GPS positioning has represented a revolution in gravimetry by obtaining the necessary precise vertical coordinates. Earlier works were limited to existing leveling lines or venturing into other areas using low levels of precision solutions, such as heights interpolated from topographic maps or inferred using barometric leveling. In the latter case, it was difficult to ensure height accuracies better than 10 m.

Digital mapping has brought a breakthrough in computing the gravimetric correction produced by the relief effect, which can be very important in volcanic areas. Compared with very time-consuming determinations using a Hammer abacus, the use of Digital Elevation Models (DEM) and automatic correction using computer codes allows for an instantaneous and very accurate determination of gravity corrections, with quality depending mainly on DEM quality and resolution (CAMACHO et al. 1988).

Third, the continuously improving computing capabilities of personal computers enabled development of non-subjective methodologies for linear and non-linear gravity inversion, facilitating construction of 3D models of anomalous density contrasts. The models consisted of thousands of model resolution elements (see e.g., CAMACHO et al. 2000). These approaches contrast with previous ones based on direct computations and testing. These methods started to be developed during the 1990s. The time evolution can be seen in Fig. 3.

3.2.2 Measurement of Deformation and Gravity Changes

During the 1980s and 1990s deformation and gravity variation measurements were carried out on several Islands. Moreover, the improvement of gravimetric instruments and, above all, the chances of very precise altimetry control have led to an interesting application of gravimetry in the geodetic control of volcanic phenomena: the microgravimetric detection and interpretation of gravity variations (in the sense of measuring with high precision gravity time variation and interpreting them to obtain characteristics of magmatic processes) (e.g., SIGURDSSON et al. 2000; DZURISIN 2007; BATTAGLIA et al. 2008). However, this technique would not apply in the Canary Islands until the increase of volcanic activity necessitated its use in the twenty-first century (see next section).

3.2.2.1 Tenerife Island The realization that the region around the Las Cañadas Caldera and the Teide volcano was one of the largest risk areas in the Canary Islands (ARAÑA and GÓMEZ 1995) resulted in most volcano research being focused there, particularly all the geodetic studies.

At the beginning of the 1980s a 17-benchmark classical geodetic micro-network was designed and set up in the Las Cañadas Caldera (see Fig. 4) (VIEIRA et al. 1996; SEVILLA et al. 1986; SEVILLA and MARTÍN 1986). It was observed several times between 1982

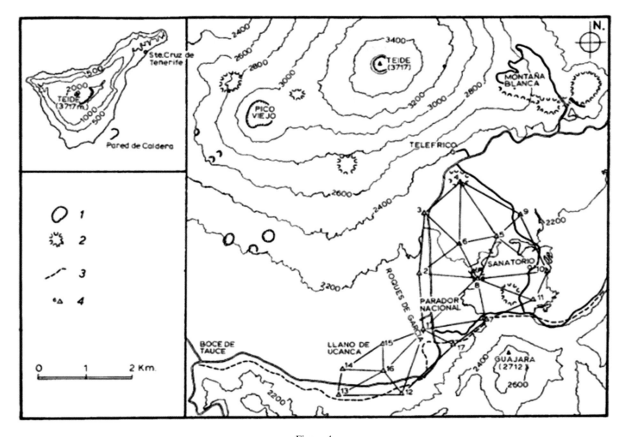

Figure 4
Location of the geodetic network in Las Cañadas, Tenerife Island (SEVILLA *et al.* 1986; YU *et al.* 2000). Meanings are: *1* Teide and Pico Viejo craters and basaltic cones; *2* salic emission centres and domes; *3* base of the caldera wall; *4* geodetic station

and 2000 to try to detect possible crustal displacements associated with volcanic reactivation within the network, and as a procedure for solving structural problems (VIEIRA *et al.* 1996; SEVILLA and MARTÍN 1986; SEVILLA and SÁNCHEZ 1996). The network is located in the southern part of the Caldera and was observed using triangulation and trilateration. In addition, in 1994 a levelling profile was connected to the geodetic network (SEVILLA *et al.*, 1996). It is composed of 52 levelling points and was observed several times using precision trigonometric levelling. No significant displacements were found from 1982 to 2000 in the geodetic network or the levelling profile (SEVILLA and ROMERO 1991; SEVILLA and SÁNCHEZ 1996; FERNÁNDEZ *et al.* 2003).

A sensitivity analysis of these networks revealed important limitations for volcano monitoring, the most important being the small spatial aperture (YU

et al., 2000). The lack of any external control point outside of Las Cañadas Caldera prevented checking for relative displacements. The use of observational methodologies with a precision of around 1 cm for detecting displacements between the dates of the different surveys made it very difficult or impossible to detect any relative displacements inside the caldera below that value.

3.2.2.2 Lanzarote Island The island's structural and geodynamic characteristics, together with the possibility of future activity, prompted the Instituto de Astronomía y Geodesia (IAG), in collaboration with the Lanzarote Inter-Island Council, to install a permanent Geodynamic Laboratory on Lanzarote in 1987 (VIEIRA *et al.* 1988, 1991a; FERNÁNDEZ 1993; FERNÁNDEZ *et al.* 1993; VIEIRA 1994). The Laboratory's scientific tasks were earth and oceanic tidal

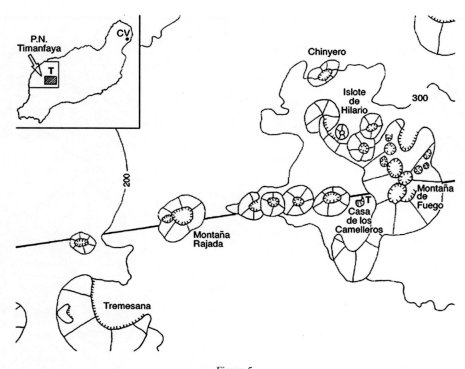

Figure 5

Inset location of geodynamic stations in Lanzarote Island; *CV* Cueva de los Verdes, *T* Timanfaya. Detailed map: plan of Timanfaya National Park with eruptive fissure of the 1730–1736 eruption, the primary volcanic cones, and location of Timanfaya station (after FERNÁNDEZ *et al* 1992)

research and geodynamic research of volcanism and associated seismicity (ROMERO *et al.* 2003). Four different modules, three located in Cueva de los Verdes (inside the La Corona volcano lava tube) and the other in Timanfaya National Park, form the Lanzarote Geodynamic Laboratory (Fig. 5). Several instruments were set up in this station since 1987 to permit the continuous observation of deformation, gravity changes, sea level, rock temperatures and different meteorological parameters (see the references in this section for more details). These modules were connected to a data centre located in the Casa de los Volcanes and from it data were transferred via modem to Madrid.

Most research in the last decades of the twentieth century in the laboratory was done in the field of earth and oceanic tides (see e.g., VIEIRA *et al.* 1991b; FERNÁNDEZ *et al.* 1991, 1992; DEL REY *et al.* 1994; VIEIRA *et al.* 1995; ARNOSO *et al.*1998; ARNOSO *et al.* 2000). However, some research also focused on studying how to apply the geodetic observation carried out in the laboratory to volcano monitoring,

developing theoretical models for interpreting geodetic observations and designing geodetic monitoring systems (see next sub-sections) (FERNÁNDEZ 1991; FERNÁNDEZ and VIEIRA 1991; FERNANDEZ 1993; FERNÁNDEZ *et al.* 1993, 1994, 1999; FERNÁNDEZ and RUNDLE 1994; FERNÁNDEZ and DIEZ 1995).

All data gathered show that no deformation or gravity change related to volcanic activity was detected between 1987 and 2000 (FERNANDEZ *et al.*1992, 1993; ARNOSO *et al.* 2000, 2001a, b; FERNÁNDEZ *et al.*, 2003).

Until the end of the year 2000 (ROMERO *et al.* 2003) the geodetic instruments installed on Lanzarote capable of providing information useful for volcano monitoring purposes only supplied information from two specific areas (Fig. 5). The lack of information about the deformation field of the whole island causes two monitoring problems (pointed out by FERNANDEZ 1993; FERNÁNDEZ and RUNDLE 1994a; FERNÁNDEZ *et al.* 1999): (a) difficulty deducing information about the characteristics of the intrusion in case of volcanic unrest; (b) difficulty identifying a process of volcanic reactivation on the island. Moreover, this situation made it extremely difficult to study the time evolution of the volcanic source, an essential factor for making civilian protection-related decisions. It should be stressed that possible future eruption scenarios in Lanzarote are not limited to a specific volcano but to an extensive active volcanic region that covers nearly the whole island (FERNÁNDEZ et al. 2003).

3.2.2.3 La Palma Island

La Palma accounts for most of the eruptions that have occurred in the archipelago over the last 500 years (historic eruptions) and all the events in this island have taken place at the Cumbre Vieja ridge to the south of the island (ROMERO 2000). Indeed, the last sub-aerial eruption in the Canary Islands occurred at the Teneguía volcano in 1971, which is why La Palma has become one of the most closely monitored islands of the Canaries in recent decades (PERLOCK et al. 2008; GONZÁLEZ et al. 2010b).

In 2000 and 2001 this island became very famous among scientists and general society due to the publication of several articles (MOSS et al. 1999; DAY et al. 1999; WARD and DAY 2001), and the consequent appearance in the media that warned a giant collapse affecting the western flank of the southern part of the island would generate a tsunami that could affect coastal areas around the North and Central Atlantic basin, although not all scientific researchers agree with this (e.g., CARRACEDO et al. 2009).

To assess the level of stability of the western flank of Cumbre Vieja (MOSS et al. 1999) a geodetic network was installed in the mid-1990s and was used three times between 1994 and 1997 utilizing electronic distance measurement techniques (EDM) and rapid-static GPS. However, these studies were limited in their spatial extension and temporal coverage. Although the results showed a coherent pattern of displacement vectors they were of the same order of magnitude as the associated errors (PERLOCK et al. 2008; GONZÁLEZ et al. 2010b) and therefore they could not be really significant.

3.3. Deformation Modeling and Designing of the Geodetic Volcano Monitoring

One of the main tasks facing volcanologists studying an active volcano area is to define the most suitable instrumental monitoring system, a task that is particularly difficult in the absence of recent activity (FERNÁNDEZ et al. 1999; YU et al. 2000). This is the case for most active volcanoes that have not erupted for decades or even longer. This task not only involves technical optimization of the monitoring systems but also their economic and scientific profitability (FERNÁNDEZ et al. 1999). In the 1980s the geodetic volcano monitoring situation was as follows: (1) Several geodetic research programs were being carried out in Lanzarote and Tenerife by Spanish research teams including volcano monitoring with a limited yet constant economic cost; (2) The volcano geodetic research in La Palma was being carried out by UK groups; (3) There was a lack of knowledge of the magnitude and pattern of deformation and gravity variations expected in case of volcanic reactivation on any of the islands based on previous observational experience (FERNANDEZ 1993); (4) There was a need to develop theoretical models and inversion techniques for monitoring use and interpretation in case of volcanic crisis (FERNANDEZ 1993); (5) It was necessary to evaluate the applicability and limitations of the geodetic techniques and methodologies that were being applied to volcano monitoring; and, (6) It was necessary also to design the geodetic monitoring system for volcanic activity on each island and the entire archipelago.

In the frame of existing international trends (BALDI and UNGUENDOLI 1987; JOHNSON and WYATT 1994; DVORAK and DZURISIN 1997; HARRIS et al. 1997; SEGALL and MATTHEWS 1997; BETTI et al. 1999; WU and CHEN 1999) it was decided to use theoretical models, develop new ones, as well as new inversion techniques (YU et al. 1998; TIAMPO et al. 2000) and to use them together with observational experience and results and knowledge about the crustal structure to define the most suitable geodetic volcano monitoring system from a theoretical study of sensitivity to ground deformation and gravity changes (FERNÁNDEZ et al. 1999). It was done that way considering aspect (3) described above but also always bearing in mind that no theoretical study can consider all the variables involved in a problem. Therefore, the results obtained would only approximate reality but could give us an useful idea of how to design geodetic observation for the purposes of monitoring volcanic unrest.

FERNÁNDEZ et al. (1993), DIÉZ-GIL et al. (1994), FERNÁNDEZ and RUNDLE (1994a) and FERNÁNDEZ and

DIEZ (1995) dealt with the issue of defining the most suitable geodetic monitoring system by studying Lanzarote, the easternmost of the Canary Islands, and the southern area of the island of La Palma. FERNÁNDEZ et al. (1999) reviewed the results for the island of Lanzarote and also applied the methodology to the Teide, the stratovolcano in Tenerife. They propose a volcano geodetic monitoring system that uses the existing facilities on each Island complemented with continuous observation or deformation and gravity networks, depending on each case. Continuous recording instruments, based on their high sensitivity, should serve as a primary tool to detect precursory geodetic signals and field observation should be used to solve the inverse problem and determine intrusion characteristics.

On account of the two main limitations of the work described by FERNÁNDEZ et al. (1999), namely the kind of source considered (spherical point magma intrusion) and the limitation in size of the zone studied, YU et al. (2000) extended it to consider a wider zone and probably a more realistic source for Tenerife Island considering its eruption history (a dike). Their study leads to the conclusion that the existing geodetic network on Tenerife Island is capable of detecting dike intrusions just below the area covered by the network and when they are very close to the surface. GPS observations covering the whole surface of the island using continuous recording stations or a very short period of re-observation would be required to detect the effects caused by the temporary rise of an intrusion towards the surface.

3.4. New Techniques

By the end of the twentieth century theoretical studies had shown a clear limitation of the observational system deployed on the islands and the need to cover the entire surface of the islands in oreder to perform operative geodetic volcano monitoring. This could not be done using classical geodetic techniques due to their characteristics and the economic and time costs (FERNÁNDEZ et al. 1999; YU et al. 2000; DZURISIN 2007). Yet the observational situation changed dramatically in the late twentieth century due to two events of great scientific and technical relief. The first was the widespread use of GNSS observation

studying deformations (e.g., DONG and BOCK 1989; BOCK 1991; DIXON et al. 1997; FERNÁNDEZ et al. 1999; SAGIYA et al. 2000; DZURISIN 2007) caused by the advance of technology that allows improved accuracies and cheaper, much smaller geodetic GPS receivers linked to permanent GNSS observations.

The second event was the appearance and subsequent rapid development and expansion of synthetic aperture radar interferometry (InSAR) (MASSONNET and FEIGL 1998; BÜRGMANN et al. 2000; HANSSEN 2001) which, in good conditions (FERNÁNDEZ et al. 2005), allowed areas of 100×100 km^2 to be covered in a single image (i.e., covering complete islands (CARRASCO et al. 2000a; FERNÁNDEZ et al. 2002) with high spatial resolution and, in after a few years, precision (FERNÁNDEZ et al. 2009; GONZÁLEZ and FERNÁNDEZ 2011). Both events drastically changed geodetic monitoring of volcanic activity worldwide and, in particular, for the Canary Islands. Both techniques are currently validated and evolved, are currently used in detecting ground movements and have brought down costs compared with classical geodetic techniques (FERNÁNDEZ et al. 2003).

A clear example of the impact of using these techniques in the Canary Islands can be seen in Tenerife Island, described below (see, for more details CARRASCO et al. 2000a, b; FERNÁNDEZ et al. 2002; RODRÍGUEZ-VELASCO et al. 2002; FERNÁNDEZ et al. 2003, 2004, 2005).

3.4.1 InSAR Application and Results

In the first InSAR study of Tenerife Island 18 radar images acquired by the European Space Agency satellites ERS-1, 2 during the period 1992–2000 were used, obtaining 21 differential interferograms. The good coherence obtained, even for time-spans beyond 7 years, permitted three important results: (a) InSAR could be applied on Tenerife for routine monitoring; (b) No deformation at Las Cañadas caldera (see Fig. 6) was detected for that period, which was in agreement with the results described previously and obtained using classical geodetic techniques; and, (c) the detection of two deformation zones outside areas typically observed, located in the region where the most recent eruptions in the island occurred (Montaña Negra 1706, Chahorra 1798, Chinyero 1909): the Garachico and Chío deformations areas

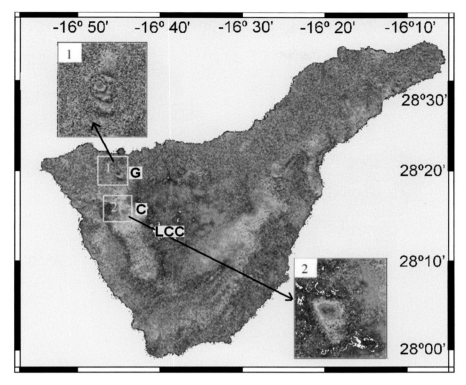

Figure 6
Differential interferogram from Tenerife Island corresponding to August 2, 1996–September 15, 2000 ERS-2 radar images. No fringe can be seen in the Las Cañadas caldera area so there is no deformation from August 2, 1996 to September 15, 2000 at the precision level attainable with one interferogram. *1* and *2* represent Garachico (*G*) and Chío (*C*) subsidence areas, respectively, from July 20, 1993 to September 15, 2000 differential interferogram ($B_\perp = 180$ m, $\Delta d = 2{,}614$ days. $B_\perp$ denotes the perpendicular baseline between the two orbits). The Garachico subsidence has 3 fringes, in other words about 9 cm of ground subsidence from 20 July 1993–15 September 2000; the Chío subsidence has 1 fringe, that is to say, about 3 cm of ground subsidence from 20 July 1993–15 September 2000. *Red circle* indicates the location of Garachico village. *LCC* Las Cañadas Caldera area (modified from FERNÁNDEZ *et al.* 2003, 2004)

(see Fig. 6). Both of them were subsidences, increasing from 1992 to 2000. By way of example Fig. 6 displays the location of these deformations and their magnitudes from 1993 to 2000.

3.4.2 GPS Network: Definition and First Observation Results

Previously described results prompted researchers to design and observe a GPS network covering all Tenerife in 2000. The main objectives were the geodetic monitoring of possible displacements associated with volcanic reactivation and corroborating the results obtained by InSAR. This GPS network has a station in deformation zone (2) (see Fig. 6) and a densification in deformation zone (1) (Fig. 6) located to the south of the village of Garachico (see Fig. 7). For details of the

observation and processing methodologies see RODRÍGUEZ-VELASCO *et al.* (2002) and FERNÁNDEZ *et al.* (2003, 2004). The precision of the obtained results is within one centimeter in height and several millimeters horizontally. Results (Fig. 8) showed subsidence of station Pinar de Chío, of the same order as obtained using InSAR in deformation zone (2), and, therefore, the two techniques confirmed one another. Results obtained in deformation zone (1) to the south of the town of Garachico were not definitive enough to confirm the displacements detected using InSAR.

4. Recent Studies

The other major change in the Canaries that has driven the development of research in volcanic

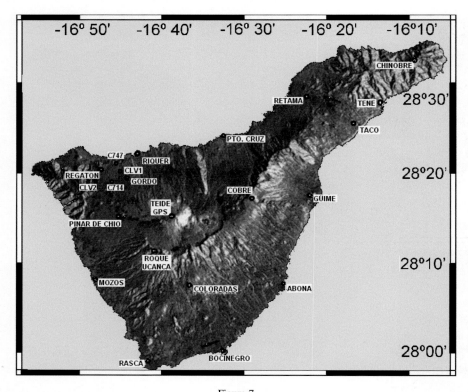

Figure 7

Global GPS network defined for Tenerife Island. It is composed of 17 stations from REGCAN95 together with the permanent station TENE, marked with *circles*. A densification network formed by two fourth-order stations (C774 and C747) and two benchmarks (CLV1 and CLV2) was installed in the zone with greater deformation near Garachico city (modified from FERNÁNDEZ *et al.* 2004)

activity monitoring in the archipelago has been the significant increase in activity coinciding with the beginning of the twenty first century. This is reflected in the seismic-volcanic crisis of Tenerife (2004–2007) that concluded without an eruption and the underwater volcanic eruption on the island of El Hierro (2011–2012) followed by various episodes of significant surface deformation and seismic activity in 2012 and 2013 (http://www.ign.es/ign/resources/volcanologia/HIERRO.html; GARCÍA *et al.* 2014). This has forced the use and development of new techniques for the improvement and optimization of geodetic monitoring, in both observational and theoretical (interpretation tools) aspects, as well as for the knowledge of the crustal structure of the different Islands. The last is a basic aspect for a correct discussion and interpretation of any observed anomaly during unrest episodes (see e.g., GOTTSMANN *et al.* 2008).

Sensors and techniques developed in recent years in the different fields of geodesy (GNSS, Earth observation from space, gravimetry, etc.) have allowed both ground deformation and gravity variations that occur before, during and after volcanic events to be measured with a precision, spatial coverage and temporal coverage unimaginable a few decades ago (FERNÁNDEZ *et al.* 1999; SIGURDSSON *et al.* 2000; GOTTSMANN *et al.* 2006; DZURISIN 2007; BATTAGLIA *et al.* 2008, CROSSLEY *et al.* 2013). These innovative observational techniques have entailed development of new methods of data processing and interpretation (theoretical models and inversion techniques), allowing rigorous and objective information to be obtained from these new observations.

4.1. Structural Studies

In the twenty first century several gravimetric works (sometimes complemented with microseismic experiments) have sought to better define gravity anomalies and show structural details on different

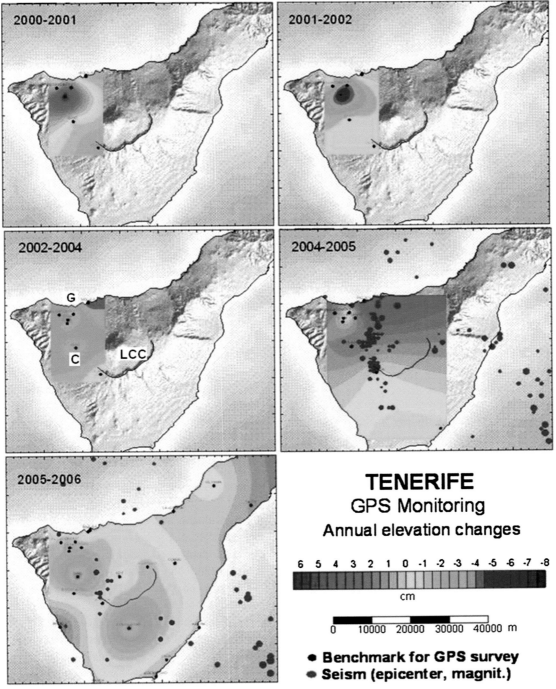

Figure 8

Elevation changes for Tenerife Island determined comparing coordinates determined in the different GNSS–GPS surveys (August 2000; July 2001; July 2002; May 2004; July 2005; January 2006). 2004–2006 campaigns were carried out in response to the volcano-tectonic crisis in Tenerife that began in 2004. *G* location of Garachico city, *C* location of the Chio deformation area, and *LCC* location of Las Cañadas Caldera

Reprinted from the journal

islands such as Lanzarote (CAMACHO *et al.* 2001; GORBATIKOV *et al.* 2004), Tenerife (CHÁVEZ-GARCÍA *et al.* 2007; GOTTSMANN *et al.* 2008; CAMACHO *et al.* 2011a, b), Fuerteventura (MONTESINOS *et al.* 2005), La Palma (CAMACHO *et al.* 2009), El Hierro (MONTESINOS *et al.* 2006; GORBATIKOV *et al.* 2013), La Gomera (MONTESINOS *et al.* 2011) and in the vicinity of the Canary Islands (LLANES 2006).

The structural pattern is similar in almost all on-shore studies for the different islands. The oldest volcanic zones have strong positive gravity anomalies interpreted in terms of large magmatic intrusive bodies and aggregation of dikes corresponding to phases of very massive shield volcanism. The areas of most recent volcanic activity (and therefore potentially more active) are characterized by negative gravity anomalies that are partly due to fracturing and slight accumulation of unconsolidated volcanic material which still not being eroded. Some of these works allow different authors to relate the determined crustal structures with displacement or other anomalies detected during the recent volcanic crisis in the archipelago (GOTTSMANN *et al.* 2008; GORBATIKOV *et al.*2013).

For example, in the case of Tenerife a large body of high density is located slightly to the SW of the center of the island (see Fig. 3), which is associated with very consolidated magmatic intrusions. Other areas of positive anomalies (high density) can be associated with ancient structures and massive rift volcanism. In the center of the island a strong significant minimum is identified revealing the low density of materials accumulated forming the Teide and the filling of the volcanic caldera. The morphology of these low density elements leads GOTTSMANN *et al.* (2008) to ensure they support a vertical collapse origin of the caldera and map the headwall of the 180-ka Icod landslide that appears to lie buried under the Pico Viejo-Pico Teide stratovolcano complex.

For the other islands models and conclusions are similar and each successive study improves the morphological definition of the structures. For example, for La Palma the following characteristics can be described (CAMACHO *et al.* 2009): a large body of high density in the northern half corresponding to ancient volcanism massive; a high-density structure elongated along the NS rift in the middle; and, lower densities in the south

and on both sides of the central rift, associated with fracturing and light recent volcanic deposits and zones of very recent activity (similar to the Teneguia volcano at the south end of the Island).

4.2. Measurement of Deformation and Gravity Changes

4.2.1 Tenerife Island

The first results from applying InSAR to Tenerife Island are described by CARRASCO *et al.* (2000a, b), ROMERO *et al.* (2002) and FERNÁNDEZ *et al.* (2002, 2003, 2005) and they were shocking, even using classical DInSAR technique and a reduced number of radar images (see previous section), because they showed clear deformation areas (Fig. 6) in areas not studied before. These results further emphasized the need to define a GPS network covering the whole island (Fig. 7), which was defined and formed (FERNÁNDEZ *et al.* 2003, 2004) by 17 vertices from the existing network, with accurate coordinates determined by the REGCAN-95 geodetic system (CATURLA 1996) plus the permanent station at Santa Cruz de Tenerife (TENE; FERNÁNDEZ *et al.* 2003). Beginning in 2000 GPS observations confirmed the DInSAR detected subsidences and the time variability of the displacement and also showed that the horizontal component displacement was very unimportant compared with the vertical one (FERNÁNDEZ *et al.* 2003, 2004, 2005). These localized subsidences were attributed, at least partially, to changes in the groundwater level (FERNÁNDEZ *et al.* 2005). Results of elevation change, comparing coordinates determined for each campaign, are shown in Fig. 8.

Background seismicity in the Canary Islands, at least during the last decades of the twentieth century, was characterized by diffuse and disperse seismicity in most of the region, concentrated in an area located between the island of Tenerife and Gran Canaria as well as in a NW–SE line where there is submarine alignment of volcanic seamounts (ROMERO RUIZ *et al.* 2000).

Anomalous low-magnitude seismicity was recorded in and around Tenerife Island from 2001 by the IGN network (IGN 2006), but more significantly during 2004 and 2005, close to a century after the last eruption in the island. A combined analysis of the observed geochemical and seismic data prior the recent seismic swarm in 2004 suggested that subsurface magma movement could be the potential mechanism for this observed seismic activity increase (PÉREZ and HERNÁNDEZ 2004; PÉREZ *et al.* 2005, 2007). This 2004–2005 anomalous seismic activity has been characterized by having a great number of epicenters located inland of Tenerife Island and some migration with time (Fig. 9) as in other volcanic areas (e.g., WAITE and SMITH 2002; WICKS *et al.* 2006). From April to December 2004 195 seismic events were located, five of them felt in May, July and August 2004; and more than 350 seismic events to February 2006. The total number of recorded seismic events during the same period, including those not located, exceeded 3,000. The model proposed by ALMENDROS *et al.* (2007) to explain the pattern of seismicity observed includes an initial deep magma intrusion under the northwest flank of the Teide volcano, and the associated stress changes produced the deep volcano-tectonic cluster. In turn, the occurrence of earthquakes permitted and enhanced the supply of fresh magmatic gases toward the surface. The gases permeated the volcanic edifice, producing lubrication of pre-existing fractures favoring the occurrence of volcano-tectonic earthquakes. On May 18 the flow front reached the shallow aquifer located under Las Cañadas Caldera and the induced instability was the driving mechanism for the observed tremor.

This reactivation produced surface gravity changes (GOTTSMANN *et al.* 2006) indicating the activity was accompanied by a sub-surface mass addition, although no widespread deformation was detected initially. While magma recharge at depth into the northwestern rift zone of Tenerife is likely to have triggered the reawakening of the central volcanic complex, the cause of the 14-month perturbation of the gravity field is most probably not related to magma flow. A more likely scenario is the migration of fluids inside the complex triggering the observed gravity changes.

Some surface deformation associated with this reactivation has been detected using GNSS (see Fig. 8) and InSAR (see Fig. 10) observations and described in different works (GONZÁLEZ *et al.* 2005; PRIETO *et al.* 2005; FERNÁNDEZ *et al.* 2006, 2007, 2008; SAMSONOV *et al.* 2008; FERNÁNDEZ *et al.* 2009; GONZÁLEZ *et al.* 2010a; TIZZANI *et al.* 2010).

SAMSONOV *et al.* (2008) present results for the three-dimensional displacement field on Tenerife Island calculated from GPS campaigns and ascending and descending ENVISAT DInSAR interferograms. The goal of their work is to provide an example of the flexibility of the technique by fusing together new varieties of geodetic data, and to observe surface deformations and study precursors of potential activity in volcanic regions. Interferometric processing of ENVISAT data was performed with GAMMA software. All possible combinations were used to create interferograms and then stacking was used to increase signal-to-noise ratio. Decorrelated areas were widely observed, particularly for interferograms with a large perpendicular baseline and large time span. Tropospheric signals were also observed, which significantly complicated the interpretation. Subsidence signal was observed in the NW part of the island and around Mount Teide and agreed in some regions with GPS data.

FERNÁNDEZ *et al.* (2009) study the state of deformation of Tenerife (Canary Islands) using Advanced Differential Synthetic Aperture Radar Interferometry (A-DInSAR). They apply the Small BAseline Subset (SBAS) (BERARDINO *et al.* 2002) DInSAR algorithm to 55 radar images acquired from descending orbits (Track 352, Frame 3037) by the ERS-1/2 satellites during 1992–2005. Their analysis reveals (Fig. 10) the summit area of the volcanic edifice is characterized by rather continuous subsidence extending well beyond Las Cañadas caldera rim and corresponding to the dense core of the island. These results were undetectable using classical DInSAR or any GPS network formed by isolated stations and observed on a survey basis. These results coupled with GPS, structural and geological information and deformation modeling suggest an interpretation based on the gravitational sinking of the dense core of the island into a weak lithosphere and that the volcanic edifice is in a state of compression. FERNÁNDEZ *et al.* (2009) also detected more localized deformation patterns, some of them previously detected using classical DInSAR (see Figs. 6, 10). They were also able to a determine deformation time series for the coherent pixels (see Fig. 10) and it is easy to see variations in the deformation time series associated with the seismic crisis in 2004.

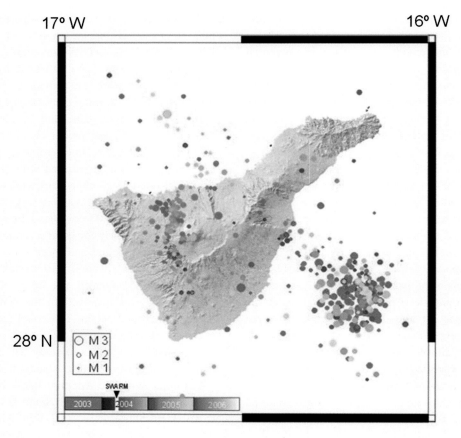

Figure 9
Epicentral locations of anomalous seismicity registered from 2003 to 2006 for Tenerife Island and surroundings (IGN 2006)

Apart from the previously described short-term surface deformation, characterized by a broad subsidence pattern with maximum ground velocities of about 4 mm/year detected via space-based geodetic observations (FERNÁNDEZ *et al.* 2009), different geophysical research (WATTS 1994; WATTS *et al.* 1997; WATTS and ZHONG 2000; COLLIER and WATTS 2001; MINSHULL and CHARVIS 2001) concluded the Tenerife volcanic complex is affected by crustal deformation processes occurring on timescales of millions of years. With the purpose of studying the relationship between these long-term and short-term deformation processes, TIZZANI *et al.* (2010) performed an advanced fluid dynamic analysis (FDA). Their results interpreted the recent surface deformation as mainly caused by a progressive sagging of the denser (less viscous) core of the island onto the weaker (but more viscous) lithosphere. Moreover, over periods comparable to the hypothesized age of loading of the oceanic

Figure 10
SBAS-DInSAR results obtained using 55 radar images acquired from 1992 to 2005 by ERS sensors at descending orbits. See text and FERNÁNDEZ *et al.* (2009) for details. **a** Geocoded mean deformation rate map computed in correspondence to coherent pixels only, and superimposed on the DEM of the island; the reported SAR azimuth and range directions (*black arrows*) are indicative. *Blue arrows* show horizontal displacement measured with error ellipses determined using GPS observations between 2000 and 2006. The *white stars*, labeled as *b, c, d, e* and *f*, identify the pixels whose DInSAR LOS deformation time series are shown in panels (**b–f**); note that in panel **f** the deformation associated with the 2004 seismic crisis has been highlighted in *orange*. **g** Plot of the mean deformation rate values (for the pixels located in coherent areas) versus topography with the locations of the areas (*black letters* from "**b**" to "**f**") affected by localized deformation (FERNÁNDEZ *et al.* 2009)

crust beneath Tenerife, this tendency would result in a total flexure of about 3–4 km, which is in agreement with independent estimations based on geophysical analyses. They use a unitary physical model to explain both the deformation recorded in deep geological

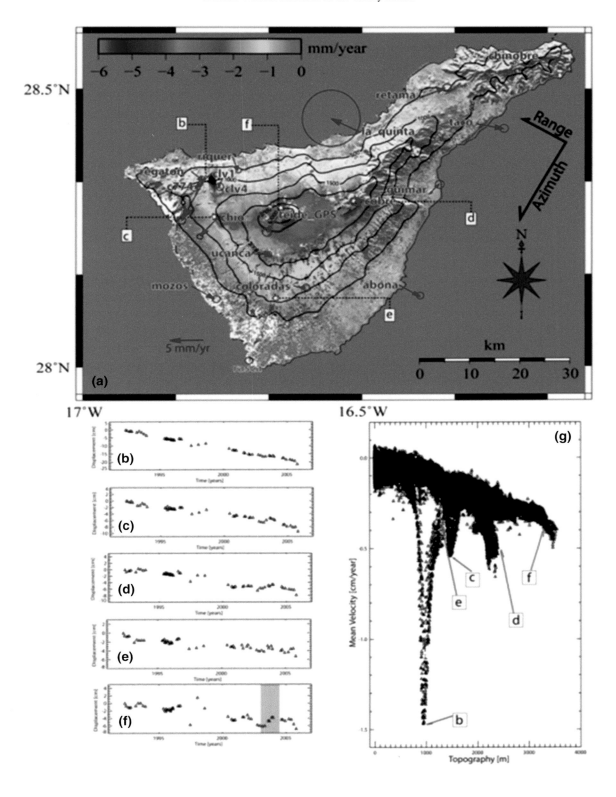

structures and the current active ground deformation processes occurring at the Tenerife volcano.

A more recent GPS and SBAS-DInSAR data combination was performed for Tenerife Island (GONZÁLEZ *et al.* 2010a). In this study the same GPS dataset was correlated for inner consistency against some previous SBAS-DInSAR results (FERNÁNDEZ *et al.* 2009). Some agreements and disagreements were found among the different regions covered by both datasets. Significant positive correlation was obtained for the NW rift zone; indeed, it was expect because GPS surveys are denser in space and time in this part of the island.

Considering some of the previous results, the related volcanic hazard and potential risk at Tenerife as well as the recent anomalous seismic activity from April 2004, a permanent GPS network of 7 stations on Tenerife was installed by scientists from ITER, the Institute of Astronomy and Geodesy (IAG), the Research Center for Seismology, Volcanology, and Disaster Mitigation of the Nagoya University, Japan and the Technological University of Madrid (UPM) (GONZÁLEZ *et al.* 2005; PRIETO *et al.* 2005; FERNÁNDEZ

et al. 2006). The spatial distribution can be seen in Fig. 11. The network sought to provide the best possible coverage of the Teide volcanic cone, deformation areas found by InSAR in that epoch (FERNÁNDEZ *et al.* 2005) and a pre-estimated non deformation area located at the south part of the island, at the ITER facilities.

Results were obtained using Bernese 4.2 software (HUGENTOBLER *et al.* 2001; FERNÁNDEZ *et al.* 2006) for the period June 2004–January 2006 (Fig. 11) showing a significant positive elevation changes for TEIT (4.4 mm/year) and PORT (9.5 mm/year) and LANO (−2.4 mm/year) in the western sector of the network. Conversely a subsidence effect is observed in the north-east station LANO (−1.9 mm/year). On the other hand, significant EW displacements are observed in the eastern half of the network.

Taking into account that seismic activity has been one of the main indicators of geological unrest in Tenerife, correlation between GPS results and seismic activity was tried as a primary search for the GPS information. Some significant correlations were detected for stations PORT ($R = -0.31$) and NORD

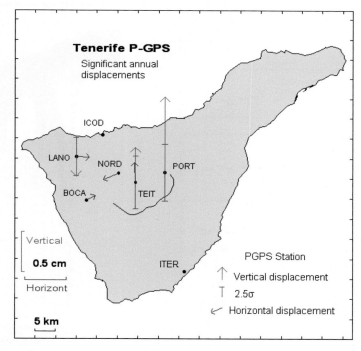

Figure 11
Spatial distribution of the stations of the Tenerife Permanent GPS network and representation of the significant annual trend values (larger than 2.5 times their standard deviations) for the ITER-PGPS stations in Tenerife. See text for exact values and discussion (PRIETO *et al.* 2005)

($R = 0.28$). Both stations and components present trends (9.5 mm/year for PORT-U and − 1.86 ±0.37 mm/year for NORD-E) and correlation ($R = -0.33$) between them. This suggests that the area close to these stations (mainly N–W of Teide) could have some deformations related with the seismicity (FERNÁNDEZ et al. 2006).

In a way unrelated to the previously described GNSS networks, another one, the TEGETEIDE network, was set up in 2005 and re-observed each year (BERROCOSO et al. 2010). It is composed of seven GNSS–GPS stations scattered throughout the island. They presented, based only on the variation in these 7 stations' coordinates, a horizontal deformation model for the whole surface of the island in order to explain the observed island displacement pattern in the geodynamic context of the Nubian plate. They concluded that the most important geologic structures, such as the volcanic rifts and the caldera, determine the current deformation pattern of Tenerife. The geodynamics of the most stable areas of the island behave similarly to that observed from the permanent GNSS–GPS reference stations located in La Palma and Gran Canaria Islands. They present anomalous geodynamic behaviour related to volcano-tectonic activity of the island and its surroundings detected in two zones of Tenerife that configure an NW–SE axis crossing the central sector of the island.

The Island of Tenerife is also home to a world-class astronomical observatory, the El Teide Observatory, where four tiltmeters, two aligned in the North–South direction and the other two in the East–West direction, are monitoring movements of the solar telescope THEMIS (Heliographic Telescope for the Study of Solar Magnetism and Instabilities). Considering THEMIS is located a few kilometers from the El Teide–Pico Viejo stratovolcano, and the precision of the inclinometers is comparable to those used in geophysical studies, EFF-DARWICH et al. (2008) carried out the analysis of the tilt measurements for the period 1997–2006. The THEMIS tiltmeters are located on the seventh floor of a tower, and hence they are less sensitive to geological processes than geophysical installations. However, THEMIS measurements are the only terrestrial data available in Tenerife for such a long period of observations that include the sustained increase in seismic activity that started in 2001. They found a significant change in the East–West tilt of approximately 35 μ-radians between the years 2000 and 2002. Some theoretical models were calculated and it was concluded that such tilt variation could not be due to dike intrusions, nor a volcanic reactivation below the El Teide–Pico Viejo volcano. The most likely explanation (EFF-DARWICH et al. 2008) comes from dislocations produced by a secondary fault associated with a major submarine fault off the eastern coast of Tenerife. They also conclude that, taking into account the nearly permanent data recording at THEMIS, the inclinometers could be considered as a complement for any ground deformation monitoring system in the island.

During and after the 2004 volcano-tectonic crisis the IGN, which is officially responsible for volcano monitoring in Spain, developed and deployed a geodetic volcano monitoring system based on classical geodesy such as precision leveling and measuring electromagnetic distances and space techniques, especially GPS. A permanent network has been designed with continuous data logging and acquisition in real-time. It consists of permanent GPS stations and tide gauges (http://www.fomento.gob.es/Contraste/MFOM/LANG_CASTELLANO/DIRECCIONES_GENERALES/INSTITUTO_GEOGRAFICO/Geofisica/volcanologia/B70_geodesia.htm).

The A-DInSAR CPT technique (MALLORQUÍ et al. 2003; BLANCO-SÁNCHEZ et al. 2008) has also been applied to study surface deformation in Tenerife. ARJONA et al. (2009) studied two sets of ESA radar images, a set of ENVISAT ascending images for 2003–2008 period and a set of ENVISAT descending images for the period 2004–2008. Their results are consistent with the ones obtained by FERNÁNDEZ et al. (2009). They did not carry out any interpretation of the detected displacements or conduct any detailed study in connection with the volcano-tectonic crisis.

Many of the previous works, mainly in relation with the 2004 volcano-tectonic crisis, present clear limitations. Not a single study has coverage from 1992 to any time after the end of the crisis (about 2006). There are no detailed studies of the advanced DInSAR deformation time series before, during and after the crisis, including an interpretation, or a GPS and DInSAR data-combined analysis covering the entire crisis period.

4.2.2 Lanzarote Island

During the first decade of this century, the Lanzarote Geodynamics facilities were being used to carry out many earth tide works, sometimes looking for possible application of tidal observations to volcano monitoring but without any clear result in this aspect or in the detection of displacements or gravity changes (see e.g., KALININA *et al.* 2004; ARNOSO *et al.* 2001a, b, c; VENEDIKOV *et al.* 2006; ARNOSO *et al.* 2011). GNSS observations were also carried out, serving to support the tide gauges installed in the laboratory (GARCÍA-CAÑADA and SEVILLA 2006) and using a single permanent GNSS–GPS station.

The lack of geodetic monitoring networks covering all of Lanzarote Island motivated the use and testing of applicability of classical DInSAR for monitoring volcanic deformation on the Island (ROMERO *et al.* 2003). They used six radar images acquired by the ERS-1 and ERS-2 satellites during the period 1992–2000. The analysis of these images confirmed the existence of long-term stability coherence across most of the island due to dry climate conditions and the large extension of recent lava flows. The analysis of 15 (redundant) interferograms allowed them to distinguish relatively important atmospheric contributions in the differential interferograms on Lanzarote, and to conclude that there has been no displacement >3 cm on the island during the period studied. These results show a clear need for a multitemporal analysis of interferometric products and the assessment of errors to look for possible small displacements below the 3 cm level.

GONZÁLEZ and FERNÁNDEZ (2011) applied a new error estimation multitemporal method on Lanzarote Island. They used a set of 14 SAR scenes from the European Space Agency (ESA) of satellites ERS-1 and ERS-2 (descending orbits) acquired in the period 1992–2000. See GONZÁLEZ and FERNÁNDEZ (2011) for details about the InSAR processing. Their results are shown in Fig. 12. Figure 12a shows the linear

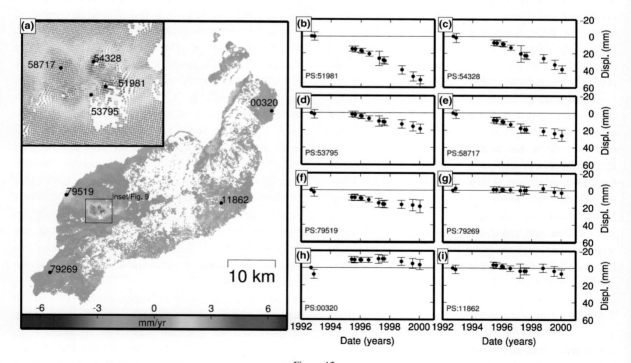

Figure 12

a Estimated descending linear deformation rate between September 2, 1992 and January 8, 2000. The *black rectangle* shows the location of the figure *inset*, which is a zoom into the Montañas del Fuego area (Timanfaya eruptive centers); **b–i** Time series of displacements and associated estimated errors of 8 selected points. See text for details. (GONZÁLEZ and FERNÁNDEZ 2011) velocity map from the estimated time series of displacements for each coherent pixel in mm/year. In addition, the mean square root of the residuals with respect to the linear model for all times series is of the order of $\sim \pm 1$ mm/year. Most of the area of Lanzarote (and northern islets) is stable at the level of $\sim \pm 1$ mm/year during the studied period. Two areas (central part and northwestern coast) show significant lengthening displacement rates (most likely a subsidence signal). The largest deformation rates are associated with the Timanfaya eruption area (Montañas del Fuego) with linear velocities of 4–6 mm/year, and affecting an area of about 7 km^2. The second deformation area is smaller in magnitude (3–4 mm/year) and located on the northwestern coast. Despite the low magnitude of the deformation rate measured along the points between both deformation areas, points indicate systematic positive (subsidence) deformation rates. It could indicate spatial continuity and a possible connection between both areas (and the generating sources). Figure 12a also shows the location of some selected time series. Time series (Fig. 12b–e) represent the estimated displacement evolution and associated estimated error (displayed as 2-sigma error bars) for 8 coherent pixels.

The time series illustrate that the estimated errors are in the range of 5–8 mm, although the repeatability is slightly higher (~ 1 cm). This observation is in accordance with previous results about the reliability of SB techniques (CASU et al. 2006). Deformation closely follows the surface temperature anomalies indicating (GONZÁLEZ and FERNÁNDEZ 2011) that magma crystallization (cooling and contraction) of the 300-year shallow magmatic body under Timanfaya volcano is still ongoing. Unfortunately, no independent ground deformation estimates were available for comparison, even considering the proximity of deformation areas to Timanfaya module of the Lanzarote Geodynamics Laboratory.

4.2.3 La Palma Island

The tsunami caused by the earthquake in SW Asia on the 26th of December 2004, which cost an enormous amount both in human lives and economic terms, spotlighted the catastrophic works published on the likelihood and possible effects of a landslide on the Island of La Palma (MOSS et al. 1999; DAY et al. 1999; WARD and DAY 2001), making it even more interesting to determine if there are any displacements on the island that might be associated with possible landslides (GONZÁLEZ et al. 2010b).

InSAR results obtained using ERS-1 and 2 ESA satellite images from 1992 to 2000 and using three different InSAR phase analysis techniques (coherent pixel time series technique, coherent target modeling method, and stacking) clearly show subsidence on the Teneguía volcano where the last eruption on La Palma took place in 1971. The stacking technique also shows a mild long wavelength signal of subsidence in the western part of Cumbre Vieja. The linear velocities of subsidence are between 4 and 9 mm/year. No deformation, at the measured level of precision, along the coherent pixels in the northern part of the island has been detected (PERLOCK et al. 2008).

FERNÁNDEZ et al. (2008) and PRIETO et al. (2009) projected and surveyed a GPS network covering La Palma Island using geodetic infrastructure installed by the IGN in the 1990s. This network includes control stations along the Cumbre Vieja rift and its flank, Cumbre Nueva slope, also inside and outside Taburiente Caldera and two stations at Teneguia volcano (Fig. 13). The geodetic network was observed in 2006, 2007 and 2008. The same equipment and observation methodology was used for all campaigns. See FERNÁNDEZ et al. (2008) and PRIETO et al. (2009) for details about observations and data processing.

Widespread subsidence was observed (Fig. 13b) (PRIETO et al. 2009) to be more emphasized in the western flank of the island varying from about 7 cm in the north to approximately 4 cm in the south, with standard deviations of about 1 cm. The effect is smaller in the east flank. Therefore, in the south they obtain a velocity of subsidence close to 4 mm/year, consistent with results previously obtained using DInSAR (PERLOCK et al. 2008). Horizontal displacements from 1994 to 2007 are shown in Fig. 13a, showing for the western flank a 2 cm displacement to the south with formal errors for all those markers under the 7 mm level. Between 2006 and 2007 (Fig. 13d) subsidence of the western flank still

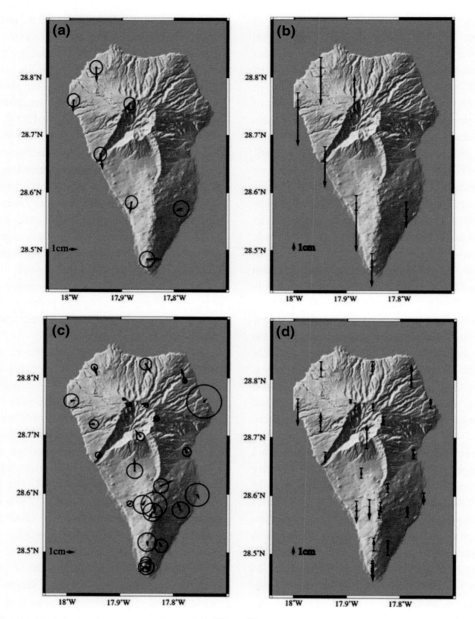

Figure 13
Horizontal (**a**) and vertical (**b**) displacements at the GPS geodetic control points between the 1994 and 2007 GPS campaigns; **c** Horizontal and **d** vertical displacements between 2006 and 2007 campaigns. The results were obtained using GAMIT software (PRIETO *et al.* 2009)

remains, but some points do not agree. The Caldera de Taburiente and its water-course to the sea describe a movement in the opposite direction. The northeast part of the island, with no markers on the previous study, also describes a displacement in the same direction as for the caldera. Displacements on the southeast flank are minor, following the same trend as

seen for the 1994–2007 period. Displacements vectors in Cumbre Vieja seem to follow an erratic trend, being of the same magnitude as their formal errors. There are no significant displacements detected in that area. Markers on the south of the island close to or in the area where the last eruption of the Teneguia volcano occurred follow the same subsidence as the western flank. Therefore, the tendency is consistent with DInSAR results again. For the 2006–2007 period (Fig. 13c) stations located on the western flank show southward displacement, but with only a 6 month interval and their magnitudes are similar to their standard deviations. The points on the caldera surroundings fit displacements in a northwest direction, as do the markers placed on the east flank. Some points on the western flank of Cumbre Vieja show eastward displacements together with the majority of the points at its south end.

All the observed significant displacements affect stations located outside the large central high density body obtained by the inverse gravimetric approach (CAMACHO et al. 2009). Therefore, detected displacements are located in younger areas and with more recent activity.

ARJONA et al. (2010) processed a stack of 15 ascending images from the period 2004–2007, a second set formed by 18 descending covering 2006 up to 2008, and a final data set formed by 16 descending images covering 1992 up to 2000 from ERS sensors using the CPT A-DInSAR technique. Their results were consistent with results described by PRIETO et al. (2009). No interpretation work was done in this study.

GONZÁLEZ et al. (2010b) analyzed 25 SAR images acquired by the European Remote Sensing (ERS1/2) satellites between May 1992 and September 2000 and 19 ASAR images acquired by the ENVISAT satellite from March 2003 to February 2008, archived by the European Space Agency (ESA). They selected descending SAR images, roughly sampled regularly over the period studied (1992–2008), to obtain a detailed image map of the ground deformation at Cumbre Vieja volcano. A similar analysis using ascending data was not feasible due to severe foreshortening that would result from the steep slopes of the western flank of Cumbre Vieja. See GONZÁLEZ et al. (2010b) for details on InSAR processing and atmospheric effect mitigation. They concentrated their ground deformation analysis in the active rift zone of Cumbre Vieja using average LOS velocity maps from the 1992–2000 (Fig. 14a) and 2003–2008 (Fig. 14b) periods. Their results show two clear subsidence signals at the Teneguia volcano area and on the western slopes of the Cumbre Vieja volcano (Fig. 14). The detected deformations were calculated using the descending

orbit pass, so it could be either subsidence, westward motion or a combination of the two. The results by PRIETO et al. (2009) suggest a large part of the ground motion could be vertical.

GONZÁLEZ et al. (2010b) modelled ground deformation using rectangular dislocation with free dip-slip motion on the fault plane, simulating in a homogeneous, isotropic and elastic half-space a normal fault mechanism (OKADA 1985). Creeping processes with associated non-volcanic tremor may release the stress on the sliding surfaces, a process that potentially can remain undiscovered without dedicated seismic observations. The smoothness and spatial distribution of our geodetic results (Fig. 14) suggests that the detachment fault is slow stable-sliding at depth beneath the western flank of the edifice in the on- and off-shore region close to the shoreline, on a fault segment with creeping friction properties. This behaviour is likely steady-state or might be punctuated by unobserved slow-slip events in the transition zone between frictionally different segments of the developing fault surfaces (BROOKS et al. 2006; SEGALL et al. 2006). Only during eruptions can the slip assumption of homogeneity be accepted.

Elastic modelling of the radar data could explain the observed deformation with slip on an active creeping detachment surface that fits the contour of the low density zone. Spatial coalescence of (a) a prominent volcano edifice underlain by a ductile layer (old sediments or debris avalanche deposits), (b) a buried buttress structure in the eastern flank and (c) concentrated westwards dilatational magmatic stresses due to repeated N–S rift intrusions in the last 7 ka results in the initiation and progressive development of an active aseismic mobilisation of the western flank. They concluded that the detachment layer may act as an efficient boundary for aseismic stress release due to gravitational loading during intereruptive periods (present activity), encouraging dike intrusions at the N–S rift zone. Release of dilatational magmatic stresses during these N–S rift intrusions encourages slip and promotes high-angle normal faulting at the border of the slip area (1949 eruption), reorienting the stress field around the volcano and resulting in near E-W fissural eruptions in the western flank. This simple model (GONZÁLEZ et al. 2010b) satisfactorily explains the ground deformation data and also complements the

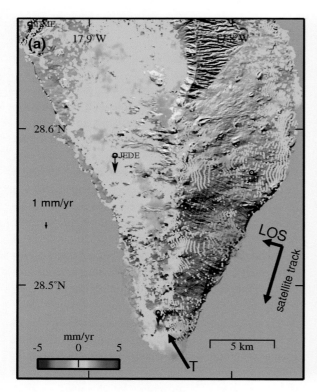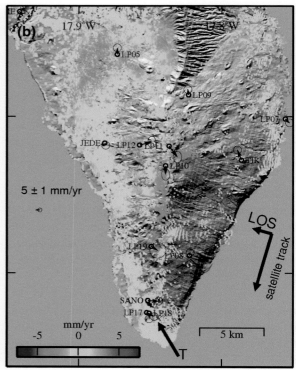

Figure 14

Stack results shown on a shaded DEM. Differential interferograms were corrected for atmospheric elevation-phase dependence. Results are only from coherent points (pixels) that exhibit LOS linear velocity (positive away from satellite, indicated with the *arrows*). **a** Stacking of 82 long temporal separation ERS differential interferograms for the period 1992–2000, accompanied by vertical GPS linear velocity between 1994 and 2007; **b** Stacking of 18 long temporal separation ENVISAT differential interferograms for the period 2003–2008. We also show estimates of horizontal GPS linear velocities between 2006 and 2007. Note the linear rate from the 2003 to 2008 results is noisier than the 1992–2000 results, mainly due to the smaller dataset, so these results should be considered with caution. The largest magnitude subsidence signal corresponds to the Teneguia volcano (*T* symbol) (González *et al.* 2010b)

geological and geophysical evidence that the Cumbre Vieja volcano is in an early state of an immature collapsing process (<20–7 ka). They speculate that ongoing creeping beneath the western flank of Cumbre Vieja tends to stabilize the flank through reorganization and a decrease of gravitational potential forces. This conclusion does not preclude that any sudden and/or unusual change in the stress field, such as a dike intrusion or groundwater pressurization, could trigger a catastrophic collapse (González *et al.* 2010b).

4.2.4 El Hierro Island

Ground tilt and gravity measurements were carried out at stations on El Hierro Island from 2004 to 2010. Tilt variations associated with seasonal temperature effects and periods of heavy rain, sometimes associated to landslides, were detected (Arnoso *et al.* 2008). Gravity tide records have been analyzed and compared to the DDW theoretical body tide model (Arnoso *et al.* 2011). No precursory deformation or gravity change for the 2011 volcanic crisis was published for those instruments.

Starting in July 2011 anomalous seismicity was observed at El Hierro Island. On the 12th of October 2011 the process led to the beginning of a submarine NW–SE fissural eruption ∼15 km from the initial earthquake loci, indicative of significant lateral magma migration. A description of the different phases of activity during that time period can be found in Carracedo *et al.* (2012), López *et al.* (2012), Ibáñez *et al.* (2012), Martí *et al.* (2013) and González *et al.* (2013). Deformation and gravity changes were measured before, during and after the eruption (López et al. 2012; Arnoso et al. 2012; Sagiya et al. 2012; González et al. 2013; García et al. 2014).

López et al. (2012) describe the multiparametric monitoring network deployed over the El Hierro Island by the IGN from July 2011 to the eruption onset on October 2011. They cover the recording of seismicity, geochemistry, geomagnetism, gravimetry, GNSS, InSAR and other additional measurements (e.g., temperature, pH and electric conductivity at wells). They describe the different parameter variations during different phases of the pre-eruption time period. No data inversion for interpretation is carried out.

Prates et al. (2013) present a processing strategy to achieve millimeter-level half-hourly positioning solutions using GPS–GNSS data and describe deformation results for July–November 2011. Some interpretation using a Mogi point source is done. Months after the eruption a new deformation was measured, of greater magnitude than before the eruption.

Martí et al. (2013) combined geological, geophysical, geodetic and petrological data and numerical modeling to propose a volcanological model of the causes and mechanisms of the El Hierro eruption. They conclude that the stress distribution in the crust beneath the Island, influenced by rheological contrast, tectonic stresses and gravitational loading, controlled the movement and eruption of magma. They do not carry out a geodetic data inversion for their interpretation, using speculative modeling based on the available data. They consider seismicity, deformation and petrological data indicate that a bath of basanitic magma coming from around a depth of 25 km was emplaced at 10–12 km below grade (discontinuity mantle/crust beneath El Hierro) creating a new reservoir where magma evolved until the initiation of the eruption. After about 2 months magma migrates laterally to the SE for nearly 20 km, always keeping the same depth and following a path controlled by stress barriers created by tectonic and rheological contrast in the upper lithosphere, ending with the submarine eruption.

González et al. (2013) complement the ground-based geodetic network with comprehensive processing, analysis and modeling of multiple space-based radar interferometric data sets [RADARSAT-2, ASAR-ENVISAT, and COSMO-SkyMed (Constellation of Small Satellites for Mediterranean Basin Observation)] in order to understand the dynamics of the magmatic system, with a temporal sampling of ~10 days during the eruption. Their results demonstrate applicability of radar observations to study off-shore eruptions, if occurring close to the coast. The data fully captures both the pre-eruptive and coeruptive phases. Elastic modeling of the ground deformation is employed to constrain the dynamics associated with the magmatic activity. This study represents the first geodetically constrained active magmatic plumbing system model for any of the Canary Islands volcanoes, and one of the few examples of submarine volcanic activity to date. Geodetic results reveal two spatially distinct shallow (crustal) magma reservoirs, a deeper central source (9.5 ±4.0 km) and a shallower magma reservoir at the flank of the southern rift (4.5 ±2.0 km) (see Fig. 15). The deeper source was recharged, explaining the relatively long basaltic eruption, contributing to observed island-wide uplift processes, and validating proposed active magma underplating. The shallowest source may be an incipient reservoir that facilitates fractional crystallization as observed on other Canary Islands. Data from this eruption supports a relationship between the depth of the shallow crustal magmatic systems and the long-term magma supply rate and oceanic lithospheric age. Such a relationship implies that a factor controlling the existence/depth of shallow (crustal) magmatic systems in oceanic island volcanoes is lithospheric thermomechanical behavior.

Gorbatikov et al. (2013) identified in the structural model for El Hierro an intrusive model at 15–25 km, suggesting it could be associated with the submarine 2011–2012 eruption.

García et al. (2014) study seismic data and displacements determined using GPS–GNSS observations from 2011 to 2013. They suggest that several magma displacement processes occurred at depth from the beginning of the unrest in July 2011. The first one culminated with the submarine eruption (October 2011). For this process they obtain via

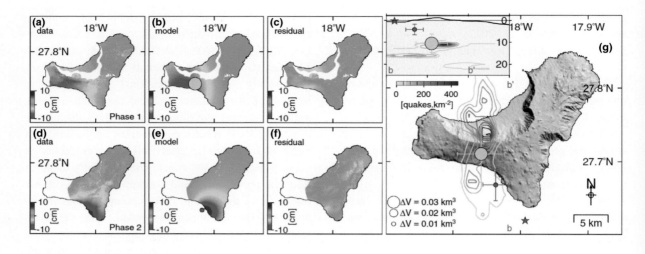

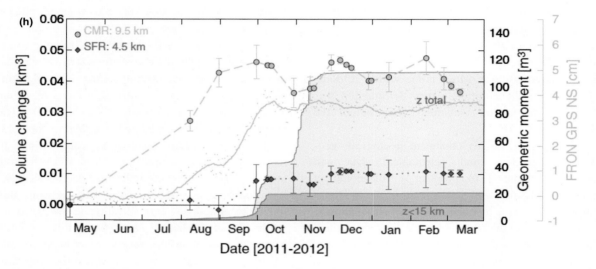

Figure 15

a Observed ground deformation (May 4, 2011 to August 8, 2011) using ascending RADARSAT-2 images, with a maximum motion toward the satellite of ∼9 cm; **b** Simulated ground deformation predicted by the best-fitting single spherical source model (see **g**) of **a**; **c** Residual of **a** and **b**; **d** Observed ground deformation (August 31, 2011 to October 30, 2011) using descending ENVISAT images from Track 109, with a maximum motion toward the satellite of ∼10–12 cm; **e** Simulated ground deformation predicted with the best-fitting single spherical source model (see **g**) of **d**; **f** Residual of **a** and **b**; **g** Location of the best-fitting spherical point sources: *orange* deep crustal source (**b**); and *dark red* the shallower crustal reservoir (**e**). Seismicity flux (events/km²), which represents the 2-D clustering of background seismicity, is shown in the background. *The Inset* shows the vertical cross section *b–b'*; **h** Time series of the volume change between May 2011 and March 2012 for the two spherical point sources (*orange circles* crust-mantle reservoir (CMR) deep source; *dark red diamonds* the SFR shallow rift reservoir) from the TSVD inversion. Cumulative seismic geometric movement, in cubic meters, is indicated as a *shaded gray* area for all events, and *shaded blue* for earthquakes shallower than 15 km. In addition, the time series of the NS component from the FRON GPS station is shown in *gray*. Applying a moderate temporal filter (*boxcar* with width 0.05 years), the time series shows minor oscillations after September 2011 which seems to correlate with the behavior of the CMR reservoir (∼9.5 km). However, these oscillations are well below the time series scatter and will require a more detailed analysis and additional GPS time series to extract robust conclusions (after GONZÁLEZ *et al*. 2013)

inversion results consistent with the InSAR data inversion results by GONZÁLEZ *et al*. (2013). They also invert deformation data for the main processes of magma injection between August 2011 and April 2013, showing a stepwise magma migration process controlled by the distribution of the maximum differential stress induced by each of the new emplaced magma pressure sources.

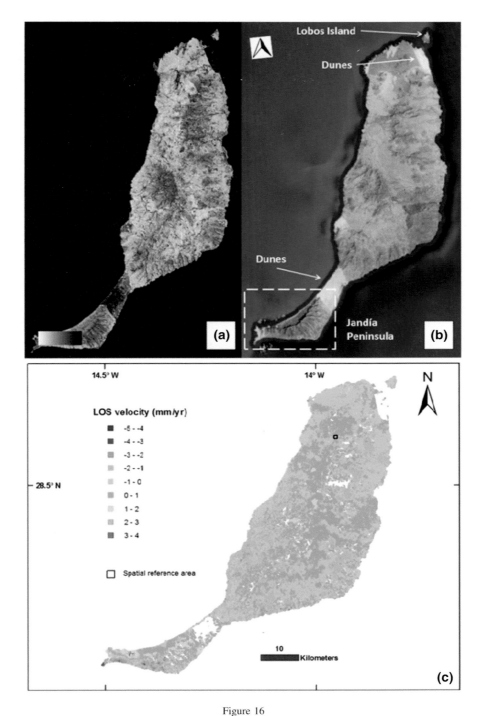

Figure 16

a Coherence map for interferogram 20061221_20070301. Figure is in radar adquisition coordinate system; **b** Google maps image of Fuerteventura. It can be noticed how the sand dune areas are completely decorrelated; **c** Geocoded mean LOS velocity map computed by stacking

Table 1

List of interferograms, with their spatial baselines, used in the Fuerteventura Island A-DInSAR study

List of interferograms	Spatial baselines (m)
20031023–20050505	111.7
20031023–20060629	143.6
20041111–20061116	−58.7
20050120–20060525	−394.5
20050120–20061221	233.0
20050120–20070301	301.8
20050505–20060629	31.9
20050505–20060803	326.9
20050922–20060420	−70.7
20050922–20060907	124.1
20050922–20061221	−68.8
20050922–20070301	0.0
20050922–20090827	−8.7
20060316–20060420	14.8
20060316–20061221	16.7
20060316–20091105	−5.1
20060420–20060907	194.8
20060420–20061221	1.9
20060420–20070301	70.7
20060420–20090827	62.0
20060420–20091105	−19.9
20060525–20061116	−342.0
20060525–20070125	81.4
20060525–20080424	253.5
20060525–20090129	143.0
20060629–20060907	−449.0
20060907–20061221	−192.9
20060907–20070301	−124.1
20060907–20090618	15.6
20060907–20090827	−132.8
20061012–20070125	−138.4
20061012–20080424	33.7
20061012–20090129	−76.8
20061221–20070301	68.8
20061221–20080424	−374.0
20061221–20090827	60.1
20061221–20091105	−21.8
20061221–20100114	−116.1
20070125–20090129	61.6
20070301–20090618	139.7
20070301–20090827	−8.7
20070301–20091105	−90.6
20080424–20090129	−110.5
20090827–20091105	−81.9
20091105–20100114	−94.3
20100114–20100218	−137.5

4.2.5 Fuerteventura Island

A set of 23 SAR descending orbit ENVISAT satellite radar images from the ESA archive, acquired in the period 2003–2010, were used for an A-DInSAR analysis. All scenes were co-registered to a common master geometry (October 12, 2006) and differential interferograms were computed using DORIS software (KAMPES *et al.* 2003). A 25-m resolution DEM from the IGN was used to remove the topographical contribution. A spatial multilooking factor of 4×20 (range $\times$ azimuth) was applied, producing pixel sizes of about 80×80 m in the ground surface. The average spatial coherence threshold used was 0.25 and, therefore, a total subset of 183,161 pixels was selected. Coherence maps were also computed for each interferogram, showing little temporal decorrelation on most of the island, except for the dune areas located in the NW and at the isthmus connecting the Jandía Peninsula (Fig. 16). A total of 46 interferograms (Table 1) out of the 78 possible combinations were selected to retrieve the linear velocity by stacking (average unwrapped interferograms). Linear trends of the orbital effects were removed on the azimuth and range directions. Unwrapping of Lobos Island was unreliable as there was no connection to the seed.

The linear velocity map of Fuerteventura for each pixel (in mm/year) is shown in Fig. 16. Linear velocity is around 1 mm/year for most of the island. Results at the Jandía Peninsula are different. A net horizontal limit can be noticed at the western part, which coincides with the Jandía Peninsula hill range (where the highest island altitude is reached). However, it is probably due to the bad quality of phase unwrapping in the area produced by the poor connections with the rest of the island (limited amount of pixels at the isthmus). Regardless, further investigation could concentrate on that area, using radar observation in another band (e.g., X-band) to study whether or not there are small current displacements. We can conclude that during the studied period 2003–2010 Fuerteventura Island was stable overall, which is consistent with the lack of recent volcanism of the area.

4.3. Deformation Modeling and Designing of the Geodetic Volcano Monitoring

Over the last 14 years constant deformation modelling work has been done to improve direct models considering different characteristics of the media (topography, structure, rheology,…). CHARCO *et al.* (2002) compute the effects on the geoid and vertical deflection produced by magmatic intrusions. It is important to estimate the change in the surface of reference for geodetic measurements. Using the elastic-gravitational model (RUNDLE 1982; FERNÁNDEZ and RUNDLE 1994) it is determined that big magmatic intrusions are necessary to produce non negligible effects on the geoid.

CHARCO *et al.* (2007) investigated the effects of topography on surface deformation and gravity changes caused by a magma intrusion in the Earth's crust. They develop a three-dimensional (3-D) indirect boundary element method (IBEM) that incorporates realistic topographic features and show that relevant topography alters both the magnitude and pattern of the deformation and gravity signal. As an example of realistic topography they consider a spherical source of dilatation located at 4 km depth below the Teide volcano summit in order to simulate deformation and gravity changes that could be observed at Tenerife if a hypothetical intrusion occurred in the volcanic system.

Their approach gives a picture of the 3-D topographic effect at Teide that can provide insight in order to improve the geodetic monitoring of the volcano. FERNÁNDEZ *et al.* (1999) and YU *et al.* (2000) performed a theoretical study of sensitivity to ground deformation and gravity changes to define the most suitable geodetic monitoring system. They considered a flat surface in their methodology, although they pointed out that the results were preliminary and needed corrections for topography. CHARCO *et al.* (2007c) employed the IBEM numerical technique in order to perform a theoretical study of sensitivity of the permanent ITER's GPS network installed in the vicinity of Teide volcano (GONZÁLEZ *et al.* 2005; PRIETO *et al.* 2005; FERNÁNDEZ *et al.* 2006). Their study assumes that the displacements and gravity changes are caused by the presence of a shallow magmatic system. They propose some improvements in order to discriminate between different geometries and processes of the magma system considering the related volcanic hazard. It is also shown that microgravity techniques would be a suitable method for monitoring the Teide stratovolcano. They suggest installation of a continuously recording gravimeter at some of the GPS locations, particularly in stations located on the volcano flanks where a shallow intrusion beneath the summit of the volcano could cause the maximum deformation and changes in gravity.

CHARCO and GALÁN DEL SASTRE (2014) used a finite element model combined with explorative inversion schemes. Their numerical methodology is applied by way of example to Tenerife Island. They study inversion results for synthetic data as observed in the 17 GPS–GNSS network stations installed in the Island by the IGN (http://www.ign.es), the Instituto Tecnológico y de Energías Renovables (ITER; http://www.iter.es) and Cartografía de Canarias (GRAFCAN; http://www.grafcan.es) using different crustal structures. Also, real topography is considered. The inversion needs around 24 h of computer code running to be obtained.

FERNÁNDEZ and LUZÓN (2002) complete the work done by FERNÁNDEZ *et al.* (1999) and YU *et al.* (2000), considering the last observational and theoretical results until that date and introducing the use of the new space geodetic techniques, permanent GPS observation and InSAR (see previous sections). They propose an updated geodetic monitoring system including those methodologies, gravity and GNSS networks basically for Tenerife, La Palma and Lanzarote, and a minimum version of it for the rest of the Islands. Obviously their conclusions and proposed geodetic volcano monitoring system for the archipelago should be revised and updated considering the recent volcanic activity in Tenerife and El Hierro and the evolution of all the involved techniques, deformation models and interpretation inversion techniques.

5. Discussion and Conclusions

We have reviewed most of the geodetic works carried out in the Canary Islands during the twentieth century and the beginning of the twenty first with a focus on volcanology and considering the technical limitations. The results obtained during the last decades point to some important issues regarding the potential and limitations of a geodetic volcano monitoring system. We can also summarize some perspectives for the future. The last two decades

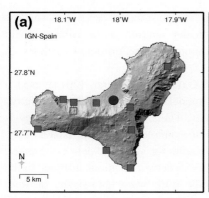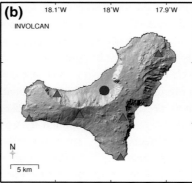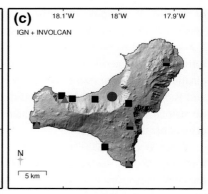

Figure 17
Fast response continuous GPS networks of the El Hierro eruption: **a** IGN; **b** INVOLCAN (ITER and Nagoya University); and, **c** IGN and INVOLCAN networks. Both networks use data provided by a publicly available GPS station, FRON (installed by GRAFCAN, *blue circle*). As we can see, none of the initiatives decided to install a station close to a publicly available one and meanwhile decided to install stations in close proximity to other locations, which do not contribute to maximize the overall monitoring capabilities

were marked by a clear advance in geodetic techniques (and monitoring possibilities) using satellite technology, continuous GPS–GNSS observation and DInSAR. The latter has evolved into the multitemporal DInSAR techniques and their use with error estimation (see e.g., GONZÁLEZ and FERNÁNDEZ 2011) together with the new generation of radar satellites help to improve precision and spatial resolution (JOYCE *et al.* 2010; SANSOSTI *et al.* 2014). Furthermore, multi-satellite and multi-band radar observation opens new options for minimizing revisit time and extracting much more information for volcano deformation studies (GONZÁLEZ *et al.* 2013), and elsewhere using MSBAS (SAMSONOV *et al.* 2014). We envision more frequent use of deformation data along with microgravity measurements as a powerful tool in volcano monitoring and for interpretation of results in the Canary Islands (GOTTSMANN *et al.* 2006; BATTAGLIA *et al.* 2008; CAMACHO *et al.* 2011c).

InSAR application has clearly shown the need for considering and properly correcting atmospheric delay, as has been clearly seen for the El Hierro study (CONG *et al.* 2012; GONZÁLEZ *et al.* 2013). Different methodologies can be used to do it: GNSS observations (CONG *et al.* 2012); mesoscale meteorological models (EFF-DARWICH *et al.* 2012; BEKAERT *et al.* 2013); or a strategy based on an empirical polynomial relationship between atmospheric phase noise and topography (GONZÁLEZ *et al.* 2013).

Corrections must be systematically applied on the multitemporal DInSAR application to islands with steep topography.

Another clear aspect that should be considered is data integration, mainly using GNSS and multitemporal DInSAR data (SAMSONOV *et al.* 2008; GONZÁLEZ *et al.* 2010a). Data integration should provide a better definition of the 3D deformation field, ensuring integrity and validation of the observed deformation, and eventually improving results precision.

Efficiently performing geodetic volcano monitoring involves knowing the background deformation in order to be able to clearly detect any significant anomaly in displacement or gravity change. This determination for the entire archipelago is being carried out in the framework of an ESA Category-1 project (FERNÁNDEZ 2012) for the period 1992–2010 using more than 1,500 radar images. The first results obtained for Fuerteventura Island have been described in this work, this being the first time this island has been studied to search for surface displacements.

Other important and basic tools for interpreting any detected anomaly are crustal structural models, deformation models and inversion techniques. In crustal models it is important to improve the knowledge of the 3D structure in a wider area, covering not only particular Islands, and it can be estimated by combining elevation, gravity, gradiometric, geoid, surface heat flow and seismic data in regional lithospheric models (e.g., FULLEA *et al.* 2013). In direct models and inversion techniques there are two main needs. First, to obtain a more advanced model capable of considering more realistic media and sources. In this case, new mathematical tools can help to develop more powerful analytical models (e.g., ARJONA *et al.* 2008). Numerical methodologies as those developed by CHARCO and GALÁN DEL SASTRE (2014) can also play an important role. Their use in between eruptions can be very useful, considering that if we know a priori information on the site, numerical modeling provides in a relatively short time a solution closer to reality. However, this kind of inversion methodology is not useful for near real-time inversion during volcanic crises due to the codes' running time (at least serveral hours). In these cases, analytical models can be the solution to invert geodetic deformation data obtained in near real-time (e.g., about 30 min, GARCÍA *et al.* 2014; or <1 min, CAMACHO *et al.* 2011c; CANNAVÒ *et al.* 2012), even considering the results obtained are often only first approximations of reality.

Since early 2000 no dedicated strategy has been discussed about the geodetic volcano monitoring for the Canary Islands (FERNÁNDEZ and LUZÓN 2002). Since then new research groups and public institutions have started different activities, shaping a complex landscape. A critical aspect to which more attention should be paid is the optimization of scarce resources. In particular, when geodetic networks are being deployed it must be acknowledged that multiple partners who are pursuing similar objectives could be involved so as to avoid overlapping infrastructure. Clear examples are GPS–GNSS networks on the islands of Tenerife (see above and previous sections) and El Hierro. During the El Hierro eruption crisis two rapid response GPS networks clearly expanded the capability to measure deformation on this island, but at the cost of revealing the worse inefficiencies in this complex scenario. As we can see in Fig. 17, at best a lack of communication ended in at least 10 GPS receivers being co-located at only 5 sites, which in the end did not contribute any new significant ground deformation information, while other parts of the island remained unmonitored, particularly the northeast and the far west. Clearly, the problems were not only logistical, as whenever possible the designers of two GPS networks avoided duplicating the resources

available to them, by placing their stations far away from a publicly available one (FRON station from the regional cartographic survey agency), located in an easy-to-access town, Frontera.

Situations like the one described for El Hierro Island reflect the need to keep updating the design of a modern and flexible geodetic volcano monitoring system that considers all the previously described results thereby tackling the challenges of maximizing resources. Rather than competing for the best sites, a smarter solution would be to spend some time defining joint data sharing policies that could satisfy all the parties involved while maximizing the scientific return of larger datasets to track activity, test hypotheses and models, etc. This will benefit not only fair competition among researchers but also foster the knowledge required for improving our understanding volcanism of the Canary Islands, and let the responsible organizations conduct improved monitoring. We propose a strategy similar to other successful consortiums where researchers and organizations merge resources to keep instrumentation pools updated and (cyber-) infrastructures that collectively maximize the strengths of a geodetic community. Examples to follow could be the geodesy consortium of UNAVCO in the USA (http://www.unavco.org), or the Supersites or Natural laboratories concepts sponsored by GEO (Group on Earth Observations) (http://supersites. earthobservations.org/). Moreover, a starting point would be an agreement to redistribute existing GPS stations and a fair access data policy for research purposes, involving at least the most relevant partners (CSIC, Grafcan, IGN, ITER-INVOLCAN, Universidad Complutense de Madrid, Universidad de Cádiz, University of Leeds, University of Nagoya, Universidad Politécnica de Madrid, etc).

Acknowledgments

This work was supported by the MINECO research project AYA2010-17448 and ESA CAT1 project 11021. We thank P. Suarez from Cartografía de Canarias SA for the information provided on the classical geodetic networks of the Canary Islands. We thank A. Manconi and one anonymous reviewer for their useful comments and suggestions to improve the manuscript. It is a contribution for the Moncloa Campus of International Excellence.

REFERENCES

ABLAY, G.J., and KEAREY, Ph., (2000), *Gravity constraints on the structure and volcanic evolution of Tenerife, Canary Islands,* J. Geophys. Res., *105*, 5783–5796.

ALMENDROS, J., IBAÑEZ, J. M., CARMONA, E., and ZANDOMENEGHI, D., (2007), *Array analyses of volcanic earthquakes and tremor recorded at Las Cañadas caldera (Tenerife Island, Spain) during the 2004 seismic activation of Teide volcano,* J. Volcanol. Geotherm. Res., *160*, 285–299.

ANCOCHEA, E., BRÄNDLE, J.L., HUERTAS, M.J., CUBAS, C.R., and HERNÁN, F. (2003), *The felsic dikes of La Gomera (Canary Islands): identification of cone sheet and radial dike swarms,* Journal of Volcanology and Geothermal Research, *120*(3–4), 197–206. doi:10.1016/S0377-0273(02)00384-0.

ANGUITA, F., GARCIA CACHO, L., COLOMBO, F., CAMACHO, A.G., and VIEIRA, R. (1991), *Roque Nublo caldera: a new stratocone caldera in Gran Canaria, Canary Islands,* J. Volcanol. Geotherm. Res., *47*, 45–63.

ANGUITA, F., and HERNAN, F. (2000), *The Canary Islands origin; a unifying model,* Journal of Volcanology and Geothermal Research, *103*, 1–26. doi:10.1016/S0377-0273(00)00195-5.

ARAÑA, V., CAMACHO, A. G., GARCÍA, A., MONTESINOS, F. G., BLANCO, I., VIEIRA, R. and FELPETO, A. (2000), *Internal structure of Tenerife (Canary Islands) based on gravity, aeromagnetic and volcanological data,* J. Volcanol. Geotherm. Research, *103*, 43–64.

ARAÑA, V. and FÚSTER, J. M. (1974), *La erupción del volcán Teneguía, La Palma, Islas Canarias [The eruption of the Teneguía volcano, La Palma, Canary Islands],* Estudios Geol., Vol. Teneguía, 15–18.

ARAÑA, V., and GÓMEZ, F. (1995), *Volcanic hazards and risks of Teide volcano (Tenerife, Canary Islands),* Per. Mineral., *64*, 23–24.

ARJONA, A., DÍAZ, J.I., FERNÁNDEZ, J. and RUNDLE, J.B. (2008), *On the Mathematical Analysis of an Elastic-gravitational Layered Earth Model for Magmatic Intrusion: The Stationary Case,* Pure appl. geophys., *165*, 1465–1490.

ARJONA, A., MONELLS, D., FERNÁNDEZ, J., DUQUE, S., and MALLORQUÍ, J. Deformation analysis employing the Coherent Pixel Technique and ENVISAT and ERS images in Canary Islands, In Proc. of 'Fringe 2009' (ed. Lacoste, H.) (ESRIN, Frascati, ESA, SP-677, March 2010) 8 pp. (CD).

ARNOSO, J., WEIXIN, C., VIEIRA, R., SHILING, T., and VELEZ, E. Tidal tilt and strain measurements in the Geodynamics Laboratory of Lanzarote. Proc. 13th Int. Symp. Earth Tides, Brussels, 1997. (ed. Ducarme, B. and Pâquet, P.) (Obs. Royal de Belgique, Série Géophysique, Bruxelles 1998) pp. 149–156.

ARNOSO J., FERNANDEZ J., VIEIRA R., and VAN RUYMBEKE M. (2000), *A preliminary discussion on tidal gravity anomalies and terrestrial heat flow in Lanzarote (Canary Islands),* Marees Terrestres Bull Inform, *132*, 10271–10282.

ARNOSO, J., FERNÁNDEZ, J. and VIEIRA, R. (2001a), *Interpretation of tidal gravity anomalies in Lanzarote, Canary Islands,* J. Geodyn., *31*, 341–354.

ARNOSO J., VIEIRA R., VELEZ E., WEIXIN C., SHILING T., JUN J., and VENEDIKOV A.P. (2001b), *Monitoring tidal and non-tidal tilt*

variations in Lanzarote island (Spain), J. Geod. Soc. Jpn., *47*, 456–462.

ARNOSO, J., VIEIRA, R., VELEZ, E., VAN RUYMBEKE, M., and VENEDIKOV, A.P., (2001c), *Studies of tides and instrumental performance of three gravimeters at Cueva de los Verdes (Lanzarote, Spain),* J. Geod. Soc. Jpn., *47*, 70–75.

ARNOSO, J., MONTESINOS, F.G., and BENAVENT, M. (2008), *Analysis of ground tilt measurements made in El Hierro (Canary Islands).* Geophysical Research Abstracts, *10*, EGU2008-A-01245, 2008, SRef-ID: 1607-7962/gra/EGU2008-A-01245.

ARNOSO, J., BENAVENT, M., BOS, M.S., MONTESINOS, F.G., and VIEIRA, R. (2011) *Verifying the body tide at the Canary Islands using tidal gravimetry observations.* J. Geodyn., *51*, 358–365. doi:10.1016/j.jog.2010.10.004.

ARNOSO, J., MONTESINOS, F. G., BENAVENT, M., and VÉLEZ, E. J. (2012) *The 2011 volcanic crisis at El Hierro (Canary Islands): monitoring ground deformation through tiltmeter and gravimetric observations,* Geophys. Res. Abstr., *14*, EGU2012-5373.

BALDI, P., and UNGUENDOLI, M., Geodetic networks for crustal movements studies, In Lecture Notes in Earth Sciences, 12. Applied Geodesy (ed. Turner, S.) (Springer, Berlin 1987) pp. 135–161.

BANDA, E., DAÑOBEITIA, J.J., SURIÑACH, E., and ANSORGE, J. (1981) *Features of crustal structure under the Canary Islands,* Earth Planet. Sci. Lett., *55*, 11–24.

BARBADILLO-FERNÁNDEZ, A., and QUIRÓS-DONATE, R. (1996) *Proyecto REGENTE. Una nueva red geodésica nacional [REGENTE Project. A new national geodetic network],* Física de la Tierra, *8*, 23–28.

BATTAGLIA, M., GOTTSMANN, J., CARBONE, D. and FERNÁNDEZ, J. (2008), *4D volcano gravimetry,* Geophysics, *73*, 6, WA3–WA18. doi:10.1190/1.2977792.

BEKAERT, D.P., HOOPER, A.J., WRIGHT, T.J. and WALTERS, R.J. (2013) *Robust corrections for topographically-correlated atmospheric noise in InSAR data from large deforming regions,* G23C-01. 2013 Fall Meeting, AGU, San Francisco, Calif., 9–13 Dec.

BERARDINO, P., G. FORNARO, R. LANARI, and E. SANSOSTI (2002), *A new algorithm for surface deformation monitoring based on small baseline differential SAR interferograms,* IEEE Trans. Geosci. Remote Sens., *40*, 2375–2383. doi:10.1109/TGRS.2002. 803792.

BERROCOSO, M., CARMONA, J., FERNÁNDEZ-ROS, A., PÉREZ-PEÑA, A., ORTÍZ, R., and GARCÍA, A. (2010), *Kinematic model for Tenerife Island (Canary Islands, Spain): Geodynamic interpretation in the Nubian plate context.* J. Afr. Earth Sci., *58*, 721–733. doi:10.1016/j.afrearsci.2010.04.005.

BETTI, B., BIAGI, L., CRESPI, H., and RIGUZZI, F. (1999), *GPS sensitivity analysis applied to non-permanent deformation control networks,* J. Geodesy *73*, 158–167.

BLANCO-SÁNCHEZ, P., MALLORQUÍ, J. J., DUQUE, S., and MONELLS, D. (2008), *The Coherent Pixels Technique (CPT): An Advanced DInSAR Technique for Nonlinear Deformation Monitoring,* Pure Applied Geophysics, *165*, 6, 1167–1193.

BOCK, Y. (1991), *Continuous monitoring of crustal deformation,* GPS World, 2, 40–47.

BOSSHARD, E., and MACFARLANE, D.J. (1970), *Crustal structure of the western Canary Islands from seismic refraction and gravity data.* J. Geophys. Res., *75*, 4901–4918.

BROOKS, B.A., FOSTER, J.H., BEVIS, M., FRAZER, L.N., WOLFE, C.J., and BEHN, M. (2006), *Periodic slow earthquakes on the flank of*

Kīlauea volcano, Hawai'i, Earth Planet. Sci. Lett. *246,* 207–216. doi:10.1016/j.epsl.2006.03.035.

BÜRGMANN, R., ROSEN, P.A., and FIELDING, E.J. (2000), *Synthetic aperture radar interferometry to measure Earth's surface topography and its deformation,* Annu. Rev. Earth Planet. Sci., *28,* 169–209.

CAMACHO, A.G., TORO, C., and FERNANDEZ, J. Cálculo de la corrección topográfica a las observaciones gravimétricas en la caldera del Teide obtenidas a partir del modelo topográfico digital de la isla de Tenerife [Computation ot the topographic correction for the gravimetric observations in the Teide caldera obtained from the digital topographic model of the Island of Tenerife] (Publ. Semin. Astron. y Geodesia, 162, Madrid, Spain, 1988).

CAMACHO, A.G., VIEIRA, R., and TORO, C. (1991) *Microgravimetric model of the Las Cañadas caldera (Tenerife),* J. Volcanol. Geotherm. Res., *47,* 75–80.

CAMACHO, A.G., MONTESINOS, F.G., and VIEIRA, R. (2000), *A 3-D gravity inversion by means of growing bodies,* Geophysics, *65,* 95–101.

CAMACHO, A.G., MONTESINOS, F.G., VIEIRA, R., and ARNOSO, J. (2001), *Modellig of crustal anomalies for Lanzarote (Canary Islands) in light of gravity,* Geophys. J. Int., *47,* 403–414.

CAMACHO, A.G., FERNÁNDEZ, J., GONZÁLEZ, P.J., RUNDLE, J.B., PRIETO, J.F., and ARJONA, A. (2009), *Structural results for La Palma Island using 3D gravity inversion.* J. Geophys. Res., *114,* B05411. doi:10.1029/2008JB005628.

CAMACHO, A.G., FERNÁNDEZ, J., and GOTTSMANN, J. (2011a), *A new gravity inversion method for multiple subhorizontal discontinuity interfaces and shallow basins.* J. Geophys. Res., *116,* B02413. doi:10.1029/2010JB008023.

CAMACHO, A.G., FERNÁNDEZ, J., and GOTTSMANN, J. (2011b), *The 3-D gravity inversion package GROWTH2.0 and its application to Tenerife Island, Spain,* Comput. Geosci., *37,* 621–633. doi:10.1016/j.cageo.2010.12.003.

CAMACHO, A.G., GONZÁLEZ, P.J., FERNÁNDEZ, J., and BERRINO, G., (2011c), *Simultaneous inversion of surface deformation and gravity changes by means of extended bodies with a free geometry: Application to deforming calderas,* J. Geophys. Res., *116,* B10401. doi:10.1029/2010JB008165.

CANALES, J.P., and DAÑOBEITIA, J.J. (1998), *The Canary Islands swell: a coherence analysis of bathymetry and gravity,* Geophys. J. Int., *132,* 479–488.

CANNAVÒ, F., CAMACHO, A.G., SCANDURA, D., ALOISI, M., BRUNO, V., GONZÁLEZ, P.J., PALANO, M., MATTIA, M., and FERNÁNDEZ, J. (2012), *A time evolving distributed model for the ground deformation associated to the May 13th 2008 eruption on Mt. Etna (Italy),* 2012 Fall Meeting, AGU, San Francisco, Calif., 3–7 Dec.

CARRACEDO, J.C., PÉREZ-TORRADO, F.J., PARIS, R., and RODRÍGUEZ BADIOLA, E. (2009), *Megadeslizamientos en las Islas Canarias,* Enseñanza de las Ciencias de la Tierra, *17,* 44–56.

CARRACEDO, J. C., PÉREZ TORRADO, F., RODRÍGUEZ GONZÁLEZ, A., SOLER, V., FERNÁNDEZ TURIEL, J. L., TROLL, V. R., and WIESMAIER, S. (2012), *The 2011 submarine volcanic eruption in El Hierro (Canary Islands),* Geol. Today, *28,* 53–58. doi:10.1111/j.1365-2451.2012.00827.x.

CARRASCO, D., FERNANDEZ, J., ROMERO, R., MARTINEZ, A., MORENO, V., and ARAÑA, V., Operational volcano monitoring for decision support demonstration. In Proceedings of the Fringe '99 Workshop Advancing ERS SAR Interferometry from Applications towards Operations, November 1999. (ed. European Space

Agency) (ESA Publications Division, SP-478, Liege, Belgium, 2000a).

CARRASCO, D., FERNÁNDEZ, J., ROMERO, R., ARAÑA, V., MARTÍNEZ, V., MORENO, V., APARICIO, A., and PAGANINI, M.. First results from operational volcano monitoring in the Canary Islands. In ESA ERS-ENVISAT Symposium "Looking down to Earth in the New Millennium" (ed. European Space Agency) (Chalmers University of Technology, Gothenburg, Sweden, 2000).

CASU, F., MANZO, M., and LANARI, R. (2006), *A quantitative assessment of the SBAS algorithm performance for surface deformation retrieval from DInSAR data,* Remote Sens. Environ., *102,* 3–4, 195–210. doi:10.1016/j.rse.2006.01.023.

CATURLA, J. L.,Compendio de los Sistemas Geodésicos de España. Publicación técnica núm. 1/1978 [Compendium of Geodetic Reference Systems in Spain] (Instituto Geográfico Nacional, Madrid, Spain, 1978) 30 pp.

CATURLA, J. L., REGCAN95, Nueva Red Geodésica de las Islas Canarias [REGCAN95, New Geodetic Network for the Canary Islands] (Instituto Geográfico Nacional, Área de Geodesia. Internal Report, Madrid, Spain, 1996) 45 pp.

CATURLA, J. L., and PRIETO, J., Utilización del Sistema GPS en Proyectos Cartográficos [Using the GPS system in Carthographic projects] (Instituto Geográfico Nacional, Las Palmas de Gran Canaria, Spain, 1996) 376 pp.

CGMW, Commission for the Geological Map of the World. (2010), Tectonic Map of Africa [Map], retrieved from http://ccgm.free.fr/Africa_Tecto_GB.html.

CHARCO, M., FERNÁNDEZ, J., SEVILLA, M.J., and RUNDLE, J.B. (2002), *Modeling magmatic intrusion's effects on the geoid and vertical deflection. Application to Lanzarote, Canary Islands, and Long Valley Caldera, California,* Física de la Tierra, *14,* 11–31.

CHARCO, M., LUZÓN, F., FERNÁNDEZ, J., TIAMPO, K.F., and SÁNCHEZ-SESMA, F.J. (2007), *Three-dimensional indirect boundary element method for deformation and gravity changes in volcanic areas. Application to Teide volcano (Tenerife, Canary Islands).* Journal of Geophysical Research, *112,* B08409. doi:10.1029/2006JB004740.

CHARCO, M., and GALÁN DEL SASTRE, P. (2014), *Efficient inversión of three-dimensional finite element models of volcano deformation.* Geophys. J. Int., *196,* 3, 1441–1454. doi:10.1093/gji/ggt490.

CHÁVEZ-GARCÍA, F.J., LUZÓN, F., RAPTAKIS, D., and FERNÁNDEZ, J., (2007), *Shear-wave velocity structure around Teide volcano: results using microtremors with the SPAC method and implications for interpretation of geodetic results.* Pure appl. geophys., *164,* 697–720.

COLLIER, J. S., and WATTS, A. B. (2001), *Lithospheric response to volcanic loading by the Canary Islands: Constraints from seismic reflection data in their flexural moats,* Geophys. J. Int., *147,* 660–676.

CONG, X., EINEDER, M., and FRITZ, T. Atmospheric delay compensation in differential SAR Interferometry for volcanic deformation monitoring-Study case: El Hierro, In Proc. 2012 IEEE International Geoscience and Remote Sensing Symposium (IGARSS), 3887–3890 (ed: Geoscience and Remote Sensing Society) (Institute of Electrical and Electronics Engineers, Munich, Germany, 2012).

CROSSLEY, D., HINDERER, J., and RICCARDI, U. (2013), *The measurement of surface gravity,* Rep. Prog. Phys., *76,* 046101, 47. doi:10.1088/0034-4885/76/4/046101.

DASH, B.P., and BOSSHARD, E. (1969), *Seismic and gravity investigations around the western Canary Islands,* Earth Planet. Sc. Lett., *7,* 169–177.

DAY, S., CARRACEDO, J.C., GUILLOU, H., and GRAVESTOCK, P. (1999), *Recent structural evolution of the Cumbre Vieja Volcano, La Palma, Canary Islands: volcanic rift zone reconfiguration as a precursor to volcanic flank instability?* J. Volcanol. Geotherm. Res. *94*, 1–4, 135–167.

DEL REY, R., ORTÍZ, R., and FERNÁNDEZ, J. (1994), *Sismicidad en Timanfaya. Primeros estudios de su relación con la marea [Seismicity on Timanfaya. Firsts studies of its relationship with tides].* Serie Casa Volcanes, Cabildo Insular Lanzarote, *3*, 87–100.

DIÉZ-GIL, J.L., FERNÁNDEZ, J., VIEIRA, R., and RUNDLE, J.B. (1994), *Modelos de deformación para el diseño de la vigilancia geodésica de actividad volcánica en Lanzarote [Deformation models for the design of geodetic monitoring of volcanic activity in Lanzarote].* Serie Casa Volcanes, Cabildo Insular Lanzarote, *3*, 61–78.

DIXON, T.H., MAO, A., BURSIK, M., HEFLIN, M., LANGBEIN, J., STEIN, R., and WEBB, F. (1997), *Continuous monitoring of surface deformation at Long Valley caldera, California, with GPS,* J. Geophys. Res., *102*, 17–34.

DONG, D., and BOCK, Y, (1989), *Global Position System network analysis with phase ambiguity resolution applied to crustal deformation studies in California,* J. Geophys. Res., *94*, 3949–3966.

DVORAK, J.J., and DZURISIN, D. (1997), *Volcano geodesy; the search for magma reservoirs and the formation of eruptive event,* Rev. Geophys. *35*, 343–384.

DZURISIN, D., (2007) *Volcano deformation: Geodetic monitoring techniques.* (Springer-Praxis Books in Geophysical Sciences. Praxis Publishing Ltd., Chichester, UK., 2007).

EFF-DARWICH, A., GRASSIN, O., and FERNÁNDEZ, J. (2008), *An upper limit to ground deformation in the Island of Tenerife, Canary Islands, for the period 1997–2006,* Pure and Applied Geophysics, *165*, 6, 1049–1070. doi:10.1007/s00024-008-0346-7.

EFF-DARWICH, A., PÉREZ-DARIAS, J.C., FERNÁNDEZ, J., GARCÍA-LORENZO, B., GONZÁLEZ-FERNÁNDEZ, A., and GONZÁLEZ, P.J. (2012), *Using a Mesoscaale metheorological model to reduce the effect of tropospheric water vapour from DInSAR data: A case study for the Island of Tenerife, Canary Islands,* Pure appl. Geophys., *169*, 1425–1441. doi:10.1007/s00024-011-0401-4.

FERNÁNDEZ, J. (1991), *Investigations on volcanic risk in Lanzarote. I- Model for gravity variations and deformations originated by a magmatic intrusion in the crust.* Comptes Rendus, JLG. Conseil de L'Europe, *72*, 49–55.

FERNÁNDEZ, J., and VIEIRA, R. (1991), *Investigations on volcanic risk in Lanzarote. II- Some observational methods,* Comptes Rendus, JLG. Conseil de L'Europe, *72*, 56–59.

FERNÁNDEZ, J., VIEIRA, R., DÍEZ, J.L., and TORO, C. (1991), *Study about the relationship among crustal thickness, heat flow and gravimetric tide in the island of Lanzarote,* Comptes Rendus, JLG. Conseil de L'Europe, *72*, 17–22.

FERNÁNDEZ, J., VIEIRA, R., DIEZ, J.L., and TORO, C. (1992), *Investigations on crustal thickness, heat flow and gravity tide relationship in Lanzarote island.* Phys Earth Planet Interiors, *74*,199–208.

FERNANDEZ, J., Técnicas geodésicas y geodinámicas aplicadas a la investigación en riesgo volcánico en la isla de Lanzarote [Geodetic and geodynamic techniques applied to research on volcanic risk in Lanzarote] (Ph. D. Thesis,Universidad Complutense de Madrid, Madrid, Spain, 1993) 149 pp.

FERNÁNDEZ, J., ARNOSO, J., and VIEIRA, R. (1993), *Investigación en riesgo volcánico en Lanzarote [Volcanic risk research in Lanzarote],* Revista de la Real Academia de Ciencias Exactas, Físicas y Naturales de Madrid, Tomo LXXXVIII, Cuad. 2° y 3°, 479–484.

FERNÁNDEZ, J., and RUNDLE, J.B. (1994), *Gravity Changes and deformation due to a magmatic intrusion in a two-layered crustal model,* J. Geophys. Res. *99*, 2737–2746.

FERNÁNDEZ, J., VIEIRA, R., VENEDIKOV, A.P., and DÍEZ, J.L. (1994), *Vigilancia de riesgo volcánico en Canarias. Isla de Lanzarote [Monitoring volcanic risk in the Canary Islands. Lanzarote Island],* Física de la Tierra, *5*, 77–88.

FERNÁNDEZ, J., and DIEZ, J.L. (1995), *Volcano monitoring design in Canary Islands by deformation model,* Cahiers Centre Eur. Geodynam. Seismol., *8*, 207–217.

FERNÁNDEZ, J., CARRASCO, J. M., RUNDLE, J. B. and ARAÑA, V. (1999), *Geodetic methods for detecting volcanic unrest: A theoretical approach,* Bull. Volcanol., *60*, 534–544.

FERNÁNDEZ, J., and LUZÓN, F. (2002), *Geodetic monitoring in Canary Islands. Present and new perspectives,* Física de la Tierra, *14*, 109–126.

FERNÁNDEZ, J., ROMERO, R., CARRASCO, D., LUZÓN, F. and ARAÑA, V. (2002), *InSAR Volcano and Seismic Monitoring in Spain. Results for the Period 1992–2000 and Possible Interpretations,* Opt. Laser Eng., *37*, 285–297.

FERNÁNDEZ, J., YU, T. T., RODRIGUEZ-VELASCO, G., GONZÁLEZ-MATESANZ, J., ROMERO, R., RODRÍGUEZ, G., QUIRÓS, R., DALDA, A., APARICIO, A. and BLANCO, M. J. (2003), *New geodetic monitoring system in the volcanic island of Tenerife, Canaries, Spain. Combination of InSAR and GPS techniques,* J. Volcanol. Geotherm. Res., *124*, 241–253.

FERNÁNDEZ, J., GONZÁLEZ-MATESANZ, F. J., PRIETO, J. F., RODRÍGUEZ-VELASCO, G., STALLER, A., ALONSO-MEDINA, A. and CHARCO, M. (2004), *GPS Monitoring in the N-W Part of the Volcanic Island of Tenerife, Canaries, Spain: Strategy and Results,* Pure appl. Geophys., *161*, 1359–1377.

FERNÁNDEZ, J., ROMERO, R., CARRASCO, D., TIAMPO, K. F., RODRÍGUEZ-VELASCO, G., APARICIO, A., ARAÑA, V. and GONZÁLEZ-MATESANZ, F. J. (2005), *Detection of displacements on Tenerife Island, Canaries, using radar interferometry,* Geophys. J. Int., *160*, 33–45.

FERNÁNDEZ J., CAMACHO, A.G., PRIETO, J.F., GONZÁLEZ, P., RODRÍGUEZ-VELASCO, G., TUNINI, L., WILLERT, V., CALVO, D., SAGIYA, T., FUJII, N., CHARCO, M., HERNÁNDEZ, P.A., PÉREZ, N.M., MALLORQUÍ, J., GONZÁLEZ-MATESANZ, J., VALDÉS, M., LÓPEZ.TURBAY, A.A., and CARRASCO, D. (2006), *Geodetic Observation of Tenerife volcanic unrest. Results and interpretation.* GARAVOLCAN 2006, 300th Anniversary Volcano Conference Commemorating the 1706 Arenas Negras Eruption, Garachico, Tenerife, Canary Islands, 22–27 May, 2006.

FERNÁNDEZ, J., CAMACHO, A.G., GONZÁLEZ, P.J., SAMSONOV, S., PRIETO, J.F., TIAMPO, K.F., GOTTSMANN, J., PUGLISI, G., GUGLIELMINO, J., MALLORQUÍ, J.J., TUNINI, L., WILLERT, V., RODRÍGUEZ-VELASCO, G., CHARCO, M., NAVARRETE, D., DUQUE, S., CARRASCO, D., BLANCO-SÁNCHEZ, and P (2007), Tenerife island (Canaries, Spain) unrest, 2004–2006, studied via integrated geodetic observations. In Proceedings of the 2007 International Geohazards Week, 5–9 November 2007,. (ed. ESA-ESRIN) (ESA Publications Division, Frascati, Italy, 2007).

FERNÁNDEZ, J., SAMSONOV, S., CAMACHO, A.G., GONZÁLEZ, P.J., PRIETO, J.F., TIAMPO, K.F., RODRIGUEZ-VELASCO, G., TUNINI, L., WILLERT, V., CHARCO, M., MALLORQUÍ, J.J., and CARRASCO, D. (2008), *Integration of two line-of-sights classical DInSAR and*

GPS data to study the 2004–2006 Tenerife volcanic unrest, Geophys. Res. Abstracts, 10, EGU2008-A-10611.

FERNÁNDEZ, J., TIZZANI, P., MANZO, M., BORGIA, A., GONZÁLEZ, P.J., MARTÍ, J., PEPE, A., CAMACHO, A.G., CASU, F., BERARDINO, P., PRIETO, J.F. and LANARI, R. (2009), Gravity-driven deformation of Tenerife measured by InSAR time series analysis Geophys. Res. Lett., 36, L04306. doi:10.1029/2008GL036920.

FERNÁNDEZ, J. [PI], (2012), Determination of the 1992–2010 deformation field in the Canary Islands by means of ERS-1/2, ENVISAT and ALOS radar images: Implications in the volcano monitoring system definition. (Cat.-1 11021) European Space Agency (ESA) (04/2012-03/2015).

FILMER, P.E., McNUTT, and M.K. (1989), Geoid anomalies over the Canary Islands group. Mar. Geophys. Res., 11, 77–87.

FOLGER, D.W., McCULLOUGH, J.R., IRWIN, B.J., and DODD, J.E. (1990), Map showing free-air gravity anomalies around the Canary Islands, Spain [Map]. U.S. Geological Survey Miscellaneous Field Studies Map MF-2098-B, scale 1:75,000.

FULLEA, J., CAMACHO, A.G., and FERNÁNDEZ, J. (2013), 3D Coupled geophysical-petrological modelling of the Canary Islands and North-Western african margin lithosphere. 15th Annual Conference of the Int. Ass. Mathematical Geosciences, Madrid, Septiembre 2–6, 2013.

GARCÍA, A., FERNÁNDEZ-ROS, A., BERROCOSO, M., MARRERO, J.M., PRATES, G., DE LA CRUZ-REYNA, S., and ORTIZ, R. (2014), Magma displacements under insular volcanic fields, applications to eruption forecasting: El Hierro, Canary Islands, 2011–2013, Geophys. J. Int., 197, 1, 322–334. doi:10.1093/gji/ggt505.

GARCÍA-CAÑADA, L., and SEVILLA, M.J., Monitoring crustal movements and sea level in Lanzarote. In Geodetic deformation monitoring: from geophysical to engineering roles (Ed. Sansò, F., and Gil, A.J.) (International Association of Geodesy Symposia, 131, Springer-Verlag, Berlin, Germany, 2006) pp. 160–165.

GELDMACHER, J., HOERNLE, K., VAN DEN BOGAARD, P., DUGGEN, S., WERNER, R. (2005), New 40Ar/39Ar age and geochemical data from seamounts in the Canary and Madeira volcanic provinces: support for the mantle plume hypothesis, Earth Planet Sci Lett 237, 85–101.

GIL MONTANER, F, (1929a). Trabajos geodésicos de primer orden en Canarias y Marruecos [First order geodetic campaigns in the Canaries and Morocco]. Boletín de la Real Sociedad Geográfica, LXIX, 141–156.

GIL MONTANER, F., (1929b) Últimos trabajos geodésicos en Canarias y Marruecos [Latest geodetic campaigns in the Canaries and Morocco] (Instituto Geográfico y Catastral, Madrid, Spain, 1929b) 23 pp.

GONZÁLEZ, P.J., PRIETO, J.F., FERNÁNDEZ, J., SAGIYA, T., FUJII, N., HERNÁNDEZ, P.A., and PÉREZ, N.M. (2005), Permanent GPS observation in Tenerife Island for volcano monitoring. Results obtained from May 2004 to present. Geophysical Research Abstracts, 7, 09545.

GONZÁLEZ, P.J., SAMSONOV, S., MANZO, M., PRIETO, J.F., TIAMPO, K.F., TIZZANI, P., CASU, F., PEPE, A., BERARDINO, P., CAMACHO, A.G., LANARI, R., and FERNÁNDEZ, J. (2010a), 3D volcanic deformation fields at Tenerife Island: integration of GPS and Time Series of DInSAR (SBAS), Cahiers Centre Eur. Geodynam. Seismol., 29, 44–50. ISBN: 978-2-91989-708-7.

GONZÁLEZ, P. J., TIAMPO, K. F., CAMACHO, A. G. and FERNÁNDEZ, J. (2010b), Shallow flank deformation at Cumbre Vieja volcano (Canary Island): Implications on the stability of steepsided volcano flanks at oceanic islands, Earth Planet. Sc. Lett., 297, 545–557.

GONZÁLEZ, P. J. and FERNÁNDEZ, J. (2011), Error estimation in multitemporal InSAR deformation time series, with application to Lanzarote, Canary Islands. J. Geophys. Res., 116, B10404.

GONZÁLEZ, P. J.,. SAMSONOV, S. V, PEPE, S., TIAMPO, K. F., TIZZANI, P., CASU, F., FERNÁNDEZ, J., CAMACHO, A. G., and SANSOSTI, E. (2013), Magma storage and migration associated with the 2011–2012 El Hierro eruption: Implications for crustal magmatic systems at oceanic island volcanoes, J. Geophys. Res. Solid Earth, 118, 4361–4377. doi:10.1002/jgrb.50289.

GORBATIKOV, A.V., KALININA, A.V., VOLKOV, V.A., ARNOSO, J., VIEIRA, R., and VÉLEZ, E. (2004), Results of analysis of the data of microseismic survey at Lanzarote Island, Canary, Spain, Pure appl. geophys., 161, 1561–1578.

GORBATIKOV, A.V., MONTESINOS, F.G., ARNOSO, J., STEPANICA, M.Y., BENAVENT, M., and TSUKANOV, A.A. (2013), New features in the subsurface structure model of El Hierro Island (Canaries) from low-frecuency microseismic sounding: an insight into the 2011 seismo-volcanic crisis, Surv. Geophys., 34, 463–489. doi:10.1007/s10712-013-9240-4.

GOTTSMANN, J., WOOLLER, M., MARTÍ, J., FERNÁNDEZ, J., CAMACHO, A., GONZÁLEZ, P., GARCÍA, A., and RYMER, H. (2006), New evidence for the reawakening of Teide volcano, Geophys. Res. Lett., 33, L20311. doi:10.1029/2006GL027523.

GOTTSMANN, J., CAMACHO, A.G., MARTÍ, J., WOOLLER, L., FERNÁNDEZ, J., GARCÍA, A. and RYMER, H. (2008), Shallow structure beneath the Central Volcanic Complex of Tenerife from new gravity data: Implications for its evolution and recent reactivation. Phys. Earth Planet. Int., 168, 212–230.

HANSSEN, R. F., (2001) Radar Interferometry. Data interpretation and error analysis, (Kluwer Academic publ., Dordrecht, The Netherlands, 2001) 308 pp.

HARRIS, A.J.L., BUTTERWORTH, A.L., CARLTON, R.W., DOWNEY, I., MILLER, P., NAVARRO, P., and ROTHERY, D.A. (1997), Low-cost volcano surveillance from space: case studies from Etna, Krafla, Cerro Negro, Fogo, Lascar and Erebus, Bull. Volcanol., 59, 49–64.

HAYFORD, J. F., Supplementary Investigation in 1909 of the Figure of the Earth and Isostasy, (Separate publications, U.S. Coast and Geodetic Survey, U.S. Government Printing Office, Washington, D.C., 1910) 80 pp.

HERNÁNDEZ-PACHECO, A., and VALS, M. C. (1982), The historical eruptions of La Palma Island (Canarias), Arquipelago. Rev. Univ. Azores, Ser. C. Nat., 3, 83–94.

HOERNLE., K, and CARRACEDO, J.C., Canary Islands geology, In Encyclopedia of Islands (ed. Gillespie, R. D., and Clague, D. A.) (University of California Press, Berkeley, California, USA, 2009), pp. 133–143.

HUGENTOBLER, U., SCHAER, S., and FRIDEZ, P. [Eds.], Bernese GPS software version 4.2 (Astronomical Institute, University of Berne, 2001) 515 pp.

IBÁÑEZ, J.M., DE ANGELIS, S., DÍAZ-MORENO, A., HERNÁNDEZ, P., ALGUACIL, G., POSADAS, A., and PÉREZ, N. (2012), Insights into the 2011–2012 submarine eruption off the coast of El Hierro (Canary Islands, Spain) from statical analysis of earthquake activity. Geophys. J. Int., 191, 2, 659–670. doi:10.1111/j.1365-246X.2012.05629.x.

IGC, Instituto Geográfico y Catastral., Espagne. Rapports sur les travaux géodésiques, astronomiques et gravimétriques exècutes par l'Institut Géographique de 1933 a 1936 présentés à la VI Assemblée Genèrale de l'Association de Géodésie de l'Union Géodésique et Géopysique Internationale (Ateliers de l'Institut Géographique et Cadastral, Madrid, Spain, 1938), 33 pp.

IGN (2006), Boletín de sísmos próximos [Data archive], retrieved from http://www.geo.ign.es.

JOHNSON, H.O., and WYATT, F.K. (1994), *Geodetic network design for fault-mechanics studies,* Manuscripta Geodaetica *19,* 309–323.

JOYCE, K., WRIGHT, K., AMBROSIA, V., and SAMSONOV, S. (2010), *Incorporating remote sensing into emergency management,* The Australian Journal of Emergency Management, *25,* 4, 14–23.

JUNG, W.Y., and RABINOWITZ, P.D. (1986), *Residual geoid anomalies of the north Atlantic ocean and their tectonic implications.* J. Geophys. Res., *91,* 10383–10396.

KALININA, A.V., VOLKOV, V.A., GORBATIKOV, A.V., ARNOSO, J., VIEIRA, R., and BENAVENT, M. (2004), *Tilt observations in the nomal mode frecuency band at the geodynamic observatory Cueva de los Verdes, Lanzarote,* Pure appl. geophys., *161,* 1597–1611. doi:10.1007/s00024-004-2423-4.

KRÖCHERT, J., MAURER, H., and BUCHNER, E. (2008), *Fossil beaches as evidence for significant uplift of Tenerife, Canary Islands,* Journal of African Earth Sciences, 51, *4,* 220–234. doi:10.1016/j.jafrearsci.2008.01.005.

LLANES, M.P., Estructura de la litosfera en el entorno de las Islas Canarias a partir del análisis gravimétrico e isostático: implicaciones geodinámicas [Lithosphere structure on the Canary Islands from gravimetric and isostatic analysis: geodynamic implications] (Ph.D. Thesis., Univ. Complutense de Madrid, Madrid, Spain, 2006).

LÓPEZ, C., BLANCO, M. J., ABELLA, R., BRENES, B., CABRERA RODRÍGUEZ, V. M., CASAS, B., DOMÍNGUEZ CERDEÑA, I., FELPETO, A., FERNÁNDEZ DE VILLALTA, M., DEL FRESNO, C., GARCÍA, O., GARCÍA-ARIAS, M. J., GARCÍA-CAÑADA, L., GOMIS MORENO, A., GONZÁLEZ-ALONSO, E., GUZMÁN PÉREZ, J., IRIBARREN, I., LÓPEZ-DÍAZ, R., LUENGO-OROZ, N., MELETLIDIS, S., MORENO, M., MOURE, D., PEREDA DE PABLO, J., RODERO, C., ROMERO, E., SAINZ-MAZA, S., SENTRE DOMINGO, M. A., TORRES, P. A., TRIGO, P., and VILLASANTE-MARCOS, V. (2012), *Monitoring the volcanic unrest of El Hierro (Canary Islands) before the onset of the 2011–2012 submarine eruption,* Geophys. Res. Lett., *39,* L13303. doi:10.1029/2012GL051846.

MACFARLANE, D.J., and RIDLEY, W.I. (1968), *An interpretation of gravity data for Tenerife, Canary Islands,* Earth Planet. Sc. Lett., *4,* 481–486.

MACFARLANE, D.J., and RIDLEY, W.I. (1969), *An interpretation of gravity data for Lanzarote, Canary Islands,* Earth Planet. Sc. Lett., *6,* 431–436.

MALLORQUÍ, J.J., MORA, O., BLANCO, P,.and BROQUETAS, A., Linear and non-linear long-term terrain deformation with DInSAR (CPT: Coherent Pixels Tehcnique), In Proc. of FRINGE 2003 Workshop(ed. European Space Agency) (ESA Publications Division, 2003) pp. 1–8.

MARTÍ, J., PINEL, V., LÓPEZ, C., GEYER, A., ABELLA, R., TÁRRAGA, M., BLANCO, M.J., CASTRO, A., and RODRÍGUEZ, C. (2013), *Causes and mechanisms of the 2011–2012 El Hierro (Canary Islands) submarine eruption,* J. Geophys. Res. Solid Earth, *118,* 8123–839. doi:10.1002/jgrb.50087.

MASSONNET, D., and FEIGL, K. (1998), *Radar interferometry and its application to changes in the Earth's surface,* Rev. Geophysics, *36,* 441–500.

MEZCUA, J., BUFORN, E., UDÍAS, A. and RUEDA, J. (1992), *Seismotectonics of the Canary Islands,* Tectonophysics, *208,* 4, 447–452.

MINSHULL, T. A., and CHARVIS, P. (2001), *Ocean island densities and models of lithospheric flexure,* Geophys. J. Inter., *145,* 731–739. doi:10.1046/j.0956-540x.2001.01422.x.

Montesinos, F.G., 1999. Inversión gravimétrica 3D por técnicas de evolución. Aplicación a la isla de Fuerteventura [3D gravimetric inversion by evolution techniques. Application to the Island of Tenerife] (Ph.D. Thesis, Complutense University of Madrid, Spain, 1999).

MONTESINOS, F.G., ARNOJO, J., and VIEIRA, R. (2005), *Using a genetic algorithm for 3-D inversión of gravity data in Fuerteventura (Canary Islands),* Int. J. Earth Sci., *94,* 301–316.

MONTESINOS, F.G., ARNOSO, J., BENAVENT, M., and VIEIRA, R. (2006), *The crustal structure of El Hierro (Canary Islands) from 3-D gravity inversion,* J. Volcanol. Geotherm. Res., *150,* 283–299.

MONTESINOS, F.G., ARNOSO, J., VIEIRA, R., and BENAVENT, M. (2011), *Subsurface geometry and structural evolution of La Gomera Island based on gravity data,* J. Volcanol. Geotherm. Res., *199,* 105–117. doi:10.10161/j.volgeores.2010.10.007.

MORGAN, W. J. (1971), *Convection plumes in the lower mantle,* Nature, *230,* 5288, 42–43. doi:10.1038/230042a0.

MOSS, J. L., McGUIRE, W. J. and PAGE, D. (1999), *Ground deformation monitoring of a potential landslide at La Palma, Canary Islands,* J. Volcanol. Geotherm. Research, *94,* 251–265.

OKADA, Y. (1985), *Surface deformation due to shear and tensile faults in a half-space,* Bull. Seismol. Soc. Am. *75,* 1135–1154.

PÉREZ, N.M and HERNÁNDEZ, P.A. (2004), *Reducing Volcanic Risk in the Canary Islands: are we doing the homework.* International Symposium "Reducing Volcanic Risk in Islands", Granadilla de Abona, Tenerife, Canary islands, Spain.

PÉREZ, N. M., MELIÁN, G., GALINDO, I., PADRÓN, E., HERNÁNDEZ, P. A., NOLASCO, D., SALAZAR, P., PÉREZ, V., COELLO, C., MARRERO, R., GONZÁLEZ, Y., GONZÁLEZ, P. and BARRANCOS, J. (2005), *Premonitory geochemical and geophysical signatures of volcanic unrest at Tenerife, Canary Islands,* Geophysical Research Abstracts, *7.*

PÉREZ, N. M., HERNÁNDEZ, P. A., PADRÓN, E., MELIÁN, G., MARRERO, R., PADILLA, G., BARRANCOS, J. and NOLASCO, D. (2007), *Precursory subsurface 222Rn and 220Rn degassing signatures of the 2004 seismic crisis at Tenerife, Canary Islands,* Pure appl. geophys., *164,* 2431–2448.

PERLOCK, P.A., GONZALEZ, P. J., TIAMPO, K.F., RODRIGUEZ-VELASCO, G., and FERNANDEZ, J. (2008), *Time evolution of deformation using Time Series of differential interferograms: Application to La Palma Island (Canary Islands),* Pure appl. geophys., *165,* 1531–1554.

PRATES, G., GARCÍA, A., FERNÁNDEZ-ROS, A., MARRERO, J.M., ORTIZ, R., and BERROCOSO, M. (2013), *Enhancement of sub-daily positioning solutions for Surface deformation surveillance at El Hierro volcano (Canary Islands, Spain),* Bull. Volcanol., *75,* 724. doi:10.1007/s00445-013-0724-3.

PRIETO, J.F., FERNÁNDEZ, J., GONZÁLEZ, P.J., SAGIYA, T., FUJII, N., HERNÁNDEZ, P.A., and PÉREZ, N.M. (2005), *Permanent ITER-GPS network in Canary Islands for volcano monitoring: Design, objectives and first results,* Geophysical Research Abstracts, *7,* 09426.

PRIETO, J., GONZÁLEZ, P., SECO, A., RODRÍGUEZ-VELASCO, G., TUNINI, L., PERLOCK, P., ARJONA, A., APARICIO, A., CAMACHO, A., RUNDLE, J., TIAMPO, K., PALLERO, J., POSPIECH, S., and FERNÁNDEZ, J. (2009), *Geodetic and Structural Research in La Palma, Canary*

Islands, Spain: 1992–2007 Results, Pure appl. geophys., *166*, 1461–1484.

QUAAS, R., GONZÁLEZ, R., GUEVARA, E., RAMOS, E., and DE LA CRUZ-REYNA, S., Monitoreo Volcánico: Instrumentación y Métodos de Vigilancia [Volcanic Monitoring: Instrumentation and Surveillance Methods], In Volcán Popocatépetl, Estudios Realizados Durante la Crisis de 1994–1995 (Ed Zepeda, O., and Sánchez, T. A.) (Centro Nacional de Prevención de Desastres, Universidad Nacional Autónoma de México, México, D.F., 1996) pp. 25–76.

RANERO, C.R., TORNE, M., and BANDA, E. (1995), *Gravity and multichannel seismic reflection constraints on the lithospheric structure of the Canary swell,* Mar. Geophys. Res., *17*, 519–534.

RODRÍGUEZ-VELASCO, G., ROMERO, R., YU, T. T., GONZÁLEZ-MATE-SANZ, F. J., QUIRÓS, R., DALDA, A., APARICIO, A., CARRASCO, D., PRIETO, J. F,. and FERNÁNDEZ, J. (2002), *On the monitoring of surface displacement in connection with volcano reactivation in Tenerife, Canary Islands, using space techniques,* Física de la Tierra, *14*, 85–108.

ROMERO, C., Actividad Volcánica Histórica en las Islas Canarias [Historic volcano activity in the Canary Islands], In Curso Internacional de Volcanología y Geofísica Volcánica (Ed. Astiz, M., and García, A.) (Serie Casa de los Volcanes 7, Cabildo Insular de Lanzarote, Lanzarote, Islas Canarias, 2000) pp. 115–128.

ROMERO, R., FERNÁNDEZ, J., CARRASCO, D., LUZÓN, F., MARTÍNEZ, F., RODRÍGUEZ-VELASCO, G., MORENO, V., ARAÑA, V., and APARICIO, A. (2002), *Synthetic Aperture Radar Interferometry (InSAR): Application to ground deformations studies for volcano and seismic monitoring,* Física de la Tierra, *14*, 55–84.

ROMERO, R., CARRASCO, D., ARAÑA, V. and FERNÁNDEZ, J. (2003), *A new approach to the monitoring of deformation on Lanzarote (Canary Islands): an 8-year perspective,* Bull. Volcanol., *65*, 1–7.

ROMERO RUIZ, C., GARCÍA-CACHO, L., ARAÑA, V., YANES LUQUE, A., and FELPETO, A. (2000), *Submarine volcanism surrounding Tenerife, Canary Islands: implications for tectonic controls, and oceanic shield forming processes,* J. Volcanol. Geotherm. Res., *103*, 105–119.

RUNDLE, J.B. (1982), *Deformation, gravity, and potential changes due to volcanic loading of the crust,* J. Geophys. Res. *87*, 10729–10744 (correction: J. Geophys. Res., 88: 10647–10652, 1983).

SAGIYA, T., MIYAZAKI, S. and TADA, T. (2000), *Continuous GPS array and present-day crustal deformation of Japan,* Pure appl. geophys., *157*, 2303–2322.

SAGIYA, T., BARRANCOS, J., CALVO, D., PADRÓN, E., HERNÁNDEZ, G. H., HERNÁNDEZ, P. A., PÉREZ, N., and SUÁREZ, J. M. P. (2012), *Crustal deformation during the 2011 volcanic crisis of El Hierro, Canary Islands, revealed by continuous GPS observation,* Geophys. Res. Abstr., *14*, EGU2012–10243.

SAMSONOV, S., TIAMPO, K., GONZALEZ, P. J., PRIETO, J., CAMACHO, A. G., and FERNANDEZ, J. (2008), *Surface deformation studies of Tenerife Island, Spain from joint GPS-DInSAR observations,* In Proc. of Second Workshop on USE of Remote Sensing Techniques (USEReST) for Monitoring Volcanoes and Seismogenic Areas, Naples, Italy, 11–14 Nov, (IEEE Xplore 10443679, Naples, Italy, 2008) pp. 1–6.

SAMSONOV, S.V., GONZÁLEZ, P.J., TIAMPO, K.F., CAMACHO, A.G., and FERNÁNDEZ, J., Spatiotemporal analysis of ground deformation at Campi Flegrei and Mt Vesuvius, Italy, observed by Envisat and Radarsat-2 InSAR during 2003–2013, In Mathematics of Planet Earth. Proc. 15th Annual Conference of the Int. Ass. Mathematical Geosciences. Lecture Notes in Earth Systen Sciences (ed. Pardo-Igúzquiza, E., Guardiola-Albert, C., Heredia, J., Moreno-Merino, L., Durán, J.J., and Vargas-Guzmán, J.A.) (Springer-Verlag Berlin Heidelberg, ISBN: 978-3-642-32407-9. doi:10. 1007/978-3-642-32408-6_84, 2014) pp. 377–382.

SANSOSTI, E., BERARDINO, P., BONANO, M., CALÒ, F., CASTALDO, R., CASU, F., MANUNTA, M. MANZO, M., PEPE, A., PEPE, S., SOLARO, G., TIZZANI, P., ZENI, G., and LANARI, R. (2014), *How second generation SAR systems are impacting the analysis of ground deformation.* International Journal of Applied Earth Observation and Geoinformation, *28*, 1–11. doi:10.1016/j.jag.2013.10.007.

SEGALL, P., and MATTHEWS, M. (1997), *Time dependent inversion of geodetic data,* J. Geophys. Res. *102*, 22931–22409.

SEGALL, P., DESMARAIS, E.K., SHELLY, D., MIKLIUS, A., and CER-VELLI, P. (2006), *Earthquakes triggered by silent slip events on Kilauea volcano, Hawaii,* Nature *442.* doi:10.1038/nature04938.

SEVILLA, M.J., and PARRA, R. (1975), *Levantamiento gravimétrico de Lanzarote [Gravimetric survey of Lanzarote],* Revista de la Real Academia de Ciencias Exactas, Físicas y Naturales de Madrid, Tomo LXIX, Cuad. 2°, 257–284.

SEVILLA, M. J., and MARTÍN, M. D. (1986), *Geodetic network design for crustal deformation studies in the Caldera of Teide area.* Tectonophysics, *130*, 235–248.

SEVILLA, M. J., MARTÍN, M. D., and CAMACHO, A. G. (1986), *Data analysis and adjustment of the first geodetic surveys in the Caldera de Teide, Tenerife, Canary Islands.* Tectonophysics, *130*, 213–234.

SEVILLA, M. J. and ROMERO, P, (1991), *Ground deformation control by statical analysis of a geodetic network in the Caldera of Teide,* J. Volcanol. Geotherm. Res., *47*, 65–74.

SEVILLA, M. J., and SÁNCHEZ, F. J., (1996) Geodetic network for deformation monitoring in the Caldera of Teide, In Proc. 2nd Workshop on European Laboratory Volcanoes, Santorini, Greece, 2–4 May, 1996. (Ed. Casale R., Fytikas M., Sigvaldasson G. and Vougioukalakis G.) (Publ. Europ. Comm. DGXII Environment and Climate Res. Progr., 1996) pp. 615–636.

SEVILLA, M.J., VALBUENA, J.L., RODRÍGUEZ-DÍAZ, G., and VARA, M.D. (1996), *Trabajos altimétricos en la caldera del Teide [Levelling campaigns on the Teide caldera].* Física de la Tierra, *8*, 117–130.

SIEBERT, L., SIMKIN, T., and KIMBERLY, P., (2011) *Volcanoes of the World* (University of California Press. University of California Press, California, 2011).

SIGURDSSON, H., HOUGHTON, B., McNUTT, S.R., RYMER, H., and STIX, J., *Encyclopedia of Volcanoes* (Academic Press, San Diego, California, 2000).

SURIÑACH, E. (1986), *La estructura cortical del archipiélago canario. Resultados de la interpretación de perfiles sísmicos profundos [Crustal structure of the Canarian Archipelago. Results of deep seismic profiles interpretation],* Anales de Física, Serie B, *82*, 62–77.

TIAMPO, K., RUNDLE, J.B., FERNANDEZ, J., and LANGBEIN, J. (2000), *Spherical and ellipsoidal volcanic sources at Long Valley Caldera, California, using a Genetic Algorithm inversion technique,* J. Volcanol. Geotherm. Res. *102*, 189–206.

TIZZANI, P., MANCONI, A., ZENI G., PEPE, A., MANZO, M., CAMACHO, A., and FERNÁNDEZ, J. (2010), *Long-term versus short-term deformation processes at Tenerife (Canary Islands),* J. Geophys. Res., *115*, B12412. doi:10.1029/2010JB007735.

TORROJA, J.M. (1926), *Nota sobre la triangulación geodésica de primer orden en las Islas Canarias [Note on the first order triangulation network on the Canary Islands],* Revista de la Real Academia de Ciencias Exactas, Físicas y Naturales de Madrid, XXIII, pp. 321–322.

VENEDIKOV, A.V., ARNOSO, J., (2006) CAI, W., VIEIRA, R., TAN, S., and VÉLEZ, E.J. (2006), *Separation of the long-term termal effects from the strain measurements in the Geodynamics Laboratory of Lanzarote.* J. Geodyn., *41*, 213–220. doi:10.1016/j.jog.2005.08.029.

VIEIRA, R., TORO, C., ARAÑA, V. (1986), *Microgravimetric survey in the Caldera of Teide, Tenerife, Canary Islands,* Tectonophys, *130*, 249–257.

VIEIRA, R., FERNÁNDEZ, J., and TORO, C. (1988), *La estación geodinámica de la Cueva de los Verdes (Lanzarote): primeros resultados de las experiencias realizadas [The geodynamic station of Cueva de los Verdes (Lanzarote): first results of the conducted experiencies].* Revista de la Real Academia de Ciencias Exactas, Físicas y Naturales de Madrid, Tomo LXXXII, Cuad. 2, 309–312.

VIEIRA, R., VAN RUYMBEKE, M., FERNANDEZ, J., ARNOSO, J., and TORO, C. (1991a), *The Lanzarote underground laboratory,* Cahiers Centre Eur. Geodynam. Seismol. *4*, 71–86.

VIEIRA, R., FERNÁNDEZ, J., TORO, C., (1991) and CAMACHO, A.G. Structural and oceanic effects in the gravimetric tides observations in Lanzarote, In Proc. XIth Int. Symp. Earth Tides (ed. Kakkuri, J.). (E. Schweizerbart'sche Verlagsbuchhandlung, Stuttgart, Germany, ISBN: 3-510-65148-0, 1991b) pp. 217–230.

VIEIRA, R. (1994), *La estación geodinámica de Lanzarote [The geodynamical station of Lanzarote].* Serie Casa Volcanes, Cabildo Insular Lanzarote, *3*, 31–40.

VIEIRA, R., VAN RUYMBEKE, M., ARNOSO, J., D'OREYE, N., FERNÁNDEZ, J., and TORO, C., Comparative study of the tidal gravity parameters observed in Timanfaya, Jameos del agua and Cueva de los Verdes stations at Lanzarote Island, In Proc. XII Int. Sym. Earth Tides (Ed. Hsu, H.T.) (Science Press, Beijing, New York, 1995) pp. 41–52.

VIEIRA, R., SEVILLA, M.J., CAMACHO, A.G., TORO, C., and MARTÍN, M.D. (1996), *Geodesia de precisión aplicada al control de movimientos y deformaciones en la Caldera del Teide [Precision geodesy applied to the monitoring of movements and deformations in the Teide Caldera],* Anales de Física, Serie B, *82*, 110–126.

WAITE, G.O., and SMITH R.B. (2002), *Seismic evidence for fluid migration accompanying subsidence of the Yellowstone caldera.* J. Geophys. Res., *107*, B9, 2177. doi:10.1029/2001JB000586.

WARD, S.N., and DAY, S.J. (2001), *Cumbre Vieja volcano-potencial collapse and tsunami at La Palma, Canary Islands,* Geophys. Res. Lett. *28*, 17, 3397–3400.

WATTS, A.B., PEIRCE, C., COLLIER, J., and HENSTOCK, T. R.R.S. Charles Darwin CD82. Cruise Report (Southampton Oceanography Centre, Southampton, UK, 1993).

WATTS, A.B. (1994), *Crustal structure, gravity anomalies and flexure of the lithosphere in the vicinity of the Canary Islands,* Geophys. J. Int. *119*, 648–666.

WATTS, A.B., PEIRCE, C., COLLIER, J., DALWOOD, R., CANALES, J.P., and HENSTOCK, T.J. (1997), *A seismic study of lithospheric flexure in the vicinity of Tenerife, Canary Islands,* Earth Planet. Sc. Lett., *146*, 431–447.

WATTS, A.B., and ZHONG, S. (2000), *Observations of flexure and the rheology of the oceanic lithosphere,* Geophys. J. Inter., *142*, 855–875.

WICKS, C.W., THATCHER, W., DZURISIN, D., and SVARC, J. (2006), *Uplift, thermal unrest and magma intrusion at Yellowstone caldera,* Nature, *440*, 72–75. doi:10.1038/nature04507.

WU, J.C., and CHEN, Y.Q. (1999), *Inverse analysis of deformation mechanism by geodetic surveys: a case study,* J. Geodyn. 27, 553–565.

WYSESSION, M.E., WILSON, J., BARTKÓ, L., and SAKATA, R. (1995) *Intraplate Seismicity in the Atlantic Ocean Basin: a Teleseismic Catalog,* Bull. Seism. Soc. Am., *85*, 3, 755–774.

YU, T.T., FERNÁNDEZ, J., and RUNDLE, J.B. (1998), *Inverting the parameters of an earthquake-ruptured fault with a genetic algorithm,* Comput. Geosci. 24, 173–182.

YU, T.T., FERNÁNDEZ, J., TSENG, C. L., SEVILLA, M. J., and ARAÑA, V. (2000), Sensitivity test of the geodetic network in Las Cañadas Caldera, Tenerife, for volcano monitoring. J. Volcanol. Geotherm. Research, *103*, 393–407.

ZAZO, C., GOY, J. L., HILLAIRE-MARCEL, C., GHALEB, B., GILLOT, P.-Y., SOLER, V., GONZALEZ, J.A., and DABRIO, C.J. (2002), *Marine Sequences of Lanzarote and Fuerteventura Revisited—a Reappraisal of Relative Sea-Level Changes and Vertical Movements in the Eastern Canary Islands during the Quaternary,* Quaternary Science Reviews, *21*, 2019–2046.

(Received April 23, 2014, revised July 10, 2014, accepted July 22, 2014, Published online August 10, 2014)

Pure Appl. Geophys. 172 (2015), 3229–3245
© 2014 Springer Basel
DOI 10.1007/s00024-014-0915-7

❚ **Pure and Applied Geophysics**

Shallow Hydrothermal Pressurization before the 2010 Eruption of Mount Sinabung Volcano, Indonesia, Observed by use of ALOS Satellite Radar Interferometry

PABLO J. GONZÁLEZ,[1,2] KESHAV D. SINGH,[3] and KRISTY F. TIAMPO[1]

Abstract—Ground deformation in volcanic regions can be a precursor to resumption of activity. Volcanic eruptions are typically brief periods of activity punctuating very long inter-eruptive periods. This makes hazard evaluation a difficult task for volcanoes with low-recurrence eruptive activity, which often are poorly monitored. As a result, analysis of inter-eruptive periods by use of remote sensing techniques can provide important information on precursory activity and improve volcano hazard assessment. In August–September 2010 Mt Sinabung, Indonesia, reawakened after at least 400 years of dormancy. The ground deformation before this eruption was investigated by use of differential interferometric synthetic aperture radar data obtained from Japanese ALOS-PALSAR radar imagery between 05 January 2007 and 31 August 2010. Results from InSAR time series processing detected significant ground deformation (subsidence) at several locations on the Karo plateau, and uplift in the summit area of Mt Sinabung. The persistent scatterers density obtained by use of ALOS data is sufficient to enable extraction of temporal and spatial patterns of the deformation. The surface deformation at the summit can be modeled by using a spherical point-source model. Source data are consistent with a very shallow (hydrothermal) reservoir, with a linear increase in overpressure before the 2010 Mt Sinabung eruption. Hydrothermal origin is consistent with seismicity, tilt-meters, and analysis of ash products collected during and after the 2010 eruption. These results support the potential of L-band interferometry for hazard assessment in poorly monitored and highly vegetated volcanic areas and also indicate that hazard assessment for Indonesian volcanoes could potentially be improved by identification of precursory (inter-eruptive) uplift periods.

Key words: InSAR, persistent scatterer analysis, time series analysis, volcano monitoring, ground deformation.

1. Introduction

Long inter-eruptive (dormant) periods are one of the most challenging aspects of evaluation of volcanic hazards. For example, before the devastating 1982 eruption of the El Chichon volcano in southeastern Mexico, the potential activity of this volcano was not very well understood and it had not been identified as an active hazardous volcano (ESPINDOLA *et al.* 2000). Another prominent example was the Mt Pinatubo (Philippines) eruption in 1991, the second largest eruption in the 20th century. Although, in retrospect, many observations were consistent with the presence of an active magmatic system, before the June 1991 eruption no historic eruptive activity was known and, consequently, only a rapid response reduced the magnitude of the losses (NEWHALL and PUNONGBAYAN 1996). Another recent, if smaller, example is the August 2010 eruption of Mt Sinabung, Indonesia, which occurred after at least 400 years of dormancy (IGUCHI *et al.* 2011).

Remote sensing technology can be used to detect departures from the background level of volcanic activity in vast and otherwise unmonitored areas. The development of targeted wide-aperture satellite-based surveys can mitigate the lack of prior information about the hazard in many volcanic areas. Differential interferometric synthetic aperture radar (DInSAR) analysis has been used to monitor and detect numerous instances of volcanic, tectonic, or anthropogenic activity over areas larger than tens of square kilometers (ROSEN *et al.* 2000; MASSONNET and FEIGL 1998). However, most early work, e.g., that by the European ERS-1/2 and Envisat, and the Canadian Radarsat-1/2, used radar sensors in the C-band wavelength range (~3–7 cm). This work has generally been limited to relatively dry and non-vegetated areas, to prevent

---

[1] Department of Earth Sciences, University of Western Ontario, London, ON N6A 5B7, Canada. E-mail: p.j.gonzalez@leeds.ac.uk; ktiampo@uwo.ca
[2] *Present Address*: Institute of Geophysics and Tectonics, School of Earth and Environment, University of Leeds, Leeds LS2 9JT, UK.
[3] Department of Earth Sciences, Indian Institute of Technology Bombay, Mumbai 400076, India. E-mail: ksingh86@alumni.uwo.ca

Table 1

Details of ALOS PALSAR synthetic aperture radar (SAR) images acquired between 05 January 2007 and 31 August 2010

N	Acquisition date	Date (decimal years)	Designation
1	05 January, 2007	2007.013699	Slave
2	20 February, 2007	2007.138128	Slave
3	08 July, 2007	2007.521918	Slave
4	23 February, 2008	2008.146347	Slave
5	09 April, 2008	2008.274658	Slave
6	25 May, 2008	2008.401826	Slave
7	10 October, 2008	2008.777397	Slave
8	25 November, 2008	2008.901826	Slave
9	**10 January, 2009**	**2009.027397**	**Master**
10	25 February, 2009	2009.151826	Slave
11	13 July, 2009	2009.535616	Slave
12	28 August, 2009	2009.660046	Slave
13	28 November, 2009	2009.910046	Slave
14	13 January, 2010	2010.035616	Slave
15	28 February, 2010	2010.160046	Slave
16	16 July, 2010	2010.543836	Slave
17	31 August, 2010	2010.668265	Slave

Bold text represents the image corresponding to the master date to which all other images were resampled

temporal decorrelation of the backscattered radar signal (WEI and SANDWELL 2010). A longer wavelength sensor (e.g., L-band, ∼15–30 cm) enables better penetration of vegetated areas and, consequently, is less affected by temporal decorrelation. L-band radar interferometry is, however, less sensitive to differential deformation, which depends on carrier radar wavelength (BURGMANN et al. 2000).

In the work discussed in this paper we analyzed the ground deformation around Mt Sinabung, Indonesia, before the 2010 eruption. Seventeen L-band Japanese ALOS PALSAR synthetic aperture radar (SAR) images were acquired between 05 January 2007 and 31 August 2010 (Table 1). The associated deformation time series were calculated by use of the software StaMPS (HOOPER et al. 2004). This study adopted the multi-temporal InSAR approach, which combines PS interferometry and the small baseline (SB) subset procedure (HOOPER 2008). The combined PS and SB approach increases the number of available data points in the study area. The radar images spanned the period between 2007 and 2010, before the eruption in 2010. The results demonstrate the validity of using L-band differential radar interferometry in remote, unmonitored areas to evaluate signals consistent with increased activity.

2. Mt Sinabung

2.1. Regional Context

Approximately 13 % of the world's active volcanoes are located along the Sunda arc. Arc-related volcanism in Indonesia results from subduction of the Indo-Australian Plate under the Sunda Plate. The rate of convergence in the Northern part of Sumatra Island is approximately 47 mm/year (Fig. 1). Convergence is oblique relative to the normal direction of the subduction trench; this results in partition of the deformation between subduction and strike–slip components. The forearc migrates northward, leading to an independent tectonic sliver plate bounded by the right-lateral Sumatran Fault (SIEH and NATAWIDJAJA 2000). In Northern Sumatra, the Sumatran fault runs parallel to the trench in a northwest-to-southeast direction (Fig. 1). The Karo plateau, which contains the Toba caldera, is located northeast of the fault. The Toba caldera was formed during the largest Quaternary eruption, ca. 74 ka. (SIEBERT et al. 2010), and blanketed the Karo plateau with ignimbrite deposits. NNW from Lake Toba are located Mt Sinabung and Mt Sibayak, the only Holocene volcanoes in the Karo plateau (Fig. 1). Mt Sinabung and Mt Sibayak are associated with a strike–slip fault with SW–NE direction (HENDRASTO et al. 2012). However, direct evidence is difficult to collect, because of the presence of thick Pleistocene-Holocene volcanic eruptive deposits (SUTAWIDJAJA et al. 2013).

2.2. Mt Sinabung Volcano

Mount Sinabung is a Pleistocene-to-Holocene stratovolcano. Its activity began after the latest caldera-forming eruption of Toba Lake. Mt Sinabung, with an elevation of 2,460 m, is a predominantly andesitic–dacitic stratovolcano. On the basis of geological and geomorphic criteria the volcanic material can be divided into two main phases, the ancient and young stages (SATAKE and HARJONO 2012; IGUCHI et al. 2012). On the eroded western flanks, the ancient phase predominantly forms lava flows of porphyritic and two pyroxene andesite with/without hornblende composition (Fig. 2). The new edifice (young phase) was created by superposition of lava flows and pyroclastic deposits, with porphyritic two-

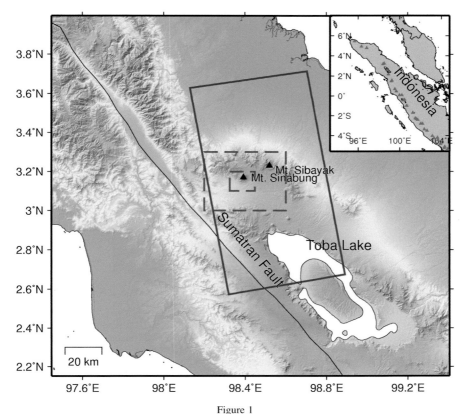

Figure 1

Location of the eruption of Mount Sinabung, Indonesia, in the Northern Sumatra region, in August and September 2010. DEM map of study area. The *blue rectangular* frame indicates the ALOS track around Mount Sinabung. The general location is shown in the *inset, top right. Gray dashed line rectangles* indicate the area covered in the other figures

pyroxene basaltic andesite to hornblende two-pyroxene andesite composition (IGUCHI *et al.* 2012). The present-day summit craters and domes are attributed to this second phase (SATAKE and HARJONO 2012). In the summit area there are four craters and/or vents aligned N–S with rather complex elongated crater geometry. The youngest is located at the southern end of the lineament and has been active, with solfatara activity, during the 20th century (SIEBERT *et al.* 2010).

The eruptive activity of Mt Sinabung is poorly understood. Recent studies have identified several Holocene eruptions (IGUCHI *et al.* 2012; SATAKE and HARJONO 2012). The youngest block-and-ash flow and associated surge deposits distributed on the southeastern flank are believed, on the basis of radiocarbon aging of charcoal, to have occurred at 1.1 ka. This deposit has been extensively described as a current density flow located approximately 5 km from the summit area. Other units in the eastern flank are 1.5 ka ash fall deposits and ca

4.2 ka pyroclastic surge deposits (IGUCHI *et al.* 2012). However, there is contradictory information on historical eruptions of Mt Sinabung. Several authors report an eruption in 1600 (OHKURA *et al.* 2012; HENDRASTO *et al.* 2012; SATAKE and HARJONO 2012). Others believe there were two, the 1,600 eruption and a second in 1881, which is unconfirmed (HENDRATNO 2010).

2.3. August–September 2010 Eruption

The 2010 eruption began on August 27 with mild phreatic explosions which caused ash fall to the E–SE of the volcano (HENDRATNO 2010). On the next day only weak signs of activity were observed, with emission of diffuse white plumes reaching 20 m. Between August 29 and 30, activity resumed with two significant explosions that produced an ash plume approximately 1.5–2 km in height. These explosions were preceded by rumblings that prompted the evacuation of

271

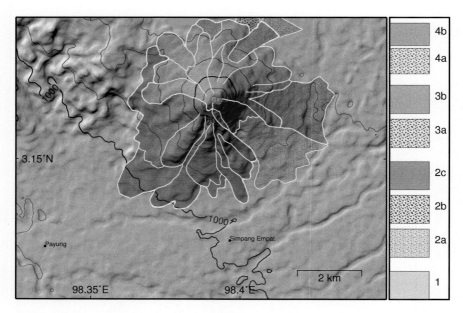

Figure 2
Geological map showing stages of activity of Mt Sinabung, modified after Iguchi *et al.* (2012). *Color and patterns* represent different units and types of product: *1* Toba Ignimbrite, *2a* debris avalanche from the Old stage, *2b* pyroclastic flow deposits from the Old stage, *2c* lava flows from the Old stage, *3a* pyroclastic flow deposits from the Young stage, *3b* lava flows from the Young stage, *4a* the most recent pyroclastic flow, and *4b* the most recent lava dome (before the 2013–2014 dome). The materials in *gray* in the W–NW area are basement rocks. *Red symbols* indicate the location of the craters

approximately 30,000 people living within a 6-km radius of the volcano (Hendratno 2010). During subsequent days, activity remained at a high level, with large explosions. On September 3, two eruptions occurred at 4:38 and 17:59 (local time, UTC +7 h), ejecting clouds of volcanic ash 2,000 and 1,000 m high, respectively. The largest explosive eruption occurred on September 7th, with an ash plume rising up to 8 km (Kusuma 2010; Hendratno 2010). The sound of the explosion was heard as far as 20 km from the volcano (Iguchi *et al.* 2011). Activity with much smaller explosions continued until September 23, when the volcano alert level was downgraded (Hendratno 2010). During the period of eruptive activity, seven eruptions were recorded. All the eruptions were phreatic type and no new magmatic material was found in volcanic ash (Iguchi *et al.* 2011).

3. Data Processing and Analysis

3.1. ALOS Dataset

To analyze the surface deformation in the Mt Sinabung region, we used differential interferometric synthetic aperture radar (DInSAR). DInSAR is a geodetic technique that measures surface ground deformation between two satellite passes over wide areas (Sansosti *et al.* 2010). DInSAR surface deformation maps are sensitive to any ground position change between ground and satellite along the satellites' line-of-sight (LoS). To minimize commonly occurring noise sources for ground deformation studies, we use multitemporal analysis in which temporal decorrelation and differential atmospheric phase delays are regarded as sources of noise affecting pairs of SAR images (differential interferograms). Multitemporal methods exploit the phase information of a (normally) reduced number of high-quality pixels. In general, two methods are used to identify pixel candidates, specifically spatially and temporally coherent pixels. Spatially coherent pixels broadly fall into the small baseline (SB) category of methods (Berardino *et al.* 2002; González and Fernández 2011a) whereas temporally coherent pixels are used in the so-called persistent scatterer methods (PS) or single master approaches (Ferretti *et al.* 2001; Werner *et al.* 2003; Kampes 2005). Hybrid methods between PS and SB have also

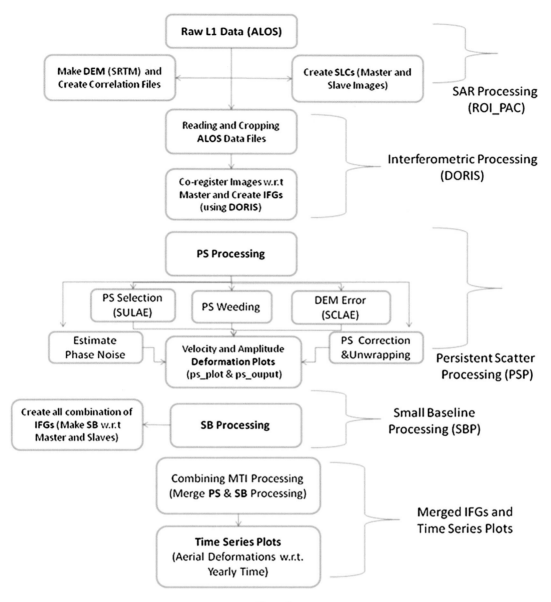

Figure 3
Processing of ALOS-InSAR data starting from ROI_PAC to StaMPS/MTI

furnished promising results in rural environments (Hooper 2008; Ferretti *et al.* 2011).

Our analysis was based on seventeen SAR scenes acquired by the PALSAR instrument aboard the Japanese ALOS satellite from January 2007 to August 2010 (Table 1). An ascending path covers the entire Karo Plateau area with Mt Sinabung approximately in the center of the scene (Fig. 1). The ALOS satellite repeat orbit cycle is 45 days.

However, the satellite did not acquired data during all passes. The ALOS satellite has an operating wavelength of 23.3 cm on a right-looking radar sensor with an approximately 34-degree look angle. The 10 January 2009 image was used as the master image. We used the Repeat Orbit Interferometry Package (ROI_PAC) (Rosen *et al.* 2004) to focus the raw images to produce single-look complex (SLC) images. All images were then coregistered to the selected master date image geometry, to form a stack of SLC images. ALOS-PALSAR data were interferometrically processed by use of Doris software (KAMPES *et al.* 2003). Interferograms were processed in two-pass differential mode, using a 90-m-resolution digital elevation model (DEM) derived from the SRTM topography mission (FARR *et al.* 2007).

3.2. Time Series Processing

We adopted a time series processing approach which entailed use of computed differential interferograms with similar but complementary approaches, namely PS, SB, and hybrid PS–SB, implemented in the software package MTI-StaMPS (http://homepages. see.leeds.ac.uk/~earahoo/stamps/). In Fig. 3, we show a hierarchical diagram of the procedure used to process the deformation time series.

First we computed the time series of phase changes, by use of a single master approach (PS method); for this we computed all the differential interferograms with respect to the selected master. The MTI-StaMPS method computes, with the time series of the deformation, the linear velocity maps, spatially-correlated look angle (SCLA) error, almost exclusively attributable to spatially-correlated DEM error, the master atmosphere, and orbit error (AOE). The PS method lacks any assumption about the nature of the time-dependent motion of the persistent scatterers (HOOPER *et al.* 2004). It searches for spatial phase statistically similar pixels in a probabilistic manner based on initial selection of pixel candidates on the basis of amplitude dispersion of the SAR images, i.e. 0.4 (HOOPER *et al.* 2004). The amplitude dispersion is a measure of the stability of SAR reflected intensity from ground scatterers; the lower the index the better. FERRETTI *et al.* (2001) showed that the SAR phase properties are correlated for pixels with low standard deviation of calibrated SAR amplitudes. The selected pixels are assumed to remain coherent over the whole stack of interferograms and can be used to estimate evolution of the deformation phase.

We next computed the time series of deformation by use of the SB method. SB methods usually work with multi-looked interferograms, which are individually phase unwrapped (BERARDINO *et al.* 2002;

GONZÁLEZ and FERNÁNDEZ 2011a). To maximize the interferometric phase correlation only small perpendicular baselines were used, while ensuring that the resulting network of image pairs contained no isolated clusters and/or subsets. The differential interferograms were formed by recombination of the resampled SLC images from the PS processing, first filtering in azimuth, to exclude the non-overlapping Doppler spectrum, and in range, to reduce the effects of geometric decorrelation. In Fig. 4, we shows all baseline combinations used in this study (36 interferometric pairs). Pixel selection in the SB method used the amplitude difference dispersion initial selection criterion, i.e., 0.6 (HOOPER 2008). For distributed scatterer pixels (those suited to the SB method), a proxy for the phase stability is given by the amplitude difference dispersion, defined as the ratio of the standard deviation of the difference in amplitude between master and slave, and the mean amplitude.

Finally, results from the PS and SB methods were merged with a hybrid PS–SB approach (HOOPER 2008). Pixels selected by both the PS and SB methods are combined after processing is completed for both previous methods. By combining PS and SB-selected pixels, the method seeks to maximize the reliability of the unwrapped phase. To combine the data sets the equivalent SB interferogram phase, $\psi_{x,i}^{SB}$, is calculated for the PS pixels by recombination of the single-master interferogram phase, $\psi_{x,i}^{SB} = W\left\{\psi_{x,s}^{SM} - \psi_{x,m}^{SM}\right\}$, where $\psi_{x,s}^{SB}$ is the single-master phase for the small baseline slave, $\psi_{x,m}^{SB}$ is the single-master phase for the small baseline master, and $W\{\cdot\}$ is the wrapping operator. Note that this equivalent SB phase is different from that extracted directly from the SB interferograms, because no spectral filtering has been applied (HOOPER 2008). The phase of the combined data set is corrected by using the estimate of spatially uncorrelated look angle error calculated in the selection steps. The resulting phase of each combined SB interferogram is then unwrapped. Phase unwrapping of SB interferograms reduces spatial aliasing when rates of deformation are high. The unwrapped phase of the SB interferograms must be inverted to derive a time series of the phase change for each pixel (HOOPER 2008).

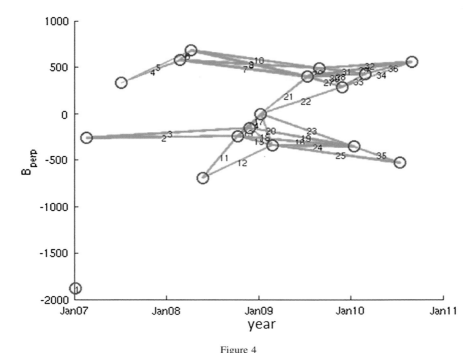

Figure 4

Baselines plot diagram, showing possible spatial and temporal baseline combinations during the SB processing, for the PS analysis the interferograms were combined with respect to the master image (10 January 2009). *Circles* represent images and *lines* represent the SB interferograms. *Bperp* perpendicular baseline

3.3. Results: Density of Pixels and Linear Velocity Maps

Because of the different approaches associated with each method, the results show some variability in the set of pixels selected (Fig. 5). In Fig. 5, we show both the distribution of the pixels and the linear LoS velocity of each selected pixel. Of the two methods, PS and SB, the PS method detects a larger number of pixels (Fig. 5a) than the SB method (Fig. 5b). However, as we can see in Fig. 5a, a larger proportion of the pixels seem to indicate shortening of the LoS (dominant blue colors). Most of the pixels present in the PS solution with shortening motion seem to be rejected during selection for the linear velocity solution in the SB method As a compromise between high density but probably "noisier" results provided by the PS method and the lower density of results obtained by use of the SB method, the hybrid PS–SB method seems to have the best characteristics, incorporating a relatively high density of pixels with apparently lower standard deviation of the spatial distribution of linear velocities (Fig. 5c).

In Fig. 6 we show that the spatial distribution of the pixels has some similarities with the land cover. Figure 6a shows a false-color Landsat image which highlights some variability in the vegetation as different shades of green. This false-color combination also presents urban areas in blue–purple colors. We computed the spatial distribution of pixels as the density of selected pixels per square km. The density is correlated with the land cover. On average for the area of the Karo plateau the hybrid PS–SB method selects approximately 100s of pixels per square km. This is likely to correspond to the characteristic rate of detection for agriculture fields, sparse roads, and small isolated towns, which dominate most of the Karo plateau region. However, approximately 1,000s of pixels per square km are selected for some areas. Some of these regions of high pixel density coincide with the blue–purple zones in Fig. 6a, which correspond to urban areas (e.g., the cities Berastagi and Kabanjahe). However, other high-pixel-density regions occur in areas of brownish to white colors, and might be interpreted as non-vegetated areas of

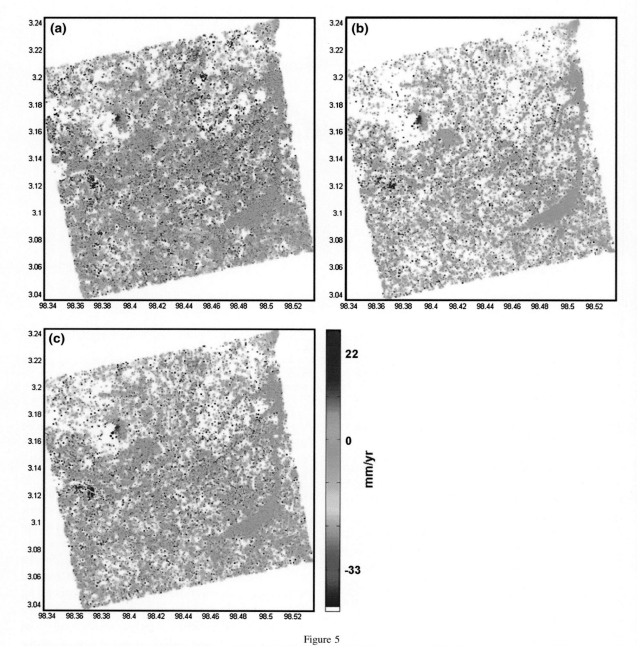

Figure 5
ALOS-InSAR time series data analysis using: **a** PS (~75,000), **b** SB (~14,500) and **c** hybrid PS–SB processing (~52,000): *Each panel* represents the pixels selected for each processing. The *color scale* denotes a change in LoS linear velocity in mm/year; *positive values* indicate motion toward the satellite (uplift) and *negative values* are indicative of deflection (subsidence). Results correspond to a cropped area around Mt Sinabung

different kinds. In this case, we are able to identify an area as being of recent pyroclastic deposits, located on the Southeast flank of Mt Sinabung and the summit areas of Mt Sinabung and Mt Sibayak, which are characterized by bare hydrothermal altered soils.

In contrast, the density of pixels drops to approximately zero to a few tens of pixels per square km in the region to the north and northwest of Mt Sinabung, and in a large part of the region between Mt Sinabung and Mt Sibayak (Fig. 6b, white). The lowest pixel

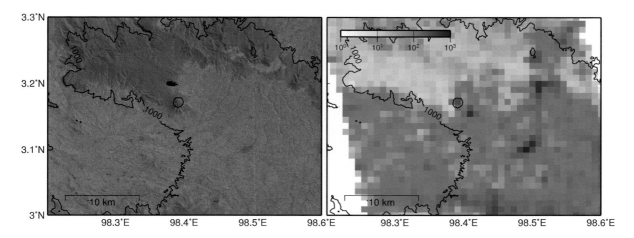

Figure 6
Illustration of the density of pixels compared with the land cover, including Landsat band combination (RGB: Landsat bands 1–2–3) of the Karo Plateau (*left panel*) and the density of pixels on a log scale (*white areas* no data, *dark blue* areas >1,000 pixels/km^2) (*right panel*). Pixels were selected by use of the hybrid PS–SB method

density regions clearly represent large decorrelation effects affecting areas covered by the rain forest vegetation. Rain forest vegetation can be clearly observed in Fig. 6a as intense dark green colors. This qualitative comparison reveals that persistent scatters L-band interferometry is adequate for sparsely populated areas in rural and/or agricultural environments, but it is still challenging to retrieve good pixel densities over rain-forest regions, which account for large areas of Indonesia. Over those regions, a potential sensor with L-band polarimetric InSAR with quad-polarization capability might improve results for deformation monitoring, as has been shown by using C-band RADARSAT-2 over vegetated areas (ALIPOUR *et al.* 2013; SAMSONOV and TIAMPO 2011).

The main results from the deformation time series analysis are presented in Fig. 7. We limit our discussion to the results obtained by use of the hybrid PS–SB method. These results span the period January 2007–August 2010. The results show relatively good spatial distribution coverage of pixels was achieved by use of the hybrid method, and its deformation time series reveals low spatial variations in its linear velocity map (Fig. 7a). The rate of deformation in the linear velocity map falls between −25 and 25 mm/year of LoS motion (positive indicates motion toward the satellite, i.e. uplift). For some pixels, however, subsidence reached

approximately 35 mm/year in an area southwest of the Mt Sinabung, as shown in the time series panels (Fig. 7d).

The 2007–2010 linear velocity map (Fig. 7a) clearly illustrates areas of ground subsidence (red pixels) located around the Southwest Karo plateau region (also south and southwest of Mt Sinabung). At this point, we cannot confirm the origin of the deformation, but considering the large extension of agricultural activity, some or all the subsidence could be the result of compaction of shallow aquifers because of groundwater extraction for irrigation purpose (GONZÁLEZ and FERNÁNDEZ 2011b). In addition, another less defined area of subsidence can be observed around the Mt Sibayak and Pinto volcanoes; this could represent active degassing processes (SIEBERT *et al.* 2010). For most of the deformation of the region, the linear velocity seems to indicate minimum to negligible motion (green pixels). Finally, a significant density of pixels with uplift motion (blue pixels) is observed in a small area corresponding to the summit of Mt Sinabung. Linear velocity values range from no significant motion for the lower parts of the volcano flanks to up to 15 mm/year of shortening in LoS at the summit (Figs. 7a, 8). The location of the deformation signal and the absence of similar uplift signals for nearby mountains indicate that the origin is related to the volcanic or hydrothermal activity at Mt Sinabung volcano.

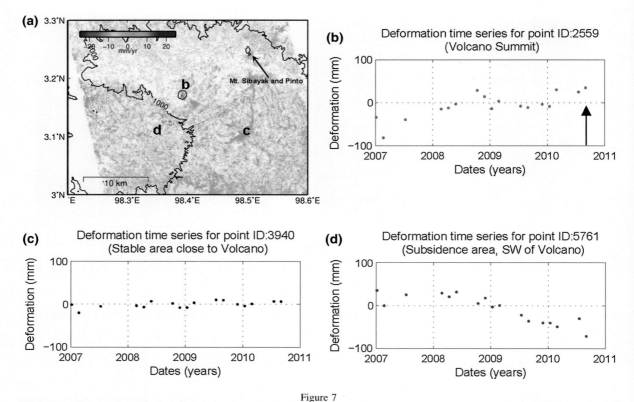

Figure 7

Time series results from ALOS-InSAR data analysis (full analyzed image frame). **a** Linear velocity of pixels selected by use of the hybrid PS–SB method (*black line* indicates 1,000 m contour lines for reference). Mt Sibayak and Pinto are indicated. *Letters* shows the location of the pixels time series displayed in **b c**, and **d**. **b** Mt Sinabung summit area (uplift); **c** pixels located at Kabanjahe city showing stable time series (no deformation); **d** subsidence area located SW of the Mt Sinabung volcano. The *black arrow* in **b** indicates the time the eruption started

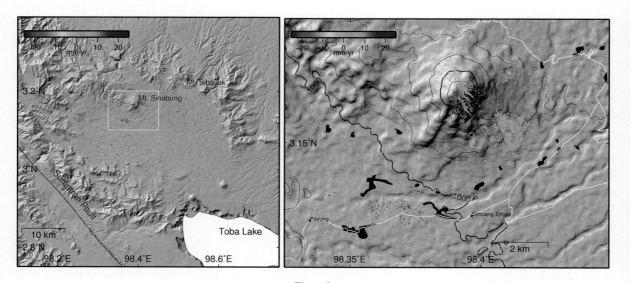

Figure 8

The velocity map showing, in the background, the topography of the area (*left panel*). The *right image* is a magnification of the area over Mt Sinabung showing some populated areas (*black polygons*), roads (*white lines*), and elevation contour lines (*black lines*). Pixel density is larger at the summit craters and the SE volcano flank, where the most recent pyroclastic deposits are located

Figure 7 also shows some relevant time series for pixels with significant uplift (Fig. 7b) stability (Fig. 7c), and subsidence (Fig. 7d). The location of the pixels is indicated by the corresponding letters in Fig. 7a. Uplift characterizes the time series of pixels from the summit area of Mt Sinabung. In Fig. 7b we show one pixel time series indicating a rather noisy but consistent uplift signal of ~20 mm/year (2 cm/year) over the 2007–2010 period (pixel ID: 2559; lat. 3.1703672, long. 98.390991). We tested those rates by use of a Monte Carlo jackknife re-estimation approach and detected a positive bias of ~0.5 cm/year that introduces the second point in the time series for the velocity of locations at the summit. If we only consider data before 2010 and exclude the second point of the time series, the uplift signal is ~1 cm/year.

As the time series of a stable point we selected that of the city of Kabanjahe, located approx. 10 km east-southeast of Mt Sinabung. The time series of LoS displacements indicate the scattering of the results can be approximately 2 cm (pixel ID: 3940; lat. 3.1556132, long. 98.412544). As a result, smaller changes could not be identified as significant temporal changes in deformation (Fig. 7c). Finally, the most significant subsidence area is located ~4 km southwest of the summit of Mt Sinabung, between the villages of Payung in the west and Simpang Empat in the east (pixel ID: 5761; lat. 3.1252968, long. 98.360863). The time series of the maximum subsidence pixel is shown in Fig. 7d, and indicates that the deformation is rather persistent and approximately constant with time. Although the short span of the time series does not enable reliable discussion of changes in deformation within the period analyzed, in the following discussion we restrict ourselves to interpreting what we regard as the more stable result, the linear velocity map (Fig. 8).

4. Modeling

The shape of the deformation pattern (Fig. 8) indicates that its cause could involve a localized volume at some depth. In addition, the approximately circular deformation pattern also is consistent with a volume and/or pressure change inside a source with axisymmetric geometry. Now, to interpret quantitatively the deformation process responsible for the observed ground displacements, we consider a simplified representation of the subsurface beneath Mt Sinabung. We select an idealized earth structure. We treat the Earth as an elastic body with homogeneous and isotropic elastic properties. There are several analytical solutions (fast computation evaluation) for simple source geometries. Considering the simplicity of the deformation pattern, we chose a spherical point source at depth, which can simulate ground displacements caused by changes in the volume and/or pressure of magma or fluids (MOGI 1958). However, this solution predicts the response for a planar surface, which is not an ideal approximation for volcano with steep topography, for example Mt Sinabung. In particular, considering that the deformation is limited to the summit area of the volcano, a very shallow source is indicated. Therefore, we approximate the effect of the topography by varying the depth of the source relative to the altitude of each surface data point (WILLIAMS and WADGE 1998). On the basis of the deformation observations, essential data for the source (location, depth, and volume) can be inverted by use of this model.

We can express the relationship between surface deformation and source data as $d = G(m) + e$, where d indicates the surface displacement, e the observation error, m the source data, and G is Green's function that relates d to m. We used a non-linear global inversion technique, bounded simulated annealing, to calculate the best-fitting model, m, which finds the minima of the misfit function in a least-squares sense (GONZÁLEZ et al. 2010). To reduce the computation burden, only points at a distance of 5 km from the summit were considered (~3,000 PSs). To avoid solutions trapped in local minima, we re-started the non-linear inversion 100 times with a homogeneously distributed random set of initial model variables with the same bounding limits. We used the improved Monte Carlo method (GONZÁLEZ et al. 2010), as applied to El Hierro volcano and a mining-related subsidence case in Canada (GONZÁLEZ et al. 2013; SAMSONOV et al. 2014). To assess the real uncertainties of the inversion, we simulated the random noise by use of an exponential and cosine theoretical covariance function. Finally, we determined the best-fitting model data as the means of the 100 best-fitting set of modeled data.

5. Discussion and Conclusions

Uplift related to the volcanic activity at Mt Sinabung began years before its 2010 eruption. In this work we proved that since 2007 uplift has been relatively steady at 1–2 cm/year in a very limited area around the summit of the volcano. Additional analysis of the time series at the summit suggests minor, although not statistically significant, acceleration before the 2010 eruption (Fig. 7b). This behavior contrasts with that of other, recently studied, volcanoes along the Sunda arc, for which no steady deformation was observed, even before large eruptions (e.g., the 2010 Merapi volcano eruption; PALLISTER et al. 2013). To provide deeper insight into its possible origin and eventual relationship with the 2010 eruption, we inverted the signal by use of a simple analytical model, the spherical point-source model (MOGI 1958). We preferred the spherical point-source model because it satisfactorily explains the simple deformation pattern observed. The main results of the inversion are shown in Fig. 9. We assumed linear velocity between 2007 and 2010, because there is no clear change in the deformation time series of the pixels in the summit area (Fig. 7b). The model (Fig. 9b), constrained with all the pixels near the volcano, gave residuals with low values and no clear patterns (Fig. 9c). The location of the best-fitting point-source model is shown in Fig. 9b by a red triangle. The inversion results indicate a very shallow source located just beneath the area of the Southern Crater (Fig. 10). Considering the realistic noise included in each Monte Carlo iteration of the inversion algorithm, we can analyze the uncertainties in the model variables, namely location (x, y, depth) and volume change (m^3). Regarding the depths of the best fit, the mean and median of scatter of the fitted models is less than 50 m, with a standard deviation of ~ 300 m for all 250 models (Fig. 10). A few models are clearly outliers (depths above the ground surface) but do not have a large effect on uncertainty. The depth is extremely shallow, $\sim 1,950$ m a.s.l. (~ 500 m beneath the surface; Mt Sinabung elevation

2,460 m). The location is also very precisely determined with estimated horizontal errors of the order of 210–170 m for latitude and longitude, respectively. The estimated source seems to be located predominantly beneath the summit Southern Crater. Finally, the volume change is estimated to increase at rates between 6×10^3 and 33×10^3 m^3/year with a 95 % confidence interval.

The shallow depth of the source seems to be incompatible with a long-lived magma reservoir (~ 300–800 m at 95 % confidence interval). Mt Sinabung lacks reported eruptions for at least the last 300 years, and its most recent eruption has been dated to 1.1 ka, according to recent geological reports (IGUCHI et al. 2012). Therefore, a magma body or remnant from its last previous eruption was very likely to be completely crystallized. As a result, we consider the deformation source to be related to a vigorous hydrothermal system with a potential argillite-rich seal lid at a depth of ~ 500 m that served to trap fluids generated deeper in the volcano structure, because of either circulation of convective meteoric waters or degassing of an undetected deeper magma reservoir.

Although the deformation data available enable study of the period before the 2010 eruption only, additional data collected during the rapid response to the 2010 eruption shed some light on the processes leading to it. Seismic monitoring was deployed soon after the activity started between August and October 2010, with up to ten seismic stations (HENDRASTO et al. 2012). According to HENDRASTO et al. (2012) seismicity accompanying the 2010 eruption was characterized mainly by *deep* (2.5–14 km) and *shallow* (0.5–2 km) events. Epicenters of the events defined a roughly NE–SW trend along a previously inferred sinistral fault zone between Mt Sinabung and Sibayak volcanoes, during the early phase (September to October 2010). Subsequently, from November 2010 to February 2011 events were located beneath the northern flank of the volcano at depths of 4–6 km, with a few events located in deeper (8–14 km) sections in the same area.

Deformation monitoring also was running from August 2010 onward, including use of tiltmeters, electro-optical distance measurements (EDM), and GPS networks (HENDRASTO et al. 2012; OHKURA

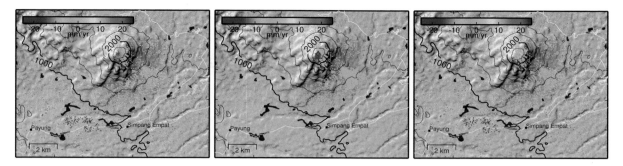

Figure 9

Left panel linear velocity (2007–2010) observed by use of ALOS images, with maximum motion toward the satellite of ∼ 30 mm/year. *Central panel*: simulated ground deformation of the *left panel* predicted by the best-fitting single spherical source model (Fig. 10). The *red triangle* marks the location of the pressure source. *Right panel*: residual between observed and simulated

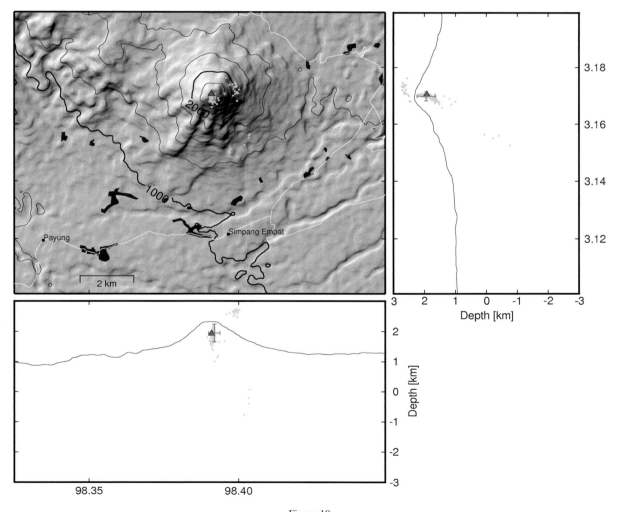

Figure 10

Location of the best-fitting spherical point sources: *orange dots* are all models obtained during the inversion; *dark red* indicates the location of the best-fitting model with the smallest residual. *Errors bars* are centered at the mean of all the best-fitting models. *Right upper panel* shows an NS profile cutting the topography through the summit of the volcano. *Lower panel* shows the same but along an EW profile

et al. 2012). In particular, ground deformation monitoring using tiltmeters indicated two distinct periods relating to the level of activity of the volcano, from September 2 to 11, and from September 12 to 25, 2010. However, the pattern of deformation was very similar, with a consistent pattern across the network. A pressure source located a 0.7 km beneath the summit and with a pressure change of 9×10^6 Pa was consistent with the tilt signals. Assuming a radius of the reservoir of 500 m, this pressure change is equivalent to 13.05×10^8 m^3 (HENDRASTO *et al.* 2012). In addition, from March to April 2011, GPS measurements were consistent with a different pressure source located SW of the summit at a depth of 6.5 km. HENDRASTO *et al.* (2012) interpreted this source as an aseismic volume (reservoir), which could explain the separation between the deep and shallow seismicity during the eruption period. This depth range also coincides with the depth of the seismicity located at the north of the volcano during the post-eruptive period (Nov. 2010–Feb. 2011).

On the basis of the mineral assemblage of the ash products, the 2010 eruption was phreatic. Ash samples consisted of altered and weathered clasts, plagioclase, clinopyroxene, orthopyroxene, hornblende, and opaque minerals (silicon oxide, iron oxide–magnetite, sodium, calcium and aluminum silicate–anorthite, pyrite, and illite). These minerals indicate the ash was formed from strongly hydrothermally altered rocks (IGUCHI *et al.* 2012; HENDRASTO *et al.* 2012). No juvenile material was recognized. This is consistent with previous knowledge of the rocks from the summit:

– domes and spines strongly altered hydrothermally;
– sulfur minerals precipitated along fine fractures; and
– plagioclase and hornblende phenocrysts replaced partly to entirely by clay and ore minerals (IGUCHI *et al.* 2012).

This suggests that the ash products were generated from the altered lava domes and spines in the crater areas, rather than from altered or non-altered rocks of the volcano edifice.

Several observations indicate that the shallow inflation source inferred by use of InSAR could represent an increasingly pressurized shallow hydrothermal system. This is supported by:

1 the co-eruptive *shallow* seismicity (0.5–2 km);
2 a pressure source, inferred from use of tiltmeters, with very shallow depth of 0.7 km beneath the summit; and, finally,
3 volcanic products clearly generated by explosion of hydrothermally altered volcanic rocks.

These products are enriched in SiO$_2$, different from the young and old volcanic products characterizing the geology of Mt Sinabung, which are of intermediate composition between young lava rocks and altered lava summit spines (IGUCHI *et al.* 2012). In particular, the altered rocks show development of argillic alterations which have been interpreted as being formed in a hydrothermal system at depth (SUTAWIDJAJA *et al.* 2013).

All these observations support our hypothesis of a sealed pressurized extremely-shallow hydrothermal system. Considering the tiltmeter-constrained volume change, it could be consistent with a possible inflation phase, at InSAR-constrained rates, of 40–220 years. Curiously, mining of the sulfur sublimates, which were economically exploited for hundreds of years, stopped during the last decade because of a decrease in sulfur content (SUTAWIDJAJA *et al.* 2013). We speculate that this time-span could be the necessary to generate small phreatic-like eruptions at Mt Sinabung under the sealed conditions in the shallow hydrothermal system. Perhaps the hydrothermal system was more effectively sealed over recent decades, eventually increasing the deformation and accelerating the pressurization, resulting in the 2010 Mt Sinabung eruption. However, the lack of apparent changes in the spatio-temporal deformation pattern of the InSAR information over the 3 year period seems to limit its potential to predict the occurrence of eruptions. Nevertheless, the eventual discovery could have raised awareness of the potential activity of this particular volcano, which was classified as dormant and/or inactive before August 2010. This result emphasizes the importance of routine processing of global spaceborne radar over volcanic areas.

The 2010 Mt Sinabung eruption could alter the complete magmatic system of the volcano, as might be indicated by activity at deeper levels during the

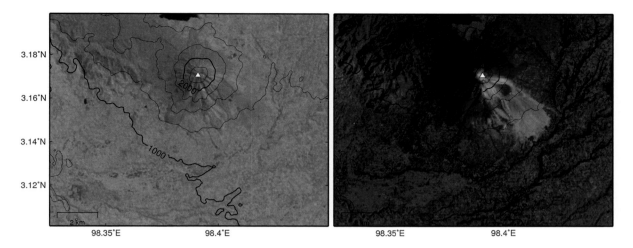

Figure 11

Same as *left panel* in Fig. 6; Landsat image acquired on May 19, 2003. NASA ALI's sensor false color combination (RGB using bands 10–6–3) on February 6, 2014. Note the *red/orange colors*, which represent intense heat radiation at the lava flow and summit areas

post-eruption phase. This activity could not be explained by rapid replenishment of magma, because, as stated previously, during the 2010 eruption no new magmatic material intruded the volcano. The change in pressure in the hydrothermal system could propagate to deeper levels and mobilize fluids or even magma along a NE–SW fracture system, which produced the deep (6–14 km) earthquakes and deformation at a source 4–6 km distant. Currently, since September 2013, Mt Sinabung is in another phase of eruption, and a new lava dome has extruded. Extrusion of the new juvenile material has formed lava flows that collapsed repeatedly in January though March 2014. Figure 11, an ALI (EO-1) false color satellite image, shows the status of the volcano in February 2014. It is characterized by the development of a massive dome and growing lava flow along its SE flank. Tragically, on February 1, 2014, fourteen people were killed by a pyroclastic flow generated by the partial collapse of the growing dome. It is worth mentioning that this scenario was identified as the most likely eruptive scenario in 2012, "*...during dome growth, partial collapse of the lava dome will generate pyroclastic flows (block-and-ash flows and surges)*" (IGUCHI *et al.* 2012; YOSHIMOTO *et al.* 2012).

Here we present results from use of ALOS-PALSAR interferometry and time series analysis to determine ground deformation for the period 2007–2010 at Mount Sinabung, Indonesia (North Sumatra). An inflation signal with a rate of ~2 cm/year was detected near the summit area. The detected signal enabled us to study the volcanic system of Mt Sinabung, and to infer that a shallow active hydrothermal system was increasing in pressure years before its 2010 eruption. Multivariable observation (seismicity, tiltmeters, and chemical and petrological analysis of erupted ash) enabled characterization of the eruption as phreatic, with the absence of juvenile material. Those observations support the extremely shallow model necessary to explain the InSAR results, and constrain the active volcanic system of Mt Sinabung. Eventually, the eruption of 2010 initiated processes that activated deeper segments of the volcano–magmatic system, leading to the 2013–2014 eruption, which clearly does involve juvenile material. Although this type of study cannot, currently, be used to directly predict the timing of future eruptions, it can serve to improve our understanding of changes in the level of activity of poorly monitored volcanoes.

Acknowledgments

The authors are grateful to the Japan Aerospace Exploration Agency (JAXA) for providing Advanced Land Observing Satellite (ALOS) InSAR data. The ALOS-PALSAR data are copyright of the Japanese Space Agency (JAXA) and the Japanese Ministry of Economy, Trade, and Industry (METI) and were made available by the US Government Research Consortium (USGRC) and the Alaska Satellite Facility (ASF). SRTM digital elevation data are provided by the Jet Propulsion Laboratory (JPL), NASA. The ROI_PAC software package was developed by JPL. Doris interferometry software (DORIS) was developed by DEOS, Delft University of Technology. We used Generic Mapping Tools (GMT) public domain software to create some figures (WESSEL and SMITH 1998). The authors would like to thank all these organizations and developers. This research was partially supported by the Natural Sciences and Engineering Research Council of Canada (NSERC) and the DALF-Canadian Commonwealth Scholarship Program (CCSP). The work of Pablo J. González was supported by a Banting Postdoctoral Fellowship. KFT is supported by an NSERC Discovery Grant.

REFERENCES

ALIPOUR, S., TIAMPO K.F., SAMSONOV S., and GONZÁLEZ P.J. (2013) *Multibaseline PolInSAR using RADARSAT-2 Quad-pol data: Improvements in interferometric phase analysis*, IEEE Geosciences and Remote Sensing Letters, *10*(6), 1280–1284. doi:10.1109/LGRS.2012.2237501.

BERARDINO, P., FORNARO, G., LANARI, R., and SANSOSTI, E., (2002). *A new algorithm for surface deformation monitoring based on small baseline differential SAR interferograms.* IEEE Trans. Geosci. Remote Sensing 40, 2375–2383.

BURGMANN, R., ROSEN, P.A., and FIELDING, E. J., (2000). *Synthetic aperture radar interferometry to measure Earth's surface topography and its deformation.* Annu. Rev. Earth Planet. Sci. 28, 169–209.

ESPINDOLA, J.M., MACIAS, J.L., TILLING, R.I., and SHERIDAN, M.F., (2000). *Volcanic history of El Chichon Volcano (Chiapas, Mexico) during the Holocene, and its impact on human activity.* Bulletin of Volcanology 62, 90–104.

FARR, T. G., et al. (2007), *The Shuttle Radar Topography Mission*, Rev. Geophys., *45*, RG2004. doi:10.1029/2005RG000183.

FERRETTI, A., PRATI,C., and ROCCA, F., (2001). *Permanent scatterers in SAR interferometry.* IEEE Trans. Geosci. Remote Sensing 39, 8–20.

FERRETTI, A., FUMAGALLI, A., NOVALI, F., PRATI, C., ROCCA, F., RUCCI, A. (2011), *A new algorithm for processing interferometric data-stacks: SqueeSAR*, IEEE Trans. Geosci. Rem. Sens., *49*(9), 3460–3470.

GONZÁLEZ P.J., and FERNÁNDEZ J. (2011a). *Error estimation in multitemporal InSAR deformation time series, with application to Lanzarote*, Canary Islands, Journal of Geophysical Research, *116*, B10404.

GONZÁLEZ P.J., and FERNÁNDEZ J. (2011b). *Drought-driven transient aquifer compaction imaged using multitemporal satellite radar interferometry*, Geology, *39*(6), 551–554.

GONZÁLEZ P.J., TIAMPO K.F., CAMACHO A.G., and FERNÁNDEZ J. (2010). *Shallow flank deformation at Cumbre Vieja volcano (Canary Islands): Implications on the stability of steep-sided volcano flanks at oceanic islands*, Earth and Planetary Science Letters, *297*(3–4), 545–557.

GONZÁLEZ P.J., SAMSONOV S., PEPE S., TIAMPO K.F., TIZZANI P., CASU F., FERNÁNDEZ J., CAMACHO A.G., and SANSOSTI E. (2013), *Magma storage and migration associated with the 2011–2012 El Hierro eruption: Implications for shallow magmatic systems at oceanic island volcanoes*, Journal of Geophysical Research—Solid Earth, *118*, 4361–4377.

HENDRASTO, M., SURONO, BUDIANTO, A., KRISTIANTO, TRIASTUTY, H., HAERANI, N., BASUKI, A., SUPARMAN, Y., PRIMULYANA, S., PRAMBADA, O., LOEQMAN, A., INDRASTUTI, N., ANDREAS, A.S., ROSADI, U., ADI, S., IGUCHI, M., OHKURA, T., NAKADA, S., and YOSHIMOTO, M., (2012) *Evaluation of Volcanic Activity at Sinabung Volcano, After More Than 400 Years of Quiet*, Journal of Disaster Research, Vol. 7, No. 1, pp. 37–47.

HENDRATNO, K., (2010), Center of Volcanology and Geological Hazard Mitigation (CVGHM), Indonesia. http://www.volcano.si.edu/world/volcano.cfm?vnum=0601-08=&volpage=var, [Accessed March 3rd, 2014].

HOOPER, A., (2008). *A multi-temporal InSAR method incorporating both persistent scatterer and small baseline approaches.* Geophys. Res. Lett. *35*, L16302.

HOOPER, A., ZEBKER, H., SEGALL, P., and KAMPES, B., (2004). *A new method for measuring deformation on volcanoes and other natural terrains using InSAR persistent scatterers.* Geophys.Res. Letters *31*, 23.

IGUCHI, M., SURONO, NISHIMURA, T., HENDRASTO, M., ROSADI, U., OHKURA, T., TRIASTUTY, H., BASUKI, A., LOEQMAN, A., MARYANTO, S., ISHIHARA, K., YOSHIMOTO, M., NAKADA, S., and N., HOKANISHI (2012) *Methods for Eruption Prediction and Hazard Evaluation at Indonesian Volcanoes*, Journal of Disaster Research, Vol. 7, No. 1, pp. 26–36.

IGUCHI, M., ISHIHARA, K., SURONO, and HENDRASTO, M, (2011) *Learn from 2010 Eruptions at Merapi and Sinabung Volcanoes in Indonesia.* Ann. Disas. Prev. Res. Inst., Kyoto Univ., No. *54 B.*, 185–194.

KAMPES, B.M., (2005). Displacement parameter estimation using permanent scatterer. Ph.D. thesis, Delft University of Technology.

KAMPES, B.M., HANSSEN, R.F., and PERSKI, Z., (2003), Radar interferometry with public domain tools, 3rd International Workshop on ERS SAR Interferometry, 'FRINGE03', Frascati, Italy, 1–5 Dec.

KUSUMA, I.N., (2010). Emergency Situation Report, ESR (4), WHO Indonesia, Mt Sinabung, Indonesia. Emergency and Humanitarian Action (EHA).

MASSONNET, D. and FEIGL, K. L., (1998). *Radar interferometry and its application to changes in the Earth's surface.* Reviews of Geophysics *36*, 4, 441–500.

MOGI, K., (1958). *Relations between the eruptions of various volcanoes and the deformations of the ground surface around them.* Bull. Earthquake Res. Inst. Univ. Tokyo 36, 99–134.

NEWHALL, C.G., and PUNONGBAYAN, A.S., (1996). Eruptions and Lahars of Mount Pinatubo, Philippines. Fire and Mud, U.S. Geological Survey, USA.

OHKURA, T., IGUCHI, M., HENDRASTO, M., ROSADI, U., (2012) Evaluation of activity of Guntur, Sinabung and Merapi volcanoes, in Indonesia based on continuous GPS observations. Japan

Geosciences Union Meeting 2012, 20–25 May, Makuhari, Chiba, Japan.

PALLISTER, J.S., SCHNEIDER, D.J., GRISWOLD, J.P., KEELER, R.H., BURTON, W.C., NOYLES, C., NEWHALL, C., and RATDOMOPURBO, A., (2013) *Merapi 2010 eruption: chronology and extrusion rates monitored with satellite radar and used in eruption forecasting.* J. Volcan. Geotherm. Res., *261*, 144–152.

ROSEN, P.A., HENLEY, S., PELTZER, G., and SIMONS, M., (2004). *Updated Repeat Orbit Interferometry Package Released.* Eos, Transactions American Geophysical Union *85*, 47.

ROSEN, P.A., HENSLEY, P., JOUGHIN, I., LI, F., MADSEN, S., RODRIGUEZ, E., and GOLDSTEIN, R., (2000). *Synthetic aperture radar interferometry.* Proc. IEEE *88*, 333–382.

SANSOSTI, E., CASU, F., MANZO, M., and LANARI R., (2010), *Spaceborne radar interferometry techniques for the generation of deformation time series: an advanced tool for Earth's surface displacement analysis,* Geophys. Res. Lett., *37*, L20305. doi:10.1029/2010GL044379.

SAMSONOV, S., and TIAMPO, K., (2011). *Polarization Phase Difference Analysis for Selection of Persistent Scatterers in SAR Interferometry.* IEEE Geoscience and Remote Sens. Letters *8*.

SAMSONOV, S.V., GONZÁLEZ P.J., TIAMPO, K.F., and D'OREYE N. (2014) *Modelling of fast ground subsidence observed in southern Saskatchewan (Canada) during 2008–2011,* Natural Hazards and Earth System Sciences, *14*, 247–257. doi:10.5194/nhess-14-247-2014.

SATAKE, K., and HARJONO, H., (2012). Multidisplinary Hazard Reduction from Earthquakes and Volcanoes in Indonesia.

Science and Technology Research Partnership for Sustainable Development, Japan.

SIEBERT, L., SIMKIN, T., and KIMBERLY, P., (2010), Volcanoes of the World, 3rd ed. Berkeley: University of California Press, 568 p.

SIEH, K., and NATAWIDJAJA, D., (2000). *Neotectonics of the Sumatran fault, Indonesia.* Journal of Geophysical Research, *105*(B12), 28,295–28,326.

SUTAWIDJAJA, I.S., PRAMBADA, O., and SIREGAR, D.A. (2013) *The August 2010 Phreatic Eruption of Mount Sinabung, North Sumatra,* Indonesian Journal of Geology, Vol. *8*, No. 1, pp. 55–61.

WEI, M., and SANDWELL, D., (2010) *Decorrelation of L-Band and C-Band Interferometry Over Vegetated Areas in California.* IEEE Transactions on Geoscience and Remote Sensing, *48*, 7, 2942–2952.

WERNER, C., WEGMULLER, U., STROZZI, T., and WIESMANN, A., (2003) Interferometric point target analysis for deformation mapping. Proc. Int. Geosci. Remote Sens. Symp.

WESSEL, P., and SMITH, W. H. F., (1998) *New improved version of the generic mapping tools released.* EOS Trans. AGU *79*, 579.

WILLIAMS, C.A., and WADGE, G., (1998) *The effects of topography on magma chamber deformation models: Application to Mt Etna and radar interferometry,* Geophys. Res. Lett., *25*(10), pp. 1549–1552. doi:10.1029/98GL01136.

YOSHIMOTO, M., NAKADA, S., HOKANISHI, N., IGUCHI, M., and OHKURA, T., (2012) Eruption Scenario of Sinabung volcano, North Sumatra, Indonesia. Japan Geosciences Union Meeting 2012, 20–25 May, Makuhari, Chiba, Japan.

(Received March 16, 2014, revised July 5, 2014, accepted July 22, 2014, Published online August 6, 2014)

Pure Appl. Geophys. 172 (2015), 3247–3263
© 2014 Springer Basel
DOI 10.1007/s00024-014-1004-7

Pure and Applied Geophysics

Retrieving the Stress Field Within the Campi Flegrei Caldera (Southern Italy) Through an Integrated Geodetical and Seismological Approach

Luca D'Auria,[1] Bruno Massa,[1,2] Elena Cristiano,[1] Carlo Del Gaudio,[1] Flora Giudicepietro,[1] Giovanni Ricciardi,[1] and Ciro Ricco[1]

Abstract—We investigated the Campi Flegrei caldera using a quantitative approach to retrieve the spatial and temporal variations of the stress field. For this aim we applied a joint inversion of geodetic and seismological data to a dataset of 1,100 optical levelling measurements and 222 focal mechanisms, recorded during the bradyseismic crisis of 1982–1984. The inversion of the geodetic dataset alone, shows that the observed ground deformation is compatible with a source consisting of a planar crack, located at the centre of the caldera at a depth of about 2.56 km and a size of about 4×4 km. Inversion of focal mechanisms using both analytical and graphical approaches, has shown that the key features of the stress field in the area are: a nearly subvertical σ_1 and a sub-horizontal, roughly NNE-SSW trending σ_3. Unfortunately, the modelling of the stress fields based only upon the retrieved ground deformation source is not able to fully account for the stress pattern delineated by focal mechanism inversion. The introduction of an additional regional background field has been necessary. This field has been determined by minimizing the difference between observed slip vectors for each focal mechanism and the theoretical maximum shear stress deriving from both the volcanic (time-varying) and the regional (constant) field. The latter is responsible for a weak NNE-SSW extension, which is consistent with the field determined for the nearby Mt. Vesuvius volcano. The proposed approach accurately models observations and provides interesting hints to better understand the dynamics of the volcanic unrest and seismogenic processes at Campi Flegrei caldera. This procedure could be applied to other volcanoes experiencing active ground deformation and seismicity.

Key words: Stress field inversion, Campi Flegrei, volcano deformation, volcanic seismicity, joint inversion.

1. Introduction

Among the various geophysical phenomena associated with the transportation of magma within the shallow crust, there is ground deformation, which can be measured using both on-site (e.g., levelling, GPS) and satellite geodetic techniques (e.g., InSAR). In the brittle crust, strains are often associated with the occurrence of volcano-tectonic earthquakes. It follows that the joint analysis of ground deformation and seismicity represents a key method for monitoring active volcanoes (Segall 2013). During the last decades, the exploitation of InSAR and GPS techniques greatly improved the spatial resolution and the continuity of geodetic data sets (Dzurisin 2006). In addition, a large amount of active volcanoes are equipped with seismic networks, able to detect and locate local microseismicity (Lee and Stewart 1981; D'Auria *et al.* 2011; Tramelli *et al.* 2013).

Volcanic eruptions are often preceded by an increment in seismicity. Generally, stress changes associated with magma migration and intrusion are believed to cause these earthquakes (Umakoshi *et al.* 2001; Pedersen and Sigmundsson 2004; Segall 2013; Cannavò *et al.* 2014). For instance, various studies have shown that the migration of seismic swarms can be related to the intrusion of a dyke (Toda *et al.* 2002; Rubin *et al.* 1998; Brandsdottir and Einarsson 1979; Patanè *et al.* 2003). In general, changes in the volcano-tectonic seismicity pattern can be quantitatively related to variations of the stress field having different origin (Dietrich 1994; Segall 2013). Geodetic or seismic measurements alone, allows putting only limited constraints to models of magma reservoir dynamics (Segall 2013). In particular, the study of ground deformation alone suffers from a limited resolution in defining the geometry of magmatic bodies, and from an intrinsic ambiguity in the possible solutions to the geodetic inverse problem (Dzurisin 2006). On the other hand, seismic study

[1] Istituto Nazionale di Geofisica e Vulcanologia, sezione di Napoli, via Diocleziano, 328, 80124 Naples, Italy. E-mail: luca.dauria@ingv.it
[2] Dipartimento di Scienze e Tecnologie, Università degli Studi del Sannio, via dei Mulini, 59a, 82100 Benevento, Italy.

alone often do not allow a quantitative estimation of the stress field magnitude (ANGELIER 1990).

Conversely, the joint inversion of ground deformation and seismicity data represents a robust and reliable approach to investigate the stress field beneath active volcanoes. For instance, joint analysis of the ground deformation and spatial pattern of *P* axes has been successfully applied to the modelling of magmatic sources at Mt. Etna (CANNAVÒ *et al.* 2014).

Following a novel approach, we propose the quantitative determination of the spatial and temporal variations of the stress field within the Campi Flegrei caldera (CFc) by joint inversion of geodetic and seismological data.

2. Geological Background

The Campi Flegrei caldera (CFc) is a collapse structure, partially submerged, and located west of

the city of Naples. Its shape is a consequence of two huge collapses due to major eruptive events: the Campanian Ignimbrite eruption (CI 40.6 ky, GEBAUER *et al.* 2014) and the Neapolitan yellow tuff eruption (NYT 14.9 ky, DEINO *et al.* 2004) (Fig. 1). After the latter event, the area underwent three eruptive periods, between 12 and 3.8 ky (DI VITO *et al.* 1999), followed by a long period of quiescence until the last eruption: Monte Nuovo, 1538 CE. The activity of CFc in the last decades consisted mostly of seismic activity, significant gas emissions and intense ground deformation (CHIODINI *et al.* 2001; D'AURIA *et al.* 2011).

In the last century, there were three major events of uplift in 1950–1952 (about 0.75 m), 1969–1972 (1.77 m) and during the period 1982–1984 (1.79 m) (DEL GAUDIO *et al.* 2010). A period of subsidence started in 1985 and was interrupted by brief episodes of uplift in the following years (D'AURIA *et al.* 2011). During the last uplift event (1982–1984), this area was also affected by intense seismicity. The aim of

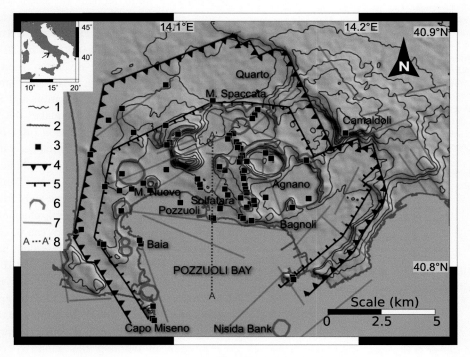

Figure 1
Map of the Campi Flegrei area with an outline of the major tectonic features. Legend (*1*) 50 m contour lines, (*2*) coast line, (*3*) seismic stations, (*4*) CI caldera rim, (*5*) NYT caldera rim, (*6*) rim of craters younger than 15 ka, (*7*) main fault traces, (*8*) trace of the N–S section discussed in Fig. 9. Data from VITALE and ISAIA (2013) and ORSI *et al.* (1999)

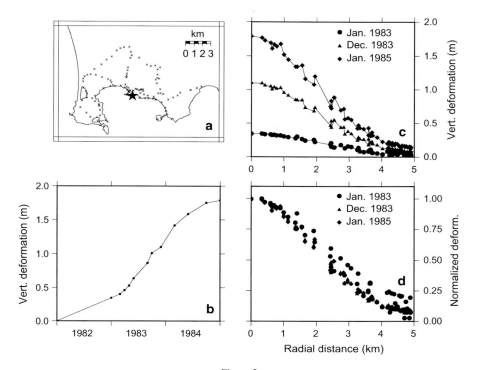

Figure 2
Optical levelling data set. **a** Map with the full distribution of benchmarks. The *star* is the position of the benchmark 25 A. **b** Vertical deformation measured at benchmark 25 A. **c** Ground deformation pattern. The *graph* represents the vertical deformation as a function of the radial distance from benchmark 25 A. We show the deformation pattern, referring to the survey of Jan. 1982 at three different times, as indicated by the three *different symbols (circle, triangle, diamond)*. **d** Same as (**c**), but with vertical deformations normalised to the value of benchmark 25 A

this work is to determine the spatial and temporal variations of the stress field within the CFc by joint inversion of geodetic and seismological data related to this last crisis.

3. Ground Deformation Data Set

The geodetic data set consists of about 1,100 optical levelling measurements referring to the interval 1982–1985. During this interval 13 surveys were performed, with a number of benchmarks ranging from a minimum of 18 (April 1983) to a maximum of 353 (January 1983). In Fig. 2a, we represent the planimetric distribution of benchmarks used between 1982 and 1985.

The ground deformation rate showed a marked increase during the first half of 1983, with an average uplift rate at benchmark 25A (black star in Fig. 2a) of about 10 cm/month. During the second half of 1984,

the rate started decreasing, reversing to subsidence since January 1985 (Fig. 2b). The ground deformation shows a roughly radial bell-shaped pattern during the entire period (Fig. 2c). However, a detailed comparison of the pattern during different intervals shows a slight change in the overall shape, in particular in the earlier phases of the crisis (Fig. 2d).

4. Inversion of the Ground Deformation Data

The study of ground deformation on active volcanoes is essential to understand the dynamics related to both the magmatic and hydrothermal systems (SEGALL 2010). Ground deformation at CFc has been the subject of various studies. In recent years, topics focused on the determination of the roles of the magmatic system and of the hydrothermal system as sources of ground deformation. BATTAGLIA *et al.* (2006) evidenced that the most likely source for the

1982–1984 uplift was a penny-shaped crack with a radius of 2.4 km and at a depth of 2.6 km, probably filled with magmatic fluids. Similar conclusions were also drawn by WOO and KILBURN (2010).

On the other hand, the role of the hydrothermal system has been invoked as well, using both analytical (BONAFEDE 1991) and numerical (CHIODINI et al. 2003) means. In particular, limited uplift episodes occurring at CFc since 1985 have been ascribed to migration of hydrothermal fluids (GAETA et al. 2003; BATTAGLIA et al. 2006; D'AURIA et al. 2011; CAMACHO et al. 2011).

Recently, D'AURIA et al. (2012) evidenced, though a geodetic imaging technique, the migration of hydrothermal fluids as a likely source for the recent episodes of limited ground uplift at CFc (2000–2007). Those results are in agreement with the results that CAMACHO et al. (2011) obtained with a similar geodetic imaging technique.

In this work, the inversion of the ground deformation data set has been performed using a hybrid non-linear technique. The ground deformation model used for the inversion is:

$$z_i^k = s_k a_i, \qquad (1)$$

where the index k refers to the benchmarks and the index i to the optical levelling surveys; the term s_k expresses the vertical deformation at the k th benchmark for a given source with normalised amplitude, and a_i is the source amplitude at the time of the i th survey. The relationship between the source amplitudes and the data is linear. However, the term s_k has a highly non-linear dependence from the source, even for very simple models (e.g., the Mogi point source).

We have addressed this inverse problem by splitting the whole problem into a linear inversion procedure embedded within a non-linear optimization algorithm.

The linear inversion is aimed at determining the value a_i given a source model s_k. To reduce the effect of local biases on the data set, we consider all the possible differences at each benchmark k:

$$d_{ij}^k = z_i^k - z_j^k. \qquad (2)$$

Substituting Eq. 2 in 1 and considering a specific source model s_k, we obtain a system of linear equations in the form: $\boldsymbol{d} = \boldsymbol{G a}$, which we solve by

Table 1

Source models used for the inversion

Model	GDOF	DOF	RSS (cm^2)	AICc
MOGI (1958)	0	15	2.935	−6467.680
MCTIGUE (1987)	1	16	2.731	−6544.621
FIALKO et al. (2001)	1	16	0.966	−7684.601
YANG et al. (1988)	4	19	0.865	−7800.115
OKADA (1985)	4	19	0.773	−7923.076

The second column (GDOF) indicates the number of geometric degrees of freedom of the model, while the third is the total number of degrees of freedom (DOF). The fourth (RSS) is the residual sum of squares (Eq. 3), while the fifth (AICc) is the corrected Akaike Information Criterion

computing a pseudo-inverse ($\boldsymbol{G}^+$) using the singular value decomposition: $\boldsymbol{a}^{\mathrm{est}} = \boldsymbol{G}^+ \boldsymbol{d}$ (ASTER et al. 2013). Once source amplitudes have been estimated, we can easily compute the residual sum of squares for the considered source model:

$$\mathrm{RSS}(s) = \sum_{i,j,k} \left[\left(d_{ij}^k \right)^{\mathrm{obs}} - s_k \left(a_i - a_j \right) \right]^2. \qquad (3)$$

To determine source model parameters, we have optimised the previous function using a genetic algorithm (SEN and STOFFA 1995).

We have considered five different elementary source models, listed in Table 1. The selection among the candidate models has been performed using the corrected Akaike Information Criterion (AICc) (BURNHAM and ANDERSON 2002). The computation of AICc requires the knowledge of the number of degrees of freedom (DOF) associated to the model. In general, we can write: DOF = 3 + GDOF + (NS− 1), where 3 refers to the DOF related to the spatial location of the source, GDOF are the DOF related to the geometry of the source, and NS−1 (with NS being the number of optical levelling surveys) are the DOF associated to the amplitude variations of the source. The term −1 appears because we must constrain the amplitude of the source term at the time of the first survey to be null ($a_1 = 0$).

The simplest model is the point source of MOGI (1958), which does not possess any GDOF. The MCTIGUE (1987) model takes into account the source radius and hence has one GDOF. The same holds for the FIALKO et al. (2001) penny-shaped crack model. The YANG et al. (1988) and OKADA (1985) models both have four GDOF: the former considers a generally oriented ellipsoid, while the latter considers a simple rectangular crack. In the modelling, values of 2 GPa and 0.25 were assumed for the shear modulus and Poisson's ratio, respectively (D'AURIA *et al.* 2012).

The AICc values indicate that the Okada rectangular crack is the best suited from a statistical point of view (Table 1).

In Fig. 3, we summarise the results of the inversion. The red rectangle in Fig. 3d is the surface projection of the crack. It has an area of about 16 km² and its centre is at about at 2.56 km b.s.l. The crack is dipping slightly 9° toward the SW. A comparison of observed and synthetic data (Fig. 3a) shows excellent

agreement, and the average residuals for each survey show no systematic trends (Fig. 3c). The average residuals for the benchmark (Fig. 3d) show a slight increase in the area E of Pozzuoli. These residuals are probably related to the presence of the Monte Olibano lava dome, whose mechanical properties strongly differ from the surrounding lithologies (ORSI *et al.* 1996).

Given the Okada source model, it is very easy to derive the corresponding volumetric variation from the crack opening value (see Fig. 3b). The final volumetric variation of the source is about 0.35 km³, which is of the same magnitude as small to medium-sized eruptions at CFc since 12 ky (0.01–0.1 km³) (WOO and KILBURN 2010).

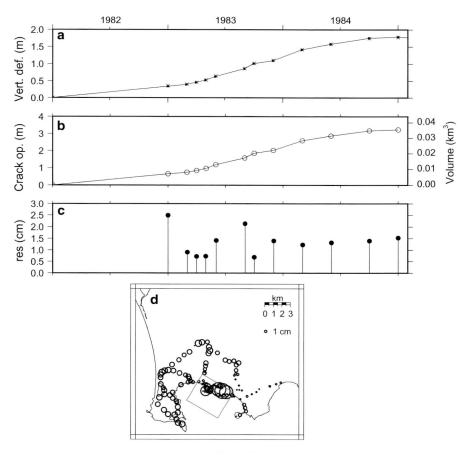

Figure 3

Results of the inversion of the geodetic data set. **a** Comparison between the observed (*crosses*) and the synthetic (*black dots*) vertical ground deformation at the benchmark 25 A. **b** Opening of the crack and its volumetric variation (*scale on the right*). **c** Average residuals for each survey. **d** Average residuals for each benchmark are indicated with *black circles*. The scale is reported on the *plot*. The *red rectangle* is the surface projection of the ground deformation source

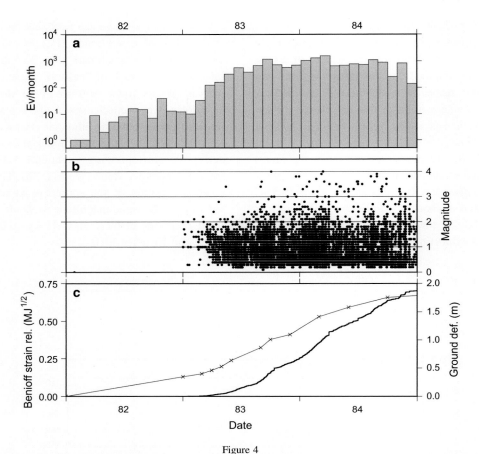

Figure 4
Time series of seismicity and ground deformation. **a** Monthly number of earthquakes since 1 January 1982. Note that the vertical scale is logarithmic. **b** Earthquakes magnitudes since 1 January 1983 (magnitudes for events occurring during 1982 are not available). **c** Comparison between the cumulative Benioff strain release (*bold black curve*) and the vertical deformation at benchmark 25 A (*solid line* with *crosses*)

5. Seismological Data Set

The bradyseismic crisis of 1982–1984 was characterised by intense seismicity, with about 16,000 recorded earthquakes and magnitudes ranging from about 0.5 to 4.2 (Fig. 4) (D'AURIA *et al.* 2011). We have analysed 222 events that occurred at CFc in the interval of 1983–1984. One hundred and ninety-two events have been relocated and their focal mechanisms have been calculated using *P* wave polarities, using the FPFIT software (REASENBERG and OPPENHEIMER 1985). Among these 192 events, 25 were previously unpublished. Each event has been relocated using a non-linear probabilistic approach (NonLinLoc software) in a three-dimensional (3D) velocity model (D'AURIA *et al.* 2008). An additional 30 mechanism data sets have been retrieved from ORSI *et al.* (1999).

The number of stations operating during the 1982–1984 intervals was highly variable. The analogue single-component network of the Osservatorio Vesuviano (OV) was progressively increased up to 21 stations. From September to November 1983, the Institut de Physique du Globe de Paris, in cooperation with OV, installed a temporary network of 18 digital three-component seismic stations. Moreover, during the interval December 1983–June 1984, the OV network was complemented by ten digital three-component stations provided by the University of Wisconsin (ASTER and MEYER 1988). In Fig. 1, we report a complete map of the seismic stations operating during the period 1982–1984.

The seismicity during the 1982–1984 bradyseismic crisis increased gradually from the second half of 1982. At the beginning of 1983, there was a marked

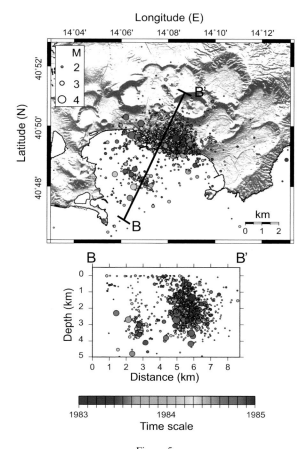

Figure 5

Earthquake hypocentres during 1983–1984. The map shows the distribution of epicentres. In the *bottom panel*, hypocentres are projected along the *B–B'* section shown on the map. Symbol size is proportional to the magnitude (as shown on the *inset* on the *top left*), while the *colour scale* indicates the time of occurrence (see *scale* on the *bottom*)

increase in both the seismicity rate and in the event magnitudes (Fig. 4a, b). The pattern of the seismicity showed an almost stationary rate during 1984 with episodes of intense seismic swarms (e.g., the 1 April 1984 swarm, with more than 400 events occurring in less than 5 h). During the entire period, the Benioff strain release pattern roughly followed those of the ground uplift (Fig. 4c), even if the former was slightly delayed (D'AURIA *et al.* 2011).

During the period 1983–1984, epicentres were located mostly in the area between the town of Pozzuoli and Agnano (Fig. 5). Another more limited cluster of events was located within the Pozzuoli Bay (Fig. 5). Hypocentre depth was generally above 3 km (D'AURIA *et al.* 2011).

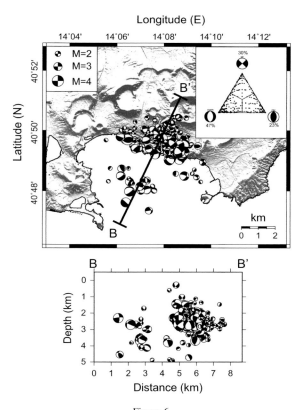

Figure 6

Focal mechanism data set. Focal mechanisms are represented both on a map (*top*) and along the B–B' cross-section (*bottom*). The size of the beachballs is proportional to the magnitude (see *inset* on the *top left*). The distribution of mechanism types is shown in the *triangular diagram* within the *inset* on the *top right* following the criterion of FROHLICH (1992)

The 192 re-analysed events have a number of *P*-wave polarities ranging between 6 and 19, with an average value of 8.6. In Fig. 6, we plot the entire focal mechanism data set. Mechanisms show a very heterogeneous distribution among the different typologies. The Frohlich triangular plot shows that normal mechanisms (47 %) prevail on inverse and strike-slip mechanisms (23 and 30 %, respectively) (Fig. 6).

In order to investigate the temporal evolution of the stress pattern in CFc, we split the focal mechanism data set into ten time intervals (A–J, see Table 2). Intervals A to I group data into time windows, while the interval J represents the entire data set and was processed in order to retrieve the overall stress field acting in the study area (see Sect. 6). The C interval partially overlaps the B one; it contains the event of 4 October 1983, the strongest of the whole crisis (*M* = 4.2) (Fig. 2b).

293

Table 2

Time intervals used for the data processing

Time intervals	Start	End	Normal (%)	Thrust (%)	Trascurrent (%)	Total
A	1983-01-17 15:24	1983-08-22 07:31	4 (57)	2 (28)	1 (15)	7
B	1983-09-01 23:04	1983-10-29 23:30	10 (42)	8 (33)	6 (25)	24
C[a]	1983-10-04 08:09	1983-10-29 23:30	9 (52)[a]	4 (24)[a]	4 (24)[a]	17[a]
D	1983-11-03 23:43	1983-12-30 23:19	9 (64)	4 (29)	1 (7)	14
E	1984-01-11 05:13	1984-03-19 23:52	10 (38)	5 (19)	11 (43)	26
F	1984-03-20 09:11	1984-03-26 20:53	12 (37)	9 (28)	11 (35)	32
G	1984-04-01 01:04	1984-04-01 04:43	23 (59)	3 (8)	13 (33)	39
H	1984-04-03 16:34	1984-04-27 15:10	25 (50)	11 (22)	14 (28)	50
I	1984-05-05 14:43	1984-12-08 22:33	11 (37)	10 (33)	9 (30)	30
J	1983-01-17 15:24	1984-12-08 22:33	104 (47)	52 (23)	66 (30)	222

[a] Note that the interval C is a subset of interval B

6. Inversion of Focal Mechanisms

A stress field acting on an intensely fractured rock causes the slip on fault planes oriented along mechanically favourable directions. Simple plotting of focal mechanisms and *P–T* axes density contouring is usually not enough to constrain the attitude of principal stress axes (McKenzie 1969).

The application of stress inversion techniques is therefore required to make a reliable reconstruction of the actual stress field responsible for the genesis of the analysed faults. Several methods of stress inversion have been proposed (Angelier and Mechler 1977; Michael 1984, 1987; Gephart and Forsyth 1984; Rivera and Cisternas 1990; Yamaji 2000; Otsubo et al. 2008).

Analytical methods are based on the Wallace-Bott hypothesis, stating that the shear traction applied on a given fault plane causes a slip in the direction and orientation of that shear traction, irrespective of the faults created in an intact rock or along a pre-existing fracture (Wallace 1951; Bott 1959; Angelier and Mechler 1977; Yamaji 2007). Following this hypothesis, the inversion of the stress field is generally done, minimizing a function of the angle between the computed shear stress and the retrieved slip vector on every analysed fault (Michael 1984; Angelier 1984, 1990; Hippolyte et al. 2012). Stress field inversion techniques allow the determination of the principal stress axes attitudes, and usually also the Bishop's ratio $\varphi_B = (\sigma_2 - \sigma_3)/(\sigma_1 - \sigma_3)$ (Bishop 1966), which is a relationship between principal stress eigenvalues that is useful to describe the shape of the stress ellipsoid.

Another class of stress inversion methods is based on a graphic approach. The simplest graphic approach to the stress inversion of a focal mechanism data set is the Right Dihedra Method (RDM; Angelier and Mechler 1977). The RDM allows the determination of a range of possible attitudes for the principal stress axes through the use of the stereographic projection technique (Ramsay and Lisle 2000; Hippolyte et al. 2012). RDM do not require an a priori selection of the real fault plane to respect to the auxiliary one.

Gaudiosi and Iannaccone (1984) performed a preliminary analysis of focal mechanisms at CFc. They processed 15 mechanisms related to the interval April–December 1983 without using a stress inversion procedure, and maintained that no regional stress components seem to dominate the stress field in Campi Flegrei area. They related the observed seismicity only to magmatic processes acting in the Pozzuoli area. Zuppetta and Sava (1991), applying an analytical stress inversion procedure to 49 focal mechanisms, retrieved for the CFc area an overall extensional stress field dominated by a sub-vertical σ_1 and a NNE-SSW trending sub-horizontal σ_3.

In this work, we performed the stress inversion procedure, applying a recent formulation of the RDM proposed by D'Auria et al. (2014). This implementation consists of an analysis of normalised moment tensors for the studied events. Following this approach, RDM can be expressed as:

$$\mathrm{RDM}(\theta, \varphi) = \frac{1}{N} \sum_{k} \mathrm{sgn}\big(\widehat{x}_k(\theta, \ \varphi)M_k\widehat{x}_k^T(\theta, \varphi)\big), \quad (4)$$

where θ and ϕ are the polar coordinates on the focal sphere, M_k is the normalised moment tensor for the event k, $\widehat{x}_k(\theta, \phi)$ is the position vector of the point on the focal sphere, and N is the number of events. The function $RDM(\theta, \phi)$ varies between -1 and 1, with negative values corresponding to high probabilities of representing the actual σ_3 (blue shades in Fig. 6) and positive values representing the actual σ_1 (red shades in Fig. 6).

In Fig. 7, we show the results of the RDM stress inversion performed on subsets of Table 2. In particular, Fig. 7j shows the result of the RDM applied to the entire data set. It shows a well-defined σ_1 attitude that appears sub-vertical, corresponding to a less defined σ_3, roughly sub-horizontal and trending NNE-SSW. RDM plot of individual subsets (Fig. 7a–i) generally show an agreement with the global pattern (Fig. 7j), indicating a substantial stability in the stress pattern during the entire considered period. The

comparison of results for each interval shows a certain degree of heterogeneity. To further constrain these results, we have applied another technique to the same dataset: the Multiple Inverse Method (MIM; YAMAJI 2000). Frequently, focal mechanism data sets are quite heterogeneous regarding stress fields responsible for their own genesis. Many methods aimed at the processing of heterogeneous data have been proposed (ANGELIER and MANOUSSIS 1980; ANGELIER 1984; YAMAJI 2000; OTSUBO et al. 2008). Operating without a priori information, the MIM is able to separate stresses within data sets related to heterogeneous stress fields, but in which it is possible to find domains where the state of stress is uniform (YAMAJI 2000; OTSUBO et al. 2008). MIM operates a resampling of the data set, through the construction of k element subsets. Every subset is inverted using a classical approach for stress tensor inversion (ANGELIER 1984; YAMAJI 2000). Results are plotted as "tadpole" symbols on the unit sphere, and significant results are identified by clusters of tadpoles (OTSUBO et al. 2006). MIM is able to manage focal mechanism

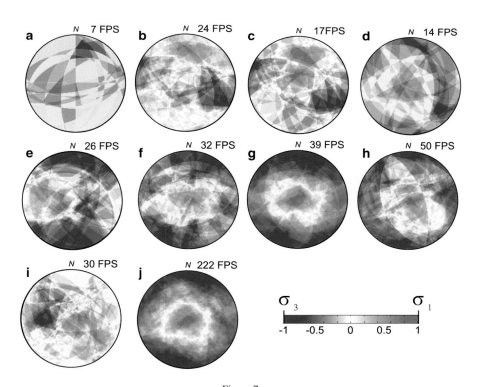

Figure 7
Results of the Right Dihedra Method (RDM). As indicated by the *colour scale*, *blue shades* indicate a high probability of containing σ_3, while *red shades* indicate high probability of containing σ_1

data, without requiring the discrimination of a real fault plane with respect to auxiliary plane (OTSUBO et al. 2006; YAMAJI and OTSUBO 2011).

In Fig. 8, we show the results of the MIM over the intervals of Table 2. An overall comparison of Figs. 7 and 8 shows that the results of RDM and MIM are in quite good agreement.

In Fig. 8j, we show the result obtained from the inversion of the entire focal mechanism data set (interval J in Table 2). It shows a strong clustering of tadpoles representing sub-vertical σ_1 attitudes (Fig. 8j, left plot), corresponding to very low clustered sub-horizontal σ_3 tadpoles (Fig. 8j, right plot) located all around the primitive circle, with a slight prevalence of NNE-SSW clusters. The prevalence of green-to-blue clusters of tadpoles allows the evaluation of the prevailing Bishop's ratio values $0 < \Phi_B < 0.5$. These results are in accordance with the RDM result (Fig. 7j), and are useful to highlight the strong heterogeneity of the analysed data set.

The A subset consists of only seven focal mechanisms. The RDM shows that a well-defined sub-horizontal σ_3, plunging toward NNE, corresponds to a poorly defined σ_1 attitude (low plunge toward ENE, medium plunge toward WSW) (Fig. 7a). The MIM results appear quite heterogeneous with a low clustering for both σ_1 and σ_3 axes (Fig. 8a). The RDM for the B subset (Fig. 7b) shows that both σ_1 and σ_3 axes are defined quite well, with a σ_1 plunging toward NNE and corresponding to a sub-horizontal σ_3 trending WNW-ESE. The MIM inversion of the B subset shows that the data set is highly heterogeneous, possibly deriving from a superposition of different stress fields. The prevailing one consists of a σ_1 medium to high plunging toward E, corresponding to a sub-horizontal N–S trending σ_3; the corresponding Φ_B (yellow to light green tadpoles) is about 0.5.

The RDM of the C subset (Fig. 7c) shows that the retrieved σ_1 and σ_3 axes attitudes is very similar to B, with σ_1 areas extended toward high plunge values. The σ_1 and σ_3 axes attitudes retrieved using MIM (Fig. 8c) are very similar to those of B; the data set is dominated by a medium–high plunging σ_1 dipping toward E, corresponding to a sub-horizontal N–S trending σ_3, and the corresponding φ_B tends to be higher than B (light green tadpoles), likely greater than 0.5.

The RDM solution for the D subset (Fig. 7d) shows a well-defined, high-plunge σ_1 corresponding to a poorly constrained σ_3 axis associated with areas extended along a great circle dipping toward ENE. The MIM applied to the same subset shows very similar results (Fig. 8d). For the E subset, the RDM result shows quite good definition for the σ_1 axis low-medium plunging toward WNW, corresponding to a less defined σ_3, with blue areas extending along an E dipping great circle, with a poor prevalence of sub-horizontal N–S trending values (Fig. 7e). Also, in this case, the agreement between RDM and MIM is good (Fig. 8e). RDM inversions for F, G and H subsets give similar results, compatible with a well-defined, sub-vertical σ_1, corresponding to a less defined low plunge σ_3 trending NNE-SSW (Fig. 7f–h). The MIM results for the same intervals show a higher variability of the σ_1, with a prevailing clustering comparable with RDM results (Fig. 8f–h). The retrieved RDM results for the I subset show a well-defined σ_3 plunging toward the west, corresponding to a poorly defined σ_1 with areas located along a wide E dipping great circle (Fig. 7i). On the other hand, the MIM results show a more clear distribution of clusters, with a prevalence of medium to high E plunging σ_1 corresponding to a sub-horizontal NNE-SSW σ_3. The prevailing clusters are characterised by high values of Φ_B.

7. Joint Inversion

The retrieved ground deformation source (Sect. 4) allows computing of the local stress field due to the inflation of the crack. In our case, this can be achieved by using an analytical formulation (OKADA 1992), but for general source models, numerical techniques (e.g., finite elements) could be used as well.

A more general model for the stress field acting in a volcanic area during unrest episodes is:

$$\sigma(x, t) = \sigma_0(x) + \sigma_M(x)\, a(t), \qquad (5)$$

where $\sigma_0(x)$ is the background stress field acting on the volcano before the unrest episode, $a(t)$ is a time-varying function describing the temporal pattern of the ground deformation source (i.e., in our case, it is

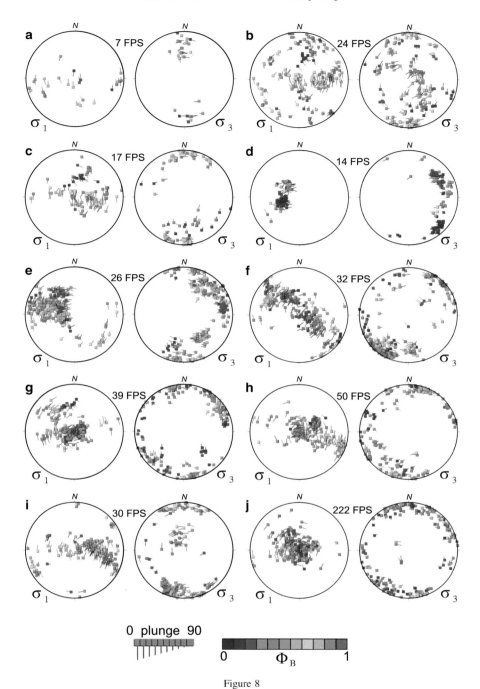

Figure 8
Results of the Multiple Inverse Method (MIM) processing of the focal mechanisms for each interval. Results are represented as *clusters* of tadpoles for both σ_1 and σ_3 axes (equal area, lower hemisphere). The *colour* of tadpoles is representative of their ϕ_B value (see *scales* on the *bottom*). Every depicted σ_1 tadpole (plot on the *left*) has a tail that allows retrieval of the corresponding σ_3 in the plot on the *right*. The σ_1 tail points toward the trend of the corresponding σ_3, and the length of each tail is inversely proportional to the σ_3 plunge (see *scale below*); the same applies to σ_3 for the plot on the *right* (YAMAJI and OTSUBO 2011)

just the crack aperture) and $\sigma_M(x)$ is the perturbation of the stress field in the volcano generated by a reference source (i.e., $a = 1$). In Eq. 5, we assume that the stress field in the volcano is modulated by a scalar function $a(t)$ according to a simple linear relationship. In detail, the background stress σ_0 can be expressed as:

$$\sigma_0 = \sigma_R + \sigma_L(x) + \sigma_M(x)a_0, \qquad (6)$$

where σ_R is a stationary background regional field, $\sigma_L(x)$ is the lithostatic load and a_0 is a constant that takes into account residual stresses deriving from previous unrest episodes arising from a similar ground deformation source. In the following, we assume that $\sigma_L(x)$ is nearly isotropic and not contributing to the deviatoric stress field pattern. Under this assumption, we can rewrite Eq. 5 as:

$$\sigma(x, t) = \sigma_M(t)[a(t) + a_0] + \sigma_R. \qquad (7)$$

In the previous expression, the function $a(t)$ is already known from the inversion of the ground deformation data (Sect. 4), and the spatial pattern $\sigma_M(x)$ can be computed easily for a given source model. The unknown regional field σ_R and the constant a_0 need to be determined from an independent data set. The focal mechanism data could be successfully used to accomplish this task.

As already mentioned in Sect. 6, the inversion of focal mechanisms provides an estimate of the stress tensor. Among the various inversion techniques, we consider the approach of ANGELIER (1990), which consists of a minimisation of the difference between the slip vector s on each fault and the direction of the maximum shear stress τ acting on it. In practice, the technique consists of searching the global minimum of the function $E = \sum_k |\delta_k|^2$, where $\delta_k = s_k - [\xi_k - (\xi_k \times n_k) \, n_k]$. In the previous expression, s_k represents the slip associated with the kth focal mechanism of the data set, n_k is the corresponding normal vector to the fault plane, and ξ_k is the stress vector acting on the fault surface (i.e., $\xi_k = \sigma n_k$). We used a non-linear optimization technique over the model parameters $m = [\alpha, \theta, \iota, \psi, \phi_B, a_0]$, with α as the magnitude of σ_1, θ the trend of σ_1, ι its plunge, ψ the rotation of σ_3 around the σ_1 axis, Φ_B as the Bishop's ratio, and a_0 is specified in Eq. 6. The first five parameters determine the regional stress tensor σ_R

Table 3

Results of the joint inversion

Parameter	Value
σ_1 Magnitude	0.39 ± 0.20 MPa
σ_1 Trend	$266° \pm 8.2°$
σ_1 Plunge	$10° \pm 5.9°$
σ_3 Trend	$0° \pm 11.5°$
σ_3 Plunge	$17° \pm 5.3°$
Φ_B	0.42 ± 0.25
a_0	0.0003 ± 0.0250 m

specified in Eq. 7. The non-linear optimization has been performed using a two-step algorithm, with the first step being a Monte Carlo search and the second an optimization using the NELDER and MEAD (1965) simplex algorithm. The results of the optimization are shown in Table 3, expressed in terms of attitudes of σ_1 and σ_3. To estimate the uncertainty of each parameter, we used a bootstrap approach (EFRON 1979; MICHEAL 1987).

The results indicate that, together with the stress field related to volcanic processes, a weak background regional field acts on the area. The orientation of field is compatible with those retrieved by D'AURIA et al. (2014) for the deeper seismogenic volume at the nearby Mt. Vesuvius volcano. The reduction in the misfit function resulting from considering the regional field is about 4 %. Nevertheless, the comparison between the modelled stress field and the results shown in Figs. 7 and 8 evidence that the contribution of σ_R is not negligible. It is worth noting that the value of the retrieved a_0 is close to zero, indicating that the effect of previous uplift events on the stress

Figure 9 ▶

Results of the joint inversion. The *left column* shows the σ_1 and σ_3 axes (respectively in *red* and *blue*) projected along a S–N cross section (A–A′ in Fig. 1) The projection of the source crack is *outlined* in *green*. *Black contours* indicate the values of the σ_1 magnitude (in MPa). The *central* and the *right columns* respectively show the stereographic projections of σ_1 and σ_3 axes within the seismogenic volume (see text for details). *Triangles* represent the regional principal stress axes (see Sect. 7). (**a**), (**b**) and (**c**) refer to the interval 1/1/1982–1/1/1983. (**d**), (**e**) and (**f**) refer to the interval 1/1/1982–1/7/1983. (**g**), (**h**) and (**i**) refer to the interval 1/1/1982–1/1/1984. (**j**), (**k**) and (**l**) refer to the interval 1/1/1982–1/1/1985. (**m**), (**n**) and (**o**) represent the stress pattern of the volcanic source alone

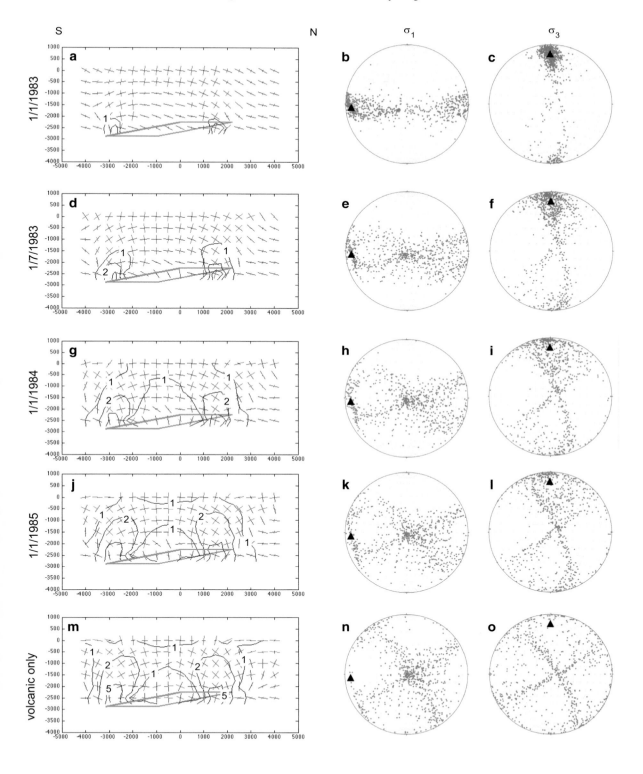

field at CFc had already almost completely vanished in 1982.

In Fig. 9, we represent the temporal evolution of the modelled stress field beneath CFc. We have computed the theoretical stress field on a regular grid of $8 \times 8 \times 4$ km, centred on benchmark 25A and with regular spacing of 250 m. In the first columns of Fig. 9, we represent the orientation of σ_1 and σ_3 along a N–S cross-section (Fig. 1), also contouring the magnitude of σ_1. In the second and third columns, the principal stress axes attitudes are plotted in stereographic projections. In the first period (Jan. 1983), the regional field is still dominant (Fig. 9a–c). However, during the first half of 1983, the subvertical dip of σ_1 becomes evident. This feature persists during the rest of the period. Moreover, the sub-horizontal N–S trending σ_3 persists until 1985. It is important to note that the stress field generated by the volcanic source alone (last row in Fig. 9) is not able to reproduce the distribution of σ_3 retrieved from the inversion of the focal mechanism data set (Figs. 7, 8).

8. Discussion and Conclusions

The inversion of the focal mechanism data set has been performed using two different approaches: the RDM and the MIM, with results summarised in Figs. 7 and 8. The overall results highlight that the stress fields acting in the CFc in the interval 1982–1984 are dominated by a well-defined, sub-vertical σ_1 corresponding to a less defined sub-horizontal σ_3 with a prevalence of NNE-SSW trends (Figs. 7j, 8j). In particular, the MIM is able to correctly identify different clusters within each subset. A comparison of results of RDM and MIM shows good agreement (Figs. 7, 8), and hence supports the reliability of the results. Moreover, we notice that there is good agreement between our results and those of ZUPPETTA and SAVA (1991).

The non-linear inversion of the optical levelling data set has shown that the best-fit ground deformation source for the 1982–1984 interval is a simple sub-horizontal crack. This source is able to model most of the observed ground deformation (Fig. 3). However, the stress field related to this source alone consists of a nearly subvertical σ_1 and a quite scattered σ_3 (Fig. 9m, n, o).

In Sect. 7, we illustrated a novel approach for the joint inversion of the ground deformation and focal mechanism data sets. The key results are that the retrieved sub-vertical σ_1 is mainly related to the effect of the ground deformation source; however, considering the pattern of σ_3 obtained by both RDM and MIM approaches, an additional background field is required. The joint inversion used in this work allows the determination of the general features of this field, shich consist of a weak extensional N–S regional stress field driving the orientation of the σ_3 axes (Fig. 9). The orientation of this field is compatible with that retrieved at the nearby Mt. Vesuvius volcano (D'AURIA et al. 2014). The regional field is dominant only in the first phases of the 1982–1984 crisis (Fig. 9a–c). During the first half of 1983, the volcanic deformation source becomes prevalent over the regional background field (Fig. 9d–f). This behaviour matches with the stress pattern inferred from focal mechanisms (Figs. 7, 8), which shows only slight variations during intervals A–C (January–October 1983). Since November 1983, the stress field pattern became quite stable with a sub-vertical σ_1 and a sub-horizontal σ_3 (intervals D–I in Table 1).

In conclusion, our results indicate that spatial and temporal variations of the stress field can be related to volcano dynamics and could be linked to the intrusion of fluids (possibly of magmatic nature) within a planar structure. Recently, MACEDONIO et al. (2014) evidenced how the intrusion of sills at a shallow depth can be a realistic model to explain volcanic unrest in calderas.

Our results also have implications for a better comprehension of seismogenic processes at CFc.

In the first column of Fig. 9, we plot the magnitude of σ_1 at different time intervals. It is evident how it is concentrated along the edges of the rectangular crack. This effect is well known for this type of source (OKADA 1992). Comparing Fig. 9 with the hypocentre locations in Fig. 5, we notice that earthquakes are grouped in two main clusters, whose positions roughly match the northern and southern edges of the ground deformation source (Fig. 3d). TROISE et al. (2003) emphasised how Coulomb stress changes in relation to magmatic processesare important in determining the location of earthquake hypocentres at CFc. However, detailed study of the relationship between ground deformation and seismogenesis at CFc must take into account the effect of hydrothermal fluids that play a fundamental role in volcanic areas (BIANCO *et al.* 2004; D'AURIA *et al.* 2011).

The technique proposed in this work could be applied to other volcanic contexts. We emphasise that the joint inversion of ground deformation and focal mechanism is more efficient than the mere comparison of the results obtained by the separate inversion. SEGALL *et al.* (2013) proposed a similar approach, which is based on the computation of Coulomb stress changes. They model the observed migration of earthquake hypocentres during dike propagation in volcanoes.

Finally, we emphasise that the CFc area is densely populated and hence subject to a high volcanic risk (ORSI *et al.* 2004). The last bradyseismic crisis of 1982–1984 led to the temporary evacuation of most of the town of Pozzuoli because of the ongoing seismic crisis (BARBERI *et al.* 1984). For this reason, a better understanding of the relationship between earthquakes and ground deformation at CFc is important not only from a scientific point of view, but also to support decision makers during the management of possible future volcanic emergencies in this area.

Acknowledgments

This work has been realised in the framework of an MED-SUV project. MED-SUV has received funding from the European Union's Seventh Programme for research, technological development and demonstration under grant agreement No. 308665.

REFERENCES

ANGELIER, J. (1984), *Tectonic analysis of fault slip data sets*, J. Geophys. Res., 89, 5835–5848.

ANGELIER, J. (1990), *Inversion of Field Data in Fault Tectonics to Obtain the Regional Stress. 3. A New Rapid Direct Inversion Method by Analytical Means, Geophys.* J. Int., *103*, 363–376.

ANGELIER, J. T., and MECHLER, P. (1977), *Sur une methode graphique de recherche des contraintes principales egalement utilisables en tectonique et en seismologie: la methode des dièdres droits.* Bulletin de la Société géologique de France (7), t.XIX, n°6, 1309–1318.

ANGELIER, J., and MANOUSSIS, S. (1980), *Classification automatique et distinction des phases superposées en tectonique de failles.* CR Acad. Sci. Paris, *290*, 651–654.

ASTER, R. C., and MEYER, R. P. (1988), *Three-dimensional velocity structure and hypocenter distribution in the Campi Flegrei caldera, Italy.* Tectonophysics, *149*(3), 195–218.

ASTER, R. C., BORCHERS, B., and THURBER, C. H. (2013), *Parameter estimation and inverse problems* (Academic Press).

BRANDSDOTTIR, B., and EINARSSON, P. (1979). *Seismic activity associated with the September 1977 deflation of the Krafla central volcano in northern Iceland,* J. Volc. Geotherm. Res., *6*, 197–212.

BARBERI, F., CORRADO, G., INNOCENTI, F. and LUONGO, G. (1984), *Phlegraean Fields 1982–1984: brief chronicle of a volcano emergency in a densely populated area.* Bull. Volcanol., *47*(2): 175–185.

BATTAGLIA, M., TROISE, C., OBRIZZO, F., PINGUE, F., and DE NATALE, G. (2006), *Evidence for fluid migration as the source of deformation at Campi Flegrei caldera (Italy).* Geophysical Research Letters, *33*(1), L01307.

BIANCO, F., DEL PEZZO, E., SACCOROTTI, G., and VENTURA, G. (2004), *The role of hydrothermal fluids in triggering the July–August 2000 seismic swarm at Campi Flegrei, Italy: evidence from seismological and mesostructural data.* Journal of Volcanology and Geothermal Research, *133*(1), 229–246.

BISHOP (1966), *The strength of solids as engineering materials,* Geotechnique, *16,* pp. 91–130.

BONAFEDE, M. (1991), *Hot fluid migration, an efficient source of ground deformation: Application to the 1982–1985 crisis at Campi Flegrei-Italy,* J. Volcanol. Geotherm. Res., *48*, 187–198.

BOTT, M. H. P. (1959), The mechanics of oblique slip faulting, Geol. Mag., 96, 109–117.

BURNHAM, K. P., and ANDERSON, D. R. (2002), Model selection and multimodel inference: a practical information-theoretic approach. Springer.

CAMACHO, A. G., P. J. GONZÁLEZ, J. FERNÁNDEZ, and G. BERRINO (2011), *Simultaneous inversion of surface deformation and gravity changes by means of extended bodies with a free geometry: Application to deforming calderas,* J. Geophys. Res., *116*, B10401, doi:10.1029/2010JB008165.

CANNAVÒ, F., SCANDURA, D., PALANO, M. and MUSUMECI C. (2014), *A Joint Inversion of Ground Deformation and Focal Mechanisms Data for Magmatic Source Modelling.* Pure and Applied Geophysics, doi:10.1007/s00024-013-0771-x.

CHIODINI, G., FRONDINI, F., CARDELLINI, C., GRANIERI, D., MARINI, L., and VENTURA, G. (2001), *CO_2 degassing and energy release at Solfatara volcano, Campi Flegrei, Italy.* Journal of Geophysical Research: Solid Earth (1978–2012), *106*(B8), 16213–16221.

CHIODINI, G., M. TODESCO, S. CALIRO, C. DEL GAUDIO, G. MACEDONIO, and M. RUSSO (2003), *Magma degassing as a trigger of bradyseismic events: The case of Phlegrean Fields (Italy),* Geophys. Res. Lett., *30*(8), 1434, doi:10.1029/2002GL016790.

D'AURIA, L., GIUDICEPIETRO, F., AQUINO, I., BORRIELLO, G., DEL GAUDIO, C., LO BASCIO, D., MARTINI M., RICCIARDI G.P., RICCIOLINO P. and RICCO, C. (2011), *Repeated fluid-transfer episodes as a mechanism for the recent dynamics of Campi Flegrei caldera (1989–2010).* Journal of Geophysical Research: Solid Earth (1978–2012), *116*(B4).

D'AURIA, L., GIUDICEPIETRO, F., MARTINI, M., and LANARI, R. (2012), *The 4D imaging of the source of ground deformation at Campi Flegrei caldera (southern Italy).* Journal of Geophysical Research: Solid Earth (1978–2012), *117* (B8).

D'AURIA, L., MARTINI, M., ESPOSITO, A., RICCIOLINO, P., and GIU-
DICEPIETRO, F. (2008), *A unified 3D velocity model for the
Neapolitan volcanic areas, In Conception, Verification and
Application of Innovative Techniques to Study Active Volcanoes*
(ed. Marzocchi W. and Zollo A.), pp. 375–390, ISBN 978-88-
89972-09-0.

D'AURIA, L., B. MASSA, and A. DE MATTEO (2014), *The stress field
beneath a quiescent stratovolcano: The case of Mount Vesuvius.*
J. Geophys. Res. Solid. Earth., *119*. doi:10.1002/2013JB010792.

DEINO, A.L., ORSI, G., PIOCHI, M. and de VITA, S. (2004), *The age of
the neapolitan Yellow Tuff caldera-forming eruption (Campi
Flegrei caldera–Italy) assessed by 40ar/39ar dating method.*
Journal of Volcanology and Geothermal Research, *133*, 157–170.

DEL GAUDIO, C., AQUINO, I., RICCIARDI, G. P., RICCO, C., and
SCANDONE, R. (2010), *Unrest episodes at Campi Flegrei: A
reconstruction of vertical ground movements during 1905–2009.*
Journal of Volcanology and Geothermal Research, *195*(1), 48-56.

DIETRICH, J. (1994), *A constitutive law for rate of earthquake
production and its application to earthquake clustering.* J. Geophys. Res. 99, 2601–2618.

DI VITO, M. A., ISAIA, R., ORSI, G., SOUTHON, J., DE VITA, S.,
D'ANTONIO, M., PAPPALARDO L. and PIOCHI, M. (1999), *Volcanism
and deformation since 12,000 years at the Campi Flegrei caldera (Italy).* Journal of Volcanology and Geothermal Research,
91(2), 221–246.

DZURISIN, D. (2006), *Volcano Deformation: Geodetic Monitoring
Techniques.* Berlin, Springer, Springer-Praxis Books in Geophysical Sciences, 441 p.

EFRON, B. (1979), *Bootstrap methods: Another look at the jackknife.* The Annals of Statistics *7*(1): 1–26.

FIALKO, Y., KHAZAN, Y., and SIMONS, M. (2001), *Deformation due to
a pressurized horizontal circular crack in an elastic half-space,
with applications to volcano geodesy.* Geophysical Journal
International, *146*(1), 181–190.

FROHLICH, C. (1992), *Triangle diagrams: Ternary graphs to display
similarity and diversity of earthquake focal mechanisms*, Phys.
Earth. Planet. Inter., *75*, 193–198.

GAETA, F. S., PELUSO, F., ARIENZO, I., CASTAGNOLO, D., DE NATALE,
G., MILANO, G., ALBANESE C. and MITA, D. G. (2003), *A physical
appraisal of a new aspect of bradyseism: The miniuplifts.* Journal
of Geophysical Research: Solid Earth (1978–2012), *108*(B8).

GAUDIOSI, G., and IANNACCONE, G. (1984), *A preliminary study of
stress pattern at Phlegraean Fields as inferred from focal
mechanisms.* Bulletin volcanologique, *47*(2), 225–231.

GEBAUER, S., SCHMITT, A.K., PAPPALARDO, L., STOCKLI, D.F., and
LOVERA, O.M. (2014), *Crystallization and eruption ages of
Breccia Museo (Campi Flegrei caldera, Italy) plutonic clasts and
their relation to the Campanian ignimbrite.* Contrib. Mineral.
Petrol., *167*: 953, doi:10.1007/s00410-013-0953-7.

GEPHART, J., and FORSYTH, D. (1984), *An improved method for
determining the regional stress tensor using earthquake focal
mechanisms data: application to the San Fernando earthquake
sequence.* J. Geophy. Res., *89*, 9305–9320.

HIPPOLYTE, J.C., BERGERAT, F., GORDON, M., BELLIER, O., ESPURT, N.
(2012), *Keys and pitfalls in mesoscale fault analysis and pa-
leostress reconstructions, the use of Angelier's methods.*
Tectonophysics, doi:10.1016/j.tecto.2012.01.012.

LEE, W.H.K., and STEWART, S.W. (1981), *Principles and applications of microearthquake networks.* In: Advances in Geophysics,
Supplement 2, Academic Press, New York, 293 pp.

MACEDONIO G., GIUDICEPIETRO F., D'AURIA L. and MARTINI M.
(2014), *Sill intrusion as a source mechanism of unrest at volcanic calderas.* J. Geophys. Res. Solid Earth, *119*, doi:10.1002/
2013JB010868.

MCKENZIE, D. P. (1969), *The relation between fault plane solutions
for earthquakes and the directions of the principal stresses*, Bull.
Seismol. Soc. Am. *59*, 591–601.

MCTIGUE, D. F. (1987), *Elastic stress and deformation near a finite
spherical magma body: Resolution of the point source paradox.*
Journal of Geophysical Research: Solid Earth (1978–2012),
92(B12), 12931–12940.

MICHAEL, A. J. (1984), *Determination of stress from slip data:
faults and folds.* J. Geophys. Res., *89*, 11517–11526.

MICHAEL, A.J. (1987), *Use of Focal Mechanisms to Determine
Stress: A Control Study.* J. Geoph. Res., *92*, 357–368.

MOGI, K. (1958), *Relations between the eruptions of various volcanoes and the deformations of the ground surfaces around them.*
Bulletin of the Earthquake Research Institute *36*, 99–134.

NELDER, J.A. and MEAD, R. (1965), *A simplex method for function
minimization*, Comput. J., *7*, pp. 308–313.

OKADA, Y. (1985), *Surface deformation due to shear and tensile
faults in a half-space.* Bulletin of the seismological society of
America, *75*(4), 1135–1154.

OKADA, Y. (1992), *Internal deformation due to shear and tensile
faults in a half-space.* Bulletin of the Seismological Society of
America, *82*(2), 1018–1040.

ORSI, G., CIVETTA, L., DEL GAUDIO, C., DE VITA, S., DI VITO, M. A.,
ISAIA, R., S.M. PETRAZZUOLI, G.P. RICCIARDI and RICCO, C. (1999),
*Short-term ground deformations and seismicity in the resurgent
Campi Flegrei caldera (Italy): an example of active block-
resurgence in a densely populated area.* Journal of Volcanology
and Geothermal Research, *91*(2), 415–451.

ORSI, G., DE VITA, S., and DI VITO, M. (1996), *The restless,
resurgent Campi Flegrei nested caldera (Italy): constraints on its
evolution and configuration.* Journal of Volcanology and Geothermal Research, *74*(3), 179–214.

ORSI, G., DI VITO, M. A., and ISAIA, R. (2004), *Volcanic hazard
assessment at the restless Campi Flegrei caldera.* Bulletin of
Volcanology, *66*(6), 514–530.

OTSUBO, M., A. YAMAJI, and A. KUBO (2008), *Determination of
stresses from heterogeneous focal mechanism data: An adaptation
of the multiple inverse method*, Tectonophysics, *457*, 150–160.

OTSUBO, M., SATO, K., YAMAJI, A. (2006), *Computerized identification of stress tensors determined from heterogeneous fault-slip
data by combining multiple inverse method and k-means clustering.* Journal of Structural Geology, *28*, 991–997, doi:10.
10106/j.jsg.2006.03.008.

PATANÈ, D., PRIVITERA, E., GRESTA, S., AKINCI, A., ALPARONE, S.,
BARBERI, G., CHIARALUCE, L., COCINA, O., D'AMICO, S., DE GORI,
P., DI GRAZIA, G., FALSAPERLA, S., FERRARI, F., GAMBINO, S.,
GIAMPICCOLO, E., LANGER, H., MAIOLINO, V., MORETTI, M., MO-
STACCIO, A., MUSUMECI, C., PICCININI, D., REITANO, D., SCARFÌ, L.,
SPAMPINATO, S., URSINO, A., and ZUCCARELLO, L. (2003), *Seis-
mological constraints for the dike emplacement of July–August
2001 lateral eruption at Mt. Etna volcano, Italy.* Annals of
geophysics, *46*(4), 599–608.

PEDERSEN, R., and SIGMUNDSSON, F. (2004), *InSAR based sill model
links spatially offset areas of deformation and seismicity for the
1994 unrest episode at Eyjafjallajökull volcano, Iceland.* Geophys. Res. Lett., *31*(14), doi:10.1029/2004GL020368.

RAMSAY J., LISLE R. (2000), *Applications of continuum mechanics in structural geology (Techniques of modern structural geology.* Vol.*3*), 701–1061.

REASENBERG, P. and OPPENHEIMER D. (1985), *FPFIT, FPPLOT and FPPAGE: Fortran computer programs for calculating and displaying earthquake fault-plane solutions*, US Geol. Surv., Open-File Rept, 85–739.

RIVERA, L. and CISTERNAS, A. (1990), *Stress tensor and fault plane solutions for a population of earthquakes*. Bull. Seismol. Soc. Am., *80*(3): 600–614.

RUBIN, A.M., GILLARD, D., and GOT, J.L. (1998), *A reinterpretation of seismicity associated with the January 1983 dike intrusion at Kilauea volcano, Hawaii*. J. geophys. Res., *103*, 10003–10015.

SEGALL, P. (2013), *Volcano deformation and eruption forecasting.* Geological Society, London, Special Publications, *380*, doi:10.1144/SP380.4.

SEGALL, P. (2010), *Earthquake and volcano deformation.* Princeton University Press.

SEGALL, P., LLENOS, A. L., YUN, S. H., BRADLEY, A. M., and SYRACUSE, E. M. (2013), *Time-dependent dike propagation from joint inversion of seismicity and deformation data.* Journal of Geophysical Research: Solid Earth *118*(11), 5785–5804.

SEN, M. K., and STOFFA, P. L. (1995), *Global optimization methods in geophysical inversion.* Elsevier.

TODA, S., STEIN, R. S., and SAGIYA, T. (2002), *Evidence from the ad 2000 Izu islands earthquake swarm that stressing rate governs seismicity.* Nature, *419*, 58–61.

TRAMELLI, A., TROISE, C., DE NATALE, G., and ORAZI, M. (2013), *A New Method for Optimization and Testing of Microseismic Networks: An Application to Campi Flegrei (Southern Italy).* Bulletin of the Seismological Society of America, *103*(3), 1679–1691, doi:10.1785/0120120211.

TROISE, C., F. PINGUE, and G. DE NATALE (2003), *Coulomb stress changes at calderas: Modeling the seismicity of Campi Flegrei*

(southern Italy), J. Geophys. Res., *108*, 2292, doi:10.1029/2002JB002006, B6.

UMAKOSHI, K., SHIMIZU, H., and MATSUWO N. (2001), *Volcano-tectonic seismicity at Unzen Volcano, Japan, 1985-1999.* Journal of Volcanology and Geothermal Research, *112*, 117–131.

VITALE, S., and ISAIA, R. (2013), *Fractures and faults in volcanic rocks (Campi Flegrei, southern Italy): insight into volcano-tectonic processes.* International Journal of Earth Sciences, 1–19.

WALLACE, R. E. (1951), *Geometry of shearing stress and relationship to faulting,* J.Geol., *59*, 111–130.

WOO, J. Y., and KILBURN, C. R. (2010), *Intrusion and deformation at Campi Flegrei, southern Italy: sills, dikes, and regional extension.* Journal of Geophysical Research: Solid Earth (1978–2012), *115*(B12).

YAMAJI, A. (2000), *The multiple inverse method: a new technique to separate stresses from heterogeneous fault-slip data.* J. Struct. Geol., *22*: 441–452.

YAMAJI, A. (2007), *An Introduction to Tectonophysics: Theoretical Aspects of Structural Geology*, 4-88704-135-7, Terrapub, Tokyo.

YAMAJI, A., and M. OTSUBO (2011), *Multiple Inverse Method Software Package User's Guide, Kyoto, Japan.* [Avialable at http://www.kueps.kyoto-u.ac.jp/web-bs/tsg/software/mim/man_e.pdf.].

YANG, X. M., DAVIS, P. M., and DIETERICH, J. H. (1988), *Deformation from inflation of a dipping finite prolate spheroid in an elastic half-space as a model for volcanic stressing.* Journal of Geophysical Research: Solid Earth (1978–2012), *93*(B5), 4249–4257.

ZUPPETTA, A., and SAVA, A. (1991), *Stress pattern at Campi Flegrei from focal mechanisms of the 1982–1984 earthquakes (Southern Italy).* Journal of Volcanology and Geothermal Research, *48*(1), 127–137.

(Received March 16, 2014, revised November 25, 2014, accepted December 3, 2014, Published online December 14, 2014)